# Interdisciplinary Cancer Research

Volume 7

The "Interdisciplinary Cancer Research" series publishes comprehensive volumes on different cancers and presents the most updated and peer-reviewed articles on human cancers. Over the past decade, increased cancer research has greatly improved our understanding of the nature of cancerous cells which has led to the development of more effective therapeutic strategies to treat cancers. This translational series is of special value to researchers and practitioners working on cell biology, immunology, hematology, biochemistry, genetics, oncology and related fields.

Nima Rezaei
Editor

# Breast Cancer Treatment: An Interdisciplinary Approach

Springer

*Editor*
Nima Rezaei
Department of Clinical Immunology
Karolinska Institutet
Stockholm, Sweden

Research Center for Immunodeficiencies,
Children's Medical Center
Tehran University of Medical Sciences
Tehran, Iran

ISSN 2731-4561 ISSN 2731-457X (electronic)
Interdisciplinary Cancer Research
ISBN 978-3-031-65826-6 ISBN 978-3-031-65827-3 (eBook)
https://doi.org/10.1007/978-3-031-65827-3



This Springer imprint is published by the registered company Springer Nature Switzerland AG
The registered company address is: Gewerbestrasse 11, 6330 Cham, Switzerland

# Preface

There are different types of treatment for breast cancer, including surgery, chemotherapy, radiotherapy, hormone therapy, immunotherapy, and targeted therapy. Deciding on the types of treatment depends on several factors, such as the stage and grade of cancer, biomarkers, and general health of patients.

The "Interdisciplinary Cancer Research" series publishes comprehensive volumes on different cancers. It plans to present the most updated and peer-reviewed interdisciplinary chapters on cancers. This interdisciplinary book series is of special value to researchers and practitioners working on cell biology, immunology, hematology, biochemistry, genetics, oncology and related fields. This is the main concept of Cancer Immunology Project (CIP) and Network of Immunity in Infection, Malignancy and Autoimmunity (NIIMA), which are two active interest groups of the Universal Scientific Education and Research Network (USERN).

The seventh volume of the book entitled *Breast Cancer Treatment: An Interdisciplinary Approach* starts with a general chapter on treatment of triple-negative breast cancer (TNBC) and ductal carcinoma in situ (DCIS), followed by chapters on ablative breast cancer surgery, radiotherapy, and percutaneous breast cancer treatment. Revolutionizing breast cancer care is the subject of Chapter 6. Hormone therapy and hormone immunotherapy combination are discussed in Chapters 7–10. The other Chapters (11–17) present potential targeted therapies for breast cancer. Finally, the last chapter presents a case study on holistic nursing care for a postoperative breast cancer patient with lymphedema.

I hope that this interdisciplinary book will be comprehensible, cogent, and of special value for researchers and oncologists who wish to extend their knowledge on breast cancer treatment.

Tehran, Iran — Nima Rezaei

# Contents

## About the Editor

**Nima Rezaei** MD, PhD, gained his medical degree (MD) from Tehran University of Medical Sciences and subsequently obtained an MSc in Molecular and Genetic Medicine and a PhD in Clinical Immunology and Human Genetics from the University of Sheffield, UK. He also spent a short-term fellowship of Pediatric Clinical Immunology and Bone Marrow Transplantation in the Newcastle General Hospital. Professor Rezaei is now the Full Professor of Immunology and Vice Dean of Research and Technologies, School of Medicine, Tehran University of Medical Sciences, and the co-founder and Head of the Research Center for Immunodeficiencies. He is also the Founder of Universal Scientific Education and Research Network (USERN). Professor Rezaei has already been the Director of more than two hundred research projects and has designed and participated in several international collaborative projects. Professor Rezaei is the editor, editorial assistant, or editorial board member of more than 50 international journals. He has edited more than one hundred international books, has presented more than a thousand lectures/posters in congresses/meetings, and has published more than 1500 scientific papers in international journals.

# Signal Transducer and Activator of Transcription as a Potential Therapeutic Target in Breast Cancer

Niloofar Deravi and Nima Rezaei

**Abstract**

Signal transducer and activator of transcription (STAT) signaling pathway, connected upstream with Janus kinases (JAK) proteins, is known to be capable of the integration of inputs from various signaling pathways. Each family member exerts a particular role in signal transduction and is hence crucial to mediate different cellular responses to cytokines. STAT proteins especially STAT3 and STAT 5 are widely investigated and known to be involved in breast cancer progression. On the other hand, STAT1 and STAT4 exert opposite roles through suppression of breast tumor growth. Persistent activation of STAT3/5 can promote chronic inflammation, increasing healthy cells susceptibility to carcinogenesis. In this chapter, we discuss the structure, functional development, and activation of STAT family members. We also summarize the roles of different STAT proteins in breast cancer and its different subtypes. We also discuss STAT signaling pathway inhibitors used to treat breast cancer.

N. Deravi
Students Research Committee, Shahid Beheshti University of Medical Sciences, Tehran, Iran

Systematic Review and Meta-analysis Expert Group (SRMEG), Universal Scientific Education and Research Network (USERN), Tehran, Iran

N. Rezaei (✉)
Research Center for Immunodeficiencies, Children's Medical Center, Tehran University of Medical Sciences, Tehran, Iran

Department of Immunology, School of Medicine, Tehran University of Medical Sciences, Tehran, Iran

Network of Immunity in Infection, Malignancy and Autoimmunity (NIIMA), Universal Scientific Education and Research Network (USERN), Stockholm, Sweden
e-mail: rezaei_nima@tums.ac.ir

 
Interdisciplinary Cancer Research, https://doi.org/10.1007/16833_2022_107
Published online: 31 January 2023

**Keywords**

Breast cancer · Carcinogenesis · Signal transducer and activator of transcription

## 1 Introduction

Breast cancer, ranked first in terms of new cases and deaths, is one of the leading cancers affecting women worldwide (Siegel et al. 2020). Due to the development in diagnostic tests for identifying ER expression models and amplification of human epidermal growth factor receptor 2 (HER2) in human breast cancer clinical samples, pathologists have classified breast cancer into three major subtypes: (1) HER2 amplified, (2) estrogen receptor (ER)-positive, and (3) triple-negative breast cancer (TNBC), which is ER-negative with the absence of amplification of HER2 amplification (Ross et al. 2009; Bertos and Park 2011; Guiu et al. 2012).

According to the profiling of gene expression, we can divide ER-positive subtype into luminal A and B subtypes; TNBC can also be subdivided into six other subtypes (Lehmann et al. 2011; Prat and Perou 2011; Eroles et al. 2012; Geyer et al. 2012; Koboldt et al. 2012). Additional subtyping of luminals and TNBC led to revealing various prognostic subtypes, some of which due to differences in both macro-and micro-environment and some due to intrinsic varieties in gene expression in cancer cells (Bertos and Park 2011). In TNBCs, poor prognosis has been linked with low expression of claudin in the cancer cells (Lu et al. 2013), while better prognosis is connected with higher presence of B-cell and low action in interleukin 8 (IL-8) (Rody et al. 2011). In order to establish more refined prognostic and therapeutic markers, molecular investigations have been conducted on luminal A and B ER-positive subtypes and HER2-amplified subtypes (Jönsson et al. 2010; Geyer et al. 2012).

In the early stages of breast cancer, some patients are treated via surgery; however, most of whom require postoperative chemotherapy, endocrine therapy, and radiotherapy in addition to surgery. These conventional treatments have for a long-time prolonged patient's survival and still hold a prominent position in the treatment of the disease, but they would also show inevitable adverse effects. Symptoms associated with chemotherapy, endocrine therapy, and radiotherapy make a patient's quality of life even worse. The two most common chemotherapeutic drugs used in breast cancer are anthracyclines and taxanes. Anthracyclines are the most general used and useful chemotherapy drugs for breast cancer. Anthracyclines have thus become an essential adjuvant and palliative medications. Anthracyclines can have a variety of toxic side effects, such as alopecia, mucositis, myelosuppression, and possible permanent cardiovascular effects (Cai et al. 2019).

Conversely, endocrine therapy can be an important treatment for hormone-positive breast cancer, the most general type of breast cancer. However, it can also cause adverse effects, including hot flashes, night sweats, cognitive changes, and fatigue (Mouridsen 2006).

Radiotherapy subsequent to breast-conserving surgery can increase patients' survival. It also helps patients diagnosed with lymph node-positive breast cancer in a mastectomy (McGale et al. 2014). Nonetheless, irradiation adverse events such as erythema, erosion, edema, and ulceration are some inevitable adverse effects of radiotherapy. Radiotherapy-induced tissue damage may take a long time to repair, and muscular rigidity in the area of the radiotherapy may result in persistent limb impairment (Ding et al. 2018). Breast cancer treatment with targeted therapy is a relatively recent concept. It exerts a remarkable therapeutic impact compared to the abovementioned traditional therapeutic approaches, yet it has fewer side effects and may lead to minimal damage in normal tissues. Transcription factors (TFs) have domains that can bind to the DNA of certain gene promoter regions. Many TFs exert role in the genesis and development of breast cancer. STAT family is one of the common outstanding TF families in breast cancer. STAT family consists of seven structurally related and highly conserved members, including STAT1, STAT2, STAT3, STAT4, STAT5a, STAT5b, and STAT6 (Furtek et al. 2016). STAT family members are attractive molecular targets in breast cancer therapy.

## 2 STAT Family Members Characteristics, Activation, and Functional Development

Several structurally and functionally conserved areas are found in all STAT proteins. During STAT monomer activation, the Src-homology 2 (SH2) domain, in conjunction with the N-terminal domain (ND), facilitates homo- and heterodimerization. Most STAT inhibitors target highly conserved SH2 domain. Subsequent to undergoing serine phosphorylation, C-terminal transcriptional activation domain (TAD) obtains further transcriptional activators, boosting STAT's transcriptional effect. The DNA-binding domain (DBD) determines DNA association. STAT proteins can constitute isoforms by either posttranslational processing of proteolytic or alternative splicing of mRNA; β, γ, or δ isoforms are the short isoforms, whereas α isoforms are full-length isoforms (Lim and Cao 2006).

STAT proteins, in their latent state, are inactive, situated in the unstimulated target cells cytoplasm as unphosphorylated dimers or monomers (Verhoeven et al. 2020). A variety of cytokines, such as interleukins, growth factors, interferons (IFNs), and hormones binding to membrane receptors can activate these proteins (Bromberg 2001; Schindler 2002). Various different and occasionally overlapping ligands activate STAT1, STAT3, STAT5a, and STAT5b, but only a few cytokines activate STAT2, STAT4, and STAT6. STATs can be activated in a variety of modalities. STATs activity is primarily mediated via the Janus kinase (JAK-STAT) pathway. Extracellular signaling polypeptides communicate with transmembrane receptors in this cascade, resulting in a method for transcriptional control without the utilization of secondary messengers. There is an association between interferon, interleukin, and hormone receptors and a tyrosine kinase of the JAK family (known as tyrosine kinase 2 (Tyk2)) as well as a tyrosine residue on their cytoplasmic domain. Coupling of receptor and ligand causes the receptors to dimerize or oligomerize, which activates the linked JAK proteins. JAKs subsequently can phosphorylate trans-phosphorylate receptor-associated tyrosine residues and also themselves, providing docking sites for the SH2 domains in the latent cytoplasmic STAT. Interactions of SH2 and pTyr are very selective, and identifying the specificity of the activation of STAT meditated by receptors is a crucial step. At the time of the coupling, on the receptor-associated STAT molecules, the JAKs would phosphorylate a specific Tyr residue (known as pY), leading to the SH2 domain of the pSTAT associated with one receptor and a reciprocal binding of the pY of the pSTAT associated with the other receptor, causing STAT hetero- or homodimerization. The dimer of pSTAT and pSTAT is subsequently liberated from the receptors to the nucleus quickly via using importins (Schindler 2002). The translocated dimer is now capable of binding certain DNA regulatory elements (DREs), and, when identified, the target gene's rate of transcription will vary substantially (Aaronson and Horvath 2002; Lim and Cao 2006; Konjević 2009). Even though phosphorylation causes STAT dimerization, there is solid evidence that, with just a minimum fraction as monomers and multimeric complexes known as statosomes, unphosphorylated STATs persist primarily as latent homo- or heterodimers (Stancato et al. 1996; Novak et al. 1998; Ndubuisi et al. 1999; Haan et al. 2000; Braunstein et al. 2003). These latent dimers, like their phosphorylated counterparts, can shuttle between the cytoplasm and nucleus (Liu et al. 2005; Reich 2013) and regulate critical cellular activities (Yang et al. 2007; Cheon and Stark 2009; Carbognin et al. 2016; Park et al. 2016). STATs lack tyrosine kinase activity and hence rely on other molecules. Alternative STATs activation is possible through growth factor receptors and non-receptor proteins' intrinsic tyrosine kinase activity. The epidermal growth factor receptor (EGFR) and platelet-derived growth factor receptor (PDGFR), via JAK-STAT signaling pathway, control apoptosis, proliferation, and differentiation, regulating stem cell maintenance, inflammatory response, embryonic development, and hematopoiesis response (Silvennoinen et al. 1993; Brooks et al. 2014; Thomas et al. 2015). Other ways of activation have been suggested for STATs, including adaptor proteins moving JAKs closer to STATs, activation via G-protein coupled receptors (GPCRs), and non-receptor tyrosine kinases. Further cellular pathways might enhance the original STAT-activating stimulus thanks to the complicated mechanisms of regulation given by multiple forms of activation (Decker and Kovarik 2000).

It is now known that the STAT can bind to various DREs throughout the genome, controlling various protein-coding genes transcription. The JAK-STAT pathway's targets include mostly genes involved in development, immunology, and stress and metabolic responses (Tsurumi et al. 2017). STATs bind to proximal DREs as classic transcription factors (TFs), causing transcriptional activation, repression, or neutral binding. They also engage more distal binding sites, far from protein-encoding genes, through which they control enhancer activity (regions of DNA that augment the activity of a promoter in an orientation- and position-independent manner, coordinating cell-type-specific gene expression, e.g., TFs) and (ii) epigenetic status of associated genes or (i) straightly guide the noncoding loci such as miRNAs or (ii) directly instruct non (37). STAT-induced epigenetic alterations are seen in areas that control binding of STAT and regulation of transcription. This creates an autoregulatory mechanism with functional consequences that are either suppressing (such as histone H3 trimethylation at Lys27) or potentiating (such as histone H3 trimethylation at Lys4) (Mandal et al. 2011).

Because of epigenetic changes that allow the access of the STATs to genes, the aimed repertoire for a certain STAT protein might differ from one cell type to another. STAT1, STAT3, STAT4, and STAT6 found in immune cells (Wei et al. 2010; Qiao et al. 2013; Qiao et al. 2016) and STAT5 found in mammary epithelium have all been shown to exhibit epigenetic alterations (Kang et al. 2014; Willi et al. 2016). STAT protein-generated transcriptional activity is affected by other parameters, including interaction with other TFs, selectivity of the gene, genome-wide competition (for access to the same DRE, members of the STAT family compete and mediate different outcomes), heterogenous signaling of STATs (distinct outcomes by the same molecule), and intensity or duration of the signaling of the STATs signaling (Villarino et al. 2017). The transcriptional result created via the JAK-STAT signaling pathway, and consequently a cell or tissue response, is mediated by a large array of intrinsic variables.

## 2.1 STAT Family Members Expression and Activity in Breast Cancer

STAT family members were discovered at the same time as three key subtypes of breast cancer (including HER2-amplified, TNBC, and ER-positive) and before further subtyping by profiling of the genes' expression. As a result, previous studies on the expression of STATs in breast cancer did not necessarily reveal any parallels or variations in patterns of expression across all breast cancer subtypes. Nonetheless, these studies provide fundamental information on the frequency of STAT family members' activity and expression in breast cancer. STAT family members' activity and expression are controlled at various levels, such as phosphorylation of tyrosine, modification of transcription, and posttranslational, and their effect on function of cells varies depending on their nuclear or cytoplasmic location (Clevenger 2004; Vafaizadeh et al. 2012; Furth 2014). STAT3 and STAT5a/b downregulation via various agents is correlated with decreased growth in cell lines of breast cancer modeling various breast cancer subtypes (Lim et al. 2012; Park et al. 2012), recommending that STATs could act as therapeutic aim for various subtypes of breast cancer.

## 2.2 STAT1 and Breast Cancer

Research on the mammary glands of mice showed that the expression of STAT1 is kept during lactation, pregnancy, and involution, though DNA binding and Y701's phosphorylation are only identified in early pregnancy, late involution, and virgin animals (Watson 2001). STAT1 regulation occurs in various stages of breast development (Watson 2001), yet the regular development of mammary glands occurs in

$STAT1^{-/-}$ mice. Contrary to normal untransformed breast cells, STAT1 is involved in the development of mammary cancer due to the phosphorylation of STAT1 Y701 is increased in human mammary tumors (Yu and Jove 2004) and is correlated with elevated survival regardless of the rest of the prognostic factors (Widschwendter et al. 2002). Moreover, elevated levels of mRNA in STAT1 were demonstrated to be part of a molecular signature that is correlated with greater metastatic outcome prediction of patients with triple-negative and hormone receptor-negative mammary cancers (Yau et al. 2010).

Recent independent researches in Hennighausen (Klover et al. 2010), Koromilas (Raven et al. 2011), Schreiber (Chan et al. 2012), and Sexl (Schneckenleithner et al. 2011) clarified the STAT1 role in the formation of mammary tumors. In spite of the implementation of various experimental ways, all these studies led to similar outcomes and supported the STAT1 role in tumor suppression in breast tumorigenesis. Hennighausen et al. reported the generation of mice bearing a STAT1 floxed (flx) allele (Klover et al. 2010). The $STAT1^{flx/flx}$ mouse was crossed to a mouse with ErbB2/neu oncogene expression (deemed NIC) while controlling the promoter of mouse mammary tumor virus (MMTV) (Klover et al. 2010). Originally, the MMTV-NIC mouse was for the expression of Cre recombinase while being controlled by an internal ribosome entry site (IRES) from the same dicistronic mRNA as the NIC oncogene (MMTV-NIC-IRES-Cre) (Ursini-Siegel et al. 2008). Mating of $STAT1^{flx/flx}$ mice with MMTV-NIC-IRES-Cre mice on background of FVB led to deletion of STAT1 in the same epithelium cell of breast with NIC expression (Klover et al. 2010). First, after 36 of birth, tumors could be detected in both groups, though the total latency of the disease was remarkably improved in mice with STAT1 deficiency (49.4 weeks) compared with STAT1-proficient animals (62.4 weeks). Since STAT1 was expressed in all cells in the microenvironment of tumor using this way, the STAT1 antitumor role was thus undeniably associated with the cell-intrinsic features in the epithelium of breast.

The Koromilas group used a tumorigenesis model in vivo. In this model, mice expressed ErbB2/Neu's active form controlled by the MMTV promoter (Siegel et al. 1999) and, on background of Balb/c, were crossed with $STAT1^{+/+}$and $STAT1^{-/-}$ mice (Durbin et al. 1996). Moreover, Neu NDL2–5-positive $STAT1^{-/-}$ female mice who had borne one litter of pups were observed to develop tumors nearly 6 weeks sooner in comparison to $STAT1^{+/+}$ counterparts (27 weeks vs. 33 weeks) (Raven et al. 2011). The reduced tumor formation latency was intensified in virgin females; $STAT1^{-/-}$ mice developed tumors at an average of 41.6 weeks while the $STAT1^{+/+}$ mice with tumors at an average of 49.1 weeks. Once development of tumor was initiated, no detectable difference was found in the number or size of tumors (Raven et al. 2011). Consequently, STAT1 acted as a suppressor for the formation of Neu NDL2–5 mammary gland tumor in mice (Raven et al. 2011). Elevated formation of tumor in $STAT1^{-/-}$ mice did not occur due to changes in morphology of the breast gland in the animals that were consistent with former findings showing no defects in mammary development in $STAT1^{-/-}$ mice (Watson 2001).

The Schreiber and Sexl groups investigated the spontaneous growth of breast tumors to assess the effect of STAT1 in vivo (Schneckenleithner et al. 2011; Chan et al. 2012). A greater tumor incidence above 90% was reported by the Schreiber group in the parous animals, while a tumor incidence of 55% was established by the Sexl group. Sexl et al. showed no tumor in virgin females, though Schreiber et al. reported a 65% incidence of tumor in virgin animals. Moreover, minute differences were shown in receptor expression in the tumors of both researches; in the Schreiber et al.'s study, tumors showed an estrogen receptor-positive $(ER)^{+}$ phenotype nearly resembling human progesterone receptor $(PR)^{+}/ER^{+}$ tumors, while in Sexl et al.'s study, only 50% of the $STAT1^{-/-}$ tumors were $ER^{+}$. The reason for these differences is not yet understood, though the different genetic backgrounds may be involved, i.e., 129S6/SvEV in the Schreiber et al.'s researches vs. Balb/c in the Sexl et al.'s researches.

Both researches of the Sexl and Schreiber labs revealed an improved manifestation of mammary intraepithelial neoplasias (MINs) from lack of STAT1 (Schneckenleithner et al. 2011; Chan et al. 2012). MINs denote precancerous lesions that may potentially develop carcinomas and are considered to be a consequence of intensified proliferation and/or reduced apoptosis in the epithelium cells of breast. STAT1 is involved in both procedures due to its capacity for inducing the genes expression inhibiting proliferation of cell and/or inducing apoptosis (Kim and Lee 2007). The role of STAT1 transcription in the apoptosis of the mammary gland was inferred by the Schreiber group's findings in which the transcriptionally inactive mutant form of STAT1 (Y701F) could not induce apoptosis (Chan et al. 2012). Yet, no signs of apoptosis or cell death were observed in the experimental system of the Sexl group. Contrarily, they observed improved proliferation in vivo in tumorigenic and non-tumorigenic breast epithelium of $STAT1^{-/-}$ cells (Schneckenleithner et al. 2011). Additional proof for STAT1's role as the fundamental regulator of the proliferation of breast epithelium cells was derived from three-dimensional (3D) culture researches, which was employed by the Sexl group to make a comparison of mammospheres formation from primary epithelium cells of breast in virgin $STAT1^{-/-}$ compared with $STAT1^{+/+}$ mice. It was reported that remarkably thicker mammosphere layers were formed by $STAT1^{-/-}$ cells with an elevated rate of proliferation, and, in some cases, they could fill in the acini lumen (Schneckenleithner et al. 2011). The usefulness of the 3D culture approach may be proved in the study of the MIN formation mechanisms in STAT1-deficient cells of epithelium.

The STAT1 antitumor effect is correlated with its potential to induce the tumors' immune surveillance (Dunn et al. 2006); thereby, Koromilas et al. additionally investigated the STAT1's contribution in the stromal compartment to inhibit tumorigenesis of breast meditated by Neu NDL2–5. Thus, Neu NDL2–5 $STAT1^{-/-}$ and $STAT1^{+/+-}$ isolated tumor cells of breast of mice and underwent evaluations of orthotopic transplantation in the syngeneic $STAT1^{-/-}$ or $STAT1^{+/+}$ mice. It was shown that the transplanted tumors' growth was not remarkably different between tumor cells of $STAT1^{-/-}$ and $STAT1^{+/+}$ in the $STAT1^{+/+}$ recipient mice that were observed for nearly 9 weeks (Raven et al. 2011). Yet, tumor development of Neu NDL2–5 $STAT1^{-/-}$ transplanted cells of breast tumor during the same period was remarkably promoted in comparison with $STAT1^{+/+}$ cells of breast tumor in the $STAT1^{-/-}$ recipient mice (Raven et al. 2011). Breast tumor histopathologic analysis revealed phenotypes of two tumor: a typical solid, nodular ErbB2/Neu-type tumor and a glandular papillary tumor, showing no difference between $STAT1^{-/-}$ and $STAT1^{+/+}$ mice (Raven et al. 2011). These data showed the tumor-site specific presence and effects of stroma in suppressing tumorigenesis of breast meditated Neu NDL2–5 by due to STAT1.

Additional evidence for a fundamental STAT1'role in the tumoral microenvironment was provided by the Sexl and Schreiber groups studies (Schneckenleithner et al. 2011; Chan et al. 2012). The Schreiber group described the selective STAT1 downregulation in a large cohort study of patients according to histochemistry analysis on biopsies of tumor (Chan et al. 2012). Downregulation of STAT1 was highly noticeable in cells of the tumor in comparison with the infiltrating lymphocytes and stromal surrounding. The STAT1 expression in the infiltrating lymphocytes and stroma of tumor may clarify why many studies have shown high levels of STAT1 in the samples of mammary cancer under experimental conditions, not allowing cells of tumor stroma separation from the other cells of tumor, including microarrays or western blotting of total tumors. These data underlined the fundamental role of the development of tumor microenvironment and the request for analysis of resolution of single cell to show the protein expression to a distinctive compartment of cells. It was shown by the Sexl group that deficiency of STAT1 serves to remarkably improve the frequency of spontaneous development of breast tumors in mice (Schneckenleithner et al. 2011). Also, the same group showed the dual STAT1 activity in suppression of the formation of breast tumor: however, STAT1 preserved the surveillance of tumor dependent to cytotoxic T lymphocyte (CTL) and was needed for full-fledged T cell cytotoxicity; however, STAT1 inhibited the proliferation of cells in the epithelium cells of breast in a cell-autonomous way (Schneckenleithner et al. 2011). That the STAT1 can control the immune cells' activity plays a crucial role in tumorigenesis regulation; i.e., preserved CTL activation by STAT1 leads to the suppression of breast tumors. Contrarily, STAT1 loss in the leukemic cells induces the robust natural killer (NK) cells activation by downregulation of MHC1 molecules (Kovacic et al. 2006).

To comprehend the STAT1's antitumor mechanism, Koromilas et al. investigated the phosphorylation of STAT1 regulation in signaling meditated by ErbB2. They showed that phosphorylation of STAT1 Y701 is reduced in ErbB2-positive cells of human mammary tumor following treatment with ErbB2 drug inhibitors, though not following treating with Src or JAKs inhibitors (Raven et al. 2011). Their findings showed that the activation of ErbB2 leads to elevated STAT1 Y701 phosphorylation through the mediation of signaling pathways, a notion further confirmed in STAT1's transient expression assays as well as an activated Neu form (Raven et al. 2011). They showed that STAT1 is an EGFR's substrate in vitro, though it is still unclear whether elevated phosphorylation of STAT1 Y701 in tumor cells of breast is directly affected by ErbB2 (Raven et al. 2011). Recent findings further supported ErbB2 and STAT1 functional interactions demonstrating the induced STAT1 expression via ErbB2 at the level of transcription via activation of STAT3 (Han et al. 2013).

Attempting to clarify the downstream activity of STAT1 of ErbB2/Neu, the Koromilas group investigated the impacts of phosphorylation of the STAT1's defective mutants, STAT1S727A and STAT1Y701F, in transformation meditated by ErbB2/Neu. It was shown that the STAT1 phosphorylation mutants' expression in assays of xenograft tumor of SCID mice compromised by immune efficiently impaired the Neu NDL2–5 transforming properties in $p53^{-/-}STAT1^{-/-}$ mouse embryonic fibroblasts (MEFs) (Raven et al. 2011). This conclusion proposed that STAT1 is able to inhibit mediated tumorigenesis meditated by ErbB2/Neu regardless of phosphorylation at either S727 or Y701. Since STAT1 and ErbB2/Neu signaling networks depend on context, it is still probable that the phosphorylated STAT1 has another activity in MEFs compared with its activity in epithelium cells of the breast. The study of transformation mediated by ErbB2/Neu was conducted in MEFs deficient of p53 (Raven et al. 2011). Thus, it is crucial to assess the impacts of STAT1 phosphorylation mutants in Neu NDL2–5-transformed $STAT1^{-/-}$ epithelium cells of the breast with wild-type p53 (Ursini-Siegel 2011). A fascinating study performed by the Koromilas group showed that the Neu NDL2–5 at tyrosine residues in addition to Y701 in transiently transfected cells can phosphorylate STAT1 (Raven et al. 2011). Up to now, little information is available regarding STAT1 regulation via phosphorylation at sites other than S727 and Y701. Earlier researches by the Maniatis group revealed the STAT1 phosphorylation via IKKε at S708, an alternation contributing to the promotion of transcription of a certain set of IFN-inducible genes (Tenoever et al. 2007). It is still probable that ErbB2 signaling results in the STAT1 phosphorylation at novel sites contributing to its antitumor activities in the breast gland.

The Sexl group suggested that interferon regulatory factor 1 (IRF-1) mediates the STAT1's inhibitory effects on ER-positive growth of epithelium of the breast (Schneckenleithner et al. 2011). Considering that STAT1 induces IRF-1 at the level of transcription, this description is in line with the results of the Schreiber group indicating that inhibition of tumor growth of $ER^{+}$ mouse needs a transcriptionally active STAT1 (Chan et al. 2012). IRF1 role was demonstrated experimentally by the remarkable downregulation of IRF-1 in tumors with deficiency in STAT1 as well as the structural similarities found between $IRF1^{-/-}$ and $STAT1^{-/-}$ breast tissues based on the 3D mammospheres (Schneckenleithner et al. 2011). Studies that IRF1 is often heterozygous in tissue of human mammary tumor, as well as its downregulated expression in high-grade mammary cancers, further supports IRF1's potential role in breast cancer (Schneckenleithner et al. 2011). Approval of the IRF1 activity downstream of STAT1 requires using $STAT1^{-/-}IRF1^{-/-}$ mice to assess the effect of the STAT1-IRF1 axis deficiency on the development of the mammary tumor. Moreover, it is interesting to describe whether downregulation of IRF1 in human mammary tumors is correlated with reduced expression of STAT1 or expression of a transcriptionally inactive STAT1.

## 2.3 STAT2 and Breast Cancer

Although the human breast tumors' percentage expressing STAT2 has not been determined, STAT2 has been found in MCF-7 cells, a cell line in ER+ human breast cancer (Schaber et al. 1998; Uluer et al. 2012). Uluer et al. revealed prominent cytoplasmic localization, and Schaber et al. reported the presence of STAT1-STAT2 heteromeric complexes following stimulation of type-I interferon.

## 2.4 STAT3 and Breast Cancer

Quantitative image analysis was used to look at tyrosine-phosphorylated STAT3 expression in 45 Stage III invasive breast tumors (Diaz et al. 2006). Phospho-STAT3 expression was found in 52% of the tumors, and it was linked to HER2 positive. Lower phospho-STAT3 levels were linked to a full pathogenic reaction. In a tissue microarray investigation of 346 node-negative breast tumors, STAT3 was found in the cytoplasm and nucleus of 69% and 23% of the tumors, respectively, while phospho-STAT3 (Tyr705) was found in the cytoplasm and nucleus of 23% and 44% of the malignancies, respectively (Dolled-Filhart et al. 2003). The nuclear STAT3 presence staining was linked to better survival in this study.

Phosphorylated STAT3 levels were measured in 923 specimens using a SAMBA 2050 automated equipment, which found expression of phosph-STAT3 in 34% of the samples and linked it to a worse prognosis (Charpin et al. 2009). In a study on 571 breast tumors, STAT3 expression was detected in 41% of the tumors, with no correlation to status of ER or prognosis (Yamashita et al. 2006). STAT3 ser727 phosphorylation was found to be lower in ER+ tumors than in ER malignancies in one research of 68 infiltrating ductal carcinomas, with equal expression level in HER2-positive and HER2-negative tumors (Yeh et al. 2006). In conclusion, these studies show that STAT3 is expressed in a large amount of all breast cancer subtypes; however, the prognostic importance and levels of STAT3 in various breast cancer subtypes vary.

The STAT3 signaling pathway, activated through growth factors or cytokines binding to their corresponding cell surface receptors, has been widely investigated (Qin et al. 2019; Ma et al. 2020). A short summary of the STAT3 signaling, STAT3 non-receptor tyrosine kinases, and intrinsic coactivators and inhibitors is provided in the following. Cytokine receptors when overexpressed, such as interleukin-10 receptor (IL-10R) and interleukin-6 receptor (IL-6R), as well as hyperactive growth factor receptors, such as insulin-like growth factor receptor (IGFR), fibroblast growth factor receptor (FGFR), and epidermal growth factor receptor (EGFR), always stimulate the phosphorylation cascade of tyrosine via ligands binding to these receptors (Yu et al. 2014). Once ligands bind to their cell surface receptors, mentioned receptors can form dimers and recruit JAKs and glycoprotein 130 (gp130), phosphorylating and activating JAKs (Garbers et al. 2015).

In contrast, active JAKs phosphorylate these receptors' cytoplasmic tyrosine residues, which subsequently bind with STAT3's SH2 domain, leading to JAK phosphorylation of STAT3 at Tyr705 (Johnson et al. 2018). Several non-receptor tyrosine kinases, including Src and Abl, may also phosphorylate and activate STAT3 (Karras et al. 1997). Through contact between their phosphorylated Tyr705 site and the SH2 domain, phosphorylated STAT3 (pSTAT3) can form a homodimer, prompting the difference of dimers of STAT3 in cell surface receptors and their translocation from the cytoplasm into the nucleus (Yu et al. 2007; Wang et al. 2018). The nuclear STAT3 binds to particular sequences of DNA and promotes the gene transcription that governs different cancer cells' phenotypes with the aid of coactivator proteins such as apurinic/apyrimidinic endonuclease-1/redox factor-1 (APE/Ref-1), NCOA/SRC1a, and CREB binding protein (CBP)/p300 (Yu et al. 2014; Guanizo et al. 2018).

STAT3's oncogenic ability in regulation of genes expression involved in proliferation of cancer cell, invasion, chemoresistance, angiogenesis, migration, stem cell self-renewal and maintenance, anti-apoptosis, immunological suppression, and autophagy has been extensively acknowledged (Yu et al. 2014; Guanizo et al. 2018). Notably, STAT3 is overexpressed and constitutively active in TNBC, which is linked to the disease's genesis, development, metastasis, chemotherapy resistance, and poor prognosis (Sirkisoon et al. 2018). STAT3 promotes the aggressiveness of TNBC by physically interacting with and functionally cooperating with other oncogenic and transcriptional factors, such as GLI1, to induce the expression of cancer-related genes (Sirkisoon et al. 2018). A decrease in an intrinsic inhibitor of STAT3 transcription, the gene correlated with retinoic-interferon-induced mortality 19 (GRIM-19), was also discovered in TNBC in recent research (Zhou et al. 2013). TCPTP, comprising the two splice variants TC48 and TC45, is likewise downregulated in TNBC cells in vitro and in vivo, contributing to STAT3 signaling activation (Shields et al. 2013). Indeed, STAT3 can localize in the mitochondria, where it is known as mitoSTAT3 and affects mitochondrial processes such as the chain of electron transport, ATP production, homeostasis of calcium, and accumulation of ROS (Wegrzyn et al. 2009; Yang and Rincon 2016). Furthermore, mitoSTAT3 has been demonstrated to enhance breast cancer cell proliferation, with Serine 727 phosphorylation playing a key role (Zhang et al. 2013). TNBC has a high level of acetylated STAT3, which causes inactivation and methylation of promoters of tumor suppressor gene, according to recent research (Lee et al. 2012). Importantly, mutating STAT3 at Lys685 or lowering acetylation of STAT3 with resveratrol may cause estrogen receptor gene demethylation and activation, making TNBC cells more sensitive to antiestrogens.

## 2.5 STAT4 and Breast Cancer

STAT4 expression has been observed both in MCF-7 cells (Liu et al. 2012; Uluer et al. 2012) and breast cancer cells. High expression of IL-12/STAT4 axis molecules including the IL-12 receptor genes and STAT4 were reported to significantly increase survival of breast cancer patients, particularly in the most aggressive subtypes including HER-2+, the luminal B (LumB), and basal like (Núñez-Marrero 2019). According to the profile integrative molecule, variations in STAT4 signaling are involved in the difference between low and high mammographic density (Kristensen et al. 2012). Because the researchers investigated the whole tissue, it was unclear if the differences in expression of STAT4 were observed in epithelium or stroma.

## 2.6 STAT5a/b and Breast Cancer

The role of STAT5 has been widely investigated in breast cancer (Tan and Nevalainen 2008). In 76% of 78 human breast adenocarcinomas, nuclear-localized tyrosine-phosphorylated STAT5a was identified (Cotarla et al. 2004). High levels of nuclear-localized p27 histological differentiation and expression were linked to the presence of nuclear-localized STAT5a. On tissue arrays of roughly 160, 683 and 443 specimens of breast cancer, expression of tyrosine-phosphorylated STAT5a/b was analyzed (Nevalainen et al. 2004). STAT5a/b activation was found in 19% of node-positive breast and 32% of node-negative breast cancers, and it was linked to a better prognosis. Another study found that 34% of 517 breast tumors tested positive for STAT5a/b, with favorable relationships between STAT5a/b with histologic grade, PR, and ER (Yamashita et al. 2006). Recent research on nuclear-localized phosphorylated STAT5a/b showed that 42% of tumors had increased levels of staining for STAT5a/b as opposed to low levels, and that these increased levels were associated with a better prognosis (Peck et al. 2011). Altogether, the findings show that nuclear-localized phosphorylated STAT5a/b is expressed in a considerable breast tumors' proportion and that this is a good prognostic signal in general.

## 2.7 STAT6 and Breast Cancer

Expression of STAT6 is now mainly characterized in cell lines of human breast cancer, but not in vast series of human breast cancer tissue, as is the case with STAT2 and STAT4. The ER+ cell line of breast cancer MCF-7 and ZR-75-1 has been found to express STAT6 (Gooch et al. 2002; Zhang et al. 2008; Godinho et al. 2012; van Agthoven et al. 2012).

## 2.8 ER+ Breast Cancer Subtypes and STAT Signaling

In ER+ human breast malignancies, STAT 1, STAT 3, and STAT 5a/b are all expressed, while STAT 2, STAT 4, and STAT 6 are expressed in cell lines of ER+ breast cancer, as shown previously. Except one study (Choi et al. 2013), several investigations so far have looked at ER+ breast cancer as a whole rather than dividing the samples into luminal A and B subtypes. Two researches found a connection between expression of STAT1 and positivity of ER (Magkou et al. 2012; Choi et al. 2013), whereas another study found that expression levels of STAT1 were decreased in ER+ malignancies than in ER− cancers (Chan et al. 2012). In 346 ER+ breast tumors, the effect of STAT3 and STAT5a/b expression on tamoxifen responsiveness was investigated (Yamashita et al. 2006). STAT5a/b-positive tumors had a better prognosis, according to the study, with an elevated rate of response to treatment with endocrine and longer life following relapse. STAT3 expression, on the other hand, did not appear to be linked to treatment response or prognosis. In two separate cohorts ($n = 221$ and $n = 97$) receiving antihormonal medication with or without adjuvant chemotherapy, the positive connection between expression of STAT5 and enhanced response to treatment with endocrine was established again (Peck et al. 2011). The antihormonal medicine type used in the first group was not specified, whereas tamoxifen was used in the second. In a small group of 16 patients, however, STAT5a/b expression levels did not have any prediction of response to a 6-month course of exemestane, a steroidal irreversible aromatase inhibitor (Yamashita et al. 2009). In fact, STAT5 expression increased after therapy in our study. The apparent consistency of the association between increased levels of expression of STAT5a/b nuclear and tamoxifen responsiveness justifies more investigation of the mechanism at work, especially since some tissue culture data contradicts with these clinical findings (Riggins et al. 2006). The activation of STAT5a/b and other activated STATs in relation to expression patterns of gene connected to the luminal A and B subtypes may give an overview of the differing prognostic results and responses to treatment with endocrine in ER+ breast tumors (Chia et al. 2012; Geyer et al. 2012).

## 2.9 HER2/Neu + Breast Cancer and STAT Signaling

Expression of STAT3 has been linked to HER2 overexpression (Diaz et al. 2006). In HER2 breast cancer stem cells, a HER2-STAT3 signaling network has been discovered (Duru et al. 2012), and therapy targeted HER2 lowers STAT3 gastric cancer phosphorylation (Kim et al. 2008). In cell lines, STAT3 and HER2 mechanistic linkages have leptin-induced elevates in expression of HER2 via STAT3 signaling (Giordano et al. 2013) and STAT1 expression improvement by STAT3 and HER2 (Han et al. 2013). The fact that only a small percentage of HER2-amplified breast tumors express STAT1 (Choi et al. 2013) could mean that this pathway is only active in a subset of human HER2-amplified breast cancers or that it is highly relevant in vitro than in vivo. Further research into the involvement of STAT in HER2-amplified breast cancers, particularly STAT3 and STAT5a/b, can be useful to determine if STAT have a role in prognosis or responsiveness to therapy in HER2-amplified breast cancers.

## 2.10 TNBC and STAT Signaling

According to one study, the fraction of TNBCs with expression of STAT1 in cancer cells is likely to be low, with the expression occurring more commonly in cells of stroma (Choi et al. 2013). On the other hand, two researches recommended that the expression of STAT1 in TNBC is associated with a better prognosis (Charpin et al. 2009; Yau et al. 2010); however, one study looked at STAT1 levels in TNBC cancerous cells and found a link between increased expression levels and node-positive disease (Greenwood et al. 2012). They also found that STAT2 and STAT3 were expressed at similar levels in node-negative and node-positive breast tumors, however, not in TNBCs. In vitro studies of TNBC cells show that decreasing the acetylation of STAT3 via resveratrol results in elevated ER expression and tamoxifen sensitivity's emergence (Lee et al. 2012) and that decreasing activation of STAT3 via administration of penta-*O*-galloyl-β-D-glucose (a herbal compound) decreases growth and metastases of xenograft (Lee et al. 2011), and this decrease works in tandem with metformin to slow the proliferation of cell lines of TNBC (Deng et al. 2012). After addressing the phosphatidylinositol 3-kinase (PI3K)/mammalian target of rapamycin (mTOR) pathway, JAK2/STAT5 activation can reduce the effect of this therapeutic introduction (Britschgi et al. 2012). TNBC therapy may need dual targeting of JAK2/STAT5 and PI3K/mTOR. Because gene array profiling can further classify TNBCs into subtypes (Lehmann et al. 2011), more delineation of expression patterns of STAT within these subtypes is needed before strong conclusions can be reached about the effect of various STAT proteins on prognosis and treatment response.

## 3 STAT Proteins: The Promising Targets in the Treatment of Breast Cancer

The etiology of various ranges of solid malignancies, including ovarian, breast, and prostate cancer, as well as malignant blood neoplasms, including myeloma and acute lymphoblastic leukemia, has been linked to the constitutive STAT3 and/or STAT5 activation. However, because abnormal activation of STAT3 was discovered more regularly than dysregulated STAT5, STAT5 has gotten less attention than STAT3, which is a therapeutic aim (Loh et al. 2019). Consequently, the focus of this chapter is on STAT3 as a possible therapeutic aim for breast cancer.

A member of the tyrphostin family of the inhibitors of tyrosine kinases small molecule like JAK2, AG490, demonstrated an ability to block the JAK2 tyrosine phosphorylation, which led to the STAT3 tyrosine phosphorylation inhibition. In MDA-MB 231 human breast cancer cell line, AG490 suppressed the phospho-STAT3 (Tyr705) expression as well as survivin, an anti-apoptotic protein, therefore blocking proliferation of cancer cells via induction of cell death (Yu and Deng 2006). Moreover, AG490 in MDA-MB-231 cells significantly inhibited cell adhesion, invasion, as well as metastasis (Chen et al. 2008). In another in vitro study, a modified structure of AG490, WP1066, overcame the TNBC drug resistance to Doxorubicin (Cheng et al. 2018). WP1066 is presently going through phase 1 RCTs in patients with melanoma which has spread to the brain or patients with malignant glioma which has come back to investigate the safety and tolerability and detect the maximum tolerated dose (NCT01904123). Tofacitinib is a targeted inhibitor of JAK3 and JAK1 and also has a partial inhibitory effect on JAK2 (Schwartz et al. 2016). An FDA-approved drug Tofacitinib is being used as a therapeutic medication in clinics for neoplastic and diseases inflammatory like ulcerative colitis, rheumatoid arthritis, and psoriasis (Dhillon 2017). It has been found through a phospho-proteome analysis that JAK/STAT signaling pathway exerts a role in the chemotherapy resistance of MCF7 breast cancer cells. Treatment with Tofacitinib in chemoresistant MCF7 cells was shown to increase chemosensitivity, resulting in higher rates of cancer cell death. This suggests the potential of Tofacitinib as a novel therapeutic agent to combat breast cancer cells chemoresistance (Nascimento et al. 2017). Studies have also shown that Tofacitinib could be of high efficacy in the anti-drug antibody immune responses reduction, often occurring when immunotoxins are given to study participants or testing animals with a healthy immune system (Onda et al. 2014).

Furthermore, Tofacitinib treatment in mice with MDA-MB-468 xenografts enhanced the antibody-based agent's delivery to tumor cells via cytokine signaling reducing as well as suppressing the recruitment of malignancy-related inflammatory cells. Accordingly, an increase in immunotoxin efficacy was reported in xenograft mouse models of TNBC (Simon et al. 2019). It has been also reported that, as a single agent, Tofacitinib caused no remarkable adverse effects on solid tumors in RCTs. The limited Tofacitinib effectiveness could be due to the RCT designs (only late-stage patients being chosen and most of whom previously failed to several medical interventions). This can also be explained by the fact that JAKs aberrant activation is not completely identified in solid tumors yet (Huynh et al. 2017). Therefore, Tofacitinib could still be considered as a promising sensitizer when used combined with other chemotherapy medications.

Small molecules, peptides, derivatives of natural product, and peptidomimetic molecules, oligonucleotides, and other inhibitors of STAT3 can be divided into many various types based on their chemical structures: small molecules, oligonucleotides, peptidomimetic molecules, natural product derivatives, and so on. In addition, inhibitors targeting the SH2 domain, as well as those aiming the DNA binding domain or other domains, can be classified into a few different classes based on the aimed STAT3's structure or region (Chai et al. 2016). From the basic scaffold of chemicals, many inhibitory peptidomimetic molecules and peptides of STAT3 have been produced. The first peptide used was Pro-pTyr-Leu-Lys-Thr-Lys, which can bind to natural C-terminal STAT3-SH2 domain and hence precludes hetero- or homodimerization with phosphorylated STAT3 monomer (Turkson et al. 2001). Peptidomimetics outperform peptides in terms of pharmacokinetic characteristics. Even inhibitors based on peptides can have a high affinity for STAT3, yet, because of their negative charge on the phosphotyrosine group and peptidic structure, majority of them have deficiency in permeability of cell. As a result, several novel peptidomimetic compounds have been created. CJ 1383, a new cell-permeable inhibitor, was found to inhibit STAT3 effectively, leading to MDA-MB-468 cell line death of breast in a dose- and time-dependent way (Chen et al. 2010). Small molecule drugs and STAT3-SH2 domain interaction had potent inhibitory impacts on STAT3 dimerization, translocation of nucleus, and activation of transcription, similar to peptidomimetics. Static is the first non-peptide small molecule produced to be a STAT3 inhibitor. It suppressed the development and survival of various cell lines of breast cancer, including MDAMB-231 and MDA-MB-435, via selectively disrupting STAT3 dimerization over other STAT members and in an efficient way interacting with the STAT3 SH2-domain (Song et al. 2005). Noori et al. showed that naringenin could enhance cyclophosphamide's anticancer effect against MDA-MB-231 breast cancer cells through targeting the STAT3 signaling pathway. They reported that naringenin could trigger apoptosis and significantly decrease cell viability. The coadministration of naringenin with cyclophosphamide enhanced its antitumor activity. Naringenin was also shown to decrease the expression of Bcl-2 and upregulate the expression of BAX. Naringenin activated caspases 3 and 9; this was also augmented via cyclophosphamide (Noori et al. 2020). LLL12 and LLY17 are two structural analogs of STA-21 that have been produced. In numerous cell lines of human cancer, such as U87, SK-BR-3, U373, HPAC, and PANC1, they all demonstrated a high specificity for inhibiting Tyr705 phosphorylation of STAT3 (84). In vitro cellular and in vivo animal models, cell-permeable compounds LLY17 and XZH-5 displayed efficacy on STAT3, as shown by the reduction of STAT3-induced biological procedures including migration, survival, and tumor growth (Linher-Melville et al. 2019; Liang 2020). Additional drugs targeting the SH2 domain have been created and demonstrated to decrease STAT3. In the murine model of bone pain induced by human breast cancer, for example, DR-1-55 suppresses the inflammatory cytokines' release including IL-6 and IL-1 (Linher-Melville et al. 2019). By targeting the SH2 domain, 6Br-6a revealed inhibitory impacts on volume of tumor in the MDA-MB-231 xenograft mouse model (Liu et al. 2020).

Finally, ODZ10117(3-(2,4-dichloro-phenoxy methyl)-5-trichloromethyl-[1, 2, 4] oxadiazole) is a recently produced small-molecule of STAT3 inhibitor found by structure-based computational database screening to recognize compounds aiming the STAT3 SH2 domain. By participating for the SH2 domain, it effectively blocked STAT3 phosphorylation. Breast cancers in vitro and in vivo models, ODZ10117 inhibited tumor growth, tumor migration and tumor invasion, tumor survival, and lung metastasis, indicating its effectiveness in inhibition of STAT3 (Kim et al. 2019). Some natural chemical entities have also been recognized as functional inhibitors that compete with the SH2 domain to reduce STAT3 activation. In cultured cell lines of breast cancer, curcumin, a component in turmeric, can selectively bound to the SH2 domain and abrogate phosphorylation of STAT3 phosphorylation, causing prevention of dimerization of STAT3 and inactivating STAT3 downstream cascade (Lin et al. 2009). The resveratrol-caffeic acid hybrid molecule 7d, on the other hand, is the first synthetic inhibitor to cause inhibition of acetylation and phosphorylation of STAT3. In breast cancer 4 T1 xenograft models, this novel derivative severely inhibited proliferation of cell via triggering growth of tumor and cell death in vitro (Li et al. 2016). GLG-302 (S3I-201, NSC 74859) is another novel STAT3 inhibitor. Its salicylic acid moiety binds to the STAT3-SH2 domain's pTyr binding site. In a mouse model, it exhibited activity against breast cancer cells and the ability to prevent receptor-negative mammary tumors and estrogen receptor-positive (Siddiquee et al. 2007). Following a computer-aided lead optimization, further STAT3-SH2 dimerization inhibitors were created using in silico site-directed fragment-based drug design. The selective STAT3 inhibitor C188–9 (STAT3 inhibitor XIII, TTI-101) binds to the STAT3-SH2 domain. C188–9 therapy inhibited STAT3 and lowered the programmed death-ligand 1 (PD-L1) expression in 4 T1 breast cancer mice, resulting in tumor growth and metastasis being limited (Zerdes et al. 2019). C188–9, in particular, is currently being tested in a phase I clinical trial for patients with advanced cancer, particularly those with breast cancer (NCT

03195699). All of these data point to the SH2 domain of STAT3 as a promising aim for novel breast cancer treatment approaches. Another type of STAT3 inhibitor targets STAT3's N-terminal region, also known as the transcriptional activation domain, which is necessary for binding of DNA, interaction of protein-protein, nuclear translocation, and activation of transcription.

ST3-H2A2 is a highly specific peptide inhibitor of the STAT3 N-terminal domain that efficiently inhibits transcription and dimerization effects of STAT3. ST3-H2A2 was found to cause cell death in cells of breast cancer MCF-7 and MDA-MB-231 however had no influence on the normal mammary epithelial cells of epithelium's survival MCF-10A, implying that it is cancer cell specific (Timofeeva et al. 2007). Platinum compounds including IS3 295, CPA-1, and CPA-7 have also been demonstrated to inhibit activated STAT3-mediated anti-apoptosis and progression of cell cycle in numerous cell lines of breast cancer by disrupting STAT3's capacity to bind its DNA response element (Turkson et al. 2004; Turkson et al. 2005). Furthermore, STAT3 N-terminal domain inhibitors based on nucleotides, such as STAT3-G-Quartet and STAT3-siRNA, showed significant inhibitory impacts on activity of STAT3 in a variety of in vitro cellular types. These findings imply that aiming the STAT3 N-terminal region may have therapeutic promise in the breast cancer pathogenesis caused by constitutively active STAT3 (Jing et al. 2004; Kunigal et al. 2009). Finally, pyrimethamine (also known as PYR or 2,4-diamino-5-p-chlorophenyl-6-ethyl-pyrimidine) is approved by FDA, an antimicrobial medication that has been introduced to treat toxoplasmosis and malaria. PYR therapy had anticancer and immune-stimulatory benefits in breast cancer mouse models, as well as significant inhibitory impacts on transcriptional activity of STAT3 (Legorreta-Herrera et al. 2010; Takakura et al. 2011; Khan et al. 2018). As a result, it's important to consider PYR's efficacy in the breast cancer treatment. BPMB (5,5′-(pentane-1,5′-diyl) bis (2-methyl-1,4 benzoquinone) was discovered to be another STAT3 transcriptional inhibitor. BPMB inhibited STAT3 transcription activation almost completely without changing its phosphorylation or translocation of nucleus. The mechanism of BPMB was discovered to be inactivation of the STAT3 complex via linker domain acylation (Koseki et al. 2019).

## 4 Conclusion

In both hematologic malignancies and solid tumors, STAT signaling was linked to survival, therapy resistance, metastasis, growth, and progression of cancer. STAT3 and STAT5 seem to play the most important roles in biology of cancer, whereas STAT6, STAT4, STAT2, and STAT1 play a smaller part. However, rather than being caused by a single dominating family member, their effects are currently thought to be caused by transcriptional changes mediated via a network of subtle and complicated interactions among distinct STAT proteins. The strength and length of the response to signal in a specific cell type are determined by the steady condition and signal-inducible function of the entire positive and negative regulators. JAK-STAT signaling can create a transcriptional result, and consequently the tissue response and cell response are greatly influenced by a wide range of external and intrinsic stimuli. Because even small changes in STAT's cellular concentrations may have significant phenotypic repercussions, it's no surprise that their actions are context-dependent and occasionally contradictory. As a result, it's important to remember that screening for the function of a single STAT member may not be enough to associate with disease features, but the balance of (counteracting) family members may. This is important to remember when developing diagnostic or therapeutic tools related to STAT, and further studies of the mutual interactions among STAT proteins is required.

On the other hand, many resources have been devoted to the development of STAT3 inhibitors. Because of the varied activities of STAT3 in survival of cancerous cell, STAT3-targeted treatment may be predicted to generate excellent therapeutic results in breast cancer patients by a direct suppression of tumor cell growth. Furthermore, several oncogenes have been shown to upregulate the production of cytokines in order to JAK/STAT3 pathway activation, increasing tumor development. For instance, the oncogenes BCR-ABL tyrosine kinase and p21 Ras oncoprotein activate JAK/STAT signaling by increasing IL-6 production, indicating that STAT3 inhibitors may similarly limit the oncogenes' oncogenic effects on breast carcinogenesis (Yu et al. 2014). In breast cancer, JAK/STAT signaling has been shown to interact with other consistently activated signaling pathways such as PI3K/Akt (Kim et al. 2015). As a result, it is hypothesized that inhibiting JAK/STAT signaling might at least partially inhibit carcinogenesis induced by oncogenic pathways. More significantly, as previously indicated, preclinical research demonstrates that inhibiting STAT3 activity efficiently prevents and/or reverses breast cancer treatment resistance to several chemotherapeutic medicines like doxorubicin. Blocking JAK/STAT signaling has also been found to be crucial in radiotherapy-induced suppression of sphere-forming capacity and self-renewal of stem cells of breast cancer (Kaushik et al. 2017). STAT3 inhibitors can be utilized as a straight tumor suppressor and as a sensitizer in conjunction with different chemotherapy or radiation for breast cancer, according to this research. However, further study on the JAK/STAT pathway is needed to better understand its biology in diverse cellular contexts and with different kinds of breast cancer backgrounds. Simultaneously, more intensive study on JAK2 and STAT3's current small molecule inhibitors is needed to assess their in vivo efficacies, selectivities, bioavailabilities, and potential adverse health consequences. In a nutshell, the targeted treatment of breast cancer is still fraught with difficulties, but, as oncology research advances, new targets are discovered, and targeted drugs are evaluated; targeted treatment will usher breast cancer treatment into a new personalized healthcare era and will have greater effects on patients with breast cancer.

**Acknowledgments** None.

**Compliance with Ethical Standards** The authors declare that there is no conflict of interest.

## References


Aaronson DS, Horvath CM (2002) A road map for those who don't know JAK-STAT. Science 296: 1653–1655

Bertos NR, Park M (2011) Breast cancer—one term, many entities? J Clin Invest 121:3789–3796

Braunstein J, Brutsaert S, Olson R, Schindler C (2003) STATs dimerize in the absence of phosphorylation. J Biol Chem 278:34133–34140

Britschgi A, Andraos R, Brinkhaus H, Klebba I, Romanet V, Müller U, Murakami M, Radimerski T, Bentires-Alj M (2012) JAK2/STAT5 inhibition circumvents resistance to PI3K/mTOR blockade: a rationale for cotargeting these pathways in metastatic breast cancer. Cancer Cell 22:796–811

Bromberg JF (2001) Activation of STAT proteins and growth control. BioEssays 23:161–169

Brooks AJ, Dai W, O'Mara ML, Abankwa D, Chhabra Y, Pelekanos RA, Gardon O, Tunny KA, Blucher KM, Morton CJ, Parker MW, Sierecki E, Gambin Y, Gomez GA, Alexandrov K, Wilson IA, Doxastakis M, Mark AE, Waters MJ (2014) Mechanism of activation of protein kinase JAK2 by the growth hormone receptor. Science 344:1249783

Cai F, Luis MAF, Lin X, Wang M, Cai L, Cen C, Biskup E (2019) Anthracycline-induced cardiotoxicity in the chemotherapy treatment of breast cancer: preventive strategies and treatment. Mol Clin Oncol 11:15–23

Carbognin E, Betto RM, Soriano ME, Smith AG, Martello G (2016) Stat3 promotes mitochondrial transcription and oxidative respiration during maintenance and induction of naive pluripotency. EMBO J 35:618–634

Chai EZ, Shanmugam MK, Arfuso F, Dharmarajan A, Wang C, Kumar AP, Samy RP, Lim LH, Wang L, Goh BC, Ahn KS, Hui KM, Sethi G (2016) Targeting transcription factor STAT3 for cancer prevention and therapy. Pharmacol Ther 162:86–97

Chan SR, Vermi W, Luo J, Lucini L, Rickert C, Fowler AM, Lonardi S, Arthur C, Young LJ, Levy DE, Welch MJ, Cardiff RD, Schreiber RD (2012) STAT1-deficient mice spontaneously develop estrogen receptor α-positive luminal mammary carcinomas. Breast Cancer Res 14:R16

Charpin C, Secq V, Giusiano S, Carpentier S, Andrac L, Lavaut MN, Allasia C, Bonnier P, Garcia S (2009) A signature predictive of disease outcome in breast carcinomas, identified by quantitative immunocytochemical assays. Int J Cancer 124:2124–2134

Chen HF, Deng HY, Fan WK (2008) effect of Janus kinase inhibitor AG490 on invasion and metastasis of human breast cancer cells. Sichuan Da Xue Xue Bao Yi Xue ban 39:63–65, 83

Chen J, Bai L, Bernard D, Nikolovska-Coleska Z, Gomez C, Zhang J, Yi H, Wang S (2010) Structure-based Design of Conformationally Constrained, cell-permeable STAT3 inhibitors. ACS Med Chem Lett 1:85–89

Cheng CC, Shi LH, Wang XJ, Wang SX, Wan XQ, Liu SR, Wang YF, Lu Z, Wang LH, Ding Y (2018) Stat3/Oct-4/c-Myc signal circuit for regulating stemness-mediated doxorubicin resistance of triple-negative breast cancer cells and inhibitory effects of WP1066. Int J Oncol 53:339–348

Cheon H, Stark GR (2009) Unphosphorylated STAT1 prolongs the expression of interferon-induced immune regulatory genes. Proc Natl Acad Sci 106:9373–9378

Chia SK, Bramwell VH, Tu D, Shepherd LE, Jiang S, Vickery T, Mardis E, Leung S, Ung K, Pritchard KI, Parker JS, Bernard PS, Perou CM, Ellis MJ, Nielsen TO (2012) A 50-gene intrinsic subtype classifier for prognosis and prediction of benefit from adjuvant tamoxifen. Clin Cancer Res 18:4465–4472

Choi J, Kim DH, Jung WH, Koo JS (2013) Differential expression of immune-related markers in breast cancer by molecular phenotypes. Breast Cancer Res Treat 137:417–429

Clevenger CV (2004) Roles and regulation of stat family transcription factors in human breast cancer. Am J Pathol 165:1449–1460

Cotarla I, Ren S, Zhang Y, Gehan E, Singh B, Furth PA (2004) Stat5a is tyrosine phosphorylated and nuclear localized in a high proportion of human breast cancers. Int J Cancer 108:665–671

Decker T, Kovarik P (2000) Serine phosphorylation of STATs. Oncogene 19:2628–2637

Deng XS, Wang S, Deng A, Liu B, Edgerton SM, Lind SE, Wahdan-Alaswad R, Thor AD (2012) Metformin targets Stat3 to inhibit cell growth and induce apoptosis in triple-negative breast cancers. Cell Cycle 11:367–376

Dhillon S (2017) Tofacitinib: a review in rheumatoid arthritis. Drugs 77:1987–2001

Diaz N, Minton S, Cox C, Bowman T, Gritsko T, Garcia R, Eweis I, Wloch M, Livingston S, Seijo E, Cantor A, Lee JH, Beam CA, Sullivan D, Jove R, Muro-Cacho CA (2006) Activation of stat3 in primary tumors from high-risk breast cancer patients is associated with elevated levels of activated SRC and survivin expression. Clin Cancer Res 12:20–28

Ding J, Guo Y, Li Q, Chen J, Hu P, Liu Q, Cao Y, Wu J (2018) The incidence of postoperative radiotherapy-induced acute dermatitis in breast cancer and its influencing factors for Chinese women. Onco Targets Ther 11:1665

Dolled-Filhart M, Camp RL, Kowalski DP, Smith BL, Rimm DL (2003) Tissue microarray analysis of signal transducers and activators of transcription 3 (Stat3) and phospho-Stat3 (Tyr705) in node-negative breast cancer shows nuclear localization is associated with a better prognosis. Clin Cancer Res 9:594–600

Dunn GP, Koebel CM, Schreiber RD (2006) Interferons, immunity and cancer immunoediting. Nat Rev Immunol 6:836–848

Durbin JE, Hackenmiller R, Simon MC, Levy DE (1996) Targeted disruption of the mouse Stat1 gene results in compromised innate immunity to viral disease. Cell 84:443–450

Duru N, Fan M, Candas D, Menaa C, Liu HC, Nantajit D, Wen Y, Xiao K, Eldridge A, Chromy BA, Li S, Spitz DR, Lam KS, Wicha MS, Li JJ (2012) HER2-associated radioresistance of breast cancer stem cells isolated from HER2-negative breast cancer cells. Clin Cancer Res 18: 6634–6647

Eroles P, Bosch A, Pérez-Fidalgo JA, Lluch A (2012) Molecular biology in breast cancer: intrinsic subtypes and signaling pathways. Cancer Treat Rev 38:698–707

Furtek SL, Backos DS, Matheson CJ, Reigan P (2016) Strategies and approaches of targeting STAT3 for cancer treatment. ACS Chem Biol 11:308–318

Furth PA (2014) STAT signaling in different breast cancer sub-types. Mol Cell Endocrinol 382: 612–615

Garbers C, Aparicio-Siegmund S, Rose-John S (2015) The IL-6/gp130/STAT3 signaling axis: recent advances towards specific inhibition. Curr Opin Immunol 34:75–82

Geyer FC, Rodrigues DN, Weigelt B, Reis-Filho JS (2012) Molecular classification of estrogen receptor-positive/luminal breast cancers. Adv Anat Pathol 19:39–53

Giordano C, Vizza D, Panza S, Barone I, Bonofiglio D, Lanzino M, Sisci D, De Amicis F, Fuqua SA, Catalano S, Andò S (2013) Leptin increases HER2 protein levels through a STAT3-mediated up-regulation of Hsp90 in breast cancer cells. Mol Oncol 7:379–391

Godinho MF, Wulfkuhle JD, Look MP, Sieuwerts AM, Sleijfer S, Foekens JA, Petricoin EF 3rd, Dorssers LC, van Agthoven T (2012) BCAR4 induces antioestrogen resistance but sensitises breast cancer to lapatinib. Br J Cancer 107:947–955

Gooch JL, Christy B, Yee D (2002) STAT6 mediates interleukin-4 growth inhibition in human breast cancer cells. Neoplasia 4:324–331

Greenwood C, Metodieva G, Al-Janabi K, Lausen B, Alldridge L, Leng L, Bucala R, Fernandez N, Metodiev MV (2012) Stat1 and CD74 overexpression is co-dependent and linked to increased invasion and lymph node metastasis in triple-negative breast cancer. J Proteome 75:3031–3040

Guanizo AC, Fernando CD, Garama DJ, Gough DJ (2018) STAT3: a multifaceted oncoprotein. Growth Factors 36:1–14

Guiu S, Michiels S, André F, Cortes J, Denkert C, Di Leo A, Hennessy B, Sorlie T, Sotiriou C, Turner N (2012) Molecular subclasses of breast cancer: how do we define them? The IMPAKT 2012 working group statement. Ann Oncol 23:2997–3006

Haan S, KORTYLEWSKI M, Behrmann I, MÜLLER-ESTERL W, HEINRICH PC, SCHAPER F (2000) Cytoplasmic STAT proteins associate prior to activation. Biochem J 345:417–421

Han W, Carpenter RL, Cao X, Lo HW (2013) STAT1 gene expression is enhanced by nuclear EGFR and HER2 via cooperation with STAT3. Mol Carcinog 52:959–969

Huynh J, Etemadi N, Hollande F, Ernst M, Buchert M (2017) The JAK/STAT3 axis: a comprehensive drug target for solid malignancies. Semin Cancer Biol 45:13–22

Jing N, Li Y, Xiong W, Sha W, Jing L, Tweardy DJ (2004) G-quartet oligonucleotides: a new class of signal transducer and activator of transcription 3 inhibitors that suppresses growth of prostate and breast tumors through induction of apoptosis. Cancer Res 64:6603–6609

Johnson DE, O'Keefe RA, Grandis JR (2018) Targeting the IL-6/JAK/STAT3 signalling axis in cancer. Nat Rev Clin Oncol 15:234–248

Jönsson G, Staaf J, Vallon-Christersson J, Ringnér M, Holm K, Hegardt C, Gunnarsson H, Fagerholm R, Strand C, Agnarsson BA (2010) Genomic subtypes of breast cancer identified by array-comparative genomic hybridization display distinct molecular and clinical characteristics. Breast Cancer Res 12:1–14

Kang K, Yamaji D, Yoo KH, Robinson GW, Hennighausen L (2014) Mammary-specific gene activation is defined by progressive recruitment of STAT5 during pregnancy and the establishment of H3K4me3 marks. Mol Cell Biol 34:464–473

Karras JG, Wang Z, Huo L, Howard RG, Frank DA, Rothstein TL (1997) Signal transducer and activator of transcription-3 (STAT3) is constitutively activated in normal, self-renewing B-1 cells but only inducibly expressed in conventional B lymphocytes. J Exp Med 185:1035–1042

Kaushik N, Kim M-J, Kim R-K, Kumar Kaushik N, Seong KM, Nam S-Y, Lee S-J (2017) Low-dose radiation decreases tumor progression via the inhibition of the JAK1/STAT3 signaling axis in breast cancer cell lines. Sci Rep 7:1–9

Khan MW, Saadalla A, Ewida AH, Al-Katranji K, Al-Saoudi G, Giaccone ZT, Gounari F, Zhang M, Frank DA, Khazaie K (2018) The STAT3 inhibitor pyrimethamine displays anticancer and immune stimulatory effects in murine models of breast cancer. Cancer Immunol Immunother 67:13–23

Kim HS, Lee MS (2007) STAT1 as a key modulator of cell death. Cell Signal 19:454–465

Kim SY, Kim HP, Kim YJ, Oh DY, Im SA, Lee D, Jong HS, Kim TY, Bang YJ (2008) Trastuzumab inhibits the growth of human gastric cancer cell lines with HER2 amplification synergistically with cisplatin. Int J Oncol 32:89–95

Kim MS, Lee WS, Jeong J, Kim S-J, Jin W (2015) Induction of metastatic potential by TrkB via activation of IL6/JAK2/STAT3 and PI3K/AKT signaling in breast cancer. Oncotarget 6:40158

Kim BH, Lee H, Song Y, Park JS, Gadhe CG, Choi J, Lee CG, Pae AN, Kim S, Ye SK (2019) Development of Oxadiazole-based ODZ10117 as a small-molecule inhibitor of STAT3 for targeted cancer therapy. J Clin Med 8:1847

Klover PJ, Muller WJ, Robinson GW, Pfeiffer RM, Yamaji D, Hennighausen L (2010) Loss of STAT1 from mouse mammary epithelium results in an increased Neu-induced tumor burden. Neoplasia 12:899–905

Koboldt D, Fulton R, McLellan M, Schmidt H, Kalicki-Veizer J, McMichael J, Fulton L, Dooling D, Ding L, Mardis E (2012) Comprehensive molecular portraits of human breast tumours. Nature 490:61–70

Konjević G (2009) STAT proteins in cancerogenesis and therapy of malignancies. Srp Arh Celok Lek 137:98–105

Koseki T, Suehiro N, Masuda Y, Miyoshi N, Muraoka D, Ogo N, Asai A (2019) Discovery of a new STAT3 inhibitor acting on the linker domain. Biol Pharm Bull 42:792–800

Kovacic B, Stoiber D, Moriggl R, Weisz E, Ott RG, Kreibich R, Levy DE, Beug H, Freissmuth M, Sexl V (2006) STAT1 acts as a tumor promoter for leukemia development. Cancer Cell 10:77–87

Kristensen VN, Vaske CJ, Ursini-Siegel J, Van Loo P, Nordgard SH, Sachidanandam R, Sørlie T, Wärnberg F, Haakensen VD, Helland Å, Naume B, Perou CM, Haussler D, Troyanskaya OG, Børresen-Dale AL (2012) Integrated molecular profiles of invasive breast tumors and ductal carcinoma in situ (DCIS) reveal differential vascular and interleukin signaling. Proc Natl Acad Sci U S A 109:2802–2807

Kunigal S, Lakka SS, Sodadasu PK, Estes N, Rao JS (2009) Stat3-siRNA induces Fas-mediated apoptosis in vitro and in vivo in breast cancer. Int J Oncol 34:1209–1220

Lee HJ, Seo NJ, Jeong SJ, Park Y, Jung DB, Koh W, Lee HJ, Lee EO, Ahn KS, Ahn KS, Lü J, Kim SH (2011) Oral administration of penta-O-galloyl-β-D-glucose suppresses triple-negative breast cancer xenograft growth and metastasis in strong association with JAK1-STAT3 inhibition. Carcinogenesis 32:804–811

Lee H, Zhang P, Herrmann A, Yang C, Xin H, Wang Z, Hoon DS, Forman SJ, Jove R, Riggs AD, Yu H (2012) Acetylated STAT3 is crucial for methylation of tumor-suppressor gene promoters and inhibition by resveratrol results in demethylation. Proc Natl Acad Sci U S A 109:7765–7769

Legorreta-Herrera M, Retana-Ugalde R, Ventura-Gallegos JL, Narváez V (2010) Pyrimethamine induces oxidative stress in plasmodium yoelii 17XL-infected mice: a novel immunomodulatory mechanism of action for an old antimalarial drug? Exp Parasitol 126:381–388

Lehmann BD, Bauer JA, Chen X, Sanders ME, Chakravarthy AB, Shyr Y, Pietenpol JA (2011) Identification of human triple-negative breast cancer subtypes and preclinical models for selection of targeted therapies. J Clin Invest 121:2750–2767

Li S, Zhang W, Yang Y, Ma T, Guo J, Wang S, Yu W, Kong L (2016) Discovery of oral-available resveratrol-caffeic acid based hybrids inhibiting acetylated and phosphorylated STAT3 protein. Eur J Med Chem 124:1006–1018

Liang R (2020) Letter to the editor re: pan et al. LLY17, a novel small molecule STAT3 inhibitor induces apoptosis and suppresses cell migration and tumor growth in triple-negative breast cancer. Breast Cancer Res Treat 184:653

Lim CP, Cao X (2006) Structure, function, and regulation of STAT proteins. Mol BioSyst 2:536–550

Lim EJ, Hong DY, Park JH, Joung YH, Darvin P, Kim SY, Na YM, Hwang TS, Ye SK, Moon ES, Cho BW, Do Park K, Lee HK, Park T, Yang YM (2012) Methylsulfonylmethane suppresses breast cancer growth by down-regulating STAT3 and STAT5b pathways. PLoS One 7:e33361

Lin L, Hutzen B, Ball S, Foust E, Sobo M, Deangelis S, Pandit B, Friedman L, Li C, Li PK, Fuchs J, Lin J (2009) New curcumin analogues exhibit enhanced growth-suppressive activity and inhibit AKT and signal transducer and activator of transcription 3 phosphorylation in breast and prostate cancer cells. Cancer Sci 100:1719–1727

Linher-Melville K, Sharma M, Nakhla P, Kum E, Ungard R, Park J, Rosa D, Gunning P, Singh G (2019) Inhibiting STAT3 in a murine model of human breast cancer-induced bone pain delays the onset of nociception. Mol Pain 15:1744806918823477

Liu L, McBride KM, Reich NC (2005) STAT3 nuclear import is independent of tyrosine phosphorylation and mediated by importin-α3. Proc Natl Acad Sci 102:8150–8155

Liu S, Li L, Zhang Y, Zhang Y, Zhao Y, You X, Lin Z, Zhang X, Ye L (2012) The oncoprotein HBXIP uses two pathways to up-regulate S100A4 in promotion of growth and migration of breast cancer cells. J Biol Chem 287:30228–30239

Liu Z, Ge X, Gu Y, Huang Y, Liu H, Yu M, Liu Y (2020) Small molecule STAT3 inhibitor, 6Br-6a suppresses breast cancer growth in vitro and in vivo. Biomed Pharmacother 121:109502

Loh C-Y, Arya A, Naema AF, Wong WF, Sethi G, Looi CY (2019) Signal transducer and activator of transcription (STATs) proteins in cancer and inflammation: functions and therapeutic implication. Front Oncol 9:48–48

Lu S, Singh K, Mangray S, Tavares R, Noble L, Resnick MB, Yakirevich E (2013) Claudin expression in high-grade invasive ductal carcinoma of the breast: correlation with the molecular subtype. Mod Pathol 26:485–495

Ma J-H, Qin L, Li X (2020) Role of STAT3 signaling pathway in breast cancer. Cell Communic Signal 18:33–33

Magkou C, Giannopoulou I, Theohari I, Fytou A, Rafailidis P, Nomikos A, Papadimitriou C, Nakopoulou L (2012) Prognostic significance of phosphorylated STAT-1 expression in premenopausal and postmenopausal patients with invasive breast cancer. Histopathology 60: 1125–1132

Mandal M, Powers SE, Maienschein-Cline M, Bartom ET, Hamel KM, Kee BL, Dinner AR, Clark MR (2011) Epigenetic repression of the Igk locus by STAT5-mediated recruitment of the histone methyltransferase Ezh2. Nat Immunol 12:1212–1220

McGale P, Correa C, Cutter D, Duane F, Ewertz M, Gray R, Mannu G, Peto R, Whelan T, Darby S (2014) Effect of radiotherapy after mastectomy and axillary surgery on 10-year recurrence and 20-year breast cancer mortality: meta-analysis of individual patient data for 8135 women in 22 randomised trials. Lancet 383:2127–2135

Mouridsen HT (2006) Incidence and management of side effects associated with aromatase inhibitors in the adjuvant treatment of breast cancer in postmenopausal women. Curr Med Res Opin 22:1609–1621

Nascimento AS, Peres LL, Fari AVS, Milani R, Silva RA, da Costa Fernandes CJ, Peppelenbosch MP, Ferreira-Halder CV, Zambuzzi WF (2017) Phosphoproteome profiling reveals critical role of JAK-STAT signaling in maintaining chemoresistance in breast cancer. Oncotarget 8: 114756–114768

Ndubuisi MI, Guo GG, Fried VA, Etlinger JD, Sehgal PB (1999) Cellular physiology of STAT3: Where's the cytoplasmic monomer? J Biol Chem 274:25499–25509

Nevalainen MT, Xie J, Torhorst J, Bubendorf L, Haas P, Kononen J, Sauter G, Rui H (2004) Signal transducer and activator of transcription-5 activation and breast cancer prognosis. J Clin Oncol 22:2053–2060

Noori S, Rezaei Tavirani M, Deravi N, Mahboobi Rabbani MI, Zarghi A (2020) Naringenin enhances the anti-cancer effect of cyclophosphamide against MDA-MB-231 breast cancer cells via targeting the STAT3 signaling pathway. Iran J Pharmaceut Res 19:122–133

Novak U, Ji H, Kanagasundaram V, Simpson R, Paradiso L (1998) STAT3 forms stable homodimers in the presence of divalent cations prior to activation. Biochem Biophys Res Commun 247:558–563

Núñez-Marrero A (2019) Assessing the role of the Interleukin-12/STAT4 Axis in breast cancer by a bioinformatics approach. Int J Sci Basic Appl Res 48:38–52

Onda M, Ghoreschi K, Steward-Tharp S, Thomas C, O'Shea JJ, Pastan IH, FitzGerald DJ (2014) Tofacitinib suppresses antibody responses to protein therapeutics in murine hosts. J Immunol 193:48–55

Park JH, Darvin P, Lim EJ, Joung YH, Hong DY, Park EU, Park SH, Choi SK, Moon ES, Cho BW, Park KD, Lee HK, Kim MJ, Park DS, Chung IM, Yang YM (2012) Hwanggeumchal sorghum induces cell cycle arrest, and suppresses tumor growth and metastasis through Jak2/STAT pathways in breast cancer xenografts. PLoS One 7:e40531

Park HJ, Li J, Hannah R, Biddie S, Leal-Cervantes AI, Kirschner K, Flores Santa Cruz D, Sexl V, Göttgens B, Green AR (2016) Cytokine-induced megakaryocytic differentiation is regulated by genome-wide loss of a uSTAT transcriptional program. EMBO J 35:580–594

Peck AR, Witkiewicz AK, Liu C, Stringer GA, Klimowicz AC, Pequignot E, Freydin B, Tran TH, Yang N, Rosenberg AL, Hooke JA, Kovatich AJ, Nevalainen MT, Shriver CD, Hyslop T, Sauter G, Rimm DL, Magliocco AM, Rui H (2011) Loss of nuclear localized and tyrosine phosphorylated Stat5 in breast cancer predicts poor clinical outcome and increased risk of antiestrogen therapy failure. J Clin Oncol 29:2448–2458

Prat A, Perou CM (2011) Deconstructing the molecular portraits of breast cancer. Mol Oncol 5:5–23

Qiao Y, Giannopoulou EG, Chan CH, Park SH, Gong S, Chen J, Hu X, Elemento O, Ivashkiv LB (2013) Synergistic activation of inflammatory cytokine genes by interferon-γ-induced chromatin remodeling and toll-like receptor signaling. Immunity 39:454–469

Qiao Y, Kang K, Giannopoulou E, Fang C, Ivashkiv LB (2016) IFN-γ induces histone 3 lysine 27 Trimethylation in a small subset of promoters to stably silence gene expression in human macrophages. Cell Rep 16:3121–3129

Qin JJ, Yan L, Zhang J, Zhang WD (2019) STAT3 as a potential therapeutic target in triple negative breast cancer: a systematic review. J Exp Clin Cancer Res 38:195

Raven JF, Williams V, Wang S, Tremblay ML, Muller WJ, Durbin JE, Koromilas AE (2011) Stat1 is a suppressor of ErbB2/Neu-mediated cellular transformation and mouse mammary gland tumor formation. Cell Cycle 10:794–804

Reich NC (2013) STATs get their move on. Jak Stat 2:e27080

Riggins RB, Thomas KS, Ta HQ, Wen J, Davis RJ, Schuh NR, Donelan SS, Owen KA, Gibson MA, Shupnik MA, Silva CM, Parsons SJ, Clarke R, Bouton AH (2006) Physical and functional interactions between Cas and c-Src induce tamoxifen resistance of breast cancer cells through pathways involving epidermal growth factor receptor and signal transducer and activator of transcription 5b. Cancer Res 66:7007–7015

Rody A, Karn T, Liedtke C, Pusztai L, Ruckhaeberle E, Hanker L, Gaetje R, Solbach C, Ahr A, Metzler D (2011) A clinically relevant gene signature in triple negative and basal-like breast cancer. Breast Cancer Res 13:1–12

Ross JS, Slodkowska EA, Symmans WF, Pusztai L, Ravdin PM, Hortobagyi GN (2009) The HER-2 receptor and breast cancer: ten years of targeted anti–HER-2 therapy and personalized medicine. Oncologist 14:320–368

Schaber JD, Fang H, Xu J, Grimley PM, Rui H (1998) Prolactin activates Stat1 but does not antagonize Stat1 activation and growth inhibition by type I interferons in human breast cancer cells. Cancer Res 58:1914–1919

Schindler CW (2002) Series introduction: JAK-STAT signaling in human disease. J Clin Invest 109:1133–1137

Schneckenleithner C, Bago-Horvath Z, Dolznig H, Neugebauer N, Kollmann K, Kolbe T, Decker T, Kerjaschki D, Wagner KU, Müller M, Stoiber D, Sexl V (2011) Putting the brakes on mammary tumorigenesis: loss of STAT1 predisposes to intraepithelial neoplasias. Oncotarget 2:1043–1054

Schwartz DM, Bonelli M, Gadina M, O'Shea JJ (2016) Type I/II cytokines, JAKs, and new strategies for treating autoimmune diseases. Nat Rev Rheumatol 12:25–36

Shields BJ, Wiede F, Gurzov EN, Wee K, Hauser C, Zhu HJ, Molloy TJ, O'Toole SA, Daly RJ, Sutherland RL, Mitchell CA, McLean CA, Tiganis T (2013) TCPTP regulates SFK and STAT3 signaling and is lost in triple-negative breast cancers. Mol Cell Biol 33:557–570

Siddiquee K, Zhang S, Guida WC, Blaskovich MA, Greedy B, Lawrence HR, Yip ML, Jove R, McLaughlin MM, Lawrence NJ, Sebti SM, Turkson J (2007) Selective chemical probe inhibitor of Stat3, identified through structure-based virtual screening, induces antitumor activity. Proc Natl Acad Sci U S A 104:7391–7396

Siegel PM, Ryan ED, Cardiff RD, Muller WJ (1999) Elevated expression of activated forms of Neu/ErbB-2 and ErbB-3 are involved in the induction of mammary tumors in transgenic mice: implications for human breast cancer. EMBO J 18:2149–2164

Siegel RL, Miller KD, Jemal A (2020) Cancer statistics, 2020. CA Cancer J Clin 70:7–30

Silvennoinen O, Schindler C, Schlessinger J, Levy DE (1993) Ras-independent growth factor signaling by transcription factor tyrosine phosphorylation. Science 261:1736–1739

Simon N, Antignani A, Hewitt SM, Gadina M, Alewine C, FitzGerald D (2019) Tofacitinib enhances delivery of antibody-based therapeutics to tumor cells through modulation of inflammatory cells. JCI Insight 4:e123281

Sirkisoon SR, Carpenter RL, Rimkus T, Anderson A, Harrison A, Lange AM, Jin G, Watabe K, Lo HW (2018) Interaction between STAT3 and GLI1/tGLI1 oncogenic transcription factors promotes the aggressiveness of triple-negative breast cancers and HER2-enriched breast cancer. Oncogene 37:2502–2514

Song H, Wang R, Wang S, Lin J (2005) A low-molecular-weight compound discovered through virtual database screening inhibits Stat3 function in breast cancer cells. Proc Natl Acad Sci U S A 102:4700–4705

Stancato LF, David M, Carter-Su C, Larner AC, Pratt WB (1996) Preassociation of STAT1 with STAT2 and STAT3 in separate Signalling complexes prior to cytokine stimulation (∗). J Biol Chem 271:4134–4137

Takakura A, Nelson EA, Haque N, Humphreys BD, Zandi-Nejad K, Frank DA, Zhou J (2011) Pyrimethamine inhibits adult polycystic kidney disease by modulating STAT signaling pathways. Hum Mol Genet 20:4143–4154

Tan SH, Nevalainen MT (2008) Signal transducer and activator of transcription 5A/B in prostate and breast cancers. Endocr Relat Cancer 15:367–390

Tenoever BR, Ng SL, Chua MA, McWhirter SM, García-Sastre A, Maniatis T (2007) Multiple functions of the IKK-related kinase IKKepsilon in interferon-mediated antiviral immunity. Science 315:1274–1278

Thomas SJ, Snowden JA, Zeidler MP, Danson SJ (2015) The role of JAK/STAT signalling in the pathogenesis, prognosis and treatment of solid tumours. Br J Cancer 113:365–371

Timofeeva OA, Gaponenko V, Lockett SJ, Tarasov SG, Jiang S, Michejda CJ, Perantoni AO, Tarasova NI (2007) Rationally designed inhibitors identify STAT3 N-domain as a promising anticancer drug target. ACS Chem Biol 2:799–809

Tsurumi A, Zhao C, Li WX (2017) Canonical and non-canonical JAK/STAT transcriptional targets may be involved in distinct and overlapping cellular processes. BMC Genomics 18:718

Turkson J, Ryan D, Kim JS, Zhang Y, Chen Z, Haura E, Laudano A, Sebti S, Hamilton AD, Jove R (2001) Phosphotyrosyl peptides block Stat3-mediated DNA binding activity, gene regulation, and cell transformation. J Biol Chem 276:45443–45455

Turkson J, Zhang S, Palmer J, Kay H, Stanko J, Mora LB, Sebti S, Yu H, Jove R (2004) Inhibition of constitutive signal transducer and activator of transcription 3 activation by novel platinum complexes with potent antitumor activity. Mol Cancer Ther 3:1533–1542

Turkson J, Zhang S, Mora LB, Burns A, Sebti S, Jove R (2005) A novel platinum compound inhibits constitutive Stat3 signaling and induces cell cycle arrest and apoptosis of malignant cells. J Biol Chem 280:32979–32988

Uluer ET, Aydemir I, Inan S, Ozbilgin K, Vatansever HS (2012) Effects of 5-fluorouracil and gemcitabine on a breast cancer cell line (MCF-7) via the JAK/STAT pathway. Acta Histochem 114:641–646

Ursini-Siegel J (2011) Beyond immune surveillance: Stat1 limits tumor growth in a cell-autonomous fashion. Cell Cycle 10:1348

Ursini-Siegel J, Hardy WR, Zuo D, Lam SH, Sanguin-Gendreau V, Cardiff RD, Pawson T, Muller WJ (2008) ShcA signalling is essential for tumour progression in mouse models of human breast cancer. EMBO J 27:910–920

Vafaizadeh V, Klemmt PA, Groner B (2012) Stat5 assumes distinct functions in mammary gland development and mammary tumor formation. Front Biosci (Landmark Ed) 17:1232–1250

van Agthoven T, Godinho MF, Wulfkuhle JD, Petricoin EF 3rd, Dorssers LC (2012) Protein pathway activation mapping reveals molecular networks associated with antiestrogen resistance in breast cancer cell lines. Int J Cancer 131:1998–2007

Verhoeven Y, Tilborghs S, Jacobs J, De Waele J, Quatannens D, Deben C, Prenen H, Pauwels P, Trinh XB, Wouters A (2020) The potential and controversy of targeting STAT family members in cancer, vol 60. Elsevier, Seminars in cancer biology, p 41

Villarino AV, Kanno Y, O'Shea JJ (2017) Mechanisms and consequences of Jak-STAT signaling in the immune system. Nat Immunol 18:374–384

Wang Y, Shen Y, Wang S, Shen Q, Zhou X (2018) The role of STAT3 in leading the crosstalk between human cancers and the immune system. Cancer Lett 415:117–128

Watson CJ (2001) Stat transcription factors in mammary gland development and tumorigenesis. J Mammary Gland Biol Neoplasia 6:115–127

Wegrzyn J, Potla R, Chwae YJ, Sepuri NB, Zhang Q, Koeck T, Derecka M, Szczepanek K, Szelag M, Gornicka A, Moh A, Moghaddas S, Chen Q, Bobbili S, Cichy J, Dulak J, Baker DP, Wolfman A, Stuehr D, Hassan MO, Fu XY, Avadhani N, Drake JI, Fawcett P, Lesnefsky EJ, Larner AC (2009) Function of mitochondrial Stat3 in cellular respiration. Science 323: 793–797

Wei L, Vahedi G, Sun HW, Watford WT, Takatori H, Ramos HL, Takahashi H, Liang J, Gutierrez-Cruz G, Zang C, Peng W, O'Shea JJ, Kanno Y (2010) Discrete roles of STAT4 and STAT6 transcription factors in tuning epigenetic modifications and transcription during T helper cell differentiation. Immunity 32:840–851

Widschwendter A, Tonko-Geymayer S, Welte T, Daxenbichler G, Marth C, Doppler W (2002) Prognostic significance of signal transducer and activator of transcription 1 activation in breast cancer. Clin Cancer Res 8:3065–3074

Willi M, Yoo KH, Wang C, Trajanoski Z, Hennighausen L (2016) Differential cytokine sensitivities of STAT5-dependent enhancers rely on Stat5 autoregulation. Nucleic Acids Res 44: 10277–10291

Yamashita H, Nishio M, Ando Y, Zhang Z, Hamaguchi M, Mita K, Kobayashi S, Fujii Y, Iwase H (2006) Stat5 expression predicts response to endocrine therapy and improves survival in estrogen receptor-positive breast cancer. Endocr Relat Cancer 13:885–893

Yamashita H, Takahashi S, Ito Y, Yamashita T, Ando Y, Toyama T, Sugiura H, Yoshimoto N, Kobayashi S, Fujii Y, Iwase H (2009) Predictors of response to exemestane as primary endocrine therapy in estrogen receptor-positive breast cancer. Cancer Sci 100:2028–2033

Yang R, Rincon M (2016) Mitochondrial Stat3, the need for design thinking. Int J Biol Sci 12: 532–544

Yang J, Liao X, Agarwal MK, Barnes L, Auron PE, Stark GR (2007) Unphosphorylated STAT3 accumulates in response to IL-6 and activates transcription by binding to NFkappaB. Genes Dev 21:1396–1408

Yau C, Esserman L, Moore DH, Waldman F, Sninsky J, Benz CC (2010) A multigene predictor of metastatic outcome in early stage hormone receptor-negative and triple-negative breast cancer. Breast Cancer Res 12:R85

Yeh YT, Ou-Yang F, Chen IF, Yang SF, Wang YY, Chuang HY, Su JH, Hou MF, Yuan SS (2006) STAT3 ser727 phosphorylation and its association with negative estrogen receptor status in breast infiltrating ductal carcinoma. Int J Cancer 118:2943–2947

Yu C, Deng HY (2006) Effects of a JAK inhibitor, AG490, on proliferation and survivin expression of breast cancer cell line MDA-MB-231. Ai Zheng 25:1227–1231

Yu H, Jove R (2004) The STATs of cancer--new molecular targets come of age. Nat Rev Cancer 4: 97–105

Yu H, Kortylewski M, Pardoll D (2007) Crosstalk between cancer and immune cells: role of STAT3 in the tumour microenvironment. Nat Rev Immunol 7:41–51

Yu H, Lee H, Herrmann A, Buettner R, Jove R (2014) Revisiting STAT3 signalling in cancer: new and unexpected biological functions. Nat Rev Cancer 14:736–746

Zerdes I, Wallerius M, Sifakis EG, Wallmann T, Betts S, Bartish M, Tsesmetzis N, Tobin NP, Coucoravas C, Bergh J, Rassidakis GZ, Rolny C, Foukakis T (2019) STAT3 activity promotes programmed-death ligand 1 expression and suppresses immune responses in breast cancer. Cancers (Basel) 11:1479

Zhang WJ, Li BH, Yang XZ, Li PD, Yuan Q, Liu XH, Xu SB, Zhang Y, Yuan J, Gerhard GS, Masker KK, Dong C, Koltun WA, Chorney MJ (2008) IL-4-induced Stat6 activities affect apoptosis and gene expression in breast cancer cells. Cytokine 42:39–47

Zhang Q, Raje V, Yakovlev VA, Yacoub A, Szczepanek K, Meier J, Derecka M, Chen Q, Hu Y, Sisler J, Hamed H, Lesnefsky EJ, Valerie K, Dent P, Larner AC (2013) Mitochondrial localized Stat3 promotes breast cancer growth via phosphorylation of serine 727. J Biol Chem 288: 31280–31288

Zhou T, Chao L, Rong G, Wang C, Ma R, Wang X (2013) Down-regulation of GRIM-19 is associated with STAT3 overexpression in breast carcinomas. Hum Pathol 44:1773–1779

# Triple-Negative Breast Cancer Therapy: Recent Advances, Challenges, and Future Perspective

Md Abdus Subhan

**Abstract**

Among the breast cancer subtypes, TNBC is recognized as most aggressive with poor prognosis and survival. Due to the lack of suitable markers, targeted therapy for TNBC is challenging. Chemotherapy, radiation therapy, and surgical resection are the existing treatment options for TNBC. Chemotherapy is sensitive to TNBC; however, development of resistance mechanisms and recurrence remain the major reasons of treatment failure in patient with TNBC. Recent advances with combination therapy approaches using chemotherapy, mAbs and ADC drugs, and siRNA-based nanoparticles-mediated gene therapy strategies are promising for TNBC therapy with enhanced prognosis and improved survival. Translation of these therapy approaches from preclinical to clinical settings is the new hope for TNBC patients.

**Keywords**

Breast cancer · Chemotherapy · Immunotherapy · Targeted therapy · TNBC

## 1 Introduction

Triple negative breast cancer (TNBC) is the most aggressive, highly heterogeneous subtype of breast cancer that lacks overexpression of estrogen receptor (ER), progesterone receptor (PR), and human epidermal growth factor receptor 2 (HER2) markers. TNBC represents 15–20% of all breast cancer (Abad et al. 2021; Manjunath and Choudhary 2021). Currently, targeted TNBC therapy is limited with poor prognosis and clinical outcome, leaving chemotherapy (CT) as

Md. A. Subhan (✉)
Department of Chemistry, Shahjalal University of Science and Technology, Sylhet, Bangladesh
e-mail: subhan-che@sust.edu



Interdisciplinary Cancer Research, https://doi.org/10.1007/16833_2023_128
Published online: 3 February 2023

the only therapy option. Generally, breast cancer patients are provided targeted therapy based on overexpression of ER, PR, and HER2 markers. Due to the lack of ER, PR, and HER2 TNBC, therapy is more challenging leading to the higher mortality rate among the women. TNBC demonstrates higher proliferation rate, and rapid metastasis to brain, bone-marrow, lung, and liver than other BC subtypes. The TNBC affects younger woman (under 40 years of age) than other BC subtypes. From a clinical perspective, TNBC is generally considered a single disease and treated with chemotherapy, consequently leading to poor therapy response and clinical outcome. However, clinical, histological, and molecular profiling indicated intrinsic heterogeneity in TNBC tumors. There are four subtypes of TNBC based on patterns of cytokeratins. Among them woman with luminal (20%) or basoluminal (28%) demonstrated noteworthy worse rate of survival than patient with basal A (26%) or basal B (26%) subtypes of TNBC (Ensenyat-Mendez et al. 2021).

Commonly used initial method of TNBC treatment is surgery and radiotherapy followed by chemotherapy. Taxens, anthracyclines, and cisplatin/carboplatin agents are the generally used chemotherapeutics (CT) in TNBC therapy. There are two CT approaches: applying CT before surgery (neoadjuvant CT/NACT) or after removal of tumor by surgical operation (adjuvant CT/ACT). Due to the development of multi-omics technology, targeted therapy of TNBC has emerged as novel approach of treatment such as PARP ((poly (ADP-ribose) polymerase) inhibitor, PI3K/AKT/mTOR inhibitor, RNase inhibitor, AR(androgen receptor) antagonist, antibody-drug-conjugates (ADC), and anti-PD-L1 immunotherapy (Fig. 1) (Manjunath and Choudhary 2021).

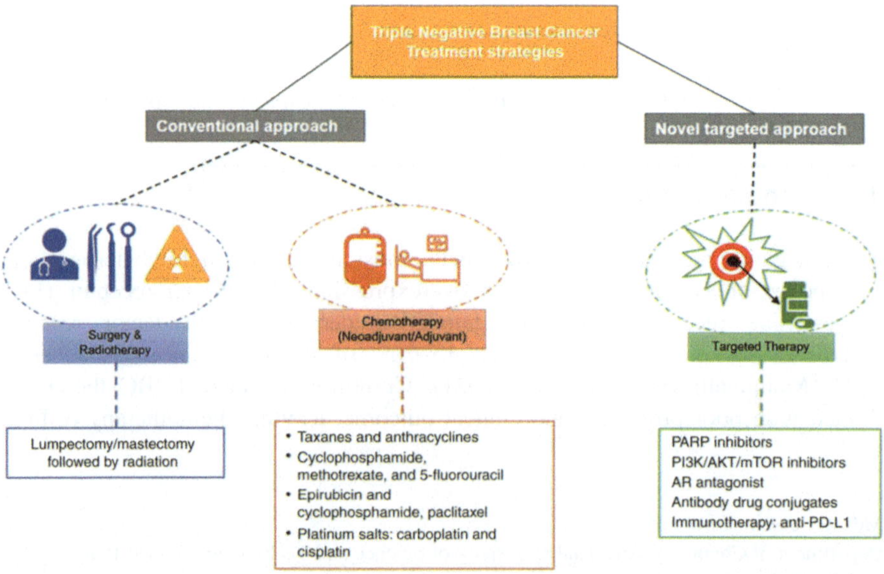


**Fig. 1** Different methods of treatment employed in TNBC therapy. (Adapted from Manjunath and Choudhary 2021)

## 2 Predictive Biomarkers, TILs, Genes and Their Roles in TNBC

The immune response can potentially and specifically inhibit tumor cells without affecting healthy cells/tissues, and generate long-term immune memory proficient for averting cancer recurrences. The immune system can either detect and eliminate tumor cells or can provide favorable TME (tumor microenvironment) for tumor growth depending on the tumor development stages. TME consists of tumor cells, inflammatory cells, and stromal cells such as cytokines. The dynamic interactions of these cells in TME dictate the environment in which tumor development occurs. The immune cells in TME can effectively promote or inhibit tumor development, which is a prognostic indicator in breast cancer (Abad et al. 2021).

In spite of the action of immune system to inhibit cancer, circumvention of immune evasion can ensue through mechanisms such as the expression of ligand of transmembrane type 1 protein, PD-L1 in tumor cells, which has an immunosuppressive function by binding to their PD-1 or B7 receptor, transmitting a repressive signal to T lymphocytes that decreases their proliferation and reduces the immune response (Voutsadakis 2016). PD-L1 is not perceived in normal breast tissue, its expression has been observed in TNBC. PD-L1 expression in tumor and existence of tumor infiltrating lymphocytes (TILs) with PD-1 expression have been associated with a high histological degree, and a greater lymphocytic infiltration of tumor (Muenst et al. 2014). PD-L1 protein expression is correlated positively with the expression of other immune regulators such as CTLA-4, IDO1 (indoleamine 2,3-dioxygenase 1), and *BRCA1* gene mutations (Wen and Leong 2019). Therapy with immune checkpoint inhibitors has the potential to enhance the prognosis of TNBC by improving the tumor-associated immune response in inhibiting breast tumor cells. There have been many investigations with PD-1 and PD-L1 inhibitors in combination with chemotherapy, radiation therapy, and other immune checkpoint inhibitors. These combination strategies may enhance immune checkpoint-based immunotherapy for TNBC in clinic (Cyprian et al. 2019).

CTLA-4 is present on the cell membrane of T lymphocytes. The activity of T lymphocyte is subdued by activation of the CTLA-4 receptor. CTLA-4 overexpression of more than 50% has been observed in TNBC (Kassardjian et al. 2018). Overexpression of CTLA-4 has been correlated with mutated BRCA1 TNBC due to increased expression of immunomodulatory genes and an increased somatic mutational load (Marra et al. 2019). Several studies with CTLA-4 inhibitors such as ipilimumab, ipilimumab +nivolumab compared to CT, combination of nivolumab with NACT (neoadjuvant CT), in patient with early TNBC are in different phases of clinical trial.

Mutation of *BRCA1/2* tumor suppressor genes that encode proteins involved in the repair of DNA double-strand breaks through the homologous recombination repair pathway, and occurs in 10–35% of TNBC patients. Mutation can occur both in germline and somatic lines (Chen et al. 2018; Furlanetto and Loibl 2020). Members of the PARP family enzymes are central to the repair of DNA single-strand breaks. About 15–20% TNBC have *gBRCA* mutation (Loibl et al. 2018). TNBC patients without *gBRCA* mutation have a somatic mutation in homologous recombination signaling pathway (BRCAness). *gBRCA* mutation and BRCAness statuses are allied with an enhanced sensitivity to CT with better outcome (Loibl et al. 2018; Pohl-Rescigno et al. 2019). Cancer cells with *gBRCA* mutation are sensitive to PARP inhibitor due to the synthetic lethality mechanism, resulting in an inability for DNA repair (Walsh 2015). *gBRCA* mutation is associated with a higher pathological complete response (PCR) rate and prognostic factors in TNBC patients.

The presence of TILs is the most contrast prognostic indicator in TNBC (Zgura et al. 2018). TNBCs are highly immunogenic as reflected by stromal TILs. TILs have a predictive and prognostic role in TNBC (Denkert et al. 2015). IDO1 is another enzyme involved in immune regulation in TNBC. IDO1 is produced primarily by activated macrophages. Overexpression of IDO1 has been allied with vascular density, and poor prognosis in breast cancer patients (Wei et al. 2018).

## 3 Drug Resistance in TNBC

Chemoresistance is one of the main reasons of treatment failure in patients with breast tumors including TNBC. There are two types of chemoresistance: primary resistance and acquired resistance (Han et al. 2019). The primary resistance is a de novo lack of therapeutic response, while acquired resistance ascends during the progression of chemotherapy. The mechanism of acquired resistance may vary based on the type of drug and patient, making it hard to envisage (Jazaeri et al. 2005).

When using signaling inhibitors to inhibit the growth and survival, cancer cells may activate new pathways that bypass the signal. As a result, many preclinical and clinical approaches using single target Abs and small molecule blockers have become unsuccessful. The following mechanisms enhance direct or indirect resistance against anticancer drugs in cancer cells: drug target alteration, drug inactivation, cell death reduction, DNA damage repair, drug efflux, epigenetic changes, and epithelial to mesenchymal (EMT) transition (Housman et al. 2014). The mechanisms of multidrug resistance (MDR) signaling are related to cell survival and growth and include the following pathways: (i) GPCR/RAS/RAF/MEK, (ii) IL6/JAK/STATs/mTOR, and (iii) HER2/PI3K/AKT. These signaling cascades activate the expression of transporter gene (ABCG2), survival genes (HER2, STAT3), transcription factors (AP-1 and NF-κB), and epigenetic genes (Histone deacetylases, HDACs) (Chen and Sikic 2012). Enhanced IL6 and IL8 secretion and STAT3 activity are crucial in development of drug resistance (Van Klompenberg et al. 2017; Liu et al. 2010; Cheng et al. 2018). Several studies have demonstrated that tumor cell dormancy can cause recurrence after a long time in patients who have received chemotherapy (Ebinger et al. 2016; Najmi et al. 2005). Tumor cell dormancy allows cancer cells to survive by acquiring mutations, gene expression alterations, slowing proliferation, and reducing metabolism, hence inhibiting the effects of anticancer drugs. HDACs can control a wide range of cellular processes and gene expressions through remodeling of chromatin (Haberland et al. 2009). Dysregulated HDAC expression and genetic mutations are involved in activating neoplastic transformation in breast and other cancers (Li and Seto 2016). HDAC inhibitors (HDACi) can suppress

HDAC activity. For example, suberoylanilide hydroxamic acid (SAHA or vorinostat) is a broad-spectrum HDACi, which can suppress all zinc-dependent HDACs at low concentration (Li and Seto 2016; Richon et al. 1998; Bradner et al. 2010). It is thus necessary to find a new method for predicting and inhibiting early drug resistance in TNBC for the success in TNBC therapy.

Doxorubicin is widely used in chemotherapy as it inhibits fast proliferating tumor cells by targeting topoisomerase-II (Trop-II) (Han et al. 2019). The DOX-resistant cells demonstrated enhanced drug-efflux rate, enhanced anchorage-dependent growth when cultured in suspension, increased tumor-forming ability in nude-mice compared to the wild-type MDA-MB-231 TNBC cells. RNA-sequencing analysis demonstrated enhanced gene expression of genes involved in membrane transport, antiapoptosis, and histone regulation. Kaplan-Meier plot analysis of TNBC patients who have received NACT, exhibited that relapse free survival (RFS) of patient with low expression of HIST1H2BK (Histone cluster 1 H2B family member K) was higher than that of patients with high HIST1H2BK expression. HIST1H2BK may be a tumor dormancy marker. qPCR analysis indicated that level of HIST1H2BK was enhanced in resistant cells. Cytotoxicity data demonstrated that DOX-resistant cancer cells were reduced by treatment with HDACi (Han et al. 2019). In DOX-resistant cancer cells, HIST1H2BK expression can be induced by the high expression of genes involved in membrane transport, antiapoptosis, and histone regulation. Drug resistance in TNBC cells, MDA-MB-231 can occur in a comparatively short period of time by DOX gradient and resistance is dependent on the specificity of resistant cell to HDAC (Han et al. 2019).

## 4 Systemic Therapy in TNBC

Although TNBC is a highly heterogeneous disease, the therapy option for early-stage patient is still taxane/anthracycline-based chemotherapy. However, high response rates to chemotherapy are not extended mainly due to early development of resistance. The 5-year survival rate of localized TNBC is 91%, in case of locally advanced TNBC 65% and only 11% in metastatic diseases (Abad et al. 2021; 11). The therapeutic approaches in localized or locally advanced TNBC are CT, surgery, and radiation therapy. The choice of TNBC therapy should be individualized and performed by a multidisciplinary approach. The selected TNBC therapy is based on the size and location of the primary tumor, the number of lesions, adenopathy involvement, age and menopausal status, and *BRCA1/2* status of the patient (Abad et al. 2021; Cardoso et al. 2019).

Since the presence of micrometastases in early-stage TNBC patient is not visible by usual imaging techniques, TNBC should be considered as systemic diseases and CT must be provided to achieve healing (Redig and Alister 2013). The generally used CT drugs that are effective in TNBC are anthracycline such as doxorubicin, epirubicin, cyclophosphamide, taxanes, platinum agents, fluoropyrimidines, eribulin, and gemcitabine (Clarke et al. 2005). NACT is a standard therapy in locally advanced TNBC. Some significant goals allied with NACT are: increased resectability by reducing tumor size (tumors less than 2 cm) allowing conservative surgery, early control of micrometastasis, and testing of chemosensitivity or chemoresistivity of the tumors in vivo (Kaufmann et al. 2003). Patient's response to NACT is a predictor of long-term response and useful to provide prognostic evidence after a short-term follow-up, unlike ACT which provides results after a long-term follow-up (Velasco et al. 2012). However, a demerit to NACT may be the surgical challenge posed by the patent who demonstrated tumor progression during this therapy. In case of NACT achieving PCR is significant after NACT as it confers a better prognosis and survival rate (Haque et al. 2018; Liedtke et al. 2008).

There are several mechanisms through which CT acts in the tumor cells such as taxanes promote microtubule polymerization and stabilization by targeting microtubule dynamics, gemcitabine function as pyrimidine nucleoside antimetabolite, 5-fluorouracil inhibit thymidylate synthase (TS), platinum agents form DNA cross-links, anthracyclines intercalate into DNA and disrupt topoisomerase-II-induced DNA repair, cyclophosphamide exerts cytotoxic effects due to cross-linking of strands of DNA and RNA and inhibit protein synthesis. Generally, most CT result in inhibition of DNA replication and transcription through cell cycle arrest and cause cancer cell death (Abad et al. 2021).

TNBC is commonly allied with a deficiency in BRCA-mediated DNA repair mechanism, leading to more sensitivity to interstrand cross-linking agents damaging the DNA such as platinum-based drugs (Abad et al. 2021; Joensuu and Gligorov 2012; Alba et al. 2012; Ando et al. 2014). For example, an enhanced PCR rate was observed in TNBC patients when carboplatin was added to anthracycline/taxane-based CT. Carboplatin was associated with toxicity, while, reduced dose showed low toxicity (von Minckwitz et al. 2014).

Immunotherapy offers innovative opportunities in the TNBC treatment. Studies with NACT plus immunotherapy with anti-PD-1 or anti-PD-L1 have shown promising results (Loibl et al. 2019). PD-1 is expressed on activated T cells, and when binds to PD-L1 on tumor cells, leads to T cell exhaustion. CTLA-4 competes with costimulatory T cell molecule, CD28 for B7 ligands, and upon stimulation, reduces T cell proliferation as well as function. CTLA-4 inhibitor (ipilimumab) and PD-1 inhibitor (pembrolizumab, nivolumab) or PD-L1 inhibitor (durvalumab, atezolizumab) activate effector T cells to harvest antitumor responses (Fig. 2) (Luo et al. 2022; Abad et al. 2021). Addition of durvalumab to anthracycline/taxane-based chemotherapy demonstrated an increased PCR rate compared to durvalumab. A significant increase in PCR was observed with higher stromal TIL in PD-L1 positive tumors (Loibl et al. 2019). Pembrolizumab increased PCR in early TNBC patients (Nanda et al. 2020; Schmid et al. 2020a, b).

PARP inhibitors trap the PARP proteins on the DNA repair intermediates, causing BRCA1/2-defect cells to accumulate more toxic DSBs (double strand breaks), thus leading to cell death. PARP inhibitors demonstrated efficacy in the metastatic TNBC (Robson et al. 2017; Litton et al. 2018a, b). Addition of veliparib to paclitaxel and carboplatin showed increased PFS (progression-free survival) compared to CT alone without altering the toxicity profile. Talazoparib has also demonstrated enhanced activity when applied as neoadjuvant monotherapy

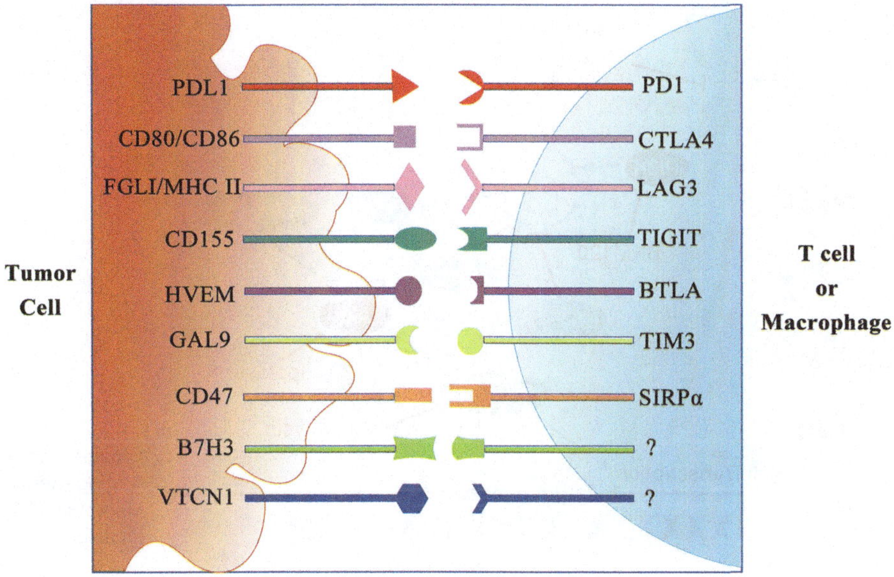


**Fig. 2** Immune checkpoint involved in T cell activation. Cancer cells evade the host immune system by the upregulation of immune checkpoints composed of the ligands on the cancer cells and the respective receptors on CD8+ T cell. (Adapted from Luo et al. 2022)

(Litton et al. 2018a, b). In metastatic TNBC both ipataseritib and capivaseritib have exhibited promising outcome particularly in PIK3CA/AKT1/PTEN altered tumors (Schmid et al. 2020a, b; Kim et al. 2017) (Fig. 3).

EGFR inhibitors such as panitumumab are being investigated for their efficacy in early TNBC (Fig. 3). A PCR rate of 42% was observed with the addition of panitumumab to NACT with nab-paclitaxel and carboplatin weekly. This study determined the role of panitumumab and validate predictive biomarkers in patient with TNBC (Matsuda et al. 2018). Some tyrosine kinase inhibitors such as apatinib (anti-VEGFR) are under investigation. The addition of apatinib to NACT remarkably enhanced PCR rate to 72.7% compared to NACT (50%) (Li et al. 2021; Liu et al. 2020).

There have been many studies to investigate efficacy of antibody-drug conjugates (ADCs) in TNBC therapy in both preclinical and clinical setup (Sahota and Vahdat 2017; Bardia et al. 2020; Subhan and Torchilin 2021). An ADC is commonly composed of a mAb, and a highly cytotoxic small molecule drug connected via a chemical linker (Sahota and Vahdat 2017; Subhan and Torchilin 2021). Sacituzumab govitecan is an ADC drug composed of an anti-Trop-2 antibody coupled to the active metabolite of irinotecan, SN-38, through a hydrolyzable linker that allows for SN-38 release intracellularly into TME (Sahota and Vahdat 2017). Sacituzumab govitecan significantly enhanced PFS and median OS (overall survival) in relapsed/refractory, metastatic TNBC (Bardia et al. 2020). ADC therapy of breast cancer, especially in TNBC, MDR, and metastatic BC are promising.

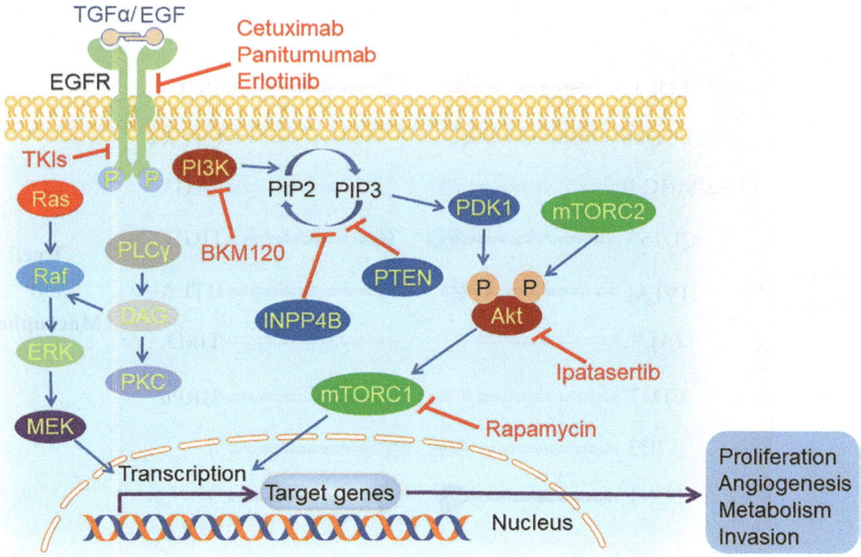


**Fig. 3** Schematic representation of EGFR and downstream signaling pathway targeted therapy in TNBC. (Adapted from Li et al. 2021)

Innovative ADC drugs such as trastuzumab emtanstine have been approved by FDA (Subhan and Torchilin 2021). Sacituzumab govitecan and trastuzumab deruxtecan are in the late phases of clinical trials for metastatic breast cancer patient including TNBC. The FDA, in recent time, has approved sacituzumab govitecan-hziy (Trodelvy) for adult TNBC patient. Several ADC drugs including ladiratuzumab vedotin, and trastuzumab duocarmizine are in the latest stages of clinical development (Subhan and Torchilin 2021).

Based on the evidences, there are several international guidelines and treatment recommendations for TNBC (Furlanetto and Loibl 2020). Guidelines recommended the utilization of CT for the patient with TNBC either with NACT or ACT settings with the exception of small tumors and low risk histology. Application of nab-paclitaxel is still controversial due to conflicting results (Untch et al. 2016, 2019; Furlanetto et al. 2017; Gianni et al. 2019). AGO (Arbeitsgemeinschaft Gynako-logische Onkologie) guidelines recommended the nab-paclitaxel instead of paclitaxel in NACT (AGO Breast Committee 2019). NCCN (National Comprehensive Cancer Network) guidelines recommended soluble taxanes as a substitute of nab-paclitaxel due to hypersensitivity (NCCN, Breast Cancer 2020). Anthracycline/taxane-based CT are the preferred option for early TNBC patients. The carboplatin is recommended for TNBC irrespective of *BRCA* status. The role of PARP and AKT inhibitors have shown promising results in early TNBC. The addition of checkpoint inhibitors in NACT for high-risk patient is looming. In patient with residual diseases after NACT, capecitabine as AD therapy is a good option (Furlanetto and Loibl 2020).

## 5 siRNA Therapy in TNBC

TNBC is the most aggressive among the breast cancers. CT, radiation therapy, and surgical operation are commonly used therapy options for TNBC. However, these approaches are commonly associated with severe side effects and chemoresistance. TNBC cells are highly sensitive to chemotherapeutic drugs. Emerged chemoresistance contribute to treatment failure and poor prognosis in TNBC (Hattab and Bakhtiar 2020). Around 90% of initially chemo-sensitive cells develop resistance. The use of siRNA-based nanoparticles as targeted approach is potential for overcoming doxorubicin and other chemoresistance (Hattab and Bakhtiar 2020). MDR evolved in breast cancer cells through the mechanisms including ABC-transporter, stimulation of antiapoptotic pathways, increase of chemotherapeutic detoxification, overexpression of β-tubulin III, KIF14, and many other signaling factors, amendments of DNA repair enzymes, chemoresistance-induced hypoxia and NF-κB signaling pathways (Navarro et al. 2012; Wu et al. 2016; Yu et al. 2018; Choi and Yu 2014). Among the different mechanisms, ABC transporter, modifications of antiapoptotic genes, and the role of hypoxia are major mechanisms of chemoresistance (Fig. 4). Two efflux pump members, namely, ABCB1 (P-glycoprotein) and ABCG2 (BCRP-breast cancer resistance protein) are recognized. ABCB1 and ABCG2 are overexpressed in TNBC tumors (Hattab and Bakhtiar 2020).

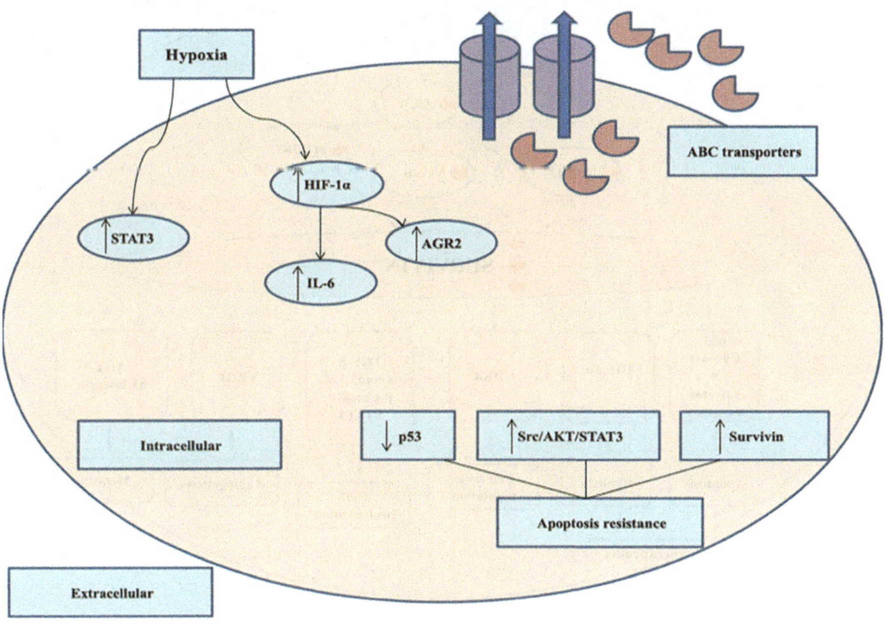


**Fig. 4** Major mechanisms of chemotherapeutics resistance in TNBC. Hypoxia prompts upregulation of HIF-1α and other signaling factors. Overexpression of ABC transporters and efflux of chemotherapeutic drugs out of the cancer cells. (Adapted from Hattab and Bakhtiar 2020)

Recent advancement in siRNA therapy may be an alternative approach for TNBC therapy. The role of siRNA in silencing the genes/proteins that are overexpressed in cancer cells exhibited potential as a TNBC therapeutic regimen. However, targeting specificity, stability of siRNA-based drugs, and delivery efficiency to tumor cells cause hurdles in application of this therapy in clinic. For efficient application of siRNA therapy in TNBC required encapsulating and transporting siRNA with high efficacy, and low toxicity profile. Recent advances in preclinical and clinical strategies utilizing engineered siRNA-loaded nanoparticles for the TNBC therapy have shown promise (Hattab and Bakhtiar 2020).

Downregulation of ABCB1 and ABCG2 by siRNA-based nanoparticles enhanced the sensitivity of cytotoxic drugs such as doxorubicin and paclitaxel. Multilayered doxorubicin-based nanoparticles enriched with alternating layers of siRNA targeted *MDR1* and poly-L-lysine enclosed by an outer shell of hyaluronic acid with high siRNA loading capacity, stability, and cellular internalization, high-efficiency, toxicity profile was developed. Administration of doxorubicin-siRNA nanoparticle complexes decreases the tumor volume (Bai et al. 2015; Deng et al. 2013).

Targeting the antiapoptotic genes/proteins is an important mechanism to reduce chemoresistance. Survivin is an inhibitor of apoptosis controlling apoptosis, tumor growth, angiogenesis, and drug resistance (Fig. 5) (Shen et al. 2013). Polymeric

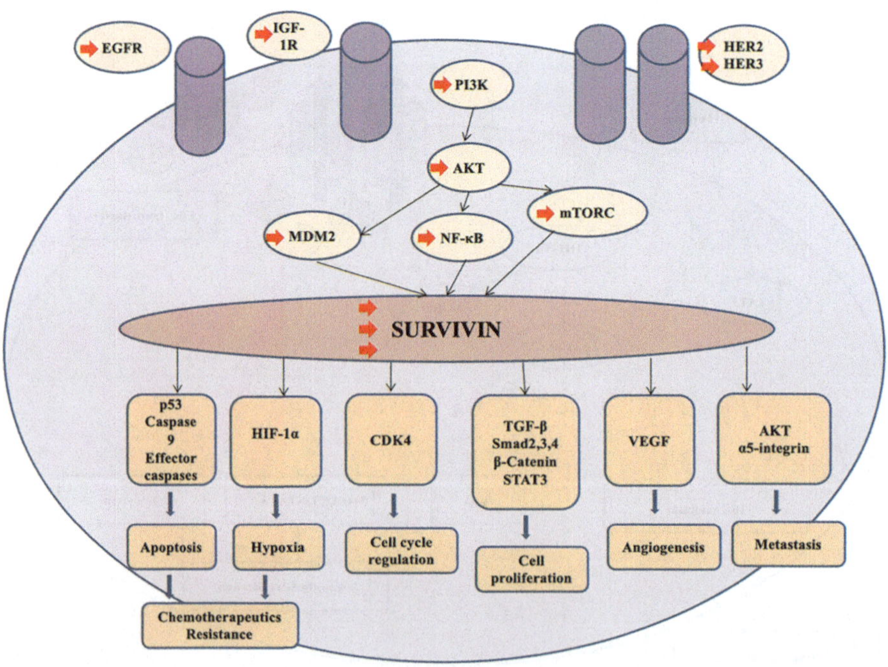


**Fig. 5** Survivin is a potential target for treatment of TNBC patients. Activation of several kinases such as HER2, HER3, IGF-1R, and EGFR prompts PI3K/AKT signaling pathways resulting in upregulation of survivin expression. Overexpression of survivin triggers numerous genes and proteins allied with cancer cell proliferation and survival, migration, angiogenesis, and promotes chemoresistance. (Adapted from Hattab and Bakhtiar 2020)

micelles (PM) of $PEG_{2000}$-PE incorporating anti-survivin siRNA can sensitize TNBC chemotherapeutics. Delivery of anti-survivin siRNA PM transfected MDA-MB-231 cells with paclitaxel demonstrated downregulation of surviving expression in resistant cancer cells causing destabilization of microtubules, and enhanced suppression of tumor growth (Salzano et al. 2014). A combination therapy with CT, gene therapy, and photothermal therapy may reveal enhanced antitumor effects.

Forkhead box protein M1(FOXM1) is a key oncogenic transcription factor (TF), which plays important role in regulating expression of genes involved in cancer cell proliferation and cell cycle progression. Administration of liposomal FOXM1 siRNA in mice implanted with MDA-MB-231 cells caused downregulation of FOXM1 expression and reduced TNBC cell growth (Hamurcu et al. 2016; Wang and Gartel 2011). Eukaryotic Elongation Factor 2 Kinase (eEF2K) is a principal regulator of tumor development and progression. TNBC patients showed overexpression of eEF2K. As the FOXM1 and eEF2K are involved in the same function in TNBC, FOXM1 regulates the expression of eEF2K. Engineered liposomal siRNA nanoparticles were able to bind to the promoter region of eEF2K and regulated its expression. Modified gold anti-eEF2K siRNA nanoparticles were highly efficient for treatment of TNBC tumors in xenografted mice, resulting in knockdown of eEF2K expression and inhibition of tumor growth and progression (Hamurcu et al. 2016; Shahbazi et al. 2017).

*MDM2* oncogene inactivates tumor suppressor p53, which is overexpressed in TNBC cells. Overexpression of *MDM2* leads to suppression of p53, resulting in tumor cell proliferation and apoptosis inhibition. Single-walled carbon nanotubes (SWNTs) functionalized with PEG efficiently delivered siRNA to target *MDM2*. PEG-modified SWNTs were effective in *MDM2* silencing, causing reduced tumor cell proliferation, and induced apoptosis in breast cancer cells (Liu et al. 2015).

Cyclin-dependent kinase (CDK)11 and casein kinase 2 (CK2) are crucial protein kinase for cancer cell survival, which are overexpressed in TNBC cells. Tenfibgen-coated nanocapsules containing anti-CDK11siRNA/anti-CK2 siRNA were effective in targeting Tenascein-C-receptors, that are found in breast cancer stroma. Anti-CDK11sirna/anti-CK2 siRNA were efficient in reducing tumor growth and proliferation (Xu et al. 2019). *C-Myc* and *CDK1* play crucial role in cell death and apoptosis. Knockdown of these two genes results in specific targeting and synthetic lethality in TNBC cells. *C-Myc* gene silencing by anti-*C-Myc* siRNA-encapsulated nanoparticles causes *C-Myc* silencing that contributes to a mild therapeutic effect on TNBC cells, while, delivery of siRNA targeting CDK1 expression resulted in silencing of TNBC cells containing higher levels of *C-Myc* expression. Thus, combination therapy with *C-Myc* and CDK1 targeting siRNA therapeutics is effective for decreased cell proliferation, apoptosis, and selective as well as specific targeting of TNBC tumor cells without any adverse effects (Xu et al. 2019; Nedeljković and Damjanović 2019).

Polo-like kinase-1 (*PLK1*) is upregulated in patients with primary or metastatic TNBC. Inhibition of *PLK1* expression is allied with enhancing tumor cell death in many metastatic cancers including TNBC (Hu et al. 2012, Maire et al. 2013). Targeted nanoplatform incorporating anti-*PLK1* siRNA encapsulated within antibody-conjugated bioreducible, cross-linked with PEI and PEG layer by layer with mesoporous silica nanoparticles (MSNP) were effective in metastatic TNBC therapy. It was found that ROS scavenging ability of MSNP contributes to the antimetastatic effect, while, interlinked anti-*PLK1* siRNA demonstrated apoptotic cell death activity (Morry et al. 2017).

TWIST is a major transcription factor that activates epithelial to mesenchymal transition (EMT), promotes stem cell like phenotyping, and inhibits apoptosis, enhancing the risk of tumor recurrence and poor prognosis in TNBC. A modified siRNA-based dendrimer, PAMAM loaded with TWIST-siRNA was fabricated and efficiently internalized into TNBC cells, which caused reduction in TWIST expression accompanied by phenotyping changes, and allied with decreasing cancer cell mobility, decreasing TWIST-induced upregulation of mesenchymal marker, attenuating dissemination and invasion of TNBC cells (Glackin 2018; Finlay et al. 2015).

MDM2, CDK11, CK2, TWIST, c-Myc, survivin, mTORC2, ATM protein, monopolar spindle 1 (MPS1) and cell division cycle protein (CDC20), Heparin-binding EGF-like growth factor (HB-EGF), EGFR, CXCR4, luciferase mRNA protein, and enhanced green fluorescent protein (eGFP) are overexpressed in TNBC cells (Fig. 6) (Chen et al. 2012; Kren et al. 2015; Liu et al. 2014; Aliabadi et al. 2011; Li et al. 2019; Werfel et al. 2018; Xu et al. 2013; Parmar et al. 2016;

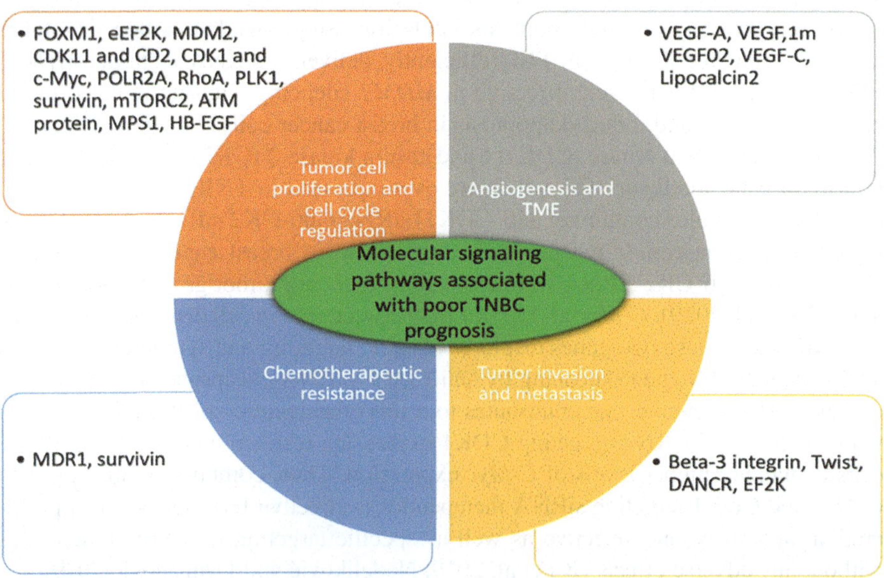


**Fig. 6** Major genes, proteins, TFs involved in significant molecular pathways associated with poor TNBC prognosis. These factors are upregulated in TNBC patients. Engineered siRNA-based anticancer drugs targeted single or multiple genes/proteins/TFs causing downregulation, suppression, and effectively improve disease progression. (Adapted from Hattab and Bakhtiar 2020)

**Table 1** Some siRNA-mediated anticancer therapeutics in different phases of clinical trials

| Drug name | Indication | Target gene/ protein | Administration route | Status | Phase |
|---|---|---|---|---|---|
| ALN-VSP02 | Solid tumors | VEGF, KSP | Systemic/i.v. infusion | Completed | I |
| Atu027 | Advanced or metastatic cancer/ solid tumors | PKN3 | Systemic/i.v. infusion | Completed | II |
| TKM-PLK1 (TKM-080301) | Solid tumors/adrenal cortical carcinoma | PLK1 | Systemic/i.v. infusion | Completed | II |

Okamoto et al. 2018; Jing et al. 2016; Gao et al. 2011; Inoue et al. 2012; Misra et al. 2015, Ho et al. 2013).

Silencing of any of these molecular factors using siRNA-mediated nanoplatforms caused enhanced tumor growth suppression. Both preclinical and clinical studies demonstrated that inhibition of any of these factors utilizing siRNA-based gene therapy is promising in inhibition of cell proliferation and survival in TNBC patients. There have been numerous preclinical studies utilizing siRNA therapeutics in nanoformulations for enhanced therapy of TNBC patients (Hattab and Bakhtiar 2020). To date, several siRNA-mediated nanoparticles have started phase I, phase II, and phase II clinical trials or have received approval (recently US-FDA approved Patisiran, Givosiran, Inclisiran, and Lumasiran) for treatment of a wide variety of solid tumors that would be suitable for patients with TNBC (Table 1).

## 6 Conclusion

Because of the lack of targeted therapy, chemotherapy, radiation therapy, and elimination of TNBC tumors by surgical operation are frequently used to date. Sensitivity of chemotherapy and ease of use have made it a frequent therapy option for TNBC tumors. Alternative strategies have emerged due to the chemoresistance and severe side effects of chemotherapy, radiation therapy, and surgery. Silencing of the molecular factors allied with TNBC tumors, using siRNA-mediated nanoplatforms trigger improved tumor growth suppression. Both preclinical and clinical studies demonstrated that inhibition of these factors using siRNA-based nanoparticle-mediated therapy is auspicious in reduction of cell proliferation and survival in TNBC patients. Silencing/downregulation of MDM2, CDK11, CK2, TWIST, c-Myc, survivin, mTORC2, ATM protein, monopolar spindle 1 (MPS1) and cell division cycle protein (CDC20), Heparin-binding EGF-like growth factor (HB-EGF), EGFR, CXCR4, luciferase mRNA protein, enhanced green fluorescent protein (eGFP) are utilized in TNBC therapy. In recent years, there have been many preclinical studies utilizing siRNA therapeutics for enhanced therapy of TNBC patients. To date, some siRNA-mediated nanoparticles have either started phase I, phase II, and phase III clinical trials or have received FDA-approval for treatment of a wide variety of solid tumors including TNBC. Finally, combination therapy approaches may be highly effective for TNBC patients.

## References


Abad MN, Calabuing-Farinas S, Lobo de Mena M et al (2021) Update on systemic treatment in early triple negative breast cancer. Ther Adv Med Oncol 13:1–18

AGO Breast Committee (2019) Diagnosis and treatment of patients with primary and metastatic breast cancer: recommendations. Available from: www.ago-online.de

Alba E, Chacon JI, Lluch A et al (2012) A randomized phase II trial of platinum salts in basal-like breast cancer patients in the neoadjuvant setting. Results from the GEICAM/2006-03, multicenter study. Breast Cancer Res Treat 136(2):487–493

Aliabadi HM, Landry B, Mahdipoor P, Uludag H (2011) Induction of apoptosis by survivin silencing through siRNA delivery in a human breast cancer cell line. Mol Pharm 8:1821–1830

Ando M, Yamauchi H, Aogi K et al (2014) Randomized phase II study of weekly paclitaxel with and without carboplatin followed by cyclophosphamide/epirubicin/5-fluorouracil as neoadjuvant chemotherapy for stage II/IIIA breast cancer without HER2 overexpression. Breast Cancer Res Treat 145(2):401–409

Bai M, Shen M, Teng Y et al (2015) Enhanced therapeutic effect of Adriamycin on multidrug resistant breast cancer by the ABCG2-siRNA loaded polymeric nanoparticles assisted with ultrasound. Oncotarget 6:43779–43790

Bardia A, Tolaney SM, Loirat D et al (2020) ASCENT: a randomized phase III study of sacituzumabgovitecan (SG) vs treatment of physician's choice (TPC) in patients (pts) with previously treated metastatic triple-negative breast cancer (mTNBC). Ann Oncol 31: S1142–S1215

Bradner JE, West N, Grachan ML et al (2010) Chemical phylogenetics of histone deacetylases. Nat Chem Biol 6(3):238–243

Cardoso F, Kyriakides S, Ohno S, ESMO Guidelines Committee et al (2019) Early breast cancer: ESMO clinical practice guidelines for diagnosis, treatment and follow-up. Ann Oncol 30: 1194–1220

Chen KG, Sikic BI (2012) Molecular pathways: regulation and therapeutic implications of multidrug resistance. Clin Cancer Res 18(7):1863–1869

Chen H, Ma X, Li Z et al (2012) Functionalization of single-walled carbon nanotubes enables efficient intracellular delivery of siRNA targeting MDM2 to inhibit breast cancer cells growth. Biomed Pharmacother 66:334–338

Chen H, Wu J, Zhang Z et al (2018) Association between BRCA status and triple-negative breast cancer: a meta-analysis. Front Pharmacol 9:909

Cheng M, Cai W, Huang W et al (2018) Histone deacetylase 6 regulated expression of IL-8 is involved in the doxorubicin (Dox) resistance of osteosarcoma cells via modulating ABCB1 transcription, European. J Pharmacol 840:1–8

Choi YH, Yu A-M (2014) ABC transporters in multidrug resistance and pharmacokinetics, and strategies for drug development. Curr Pharm Des 20:793–807

Clarke M, Collins R, Darby S et al (2005) Effects of chemotherapy and hormonal therapy for early breast cancer on recurrence and 15-year survival: an overview of the randomised trials. Lancet 365:1687–1717

Cyprian FS, Akhtar S, Gatalica Z et al (2019) Targeted immunotherapy with a checkpoint inhibitor in combination with chemotherapy: a new clinical paradigm in the treatment of triple-negative breast cancer. Bosn J Basic Med Sci 19:227–233

Deng ZJ, Morton SW, Ben-Akiva E et al (2013) Layer-by-layer nanoparticles for systemic codelivery of an anticancer drug and siRNA for potential triple-negative breast cancer treatment. ACS Nano 7:9571–9584

Denkert C, von Minckwitz G, Brase JC, Sinn BV, Gade S, Kronenwett R et al (2015) Tumor-infiltrating lymphocytes and response to neoadjuvant chemotherapy with or without carboplatin in human epidermal growth factor receptor 2-positive and triple-negative primary breast cancers. J Clin Oncol 33(9):983–991

Ebinger S, Ozdemir E, Tiedt S et al (2016) Characterization of a novel dormant, drug resistant, stem cell subpopulation in acute lymphoblastic leukemia. Eur J Cancer 30(6):849–862

Ensenyat-Mendez M, Llinàs-Arias P, Orozco JIJ et al (2021) Current triple-negative breast cancer subtypes: dissecting the Most aggressive form of breast cancer. Front Oncol 11:681476

Finlay J, Roberts CM, Lowe G, Loeza J, Rossi JJ, Glackin CA (2015) RNA-based TWIST1 inhibition via dendrimer complex to reduce breast cancer cell metastasis. Biomed Res Int 2015:382745

Furlanetto J, Loibl S (2020) Optimal systemic treatment for early triple-negative breast cancer. Breast Care 15:217–226

Furlanetto J, Jackisch C, Untch M, Schneeweiss A, Schmatloch S, Aktas B et al (2017) Efficacy and safety of nab-paclitaxel 125 mg/m2 and nab-paclitaxel 150 mg/m2 compared to paclitaxel in early high-risk breast cancer. Results from the neoadjuvant randomized GeparSepto study (GBG 69). Breast Cancer Res Treat 163(3):495–506

Gao J, Liu W, Xia Y et al (2011) The promotion of siRNA delivery to breast cancer overexpressing epidermal growth factor receptor through anti-EGFR antibody conjugation by immunoliposomes. Biomaterials 32:3459–3470

Gianni L, Mansutti M, Anton A et al (2019) Event-free survival analysis of the prospectively randomized phase III ETNA study with neoadjuvant nab-paclitaxel (nab-P) versus paclitaxel (P) followed by anthracycline regimens in women with HER2-negative high-risk breast cancer. J Clin Oncol 37:515

Glackin CA (2018) Nanoparticle delivery of TWIST small interfering RNA and anticancer drugs: a therapeutic approach for combating cancer. Enzyme 44:83–101

Haberland M, Montgomery RL, Olson EN (2009) The many roles of histone deacetylases in development and physiology: implications for disease and therapy. Nat Rev Genet 10(1):32–42

Hamurcu Z, Ashour A, Kahraman N, Ozpolat B (2016) FOXM1 regulates expression of eukaryotic elongation factor 2 kinase and promotes proliferation, invasion and tumorgenesis of human triple negative breast cancer cells. Oncotarget 7:16619–16635

Han J, Lim W, You D et al (2019) Chemoresistance in the human triple-negative breast cancer cell line MDA-MB-231 induced by doxorubicin gradient is associated with epigenetic alterations in histone deacetylase. J Oncol 2019:1345026

Haque W, Verma V, Hatch S et al (2018) Response rates and pathologic complete response by breast cancer molecular subtype following neoadjuvant chemotherapy. Breast Cancer Res Treat 170:559–567

Hattab D, Bakhtiar A (2020) Bioengineered siRNA-based Nanoplatforms targeting molecular signaling pathways for the treatment of triple negative breast cancer: preclinical and clinical advancements. Pharmaceutics 12:929

Ho EA, Osooly M, Strutt D, Masin D, Yang Y, Yan H, Bally M (2013) Characterization of long-circulating cationic nanoparticle formulations consisting of a two stage PEGylation step for the delivery of siRNA in a breast cancer tumor model. J Pharm Sci 102:227–236

Housman G, Byler S, Heerboth S et al (2014) Drug resistance in cancer: an overview. Cancers 6(3): 1769–1792

Hu K, Law JH, Fotovati A, Dunn SE (2012) Small interfering RNA library screen identified polo-like kinase-1 (PLK1) as a potential therapeutic target for breast cancer that uniquely eliminates tumor-initiating cells. Breast Cancer Res 14:R22

Inoue S, Patil R, Portilla-Arias J et al (2012) Nanobiopolymer for direct targeting and inhibition of EGFR expression in triple negative breast cancer. PLoS One 7:e31070

Jazaeri AA, Awtrey CS, Chandramouli GVR et al (2005) Gene expression profles associated with response to chemotherapy in epithelial ovarian cancers. Clin Cancer Res 11(17):6300–6310

Jing H, Cheng W, Li S et al (2016) Novel cell-penetrating peptide-loaded nanobubbles synergized with ultrasound irradiation enhance EGFR siRNA delivery for triple negative breast cancer therapy. Colloids Surf B Biointerfaces 146:387–395

Joensuu H, Gligorov J (2012) Adjuvant treatments for triple-negative breast cancers. Ann Oncol 23: 40–45

Kassardjian A, Shintaku PI, Moatamed NA (2018) Expression of immune checkpoint regulators, cytotoxic T lymphocyte antigen 4 (CTLA-4) and programmed death-ligand 1 (PD-L1), in female breast carcinomas. PLoS One 13:e0195958

Kaufmann M, Von MincNwitz G, Smith R et al (2003) Internacional expert panel on the use of primary (preoperative) systemic treatment of operable breast cancer: review and recommendations. Clin Oncol 21:2600–2608

Kim SB, Dent R, Im SA et al (2017) LOTUS investigators. Ipatasertib plus paclitaxel versus placebo plus paclitaxel as first-line therapy for metastatic triple-negative breast cancer (LOTUS): a multicentre, randomised, double-blind, placebo-controlled, phase 2 trial. Lancet Oncol 18(10):1360–1372

Kren BT, Unger GM, Abedin MJ et al (2015) Preclinical evaluation of cyclin dependent kinase 11 and casein kinase 2 survival kinases as RNA interference targets for triple negative breast cancer therapy. Breast Cancer Res 17:19

Li Y, Seto E (2016) HDACs and HDAC inhibitors in cancer development and therapy. Cold Spring Harb Perspect Med 6(10):1–34, a026831

Li F, Aljahdali I, Ling X (2019) Cancer therapeutics using survivin BIRC5 as a target: what can we do after over two decades of study? J Exp Clin Cancer Res 38:368

Li Y, Zhan Z, Yin X, Fu S, Deng X (2021) Targeted therapeutic strategies for triple-negative breast cancer. Front Oncol 11:731535

Liedtke C, Mazouni C, Hess KR et al (2008) Response to neoadjuvant therapy and long-term survival in patients with triple-negative breast cancer. Int J Clin Oncol 26:1275–1281

Litton JK, Rugo HS, Ettl J et al (2018a) Talazoparib in patients with advanced breast cancer and a germline BRCA mutation. N Engl J Med 379(8):753–763

Litton JK, Scoggins ME, Hess KR et al (2018b) Neoadjuvant Talazoparib for patients with operable breast cancer with a germline BRCA pathogenic variant. J Clin Oncol 38(5):388–394

Liu Y, Li P-K, Li C, Lin J (2010) Inhibition of STAT3 signaling blocks the anti-apoptotic activity of IL-6 in human liver cancer cells. J Biol Chem 285(35):27429–27439

Liu Y, Zhu YH, Mao CQ et al (2014) Triple negative breast cancer therapy with CDK1 siRNA delivered by cationic lipid assisted PEG-PLA nanoparticles. J Control Release 192:114–121

Liu Y, Zhang X, Han C, Rodriguez-Aguayo C, Lopez-Berestein G, Rao PH et al (2015) TP53 loss creates therapeutic vulnerability in colorectal cancer. Nature 520:697–701

Liu Y, Zhang X, Wang L et al (2020) 270P Apatinib added to taxanes and platinum neoadjuvant chemotherapy for patients with triple-negative and HER2-positive breast cancer: a multicenter, randomized, phase II, open-label trial. Ann Oncol 31:S346

Loibl S, Weber KE, Timms KM et al (2018) Survival analysis of carboplatin added to an anthracycline/taxane-based neoadjuvant chemotherapy and HRD score as predictor of response-final results from GeparSixto. Ann Oncol 29(12):2341–2347

Loibl S, Untch M, Burchardi N et al (2019) A randomised phase II study investigating durvalumab in addition to an anthracycline/taxane-based neoadjuvant therapy in early triple-negative breast cancer: clinical results and biomarker analysis of GeparNuevo study. Ann Oncol 30:1279–1288

Luo C, Wang P, He S, Zhu J, Shi Y, Wang J (2022) Progress and Prospect of immunotherapy for triple-negative breast cancer. Front Oncol 12:919072

Maire V, Némati F, Richardson M et al (2013) Polo-like kinase 1: a potential therapeutic option in combination with conventional chemotherapy for the management of patients with triple-negative breast cancer. Cancer Res 73:813–823

Manjunath M, Choudhary B (2021) Triple-negative breast cancer: a run-through of features, classification and current therapies. Oncol Lett 22:512
Marra A, Viale G, Curigliano G (2019) Recent advances in triple negative breast cancer: the immunotherapy era. BMC Med 17:90
Matsuda N, Wang X, Lim B et al (2018) Safety and efficacy of panitumumab plus neoadjuvant chemotherapy in patients with primary her2-negative inflammatory breast cancer. JAMA Oncol 4:1207–1213
Misra AC, Luker KE, Durmaz H, Luker GD, Lahann J (2015) CXCR4-targeted nanocarriers for triple negative breast cancers. Biomacromolecules 16:2412–2417
Morry J, Ngamcherdtrakul W, Gu S et al (2017) Targeted treatment of metastatic breast cancer by PLK1 siRNA delivered by an antioxidant nanoparticle platform. Mol Cancer Ther 16:763–772
Muenst S, Schaerli AR, Gao F et al (2014) Expression of programmed death ligand 1 (PD-L1) is associated with poor prognosis in human breast cancer. Breast Cancer Res Treat 146:15–24
Najmi S, Korah R, Chandra R, Abdellatif M, Wieder R (2005) Flavopiridol blocks integrin-mediated survival in dormant breast cancer cells. Clin Cancer Res 11(5):2038–2046
Nanda R, Liu MC, Yau C et al (2020) Effect of pembrolizumab plus neoadjuvant chemotherapy on pathologic complete response in women with early-stage breast cancer: an analysis of the ongoing phase 2 adaptively randomized ISPY2 trial. JAMA Oncol 6(5):676–684
National Comprehensive Cancer Network. Breast cancer (version 1.2020) (2020). Available from: https://www.nccn.org/professionals/physician_gls/pdf/breast.pdf
Navarro G, Sawant RR, Biswas S, Essex S, Tros de Ilarduya C, Torchilin VP (2012) P-glycoprotein silencing with siRNA delivered by DOPE-modified PEI overcomes doxorubicin resistance in breast cancer cells. Nanomedicine 7:65–78
Nedeljković M, Damjanović A (2019) Mechanisms of chemotherapy resistance in triple-negative breast cancer—how we can rise to the challenge. Cell 8:957
Okamoto A, Asai T, Hirai Y et al (2018) Systemic administration of siRNA with anti-HB- EGF antibody-modified lipid nanoparticles for the treatment of triple-negative breast cancer. Mol Pharm 15:1495–1504
Parmar MB, Ballesteros BEA, Fu T et al (2016) Multiple siRNA delivery against cell cycle and anti-apoptosis proteins using lipid-substituted polyethylenimine in triple-negative breast cancer and non-malignant cells. J Biomed Mater Res Part A 104:3031–3044
Pohl-Rescigno E, Hauke J, Loibl S et al (2019) Germline mutation status and therapy response in high-risk early-stage 2 breast cancer: a secondary analysis of the GeparOcto randomized clinical trial. J Clin Oncol Forthcoming 37(15):573
Redig AJ, Allister S (2013) Breast cancer as a systemic disease: a view of metastasis. J Intern Med 274:113–126
Richon VM, Emiliani S, Verdin E et al (1998) A class of hybrid polar inducers of transformed cell diferentiation inhibits histone deacetylases. Proc Natl Acad Sci U S A 95(6):3003–3007
Robson V, Im SA, Senkus E et al (2017) Olaparib for metastatic breast cancer in patients with a germline BRCA mutation. N Engl J Med 377(6):523–533
Sahota S, Vahdat LT (2017) Sacituzumabgovitecan: an antibody-drug conjugate. Expert Opin Biol Ther 17:1027–1031
Salzano G, Riehle R, Navarro G, Perche F, De Rosa G, Torchilin VP (2014) Polymeric micelles containing reversibly phospholipid-modified anti-survivin siRNA: a promising strategy to overcome drug resistance in cancer. Cancer Lett 343:224–231
Schmid P, Abraham J, Chan S et al (2020a) Capivasertib plus paclitaxel versus placebo plus paclitaxel as first-line therapy for metastatic triple-negative breast cancer: the PAKT trial. J Clin Oncol 38(5):423–433
Schmid P, Salgado R, Park YH et al (2020b) Pembrolizumab plus chemotherapy as neoadjuvant treatment of high-risk, early-stage triple-negative breast cancer: results from the phase 1b open-label, multicohort KEYNOTE-173 study. Ann Oncol 31(5):569–581

Shahbazi R, Asik E, Kahraman N et al (2017) Modified gold-based siRNA nanotherapeutics for targeted therapy of triple-negative breast cancer. Nanomedicine 12:1961–1973

Shen J, Sun H, Xu P et al (2013) Simultaneous inhibition of metastasis and growth of breast cancer by co-delivery of twist shRNA and paclitaxel using pluronic P85-PEI/TPGS complex nanoparticles. Biomaterials 34:1581–1590

Subhan M, Torchilin VP (2021) Advances with antibody-drug conjugates in breast cancer treatment. Eur J Pharm Biopharm 169:241–255

Untch M, Jackisch C, Schneeweiss A, Conrad B, Aktas B, Denkert C, German Breast Group (GBG); Arbeitsgemeinschaft Gynäkologische Onkologie—Breast (AGO-B) Investigators et al (2016) Nab-paclitaxel versus solvent-based paclitaxel in neoadjuvant chemotherapy for early breast cancer (GeparSepto-GBG 69): a randomised, phase 3 trial. Lancet Oncol 17(3):345–356

Untch M, Jackisch C, Schneeweiss A, Schmatloch S, Aktas B, Denkert C et al (2019) NAB-paclitaxel improves disease-free survival in early breast cancer: GBG 69-GeparSepto. J Clin Oncol 37(25):2226–2234

Van Klompenberg MK, Leyden E, Arnason AH, Zhang J-T, Stefanski CD, Prosperi JR (2017) APC loss in breast cancer leads to doxorubicin resistance via STAT3 activation. Oncotarget 8(61): 102868–102879

Velasco M, Martínez S, Cerdà P et al (2012) Quimioterapia neoadyuvante en el cáncer de mama localmente avanzado. Revista de Senología y Patología Mamaria 25:14–21

von Minckwitz G, Schneeweiss A, Loibl S et al (2014) Neoadjuvant carboplatin in patients with triple-negative and HER2-positive early breast cancer (GeparSixto; GBG 66): a randomised phase 2 trial. Lancet Oncol 15(7):747–456

Voutsadakis IA (2016) Immune blockade inhibition in breast cancer. Anticancer Res 36:5607–5622

Walsh CS (2015) Two decades beyond BRCA1/2: homologous recombination, hereditary cancer risk and a target for ovarian cancer therapy. Gynecol Oncol 137(2):343–350

Wang M, Gartel AL (2011) The suppression of FOXM1 and its targets in breast cancer xenograft tumors by siRNA. Oncotarget 2:1218–1226

Wei L, Zhu S, Li M et al (2018) High indoleamine 2,3-dioxygenase is correlated with microvessel density and worse prognosis in breast cancer. Front Immunol 9:724

Wen WX, Leong C (2019) Association of BRCA1- and BRCA2-deficiency with mutation burden, expression of PD-L1/PD-1, immune infiltrates, and T cell-inflamed signature in breast cancer. PLoS One 14:0215381

Werfel TA, Wang S, Jackson MA et al (2018) Selective mTORC2 inhibitor therapeutically blocks breast cancer cell growth and survival. Cancer Res 78:1845–1858

Wu Y, Zhang Y, Zhang W, Sun C, Wu J, Tang J (2016) Reversing of multidrug resistance breast cancer by co-delivery of P-gp siRNA and doxorubicin via folic acid-modified core-shell nanomicelles. Colloids Surf B Biointerfaces 138:60–69

Xu R, Huang Y, Mai J et al (2013) Multistage vectored siRNA targeting ataxia-telangiectasia mutated for breast cancer therapy. Small 9:1799–1808

Xu J, Liu Y, Li Y et al (2019) Precise targeting of POLR2A as a therapeutic strategy for human triple negative breast cancer. Nat Nanotechnol 14:388–397

Yu M, Han S, Kou Z et al (2018) Lipid nanoparticle-based co-delivery of epirubicin and BCL-2 siRNA for enhanced intracellular drug release and reversing multidrug resistance. Artif Cells Nanomed Biotechnol 46:323–332

Zgura A, Galesa L, Bratila E et al (2018) Relationship between tumor infiltrating lymphocytes and progression in breast cancer. Maedica (Buchar). J Clin Oncol 13:317–320

# Advances in Local Ablative Techniques for Breast Cancer

Billy Ho-Hung Cheung and Michael Tiong-Hong Co

**Abstract**

Breast cancer, the most prevalent malignancy in women, has witnessed a paradigm shift in its management over the years. From the historically predominant mastectomies, which often led to significant physical and psychological morbidity, the focus has shifted to breast-conserving surgery and minimally invasive ablative techniques.

This chapter provides a comprehensive review of the various ablative modalities available for breast cancer management, emphasizing their mechanisms, indications, efficacy, and potential complications.

Ablative techniques characterized by their minimally invasive nature aim to obliterate tumor cells while preserving breast tissue aesthetics. These techniques, including cryoablation, radiofrequency ablation (RFA), high-intensity focused ultrasound (HIFU), microwave ablation (MWA), and laser therapy, aim to obliterate tumor cells while maximizing breast tissue preservation and aesthetic outcomes. Moreover, due to the minimally invasive nature, they are particularly beneficial for patients unfit for traditional surgeries, as most of them can be performed without general anesthesia.

However, the non-excisional nature of these procedures raises concerns about the completeness of ablation and oncological safety. Accurate imaging guidance, primarily ultrasound and magnetic resonance imaging (MRI), is paramount for the precision of these techniques. The chapter also delves into the role of ablative surgery in palliative settings, highlighting its potential in symptom alleviation and prevention of disease-related complications.

B. H.-H. Cheung · M. T.-H. Co (✉)
Division of Breast Surgery, Department of Surgery, Queen Mary Hospital and Tung Wah Hospital, School of Clinical Medicine, Li Ka Shing Faculty of Medicine, The University of Hong Kong, Hong Kong S.A.R., China
e-mail: mcth@hku.hk



Interdisciplinary Cancer Research, https://doi.org/10.1007/16833_2023_189
Published online: 20 October 2023

As the field of breast cancer surgery continues to evolve, these ablative techniques, backed by technological advancements and interdisciplinary research, promise to play an increasingly significant role in personalized patient care.

**Keywords**

Ablative therapy · Breast oncology · Early breast cancer · Minimally invasive breast cancer treatment · Oncological safety

## Abbreviations

| | |
|---|---|
| CT | Computed tomography |
| HIFU | High-intensity focused ultrasound |
| LITT | Laser interstitial thermotherapy |
| MRI | Magnetic resonance imaging |
| MWA | Microwave ablation |
| RF | Radiofrequency |
| RFA | Radiofrequency ablation |
| US | Ultrasound |

## 1 Introduction

Breast cancer remains the most prevalent cancer in women globally (Lagendijk et al. 2018). Over the past few decades, the approach to breast cancer management has undergone a transformative evolution. Historically, mastectomies were the mainstay of treatment, often resulting in significant physical and psychological morbidity (Piot-Ziegler et al. 2010). The subsequent shift to breast-conserving surgery permitted the preservation of not only the breast tissue but also the patient's self-image and psychological well-being (Zehra et al. 2020).

## 2 Overview of Local Ablative Techniques in Breast Cancer Management

In the contemporary era, the focus has been further refined to prioritize both efficacy and quality of life. This has led to the emergence and adoption of ablative techniques. These methods, characterized by their minimally invasive nature, aim to obliterate tumor cells while maximizing the preservation of breast tissue and its aesthetic appearance (Peek et al. 2017). The attraction of these techniques lies not just in their reduced invasiveness but also in their potential to minimize hospital stays, reduce recovery times, minimize deformity and scarring, and offer treatment options to those who might be unfit for traditional surgeries (Carriero et al. 2023). In addition, adjuvant therapy may be administered faster after ablative treatment due to the less-destructive nature of the approach.

However, due to the non-excisional nature of ablative procedures, completeness of ablation and its oncological safety is of constant concern (Peek et al. 2017). In addition, due to the limitation in the instrument design, almost all ablative techniques are designed for small, early tumors (Kwong et al. 2023). In this chapter, we explore the various ablative modalities available, their indications, efficacy, and potential complications, outlining a comprehensive review of their roles in modern breast cancer management (Table 1).

**Table 1** Summary of ablative procedures

| Ablative approach | Mechanism of action | Advantages | Limitations | Ablative rate |
|---|---|---|---|---|
| Cryoablation | 1. Two-phase processes with alternating freezing and thawing cycles<br>2. Cell rupture and osmotic dehydration | 1. Minimally invasive<br>2. Visualization possible with the "ice ball" | 1. Local complications with ulceration and frostbite | 74.1 ± 28.9% |
| Radiofrequency ablation (RFA) | 1. High-frequency electrical currents produce heat to induce coagulative necrosis | 1. Minimally invasive<br>2. More evidence available<br>3. Highest overall ablative rates | 1. Potential skin burn | 87.1 ± 12.8% |
| High-intensity focused ultrasound (HIFU) | 1. Focuses ultrasound to raise the temperature and induce cell death | 1. Completely noninvasive<br>2. Precise targeting | 1. Limited by the depth of penetration<br>2. Longer procedure times | 47.6 ± 29.9% |
| Microwave ablation (MWA) | 1. Microwave energy to produce heat and destroy tumor cells | 1. Rapid heating<br>2. Larger ablation zones | 1. Heat diffusion to surrounding tissues<br>2. Ensuring clear margins can be challenging | 83.2 ± 11.6% |
| Laser ablation | 1. Uses laser energy to generate heat and damage to tumor cells | 1. Minimally invasive, rapid coagulation<br>2. Outpatient procedure possible | 1. Limited depth of penetration<br>2. Risk of burns<br>3. Lowest ablative rates | 52.2 ± 29.2% |

## 3 Localization Techniques

Accurate imaging guidance is paramount to ensure the precision and efficacy of the procedure. Ultrasound (US) emerged as the predominant choice. It was utilized across most of the published studies in all ablative modalities (Peek et al. 2017). Magnetic resonance imaging (MRI), known for its high-resolution soft tissue contrast, was predominantly used in studies focusing on High-intensity focused ultrasound (HIFU) and cryoablation (Furusawa et al. 2006; Gianfelice et al. 2003a, b, c; Khiat et al. 2006; Merckel et al. 2013, 2016; Siedek et al. 2019; Yamamoto et al. 2011). A minority of studies used computer tomography (CT) with US (Cazzato et al. 2015; Harries et al. 1994; Littrup et al. 2009) or stereotactic guidance (Bloom et al. 2001; Dowlatshahi et al. 2002).

## 4 Ablative Treatment Modalities for Curative Intent

### 4.1 Cryotherapy

**The Evolution**

While the concept of using extreme cold to treat diseases can be traced back to the nineteenth century, which mainly used ice for various conditions, the invention of the cryoprobe in the 1960s, which utilized liquid nitrogen, allowed for more targeted and deeper tissue freezing (Bald 1984; Budman et al. 1986). Among all the ablative techniques described, only cryotherapy uses freezing instead of heat (Karow and Webb 1965).

The initial focus of cryotherapy, also called cryoablation, in oncology was on prostate and liver cancers (Yilmaz et al. 2016). The ability to freeze and destroy tumor cells while sparing surrounding healthy tissue made it an attractive option. By the late twentieth to early twenty-first century, the potential of cryotherapy in breast cancer began to be explored (Pfleiderer et al. 2002; Tafra et al. 2003). Given the organ's accessibility and the desire for less invasive treatments that preserved the appearance of the breast, cryotherapy seemed a promising avenue.

The current system, such as the ProSense™ system developed by IceCure Medical Limited (Figs. 1 and 2), commonly consists of a cryoprobe (indicated by a red arrow) for the approach toward the breast tumor, which connects to a console containing a liquid nitrogen reservoir for freezing.

**Mechanism of Action**

Cryotherapy induces cell death through a two-phase process with alternating freezing and thawing cycles (Takada and Toi 2019). The rapid freezing causes intracellular ice formation, leading to cell rupture and osmotic dehydration (Chu and Dupuy 2014). The subsequent thawing results in osmotic imbalances, leading to further cell death. Freezing also damages the tumor's vasculature, depriving it of essential nutrients and indirect damage due to ischemia and immunologic response.

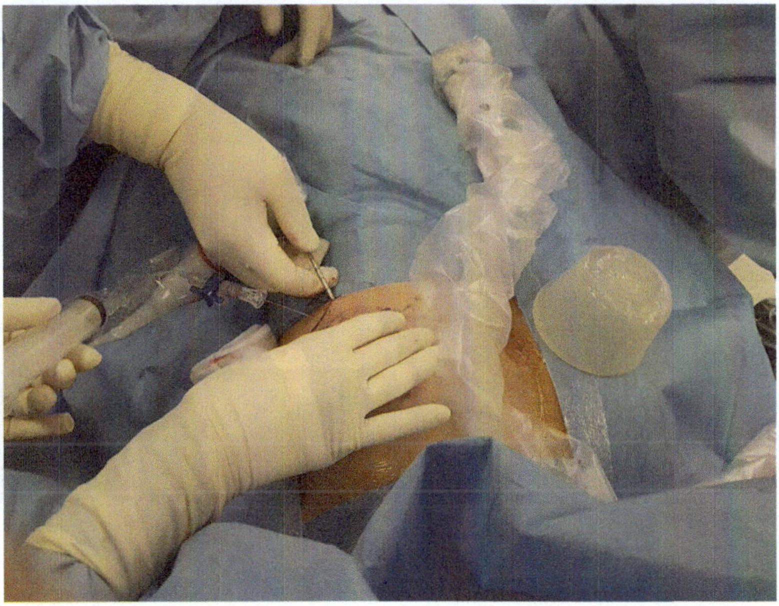

**Fig. 1** Cryoablation in action

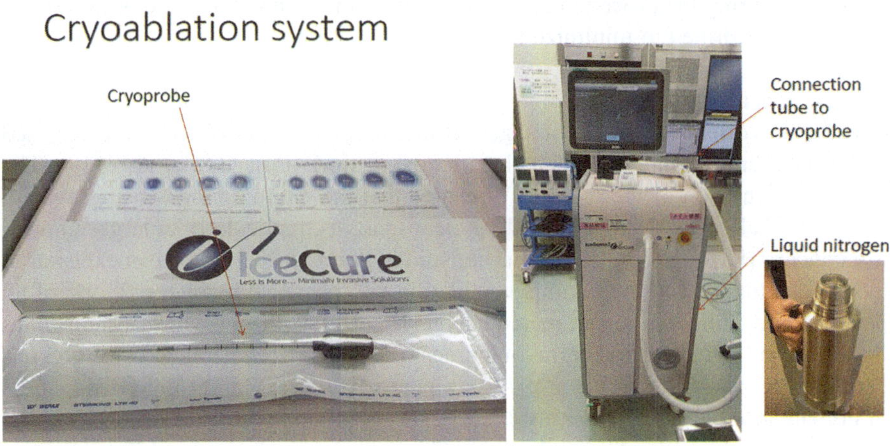


**Fig. 2** A contemporary cryotherapy system in the operation theater (the ProSense™ system, IceCure Medical Limited)

## The Procedure

Cryotherapy has been primarily explored for small breast tumors. The visualization of the tumor was performed under ultrasound to confirm the site of incision and the axis for probe insertion (Pfleiderer et al. 2005) (Fig. 3). Then the cryoprobe was inserted into the tumor under imaging guidance. Liquid nitrogen or argon gas is then

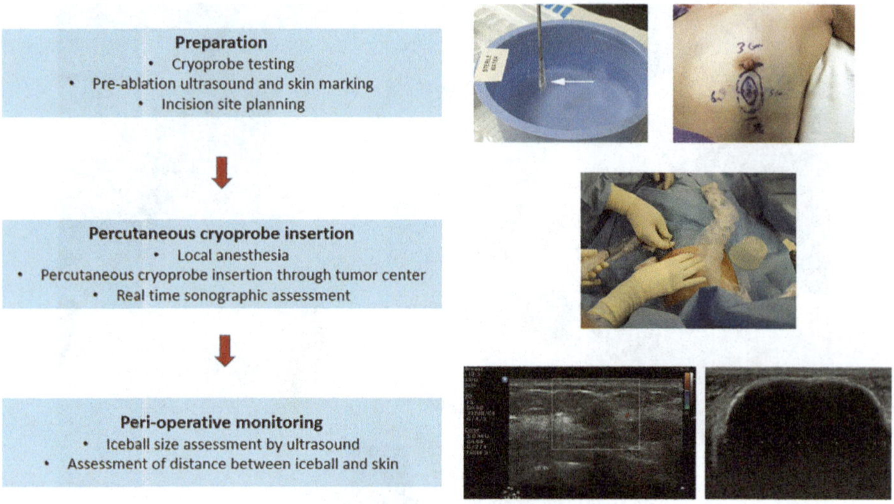


**Fig. 3** Illustration of the cryotherapy. (Kwong et al. 2023)

used to freeze the tumor, with the process monitored in the US in real time to ensure complete tumor ablation. The distance between the ice ball and the skin is constantly monitored during the procedure, and local infiltration of normal saline or tumescent should be performed to minimize the risk of frost injury.

### Advantages and Challenges

Cryotherapy offers several advantages. It is minimally invasive, can be performed under local anesthesia, and has a short recovery time (Pusceddu et al. 2019). The visual confirmation of the "ice ball" ensures that the entire tumor is being treated; however, challenges remain. Providing clear margins, particularly in larger tumors, is a concern (Kwong et al. 2023). Earlier studies commonly followed an excision of the ablated tumor, and their complete ablation was achieved in 74.1 ± 28.9% of all patients, according to a recent meta-analysis (Peek et al. 2017). Most of the studies only focused on the short-term outcomes. The American College of Surgeons Oncology Group (ACOSOG) Z1072 is a phase II trial exploring the effectiveness of cryoablation in the treatment of breast cancer (Simmons et al. 2016). Successful ablation was achieved in 80 (92%) of the 87 lesions reported when multifocal disease outside of the targeted zone was not considered a treatment failure. Overall, the best results are achieved with small (<15 mm) ductal tumors treated by the application of multiple cryoprobes (Lanza et al. 2015).

With cryoablation, 10.9% of patients developed a complication, which was primarily local complications, including skin necrosis, hematoma, seromata, bleeding, and, less commonly, skin ulceration (Peek et al. 2017). There is also the risk of frostbite to the skin and damage to surrounding structures, which is unique to cryoablation, but the incidence is very low.

The long-term oncological outcome, in particular local recurrence, needs to wait for current future reports of prior/recent trials for a comprehensive assessment, which include two major phase 2 trials, the FROST trial (NCT01992250) (*Cryoablation of Small Breast Tumors in Early Stage Breast Cancer (FROST) (NCT01992250)*, 2013) and the Ice3 trial (NCT02200705) (Cryoablation of Low Risk Small Breast Cancer- Ice3 Trial (NCT02200705) 2014), which are currently underway to evaluate complete response or local recurrence focusing on tumor size ≤1.5 cm.

**Future Directions**

Modern cryotherapy equipment allows for more precise temperature control, ensuring uniform freezing. The integration of cryotherapy with advanced imaging modalities has further enhanced its precision and safety.

As with other ablative techniques, the potential of combining cryotherapy with other treatments, such as immunotherapy or targeted therapies, is being explored (Ou et al. 2023). The idea is that the cell death induced by freezing might stimulate an immune response, which could be augmented by other therapies.

In summary, the journey of cryotherapy from its rudimentary beginnings to its current role in oncology highlights the innovative spirit of medical science. Its application in breast cancer, while still evolving, offers a glimpse into the future of personalized and minimally invasive cancer treatments.

## 4.2 Radiofrequency Ablation (RFA) in Breast Cancer Management

**The Evolution**

The earliest applications of radiofrequency (RF) energy in medicine were for the treatment of initially mainly neurological conditions for more than a century (Ihnát et al. 2014). The potential of RFA in oncology was recognized in the late twentieth century, with the initial focus on liver tumors (Rhim and Dodd 3rd. 1999). The success in liver tumors paved the way for its exploration into other malignancies, including breast cancer (Izzo et al. 2001; Jeffrey et al. 1999).

Over time, as technology advanced, the electrodes became thinner, the imaging guidance more precise, and the procedure itself more refined (Waaijer et al. 2014) (Fig. 4). Modern RF generators can monitor tissue impedance, allowing for real-time feedback and ensuring more uniform heating of the tumor (Xia et al. 2021). Another important development is the cool-tip system, which allows the circulation of water internally to cool the tissue adjacent to the electrode to limit tissue charring while permitting higher energy for a larger ablative zone (Kinoshita 2016).

**Mechanism of Action**

RFA works by thermal ablation, where high-frequency electrical currents pass through tissues with resistance, especially from tumor cells, generating heat (Nguyen et al. 2014). The heat leads to coagulative necrosis and subsequent eradication of tumor cells. The surrounding healthy tissues, with their different electrical

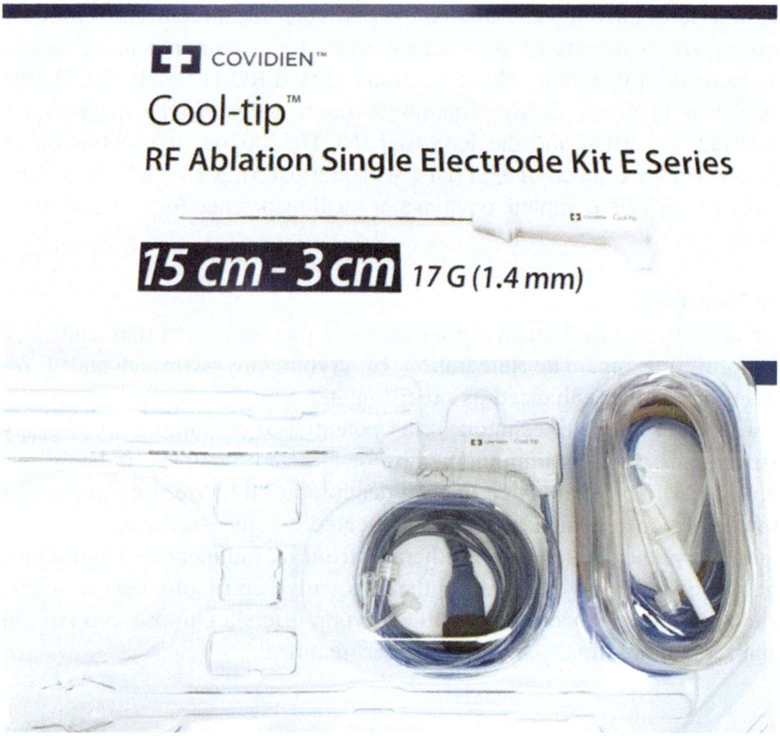


**Fig. 4** An example of radiofrequency ablation system (Covidien, Boulder, CO, USA)

properties, are less affected, allowing for a degree of selectivity in the ablation process.

**The Procedure**

Typically performed under ultrasound or MRI guidance, a thin electrode is percutaneously inserted into the tumor (Kinoshita et al. 2011) (Fig. 5). Once correctly positioned, radiofrequency energy is delivered, raising the temperature within the tumor to a point where cells cannot survive. The real-time imaging guidance of the procedure ensures precision, minimizing damage to surrounding healthy tissues.

**Advantages and Challenges**

RFA is minimally invasive, often performed under local anesthesia, making it especially suitable for patients who might be unfit or unwilling to undergo traditional surgery (Schassburger et al. 2014). It has shown the most promise for tumors less than 2 cm in size (Nguyen et al. 2014). Its efficacy diminishes with larger tumors due to the challenge of ensuring uniform heat distribution throughout the tumor mass (Nguyen et al. 2014). It is particularly beneficial for patients who are not ideal candidates for surgery due to age, comorbidities, or personal preference.

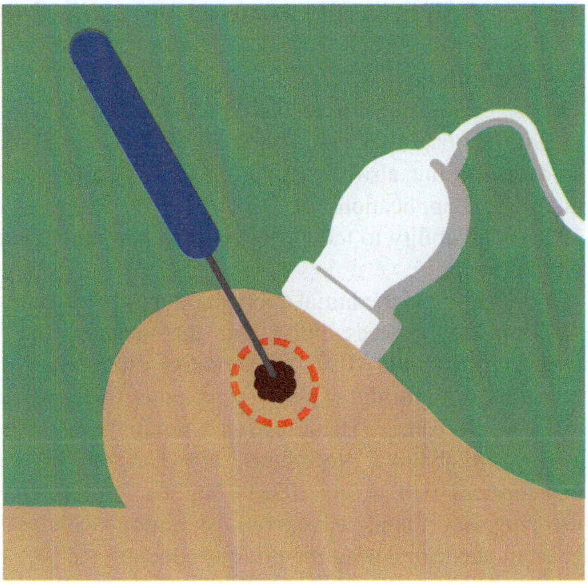

**Fig. 5** A schematic diagram for percutaneous RFA in the breast with a patient lying supine. The brownish focus representing a tumor focus, the blue handle represents an RFA probe while the procedure is done under ultrasound guidance. The red dotted circle represents the thermal ablative zone by the RFA probe with margins included

Additionally, RFA can be a valuable tool for treating recurrent tumors or those in anatomically challenging locations (Garbay et al. 2008).

For tumors less than 2 cm in size, RFA has demonstrated commendably high ablative rates. In 87.1 ± 12.8% (491/564) of all patients who underwent surgical excision, complete ablation of the tumor was achieved (Peek et al. 2017). The oncological outcome was better documented compared to the rest of other ablative modalities with a collective local recurrence rate of 3.1% at a mean follow-up time of 30.8 ± 16.9 months (Peek et al. 2017).

While RFA is generally safe, risks associated include pain, skin burns, hematoma formation, and infection. In rare instances, there is potential damage to surrounding structures, especially if the tumor is close to the skin or chest wall, with reports documented on muscle burn, pneumothorax, and fistulations (Peek et al. 2017; Vilar et al. 2012; Waaijer et al. 2014; Wiksell et al. 2010).

**Future Directions**

As technology advances, there is potential for refining RFA techniques to treat larger tumors or those in more challenging locations. Combining RFA with other therapies, such as immunotherapy, might also enhance its antitumor effects, paving the way for more integrated and holistic treatment strategies.

In conclusion, RFA represents a significant stride in the ongoing journey to optimize breast cancer treatment. By offering a blend of efficacy, reduced invasiveness, and cosmetic preservation, it holds promise as a valuable tool in individualized patient care.

## 4.3 High-Intensity Focused Ultrasound (HIFU)

**The Evolution**

The use of ultrasound in medicine began in the early twentieth century, primarily for diagnostic imaging (Campbell 2013). However, the idea of using ultrasound not only to diagnose but also to treat took a few more decades to materialize. The first therapeutic applications of HIFU were in neurosurgery in the 1950s (Fry et al. 1958). The ability to target deep-seated structures without making an incision was revolutionary.

For oncology, the initial application was in prostate cancer (Beerlage et al. 1999). The prostate's accessibility via the rectum and the challenges of surgical interventions made HIFU an attractive alternative. Its success in prostate cancer paved the way for the exploration of other malignancies, including breast cancer.

In breast cancer, HIFU offers a noninvasive approach to tumor ablation, and among all ablative procedures, HIFU is the only procedure that is genuinely incision-less (Co et al. 2022) (Fig. 6) (Group 2023). The procedure involves focusing ultrasound waves on the tumor, with real-time imaging (often MRI) guiding and monitoring the process (Bachu et al. 2021).

**Mechanism of Action**

HIFU operates by focusing ultrasound waves on a specific point, much like how a magnifying glass can focus sunlight (Izadifar et al. 2020) (Fig. 7). At this focal point,

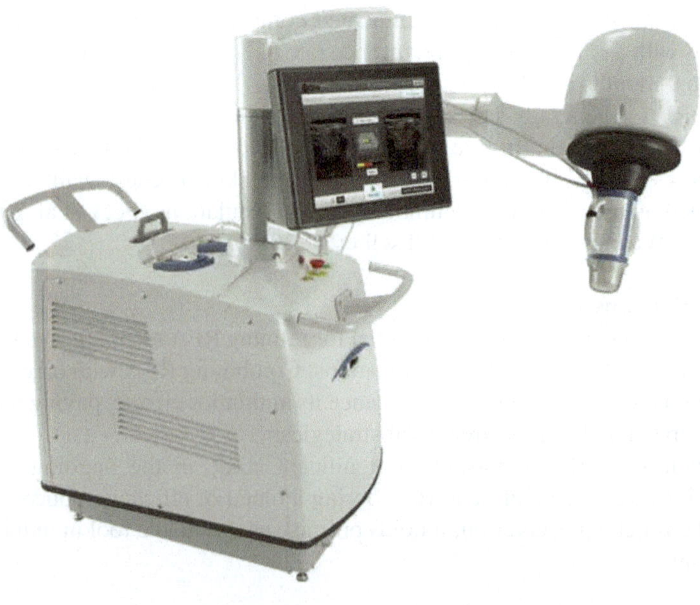

**Fig. 6** An example of the HIFU system (ECHOPULSE®). (Group 2023)

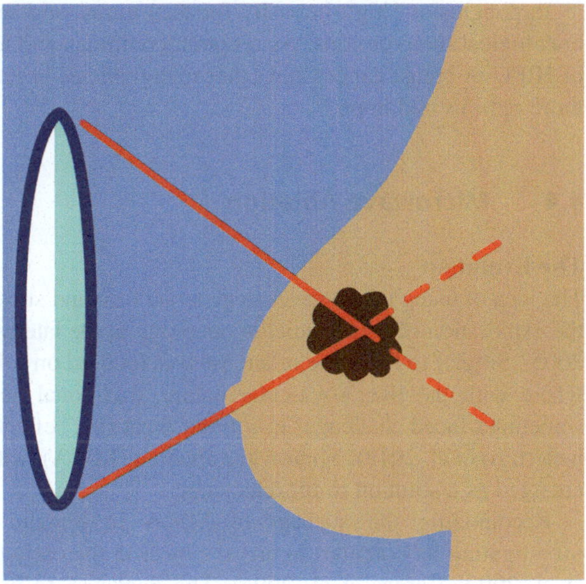

**Fig. 7** Schematic diagram showcasing the mechanism of HIFU. The brownish focus represents a breast tumor. The convex shape represents the HIFU plate. The red line shows the convergence of HIFU wave that targets the tumor

the energy is intense enough to cause rapid temperature elevation, leading to coagulative necrosis and cell death, all without affecting the surrounding tissues.

**The Procedure**

This is also the only ablative treatment for breast cancer that is entirely noninvasive and needleless (Wei et al. 2020) (Fig. 7). Most of the time, US and magnetic resonance imaging MRI are used to provide spatial targeting and real-time monitoring (Siedek et al. 2019).

**Advantages and Challenges**

With no incisions, the risk of infection and cosmetic outcomes were a substantial advantage of HIFU. The procedure can also be repeated if necessary. However, HIFU does have limitations. Treating larger tumors or those close to the skin or chest wall can be challenging. Ensuring complete ablation and clear margins is also a concern. With HIFU, complete ablation was achieved only in $47.6 \pm 29.9\%$ (60/126) of all patients (Peek et al. 2017).

Modern HIFU systems have seen significant refinements. Improved imaging integration, real-time temperature monitoring, and more precise focusing mechanisms have enhanced both safety and efficacy.

**Future Directions**

The potential of HIFU extends beyond tumor ablation. Ongoing research is exploring its role in drug delivery, as HIFU can enhance the permeability of tumor vessels (Izadifar et al. 2020). In combination with other therapies, such as immunotherapy, there is potential for augmentation in antitumor response.

In conclusion, high-intensity focused ultrasound represents a paradigm shift in oncological interventions. As research continues and technology advances, the role of HIFU in breast cancer and other malignancies is set to expand, offering patients more options and hope.

## 4.4 Microwave Ablation

**The Evolution**

The idea of using microwaves to generate heat and subsequently destroy tissue led to the exploration of its potential in oncology in the late twentieth century (Simon et al. 2005). Similar to RFA, the initial use was focused on solid organs, including the liver, where with the liver's vascular nature, traditional heat-based ablation techniques sometimes faced challenges due to the "heat sink" effect, where blood flow dissipates heat (Izzo et al. 2019). Microwave ablation (MWA), with its rapid heating capability, emerged as a solution to this challenge.

Recognizing the success of MWA in hepatic tumors, researchers began investigating its potential in breast cancer in the early twenty-first century (Gardner et al. 2002; Vargas et al. 2004). The goal was to provide a minimally invasive, effective treatment option, especially for early-stage tumors or those in patients unfit for traditional surgery.

**Mechanism of Action**

MWA operates by producing oscillating electric and magnetic fields (Zhou et al. 2014). These fields agitate water molecules within cells, generating friction and heat. This rapid, localized heating leads to coagulative necrosis and cell death. Unlike radiofrequency ablation, which relies on electrical conductivity, the efficacy of MWA is less affected by changes in tissue properties, allowing for more consistent and larger ablation zones (Izzo et al. 2019).

**The Procedure**

MWA in breast cancer involves the percutaneous insertion of a microwave antenna into the tumor, typically under ultrasound guidance (Mays et al. 2016; Pan et al. 2021). Once positioned, microwave energy is delivered, raising the temperature within the tumor to lethal levels. The real-time imaging of the procedure ensures precision and minimizes damage to surrounding tissues.

**Advantages and Challenges**

The rapid heating capability of MWA allows for shorter procedure times compared to other ablation techniques (van de Voort et al. 2021). Its ability to create larger, more consistent ablation zones is another advantage. From the reported studies in the literature, $83.2 \pm 11.6\%$ (89/107) of patients obtained complete microwave ablation (Peek et al. 2017).

Modern MWA systems have seen significant advancements. Improved antenna designs, real-time temperature monitoring, and integration with advanced imaging modalities have enhanced the safety and efficacy of the procedure (Hernández et al. 2015).

However, there are challenges. Ensuring clear margins in more extensive tumors and the potential for burns or damage to surrounding structures are concerns that practitioners must navigate.

**Future Directions**

The future of MWA in breast cancer is promising. Ongoing research explores its combination with other therapies, especially its role in immune response activation, which can be crucial for immunotherapy (Zhou et al. 2021).

In conclusion, microwave ablation represents the confluence of technological innovation and clinical need. Its journey from broader medical applications to its niche in oncology underscores the relentless pursuit of better, less invasive treatments in the ever-evolving landscape of cancer care.

## 4.5 Laser Therapy in Breast Cancer Management

**The Evolution**

The term "laser" stands for "light amplification by stimulated emission of radiation." Since its invention in the 1960s, lasers have found applications across various medical fields, from ophthalmology to dermatology (Dixon 1988). The idea of using lasers to target and destroy tumors emerged soon after their medical introduction.

Laser interstitial thermotherapy (LITT) is a variation of laser ablation where laser energy is delivered through optical fibers placed within the tumor (Kerbage et al. 2022). It allows for the treatment of slightly larger tumors.

**Mechanism of Action**

Lasers work by delivering concentrated light to specific tissues (Haraldsdottir et al. 2008). In oncology, the goal is to use this light to generate heat and cause localized damage to tumor cells, leading to coagulative necrosis. The precision of lasers allows for targeted ablation, minimizing damage to surrounding healthy tissue.

**The Procedure**

The technique involves the use of lasers to directly target and destroy tumors. Under imaging guidance, typically ultrasound, a laser fiber is inserted into the tumor, and laser energy is delivered to ablate the tumor.

**Advantages and Challenges**

The accuracy of the laser ensures that damage to surrounding tissues is minimized, which makes outpatient procedures possible and reduces hospital stays. Real-time coagulative changes can be indirectly measured through continuous temperature monitoring of the tumor both centrally and to ensure the adequacy of treatment (Dowlatshahi et al. 2002).

For laser ablation, however, the complete ablative rate is a bit disappointing at 52.2 ± 29.2% (48/92) of all patients had complete ablation posttreatment (Peek et al. 2017). This could be due to the limited depth of penetration, making them more suitable for smaller tumors. Local recurrences reported with laser ablation were up to 10.7% at a mean follow-up time of 20.5 ± 0 months (Peek et al. 2017). This stresses the importance of patient selection for such a procedure.

**Future Directions**

As technology advances, there is potential for refining laser techniques to treat larger tumors or those in more challenging locations (Kerbage et al. 2022; Nori et al. 2018; Schwartzberg et al. 2018). Combining laser therapy with other treatments, such as immunotherapy or targeted therapies, is also being explored to enhance its antitumor effects.

In conclusion, laser therapy in breast cancer represents a fusion of technology and clinical need. While it may not replace traditional surgical interventions, it offers an alternative, especially for early-stage tumors or in palliative settings. As research continues and technology evolves, the role of lasers in breast cancer management is set to expand, offering patients more therapeutic options.

## 5 Ablative Surgery in Palliative Settings

Ablative surgery, regarded for its curative potential, has increasingly been recognized for its role in palliative care. The primary objective shifts from curative intent to symptom alleviation, prevention of disease-related complications, and enhancement of the patient's overall well-being. Moreover, in the setting of oligometastatic, ablative options toward the metastatic lesion(s) may help provide better systemic control while limiting the extent and the associated risks of metastasectomies.

### 5.1 Local Therapy

Advanced or recurrent breast tumors can present with a range of symptoms. These may include pain, ulceration, bleeding, and foul-smelling discharge (Kim et al. 2023). Such symptoms can significantly impair a patient's quality of life, both physically and emotionally. Local ablative therapies can directly target and reduce the tumor burden, providing symptomatic relief.

In addition, emerging evidence has shown that due to the local destructive nature of ablative therapy modulates the immune response toward the tumor (Ou et al. 2023). Earlier studies have demonstrated that the tumor microenvironment (TME) was mainly immunosuppressive for solid tumors, which hinders the effectiveness of immunotherapy (Binnewies et al. 2018). The breakdown of tumor cells using the local ablative therapy leads to the discharge of tumor antigens and the creation of damage-associated molecular patterns. This induces immunogenic cell death (ICD) forms an immune-responsive tumor microenvironment (TME) and further activates dendritic cells (DCs), macrophages, and CD8+ cytotoxic T lymphocytes (CTLs) to carry out antitumor immune responses (Shao et al. 2019). Such a concept was proven with survival advantages in other solid tumors such as colorectal and prostate cancer (Palma et al. 2020; Phillips et al. 2020) and also in small-scale studies among breast cancer patients (McArthur et al. 2016; Ye et al. 2018).

Vigorous patient selection for ablative treatments in the setting of oligometastases is crucial. Factors to consider include the number and location of metastatic lesions, the primary tumor's biology and responsiveness to treatment, and the patient's overall health and performance status.

## 5.2 Ablative Therapies Toward Metastatic Lesion(s)

While metastatic breast cancer is an incurable disease and systemic therapy was the standard of care for them, there were long-term survivors (Tomiak et al. 1996). One of the essential characteristics of these patients is the limited metastatic disease involved, which is also known as oligometastatic disease (OMD), commonly defined as limited metastases below five (Hellman and Weichselbaum 1995). The concept behind treating metastases aggressively is that by controlling these sites of metastatic disease, mainly when the metastasis is limited in quantity, one might delay the progression to widespread metastasis, thereby prolonging survival and improving quality of life (Hortobagyi 2002). The evidence was commonly discussed among breast cancer patients with lung (Friedel et al. 2002) and liver metastases (Abbott et al. 2012; Vlastos et al. 2004)with surgical excision showing an improved survival.

However, level I evidence is still lacking, and hence it is still uncertain if metastasectomy is definitely beneficial. Surgical excision is not without risks. Thus clinicians are exploring ablative techniques in organs commonly involved in breast cancer metastases due to the better risk profile, which is more acceptable in the palliative setting.

### 5.2.1 Bone

Bone is the commonest site of metastasis involved in breast cancer patients (Kennecke et al. 2010). Traditionally treated with radiation or systemic therapies, ablative techniques such as RFA, MWA, and cryoablation are considered for pain management and local tumor control (Papalexis et al. 2022; McArthur et al. 2017; Liu et al. 2018). These minimally invasive procedures can provide immediate relief, especially in weight-bearing bones, reducing the risk of fractures and improving patients' mobility and overall quality of life.

### 5.2.2 Liver

Liver is another common site for breast cancer metastases. Historically, surgical resections were the mainstay for isolated metastases (Cheung et al. 2019). However, liver surgery always carries significant risks, and ablative therapies, including RFA, MWA, and laser, have been proven to be safe alternatives to traditional surgeries and sometimes also permit a percutaneous approach (Bergenfeldt et al. 2011; Meloni et al. 2009; Rahnemai-Azar et al. 2021; Tomita et al. 2022; Sofocleous et al. 2007). For cryotherapy, the current evidence is limited, and further studies are required (Bala et al. 2019). The usage of HIFU is continually being investigated, especially in situations where other approaches are deemed not suitable with some initial success (Yang et al. 2021). Furthermore, there is also a study evaluating the combined effect of HIFU on improving the risks of liver resection for metastases (Dupré et al. 2017).

### 5.2.3 Lung

For patients with isolated lung metastases, surgical resection has shown survival benefits (Rashid and Takabe 2012). However, for those who are not surgical candidates or would like to avoid the associated risks, ablative therapies such as RFA and cryoablation provide potential options (Kent et al. 2021; Vogl et al. 2017). These techniques focus on the targeted destruction of tumor tissue, with the added advantage of preserving maximum lung function and reducing post-procedural complications.

Ablative therapy in the thoracic cavity, despite being minimally invasive, can still lead to a wide variety of complications, and post-procedural monitoring is crucial, which includes pneumothorax, bleeding, hemothorax, and even injury to other adjacent organs (Páez-Carpio et al. 2021).

### 5.2.4 Others

Other less common sites of breast cancer metastasis include the brain, adrenal glands, ovaries, and skin. Ablative techniques for these sites are more niche but are gradually gaining traction, such as the application of RFA in adrenal metastases (Yamakado 2014). As with all treatments, the selection of an ablative method should be individualized, considering the patient's overall health, location and size of the metastasis, and potential risks and benefits.

## 6 Conclusion

The landscape of breast cancer surgery is ever-evolving, with ablative techniques offering promising results in both curative and palliative settings. As technology advances and interdisciplinary research continues, it is anticipated that these modalities will become even more refined, offering patients personalized and effective treatment options. The key to success lies in appropriate patient selection, understanding the limitations of each modality, and continuous monitoring for potential recurrences.

**Acknowledgments** None.

**Compliance with Ethical Standards** All methods were carried out in accordance with relevant guidelines and regulations. No conflict of interest was reported among all authors.

## References


Abbott DE, Brouquet A, Mittendorf EA, Andreou A, Meric-Bernstam F, Valero V, Green MC, Kuerer HM, Curley SA, Abdalla EK, Hunt KK, Vauthey JN (2012) Resection of liver metastases from breast cancer: estrogen receptor status and response to chemotherapy before metastasectomy define outcome. Surgery 151(5):710–716. https://doi.org/10.1016/j.surg.2011.12.017

Bachu VS, Kedda J, Suk I, Green JJ, Tyler B (2021) High-intensity focused ultrasound: a review of mechanisms and clinical applications. Ann Biomed Eng 49(9):1975–1991. https://doi.org/10.1007/s10439-021-02833-9

Bala MM, Riemsma RP, Wolff R, Pedziwiatr M, Mitus JW, Storman D, Swierz MJ, Kleijnen J (2019) Cryotherapy for liver metastases. Cochrane Database Syst Rev 7(7):Cd009058. https://doi.org/10.1002/14651858.CD009058.pub3

Bald WB (1984) A helium gas probe for use in cryosurgery. Cryobiology 21(5):570–573. https://doi.org/10.1016/0011-2240(84)90056-7

Beerlage HP, Thuroff S, Debruyne FMJ, Chaussy C, de la Rosette J (1999) Transrectal high-intensity focused ultrasound using the ablatherm device in the treatment of localized prostate carcinoma. Urology 54(2):273–277. https://doi.org/10.1016/s0090-4295(99)00104-1

Bergenfeldt M, Jensen BV, Skjoldbye B, Nielsen D (2011) Liver resection and local ablation of breast cancer liver metastases--a systematic review. Eur J Surg Oncol 37(7):549–557. https://doi.org/10.1016/j.ejso.2011.04.013

Binnewies M, Roberts EW, Kersten K, Chan V, Fearon DF, Merad M, Coussens LM, Gabrilovich DI, Ostrand-Rosenberg S, Hedrick CC, Vonderheide RH, Pittet MJ, Jain RK, Zou W, Howcroft TK, Woodhouse EC, Weinberg RA, Krummel MF (2018) Understanding the tumor immune microenvironment (TIME) for effective therapy. Nat Med 24(5):541–550. https://doi.org/10.1038/s41591-018-0014-x

Bloom KJ, Dowlat K, Assad L (2001) Pathologic changes after interstitial laser therapy of infiltrating breast carcinoma. Am J Surg 182(4):384–388. https://doi.org/10.1016/s0002-9610(01)00732-2

Budman H, Shitzer A, Del Giudice S (1986) Investigation of temperature fields around embedded cryoprobes. J Biomech Eng 108(1):42–48. https://doi.org/10.1115/1.3138578

Campbell S (2013) A short history of sonography in obstetrics and gynaecology. Facts Views Vis Obgyn 5(3):213–229

Carriero S, Lanza C, Pellegrino G, Ascenti V, Sattin C, Pizzi C, Angileri SA, Biondetti P, Ianniello AA, Piacentino F, Lavorato R, Ierardi AM, Carrafiello G (2023) Ablative therapies for breast cancer: state of art. Technol Cancer Res Treat 22:15330338231157193. https://doi.org/10.1177/15330338231157193

Cazzato RL, de Lara CT, Buy X, Ferron S, Hurtevent G, Fournier M, Debled M, Palussiere J (2015) Single-centre experience with percutaneous cryoablation of breast cancer in 23 consecutive non-surgical patients. Cardiovasc Intervent Radiol 38(5):1237–1243. https://doi.org/10.1007/s00270-015-1181-5

Cheung TT, Chok KSH, Chan ACY, Tsang SHY, Dai WC, Yau TCC, Kwong A, Lo CM (2019) Survival analysis of breast cancer liver metastasis treated by hepatectomy: a propensity score analysis for Chinese women in Hong Kong. Hepatobiliary Pancreat Dis Int 18(5):452–457. https://doi.org/10.1016/j.hbpd.2019.08.001

Chu KF, Dupuy DE (2014) Thermal ablation of tumours: biological mechanisms and advances in therapy. Nat Rev Cancer 14(3):199–208. https://doi.org/10.1038/nrc3672

Co M, Chen C, Lee C, Yu S, Kwong A (2022) Prospective clinical trial on the learning curve of high-intensity-focused ultrasound for the treatment of breast fibroadenoma. Surg Today 52(7):1048–1053. https://doi.org/10.1007/s00595-021-02421-3

Cryoablation of Low Risk Small Breast Cancer- Ice3 Trial (NCT02200705) (2014) Retrieved 13/8 from https://www.clinicaltrials.gov/study/NCT02200705?cond=Ice3%20trial&term=Cryoablation&rank=1

Cryoablation of Small Breast Tumors in Early Stage Breast Cancer (FROST) (NCT01992250) (2013). Retrieved 13/8 from https://www.clinicaltrials.gov/study/NCT01992250?cond=Small%20Breast%20Tumors%20in%20Early%20Stage%20Breast%20Cancer&term=Cryoablation&rank=1

Dixon JA (1988) Current laser applications in general surgery. Ann Surg 207(4):355

Dowlatshahi K, Francescatti DS, Bloom KJ (2002) Laser therapy for small breast cancers. Am J Surg 184(4):359–363., Article Pii s0002-9610(02)00942-x. https://doi.org/10.1016/s0002-9610(02)00942-x

Dupré A, Pérol D, Blanc E, Peyrat P, Basso V, Chen Y, Vincenot J, Kocot A, Melodelima D, Rivoire M (2017) Efficacy of high-intensity focused ultrasound-assisted hepatic resection (HIFU-AR) on blood loss reduction in patients with liver metastases requiring hepatectomy: study protocol for a randomized controlled trial. Trials 18(1):57. https://doi.org/10.1186/s13063-017-1801-2

Friedel G, Pastorino U, Ginsberg RJ, Goldstraw P, Johnston M, Pass H, Putnam JB, Toomes H (2002) Results of lung metastasectomy from breast cancer: prognostic criteria on the basis of 467 cases of the international registry of lung metastases. Eur J Cardiothorac Surg 22(3): 335–344. https://doi.org/10.1016/s1010-7940(02)00331-7

Fry FJ, Ades HW, Fry WJ (1958) Production of reversible changes in the central nervous system by ultrasound. Science 127(3289):83–84. https://doi.org/10.1126/science.127.3289.83

Furusawa H, Namba K, Thomsen S, Akiyama F, Bendet A, Tanaka C, Yasuda Y, Nakahara H (2006) Magnetic resonance-guided focused ultrasound surgery of breast cancer: reliability and effectiveness. J Am Coll Surg 203(1):54–63. https://doi.org/10.1016/j.jamcollsurg.2006.04.002

Garbay JR, Mathieu MC, Lamuraglia M, Lassau N, Balleyguier C, Rouzier R (2008) Radiofrequency thermal ablation of breast cancer local recurrence: a phase II clinical trial. Ann Surg Oncol 15(11):3222–3226. https://doi.org/10.1245/s10434-008-0026-8

Gardner RA, Vargas HI, Block JB, Vogel CL, Fenn AJ, Kuehl GV, Doval M (2002) Focused microwave phased array thermotherapy for primary breast cancer. Ann Surg Oncol 9(4): 326–332. https://doi.org/10.1245/aso.2002.9.4.326

Gianfelice D, Khiat A, Amara M, Belblidia A, Boulanger Y (2003a) MR imaging-guided focused ultrasound surgery of breast cancer: correlation of dynamic contrast-enhanced MRI with histopathologic findings. Breast Cancer Res Treat 82(2):93–101. https://doi.org/10.1023/B:BREA.0000003956.11376.5b

Gianfelice D, Khiat A, Amara M, Belblidia A, Boulanger Y (2003b) MR imaging-guided focused US ablation of breast cancer: histopathologic assessment of effectiveness-initial experience. Radiology 227(3):849–855. https://doi.org/10.1148/radiol.2281012163

Gianfelice D, Khiat A, Boulanger Y, Amara M, Belblidia A (2003c) Feasibility of magnetic resonance imaging-guided focused ultrasound surgery as an adjunct to tamoxifen therapy in high-risk surgical patients with breast carcinoma. J Vasc Interv Radiol 14(10):1275–1282. https://doi.org/10.1097/01.Rvl.0000092900.73329.A2

Group VE (2023) Breast fibroadenoma treatment HIFU ablation system ECHOPULSE®. https://www.medicalexpo.com/prod/theraclion/product-80971-511153.html

Haraldsdottir KH, Ivarsson K, Gotberg S, Ingvar C, Stenram U, Tranberg KG (2008) Interstitial laser thermotherapy (ILT) of breast cancer. Eur J Surg Oncol 34(7):739–745. https://doi.org/10.1016/j.ejso.2008.01.008

Harries SA, Amin Z, Smith MEF, Lees WR, Cooke J, Cook MG, Scurr JH, Kissin MW, Bown SG (1994) Interstitial laser photocoagulation as a treatment for breast-cancer. Br J Surg 81(11): 1617–1619. https://doi.org/10.1002/bjs.1800811118

Hellman S, Weichselbaum RR (1995) Oligometastases. J Clin Oncol 13(1):8–10. https://doi.org/10.1200/jco.1995.13.1.8

Hernández JI, Cepeda MF, Valdés F, Guerrero GD (2015) Microwave ablation: state-of-the-art review. Onco Targets Ther 8:1627–1632. https://doi.org/10.2147/ott.S81734

Hortobagyi GN (2002) Can we cure limited metastatic breast cancer? J Clin Oncol 20(3):620–623. https://doi.org/10.1200/jco.2002.20.3.620

Ihnát P, Ihnát Rudinská L, Zonča P (2014) Radiofrequency energy in surgery: state of the art. Surg Today 44(6):985–991. https://doi.org/10.1007/s00595-013-0630-5

Izadifar Z, Izadifar Z, Chapman D, Babyn P (2020) An introduction to high intensity focused ultrasound: systematic review on principles, devices, and clinical applications. J Clin Med 9(2). https://doi.org/10.3390/jcm9020460

Izzo F, Thomas R, Delrio P, Rinaldo M, Vallone P, DeChiara A, Botti G, D'Aiuto G, Cortino P, Curley SA (2001) Radiofrequency ablation in patients with primary breast carcinoma – a pilot study in 26 patients. Cancer 92(8):2036–2044. https://doi.org/10.1002/1097-0142(20011015)92:8<2036::Aid-cncr1542>3.0.Co;2-w

Izzo F, Granata V, Grassi R, Fusco R, Palaia R, Delrio P, Carrafiello G, Azoulay D, Petrillo A, Curley SA (2019) Radiofrequency ablation and microwave ablation in liver tumors: an update. Oncologist 24(10):E990–E1005. https://doi.org/10.1634/theoncologist.2018-0337

Jeffrey SS, Birdwell RL, Ikeda DM, Daniel BL, Nowels KW, Dirbas FM, Griffey SM (1999) Radiofrequency ablation of breast cancer – first report of an emerging technology. Arch Surg 134(10):1064–1068. https://doi.org/10.1001/archsurg.134.10.1064

Karow AM, Webb WR (1965) Tissue freezing: a theory for injury and survival. Cryobiology 2(3): 99–108. https://doi.org/10.1016/S0011-2240(65)80094-3

Kennecke H, Yerushalmi R, Woods R, Cheang MC, Voduc D, Speers CH, Nielsen TO, Gelmon K (2010) Metastatic behavior of breast cancer subtypes. J Clin Oncol 28(20):3271–3277. https://doi.org/10.1200/jco.2009.25.9820

Kent CL, McDuff SGR, Salama JK (2021) Oligometastatic breast cancer: where are we now and where are we headed? -a narrative review. Ann Palliat Med *10*(5):5954–5968. https://doi.org/10.21037/apm-20-1128

Kerbage Y, Rouillès J, Vignion AS, Delhem N, Thecua E, Deleporte P, Collinet P, Mordon S (2022) Laser interstitial thermotherapy (LITT) for breast cancer: dosimetry optimization and numerical simulation. Lasers Med Sci 37(1):489–498. https://doi.org/10.1007/s10103-021-03286-1

Khiat A, Gianfelice D, Amara M, Boulanger Y (2006) Influence of post-treatment delay on the evaluation of the response to focused ultrasound surgery of breast cancer by dynamic contrast enhanced MRI. Br J Radiol 79(940):308–314. https://doi.org/10.1259/bjr/23046051

Kim SJ, Patel I, Park C, Shin DY, Chang J (2023) Palliative care and healthcare utilization among metastatic breast cancer patients in U.S. hospitals. Sci Rep 13(1):4358. https://doi.org/10.1038/s41598-023-31404-2

Kinoshita T (2016) Non-surgical ablation therapy for early-stage breast cancer. Springer

Kinoshita T, Iwamoto E, Tsuda H, Seki K (2011) Radiofrequency ablation as local therapy for early breast carcinomas. Breast Cancer 18(1):10–17. https://doi.org/10.1007/s12282-009-0186-9

Kwong A, Co M, Fukuma E (2023) Prospective clinical trial on expanding indications for cryosurgery for early breast cancers. Clin Breast Cancer 23(4):363–368. https://doi.org/10.1016/j.clbc.2023.01.007

Lagendijk M, van Maaren MC, Saadatmand S, Strobbe LJA, Poortmans PMP, Koppert LB, Tilanus-Linthorst MMA, Siesling S (2018) Breast conserving therapy and mastectomy revisited: breast cancer-specific survival and the influence of prognostic factors in 129,692 patients. Int J Cancer 142(1):165–175. https://doi.org/10.1002/ijc.31034

Lanza E, Palussiere J, Buy X, Grasso RF, Beomonte Zobel B, Poretti D, Pedicini V, Balzarini L, Cazzato RL (2015) Percutaneous image-guided cryoablation of breast cancer: a systematic review. J Vasc Interv Radiol *26*(*11*):1652–1657.e1651

Littrup PJ, Jallad B, Chandiwala-Mody P, D'Agostini M, Adam BA, Bouwman D (2009) Cryotherapy for breast cancer: a feasibility study without excision. J Vasc Interv Radiol 20(10): 1329–1341. https://doi.org/10.1016/j.jvir.2009.06.029

Liu B, Wu Z, Mo H, He J, Lin X, Guan J, Wei C, Yuan Z (2018) Safety and efficacy of microwave ablation for breast cancer thoracic metastases. Cancer Manag Res 10:5685–5689. https://doi.org/10.2147/cmar.S176743

Mays RO, Neira LM, Luyen H, Wilke LG, Behdad N, Hagness SC (2016, 10–15 April) Advances in microwave ablation antennas for breast tumor treatment. 2016 10th European conference on antennas and propagation (EuCAP)

McArthur HL, Diab A, Page DB, Yuan J, Solomon SB, Sacchini V, Comstock C, Durack JC, Maybody M, Sung J, Ginsberg A, Wong P, Barlas A, Dong Z, Zhao C, Blum B, Patil S, Neville D, Comen EA et al (2016) A pilot study of preoperative single-dose Ipilimumab and/or cryoablation in women with early-stage breast cancer with comprehensive immune profiling. Clin Cancer Res 22(23):5729–5737. https://doi.org/10.1158/1078-0432.Ccr-16-0190

McArthur TA, Narducci CA, Lander PH, Lopez-Ben R (2017) Percutane image-guided cryoablation of painful osseous metastases: a retrospective single-center review. Curr Probl Diagn Radiol 46(4):282–287. https://doi.org/10.1067/j.cpradiol.2016.11.007

Meloni MF, Andreano A, Laeseke PF, Livraghi T, Sironi S, Lee FT Jr (2009) Breast cancer liver metastases: US-guided percutaneous radiofrequency ablation – intermediate and long-term survival rates. Radiology 253(3):861–869. https://doi.org/10.1148/radiol.2533081968

Merckel LG, Bartels LW, Kohler MO, van den Bongard H, Deckers R, Mali W, Binkert CA, Moonen CT, Gilhuijs KGA, van den Bosch M (2013) MR-guided high-intensity focused ultrasound ablation of breast cancer with a dedicated breast platform. Cardiovasc Intervent Radiol 36(2):292–301. https://doi.org/10.1007/s00270-012-0526-6

Merckel LG, Knuttel FM, Deckers R, van Dalen T, Schubert G, Peters N, Weits T, van Diest PJ, Mali W, Vaessen P, van Gorp J, Moonen CTW, Bartels LW, van den Bosch M (2016) First clinical experience with a dedicated MRI-guided high-intensity focused ultrasound system for breast cancer ablation. Eur Radiol 26(11):4037–4046. https://doi.org/10.1007/s00330-016-4222-9

Nguyen T, Hattery E, Khatri VP (2014) Radiofrequency ablation and breast cancer: a review. Gland Surg 3(2):128–135. https://doi.org/10.3978/j.issn.2227-684X.2014.03.05

Nori J, Gill MK, Meattini I, Paoli CD, Abdulcadir D, Vanzi E, Boeri C, Gabbrielli S, Giannotti E, Lucci F, Vezzosi V, De Benedetto D, Bicchierai G, Bianchi S, Sanchez L, Orzalesi L, Carmelo G, Miele V, Livi L, Casella D (2018) The evolving role of ultrasound guided percutaneous laser ablation in elderly unresectable breast cancer patients: a feasibility pilot study. Biomed Res Int *2018*:Article 9141746. https://doi.org/10.1155/2018/9141746

Ou W, Stewart S, White A, Kwizera EA, Xu J, Fang Y, Shamul JG, Xie C, Nurudeen S, Tirada NP, Lu X, Tkaczuk KHR, He X (2023) In-situ cryo-immune engineering of tumor microenvironment with cold-responsive nanotechnology for cancer immunotherapy. Nat Commun 14(1):392. https://doi.org/10.1038/s41467-023-36045-7

Páez-Carpio A, Gómez FM, Isus Olivé G, Paredes P, Baetens T, Carrero E, Sánchez M, Vollmer I (2021) Image-guided percutaneous ablation for the treatment of lung malignancies: current state of the art. Insights Imaging 12(1):57. https://doi.org/10.1186/s13244-021-00997-5

Palma DA, Olson R, Harrow S, Gaede S, Louie AV, Haasbeek C, Mulroy L, Lock M, Rodrigues GB, Yaremko BP, Schellenberg D, Ahmad B, Senthi S, Swaminath A, Kopek N, Liu M, Moore K, Currie S, Schlijper R et al (2020) Stereotactic ablative radiotherapy for the comprehensive treatment of oligometastatic cancers: long-term results of the SABR-COMET phase II randomized trial. J Clin Oncol 38(25):2830–2838. https://doi.org/10.1200/jco.20.00818

Pan H, Qian MJ, Chen H, Wang H, Yu MX, Zhang K, Wang SQ, Deng J, Xu Y, Ling LJ, Ding Q, Xie H, Wang S, Zhou WB (2021) Precision breast-conserving surgery with microwave ablation guidance: a pilot single-center, prospective cohort study. Front Oncol *11*:Article 680091. https://doi.org/10.3389/fonc.2021.680091

Papalexis N, Parmeggiani A, Peta G, Spinnato P, Miceli M, Facchini G (2022) Minimally invasive interventional procedures for metastatic bone disease: a comprehensive review. Curr Oncol 29(6):4155–4177. https://doi.org/10.3390/curroncol29060332

Peek MCL, Ahmed M, Napoli A, Usiskin S, Baker R, Douek M (2017) Minimally invasive ablative techniques in the treatment of breast cancer: a systematic review and meta-analysis. Int J Hyperth 33(2):191–202. https://doi.org/10.1080/02656736.2016.1230232

Pfleiderer SOR, Freesmeyer MG, Marx C, Kuhne-Heid R, Schneider A, Kaiser WA (2002) Cryotherapy of breast cancer under ultrasound guidance: initial results and limitations. Eur Radiol 12(12):3009–3014. https://doi.org/10.1007/s00330-002-1511-2

Pfleiderer SOR, Marx C, Camara O, Gajda M, Kaiser WA (2005) Ultrasound-guided, percutaneous cryotherapy of small (<= 15 mm) breast cancers. Investig Radiol 40(7):472–477. https://doi.org/10.1097/01.rli.0000166935.56971.ff

Phillips R, Shi WY, Deek M, Radwan N, Lim SJ, Antonarakis ES, Rowe SP, Ross AE, Gorin MA, Deville C, Greco SC, Wang H, Denmeade SR, Paller CJ, Dipasquale S, DeWeese TL, Song DY, Wang H, Carducci MA et al (2020) Outcomes of observation vs stereotactic ablative radiation for oligometastatic prostate cancer: the ORIOLE phase 2 randomized clinical trial. JAMA Oncol 6(5):650–659. https://doi.org/10.1001/jamaoncol.2020.0147

Piot-Ziegler C, Sassi ML, Raffoul W, Delaloye JF (2010) Mastectomy, body deconstruction, and impact on identity: a qualitative study. Br J Health Psychol 15(3):479–510

Pusceddu C, Paliogiannis P, Nigri G, Fancellu A (2019) Cryoablation in the management of breast cancer: evidence to date. Breast Cancer-Targets and Ther 11:283–292. https://doi.org/10.2147/bctt.S197406

Rahnemai-Azar AA, Selby LV, Lustberg MB, Pawlik TM (2021) Surgical management of breast cancer liver metastasis. Surg Oncol Clin N Am 30(1):27–37. https://doi.org/10.1016/j.soc.2020.09.003

Rashid OM, Takabe K (2012) The evolution of the role of surgery in the management of breast cancer lung metastasis. J Thorac Dis 4(4):420

Rhim H, Dodd GD 3rd. (1999) Radiofrequency thermal ablation of liver tumors. J Clin Ultrasound 27(5):221–229. https://doi.org/10.1002/(sici)1097-0096(199906)27:5<221::aid-jcu1>3.0.co;2-t

Schassburger KU, Lofgren L, Lagerstedt U, Leifland K, Thorneman K, Sandstedt B, Auer G, Wiksell H (2014) Minimally-invasive treatment of early stage breast cancer: a feasibility study using radiofrequency ablation under local anesthesia. Breast 23(2):152–158. https://doi.org/10.1016/j.breast.2013.12.007

Schwartzberg B, Lewin J, Abdelatif O, Bernard J, Bu-Ali H, Cawthorn S, Chen-Seetoo M, Feldman S, Govindarajulu S, Jones L, Juette A, Kavia S, Maganini R, Pain S, Shere M, Shriver C, Smith S, Valencia A, Whitacre E, Whitney R (2018) Phase 2 open-label trial investigating percutaneous laser ablation for treatment of early-stage breast cancer: MRI, pathology, and outcome correlations. Ann Surg Oncol 25(10):2958–2964. https://doi.org/10.1245/s10434-018-6623-2

Shao Q, O'Flanagan S, Lam T, Roy P, Pelaez F, Burbach BJ, Azarin SM, Shimizu Y, Bischof JC (2019) Engineering T cell response to cancer antigens by choice of focal therapeutic conditions. Int J Hyperth 36(1):130–138. https://doi.org/10.1080/02656736.2018.1539253

Siedek F, Yeo SY, Heijman E, Grinstein O, Bratke G, Heneweer C, Puesken M, Persigehl T, Maintz D, Grull H (2019) Magnetic resonance-guided high-intensity focused ultrasound (MR-HIFU): technical background and overview of current clinical applications (part 1). Rofo-Fortschritte Auf Dem Gebiet Der Rontgenstrahlen Und Der Bildgebenden Verfahren 191(6):522–530. https://doi.org/10.1055/a-0817-5645

Simmons RM, Ballman KV, Cox C, Carp N, Sabol J, Hwang RF, Attai D, Sabel M, Nathanson D, Kenler A, Gold L, Kaufman C, Han L, Bleznak A, Stanley Smith J, Holmes D, Fornage B, Le-Petross C, Hoda S et al (2016) A phase II trial exploring the success of cryoablation therapy in the treatment of invasive breast carcinoma: results from ACOSOG (Alliance) Z1072. Ann Surg Oncol 23(8):2438–2445. https://doi.org/10.1245/s10434-016-5275-3

Simon CJ, Dupuy DE, Mayo-Smith WW (2005) Microwave ablation: principles and applications. Radiographics 25:S69–S84. https://doi.org/10.1148/rg.25si055501

Sofocleous CT, Nascimento RG, Gonen M, Theodoulou M, Covey AM, Brody LA, Solomon SM, Thornton R, Fong Y, Getrajdman GI, Brown KT (2007) Radiofrequency ablation in the management of liver metastases from breast cancer. Am J Roentgenol 189(4):883–889. https://doi.org/10.2214/ajr.07.2198

Tafra L, Smith SJ, Woodward JE, Fernandez KL, Sawyer KT, Grenko RT (2003) Pilot trial of cryoprobe-assisted breast-conserving surgery for small ultrasound-visible cancers. Ann Surg Oncol 10(9):1018–1024. https://doi.org/10.1245/aso.2003.04.002

Takada M, Toi M (2019) Cryosurgery for primary breast cancers, its biological impact, and clinical outcomes. Int J Clin Oncol 24(6):608–613. https://doi.org/10.1007/s10147-019-01448-4

Tomiak E, Piccart M, Mignolet F, Sahmoud T, Paridaens R, Nooy M, Beex L, Fentiman IS, Muller A, vanderSchueren, E., & Rubens, R. D. (1996) Characterisation of complete responders to combination chemotherapy for advanced breast cancer: a retrospective EORTC breast group study. Eur J Cancer 32A(11):1876–1887. https://doi.org/10.1016/0959-8049(96)00189-x

Tomita K, Matsui Y, Uka M, Umakoshi N, Kawabata T, Munetomo K, Nagata S, Iguchi T, Hiraki T (2022) Evidence on percutaneous radiofrequency and microwave ablation for liver metastases over the last decade. Jpn J Radiol 40(10):1035–1045. https://doi.org/10.1007/s11604-022-01335-5

van de Voort EMF, Struik GM, Birnie E, Moelker A, Verhoef C, Klem T (2021) Thermal ablation as an alternative for surgical resection of small (<= 2 cm) breast cancers: a meta-analysis. Clin Breast Cancer 21(6):E715–E730. https://doi.org/10.1016/j.clbc.2021.03.004

Vargas HI, Dooley WC, Gardner RA, Gonzalez KD, Venegas R, Heywang-Kobrunner SH, Fenn AJ (2004) Focused microwave phased array thermotherapy for ablation of early-stage breast cancer: results of thermal dose escalation. Ann Surg Oncol 11(2):139–146. https://doi.org/10.1245/aso.2004.03.059

Vilar VS, Goldman SM, Ricci MD, Pincerato K, Oliveira H, Abud TG, Ajzen S, Baracat EC, Szejnfeld J (2012) Analysis by MRI of residual tumor after radiofrequency ablation for early stage breast cancer. Am J Roentgenol 198(3):W285–W291. https://doi.org/10.2214/ajr.10.5581

Vlastos G, Smith DL, Singletary SE, Mirza NQ, Tuttle TM, Popat RJ, Curley SA, Ellis LM, Roh MS, Vauthey JN (2004) Long-term survival after an aggressive surgical approach in patients with breast cancer hepatic metastases. Ann Surg Oncol 11(9):869–874. https://doi.org/10.1245/aso.2004.01.007

Vogl TJ, Nour-Eldin NEA, Albrecht MH, Kaltenbach B, Hohenforst-Schmidt W, Lin H, Panahi B, Eichler K, Gruber-Rouh T, Roman A (2017) Thermal ablation of lung tumors: focus on microwave ablation. Rofo-Fortschritte Auf Dem Gebiet Der Rontgenstrahlen Und Der Bildgebenden Verfahren 189(9):828–843. https://doi.org/10.1055/s-0043-109010

Waaijer L, Kreb DL, Gallardo MAF, Van Rossum PSN, Postma EL, Koelemij R, Van Diest PJ, Klaessens J, Witkamp AJ, Van Hillegersberg R (2014) Radiofrequency ablation of small breast tumours: evaluation of a novel bipolar cool-tip application. Ejso 40(10):1222–1229. https://doi.org/10.1016/j.ejso.2014.07.031

Wei FQ, Chen WY, Lin XT (2020) HIFU ablation as a therapy for breast tumor: a meta-analysis of 23 prospective feasibility studies. Breast J 26(7):1478–1480. https://doi.org/10.1111/tbj.13754

Wiksell H, Lofgren L, Schassburger KU, Grundstrom H, Janicijevic M, Lagerstedt U, Leifland K, Nybom R, Rotstein S, Saracco A, Schultz I, Thorneman K, Wadstrom C, Westman L, Wigzell H, Wilczek B, Auer G, Sandstedt B (2010) Feasibility study on the treatment of small breast carcinoma using percutaneous US-guided preferential radiofrequency ablation (PRFA). Breast 19(3):219–225. https://doi.org/10.1016/j.breast.2010.01.016

Xia LY, Hu QL, Xu WY (2021) Efficacy and safety of radiofrequency ablation for breast cancer smaller than 2 cm: a systematic review and meta-analysis. Front Oncol *11*:Article 651646. https://doi.org/10.3389/fonc.2021.651646

Yamakado K (2014) Image-guided ablation of adrenal lesions. Semin Intervent Radiol 31(2): 149–156. https://doi.org/10.1055/s-0034-1373797

Yamamoto N, Fujimoto H, Nakamura R, Arai M, Yoshii A, Kaji S, Itami M (2011) Pilot study of radiofrequency ablation therapy without surgical excision for T1 breast cancer: evaluation with MRI and vacuum-assisted core needle biopsy and safety management. Breast Cancer 18(1):3–9. https://doi.org/10.1007/s12282-010-0197-6

Yang T, Ng DM, Du N, He N, Dai X, Chen P, Wu F, Chen B, Fan X, Yan K, Zhou X, Dong M, Zheng Z, Gu L (2021) HIFU for the treatment of difficult colorectal liver metastases with unsuitable indications for resection and radiofrequency ablation: a phase I clinical trial. Surg Endosc 35(5):2306–2315. https://doi.org/10.1007/s00464-020-07644-y

Ye P, Yin H, Gu X, Ye Y, Zhao Q, Chang Z, Han B, Chen X, Liu P (2018) Improved synergetic therapy efficiency of cryoablation and nanoparticles for MCF-7 breast cancer. Nanomedicine (Lond) 13(15):1889–1903. https://doi.org/10.2217/nnm-2018-0168

Yilmaz S, Ozdogan M, Cevener M, Ozluk A, Kargi A, Kendiroglu F, Ogretmen I, Yildiz A (2016) Use of cryoablation beyond the prostate. Insights Imaging 7(2):223–232. https://doi.org/10.1007/s13244-015-0460-7

Zehra S, Doyle F, Barry M, Walsh S, Kell MR (2020) Health-related quality of life following breast reconstruction compared to total mastectomy and breast-conserving surgery among breast cancer survivors: a systematic review and meta-analysis. Breast Cancer 27(4):534–566. https://doi.org/10.1007/s12282-020-01076-1

Zhou WB, Jiang YN, Chen L, Ling LJ, Liang MD, Pan H, Wang SQ, Ding Q, Liu XA, Wang S (2014) Image and pathological changes after microwave ablation of breast cancer: a pilot study. Eur J Radiol 83(10):1771–1777. https://doi.org/10.1016/j.ejrad.2014.06.015

Zhou WB, Yu MX, Pan H, Qiu W, Wang H, Qian MJ, Che N, Zhang K, Mao XR, Li L, Wang RX, Xie H, Ling LJ, Zhao Y, Liu XA, Wang C, Ding Q, Wang S (2021) Microwave ablation induces Th1-type immune response with activation of ICOS pathway in early-stage breast cancer. J Immunother Cancer 9(4):Article e002343. https://doi.org/10.1136/jitc-2021-002343

# Radiotherapy in Breast Cancer

Mohammad Mehedi Hasan, Parvathy Mohanan, Shabana Bibi, Catherine Babu, Yohan Joe Roy, Ashlyn Mathews, Govinda Khatri, and Stavros P. Papadakos

**Abstract**

Radiotherapy or radiation therapy is a fundamental part of the multidisciplinary management of breast cancer. It is a treatment that utilizes high-energy X-rays, protons, or other unstable particles, in which hyperproliferative cells, such as cancer cells are affected more and therefore more susceptible to radiotherapy compared to normal cells. When treating early-stage breast cancer, radiotherapy can often be given before, during, or after surgery, in order to shrink tumors or to eradicate cancer cells. Postoperative radiation allows to destroy any cancer cells that may remain after surgery and lowers the risk of recurrence. Radiotherapy can also be used in cases of unresectable or metastatic breast cancers where surgery is no longer a viable option. The modality used, such as external beam therapy, brachytherapy, intraoperative therapy, and the dosage used, varies depending on the treatment plan. The dose of radiation given is split into fractions, usually

M. M. Hasan
Department of Biochemistry and Molecular Biology, Faculty of Life Science, Mawlana Bhashani Science and Technology University, Tangail, Bangladesh

P. Mohanan · C. Babu · Y. J. Roy · A. Mathews
Department of General Medicine, Medical University Sofia, Sofia, Bulgaria

S. Bibi (✉)
Yunnan Herbal Laboratory, College of Ecology and Environmental Sciences, Yunnan University, Kunming, Yunnan, China

Department of Biological Sciences, International Islamic University, Islamabad, Pakistan
e-mail: shabana_bibi@ynu.edu.cn; dazzling.grin@gmail.com

G. Khatri
Dow University of Health Sciences, Karachi, Pakistan

S. P. Papadakos
First Department of Pathology, School of Medicine, National and Kapodistrian University of Athens, Athens, Greece

 
Interdisciplinary Cancer Research, https://doi.org/10.1007/16833_2023_176
Published online: 30 August 2023

given daily, over a few days or weeks, depending on the choice of radiotherapy. A combination of chemotherapy or immunotherapy with radiotherapy is recommended for the progressive stages of breast cancer. This chapter dwells on different types of radiotherapy, methods of administration, evolution, benefits, and adverse effects of radiotherapy in breast cancer.

**Keywords**

Brachytherapy · Breast cancer · External beam therapy · Intraoperative therapy · Radiotherapy · Surgery · Tumor

## 1 Introduction

Radiotherapy or radiation therapy (RT) is a treatment modality that makes use of high-energy ionizing radiation in order to kill or slow the growth of cancer cells. All cells can be killed using such high-energy radiation; however, cancer cells are more susceptible than healthy cells due to their impaired nature of repair mechanisms. Radiation therapy works by causing damage to DNA found within the cell; normal cells are able to repair these damages, whereas cancerous cells are not able to do so (Baskar et al. 2014). The therapy modality also relies on the higher rate of division seen in cancer cells, as the damaged DNA is able to cause cell destruction faster, as well as due to radiation affecting DNA during division, and thus removing the cancerous cells from the site quicker. Radiation therapy can be used as the sole treatment or used with, before, or after other oncological treatments depending on the nature and progression of the cancer and the patient needs (Jaffray and Gospodarowicz 2015).

The ionizing energy generated to be used in radiation therapy can be created by different forms and machines ranging from a radioactive source or different machines that can accelerate particles and create this ionizing radiation. Most commonly this may be in the form of a linear accelerator or LINAC. In a linear accelerator, high-speed particles such as electrons and protons are released and made to collide with a metal target inside the linear accelerator such as tungsten. This collision generates X-ray or photon beams and is then delivered to the patient (Healy et al. 2017; Pereira et al. 2014). The generated high-energy particle can defer depending on the radiation therapy type used. In case electrons are required, the metal target is bypassed allowing electrons to be focused toward the patient.

As stated above radiation therapy has a higher impact on cancer than healthy cells. Radiation therapy relies on ionizing radiation, which causes atoms to lose an electron causing the atom to become positively charged. Radiation therapy, specifically the ionizing radiation used, affects the cells the greatest, when the cell is in the M phase of the cell division (Ingledew and Kong 2021). The increased cell division rate means that cancer cells will be in the M phase more often than healthy cells, thus they are more likely to be affected. The damage can be indirect or direct damage. Most commonly indirect damage occurs within the cell; water molecules in the cells are ionized and become "free radicals" or "superoxide radicals" (Hubenak et al. 2014; Ingledew and Kong 2021). These radicals in turn cause damage to the DNA and cause cell destruction. Less commonly, direct damage to the DNA can also occur with the use of protons or other heavier particles.

The purpose of this article is to demonstrate the use of radiologic technology in the treatment of breast cell carcinoma. This article describes breast cancer, causes of breast cancer, and different types of breast cancer, with a special emphasis on the radiotherapy aspect of breast cell carcinoma treatment and management.

## 2 History

In 1895, Wilhelm Conrad Roentgen, a German physicist discovered X-rays and thus opened up the vast potential of X-ray within medicine. Named after the Roentgen, the unit for exposure of radiation was defined as Roentgen (Cassidy et al. 2015). Very shortly, a few months after this discovery, the first-ever cancer treatment using X-ray was conducted in January 1896 (Gianfaldoni et al. 2017). In this procedure, Emil Herman Grubbe used a single treatment of X-ray that lasted an hour for breast cancer (Gianfaldoni et al. 2017). Same time as this husband and wife, Marie and Pierre Curie, as well as Antoine-Henry Becquerel, discovered Polonium and Radium as sources of radioactivity; the effects of which were reported 3 years later. Premature use of radiotherapy did not have the clear and precise planning as we know today; the treatment utilized low-energy radiation, which does not pose the same ability to penetrate the skin as high-energy radiation can. The nature of low-energy radiation meant that in the early 1900s, skin cancers were frequently treated with radiation. However, the lack of understanding regarding radiation and its effect on the human body caused the emergence of the field of Medical Physics in the early 1900s. The effect of this was seen with the introduction of an ionizing chamber which allows for the measurement of radiation doses as well as advancements in Device Engineering.

Advances in the Orthovoltage Era allowed the use of radiotherapy in more deeper cancers than the limited use that only allowed treatment limited to the skin. The Megavoltage Era introduced higher energy radiation, this further enhanced the ability to go much deeper past the skin, much more than the advances seen in the Orthovoltage Era. In the late 1900s more advanced technology such as the Linear Accelerator, computers and imaging (CT/MRI) aided radiotherapy including 3D conformal therapies. Curative efforts were seen as early as 1899, where successful skin cancer treatments were conducted (Fig. 1) (Cassidy et al. 2015).

## 3 Overview of Breast Cancer

Cancer refers to the uncontrolled division of metaplastic cells within a tissue with breast cancer being the uncontrollable growth of cancerous tissue within the breast, typically affecting the ducts and lobules (Alkabban 2021).

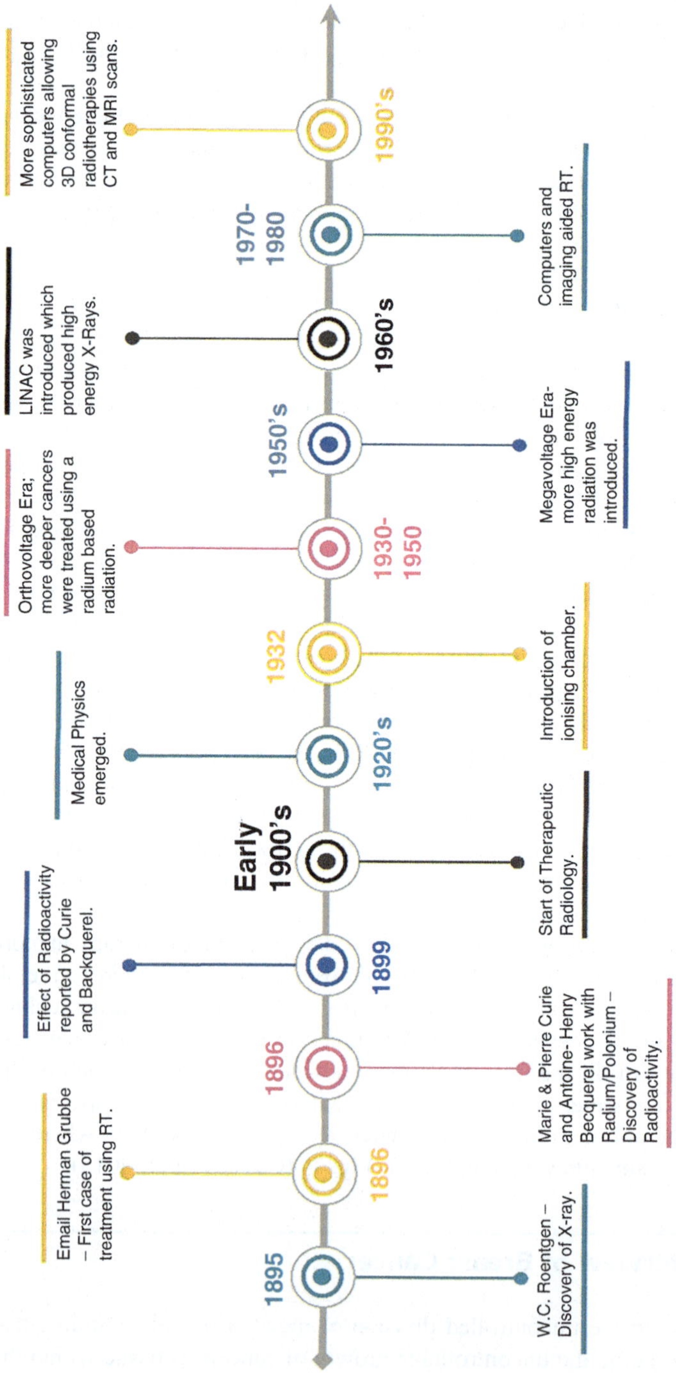


**Fig. 1** Historical perspective on the evolution of radiotherapy

## 3.1 Types

Invasive and non-invasive is the key distinction within breast carcinomas. The former refers to a neoplasm that spreads within the local breast parenchyma with the possibility of metastasizing to distant organs. Non-invasive neoplasia, as the name suggests, is relatively localized to a specific region and does not spread further than the glandular duct and lobules found within the breast. The most frequent types of non-invasive would be ductal and lobular carcinoma in situ (Sharma et al. 2010).

***Non-invasive*** Non-invasive carcinomas are further categorized into: Ductal carcinoma in situ and Lobular carcinoma in situ. The former is the most prevailing subtype of localized carcinoma originating in the glandular ducts and will not advance into any circumferential breast parenchyma and connective tissue. Lobular carcinoma in situ in contrast is classified as a non-invasive breast cancer that advances in the reach of the lobules of the female breast, specifically within the lactiferous glands at the end of breast ducts (Sharma et al. 2010).

***Invasive*** On the contrary invasive breast cancers refer to the spread of the malignant neoplasia outward from the lobular and ductal structures within the breast into the neighboring mammary tissue. They are often categorized into invasive ductal carcinoma and invasive lobular carcinomas (Alkabban 2021; Goh et al. 2019).

The most common kind of mammary carcinoma is invasive ductal carcinoma, which refers to cancer that begins in the glandular ducts of the breast and progresses to the surrounding breast tissue. If the cancer is not treated properly, metastatic progression to the lymph nodes and perhaps to other areas of the body is frequently seen as a consequence (Arps et al. 2013; Goh et al. 2019).

The next most frequent breast cancer is invasive lobular carcinoma; it refers to neoplasia that originated within specific glandular structures of the breast such as the lobules and spread into the peripheral mammary tissue. Less common invasive includes Paget's disease angiosarcoma, inflammatory breast disease, and metastatic breast cancer (Alkabban 2021; Goh et al. 2019; Sharma et al. 2010).

## 3.2 Breast Cancer Subtypes

Breast cancer can be further categorized into 3 main subtypes: hormone receptor-positive breast cancer, HER2-positive breast cancer, and triple-negative breast cancer.

### 3.2.1 Hormone Receptor-Positive Breast Cancer

This refers to cancers that require specific hormones such as estrogen or progesterone to advance and they typically make up approximately 60% of breast cancers. The neoplastic cells have these protein-derived receptors on the cell surface membrane and are then defined as hormone receptor-positive. Growth depends on the presence of hormones and typically is found in woman who are postmenopausal (Peddi 2018).

### 3.2.2 HER2-Positive Breast Cancer

This refers to breast carcinomas that require human epidermal growth factor receptor 2 (HER2) to progress and is typically present in approximately one-fifth of breast carcinomas. The HER2 polypeptide is coded for by the HER2 gene, which allows for the progression of neoplastic growth of these specific cancers (Yersal and Barutca 2014). The cancer is referred to as HER2 positive or negative based on the levels of protein and presence of genes within the patient (Loibl and Gianni 2017).

### 3.2.3 Triple-Negative Breast Cancer

This subclassification is characterized by not having any expression of the common breast cancer receptors such as estrogen, progesterone, and HER2. This cancer is not common and only makes up about 15% of invasive breast cancers (Kumar and Aggarwal 2016). Certain risk factors make this classification of cancer more likely to be expressed such as having a younger age and being of Hispanic or Black origin. The presence of the BRCA 1 mutation also increases the risk of development of malignancies (Lee et al. 2011).

## 3.3 Risk Factors and Etiology

The development of breast cancer is regarded as being multifactorial in nature and the formation of a neoplasia-causing mutation depends on multiple risk factors. The most common risk factors are seen to be genetic however there are also other major components such as age that can lead to carcinoma.

***Age*** The incidence rate of breast malignancies tends to increase as a woman advances in age, with the majority of cancers presenting in women older than 50.1 years or more (Feng et al. 2018).

***Gender*** The risk of neoplasia is significantly increased in women compared to men with studies showing that invasive breast cancer affects about 1 in 8 women who are born in the United States during their lifetime and in contrast the figure is approximately 1 in 1000 for men (Alkabban 2021).

***Previous Breast Cancer*** Studies have indicated that a previous breast malignancy can increase the chance of a new primary neoplasm in the other breast (Feng et al. 2018).

***Changes in Histology*** Cellular abnormalities such as lobular carcinoma in situ and atypical proliferative changes can lead to a significant increase in the likelihood of development of breast cancers (Ashbeck et al. 2007).

***Genetic and Familial Risk Factors*** Inherited mutations in BRCA 1 and 2 are linked to the development of neoplasia and are associated with increased risk of prostate, ovarian, and breast carcinomas (Tai et al. 2007). Consanguineous marriage that is often practiced within the Ashkenazi Jewish ethnic groups can cause a twofold increase in the risk of breast cancer development within first-degree relatives of the patient (Feng et al. 2018).

***Reproductive Risk Factors*** Certain childbearing and hormonal markers can lead to increased risk to developing malignancies. These include overall exposure to estrogen, early development of menarche or delivery post 30 years old to name a few (Anderson et al. 2014). This would increase the amount of estrogen the individual is exposed to and could lead to the development of cancers, especially ER-positive cancers.

***Exogenous Hormone Use*** This can increase the likelihood of developing carcinomas, especially hormone receptor-positive carcinomas and typically seen in use as contraceptive as well as use in postmenopausal women who use hormone replacement therapy (Peddi 2018; Sullivan et al. 2016).

## 3.4 Pathophysiology

The current general consensus regarding the development of breast cancer pathogenesis is that invasive carcinomas develop through a set of cellular transformations. These amendments lead to neoplastic cells with regenerative features and proliferative development.

Genetic testing has identified distinct subclassifications of tumors with specific pathogenesis and clinical demeanor which are thought to be influenced by mutations within the receptors as previously mentioned (Feng et al. 2018). It is commonly agreed that certain risk factors can increase the expression of a mutation, in this case DNA damage, linked to estrogen exposure. They can be explained by the following hypotheses of the tumor-forming properties of estrogen. The hormone acts as an initiator and generates free radicals on the genetic components. Estrogen acts as a promoter which induces proliferation of cancerous cells as well as the premalignant cellular complexes (Russo and Russo 2006).

A genetic approach is that there will be an inheritance of defective DNA or cancer-associated genes like BRCA1 and BRCA2 thus increased cancerous activity leading to increased risk of malignancy formation. Therefore, it is a plausible assumption that there is increased risk for malignancy formation if there is a positive family history for breast and ovarian cancers (Alkabban 2021). The immune system of a healthy person eliminates cells that contain abnormal DNA or are developing improperly. This fails in breast cancer patients, resulting in tumor development and spread. There are specific cells that are involved in this process such as tumor suppressor cells such as the TP53 antigen, however when mutations arise in these cells cancer can develop (Feng et al. 2018).

Mutations in proto-oncogenes can change their role leading to oncogene production, meaning that this transition disrupts the balance of the cell cycle regulation process, which gives rise to uncontrolled cellular development (Feng et al. 2018). Tumor suppressor genes function to inhibit cellular division and stop the progression of cells that have damaged DNA components. In patients with cancer these tumor suppressor genes are often disabled this is caused by mutagenic changes.

## 3.5 Signs and Symptoms

Patients are identified upon symptoms prior or during clinical examination or by screening. Typically, the patient presents with a noticed lump, there are abnormalities such as the following that should be investigated (Alkabban 2021; Raina et al. 2005; Sun and Lee 2020):

- Firmness
- Inconsistency
- Nodular Focus
- Asymmetry
- Lump is attached to skin or muscle

Other signs and symptoms may include the following:

- Variation in shape and size of the breasts.
- Surface changes on the skin such as denting.
- Skin changes such as change in thickness or color.
- Relatively recent nipple inversion.
- Nipple discharge, particularly if bloodstained.
- Axillary lumps or lymph node involvement.

Breast cancer especially in later stages is known to have lymphatic and distant metastases, most commonly to the bone and lung. So, the assessment should include the axilla and supraclavicular fossa, the chest and sites of skeletal pain, and abdominal and neurologic examinations (Alkabban 2021; Feng et al. 2018).

## 3.6 Diagnosis

In addition to clinical symptoms diagnosis is confirmed with imaging and biopsy. This will determine the extent and exact location of the cancer as well as possible metastasis. There are multiple modalities of imaging for cancer most commonly involving ultrasound or mammography such as the following:

Ultrasound is usually the first mode of detecting irregularities; it utilizes ultrasound waves to distinguish between alternating densities of tissues such as fluid (cysts) and solid masses which can indicate neoplasms (Evans et al. 2018). An ultrasound can distinguish between a solid tumor that could be cancer and a fluid-filled cyst that isn't usually cancerous (Gokhale 2009).

Mammogram and diagnostic mammography is used within non-dense breast parenchyma and connective tissue to detect calcifications and non-palpable defects often seen in postmenopausal women. The latter is frequently utilized if something suspect is discovered on a screening mammography, such as when a patient complains of new symptoms including nodular abnormalities and nipple discharge (Gajdos et al. 2002; Huang et al. 1999).

A breast MRI is commonly used post diagnosis to explore the proliferation of the neoplasm within the breast or to confirm further metastatic advancement. The main benefit of an MRI is that there is high resolution and quality of images produced so it gives a better outcome when determining factors such as metastasis (Menezes et al. 2014).

A biopsy is a procedure that isolates and removes tissue so that it can be tested in a laboratory. There are multiple types but most common diagnostic biopsies include fine needle aspiration biopsy and core needle biopsy. Fine needle aspiration biopsy comprises of a small sample of cells being removed via a needle (Wu and Burstein 2004).

Core needle biopsy is used for larger samples of tissue to identify the presence of tumor markers, indicating possible malignancies. Other biopsies include surgical, image-guided, and sentinel biopsy (Bilous 2010; White et al. 2001).

## 3.7 Staging

The advancement and spread of malignancy are often determined by staging. Factors that are included in staging are multiple and include lymphatic spread and dimensions of the tumor, among others. The TNM staging system is typically agreed to be the most efficient way of staging breast cancer based on common standards (Cserni et al. 2018). The system contains three segments:

***Tumor (T)*** This refers to the size of the tumor and is subclassified as T1–4 with T1 being the size less than 2 cm and T4 being spread to the chest wall and other anatomical structures. These subclassifications are also further divided into subgroups (Cserni et al. 2018).

***Node (N)*** This describes involvement of lymph nodes and further spread. It is divided into two approaches: pathological (tissue sample assed from tissue obtained during surgery) and clinical (staging after physical examination involving use of test and imaging results) (Cserni et al. 2018).

***Metastasis (M)*** This specifies if there is metastatic malignant spread to surrounding or distal structures. It is typically split into 4 different categories in breast cancer M0-pM1 (Cserni et al. 2018).

### 3.8 Prognosis

The prognosis of survival after breast cancer is determined based on many different criteria which include data from the staging systems, type of malignancy, and other variable factors such as race, sex, and age. An effective method of defining the prognosis is via the survival rate calculation and it can be categorized into different subtypes such as 5-year survival rate and 10-year survival rate. The rates can be further subdivided into categories that include metastatic cancers and non-metastatic cancers with the former having a lower percentage of survival and the latter with a higher percentage of survival (Soerjomataram et al. 2008). Women with invasive breast cancer have a 99% 5-year survival rate if the cancer is exclusively found in the breast. If the cancer has advanced to the lymph nodes in the area (regional lymph node), the 5-year survival rate is 86%. If cancer has migrated to another region of the body, the 5-year survival rate is only 28%. Sixty-three percent (63%) of breast cancer patients are diagnosed at this stage. When compared to women over the age of 65, adolescent and young adult females aged 15–39 in the United States are 47% less likely to be diagnosed with breast cancer at an early stage (68%). This might be because most breast cancer screening does not start until the age of 40 unless a person is at greater risk, therefore the eventual prognosis is drastically altered (Sung et al. 2021).

## 4 Types of Radiation Therapy

Radiation therapy is rarely used as the sole modality of treatment for breast cancer. The primary purpose of radiation is to shrink the tumor in order to protect healthy tissue or to irradiate any cancer cells that may remain after the first therapy. Radiation therapy can be used before, after, or with other modality of treatments; adjuvant, neo-adjuvant, or palliative.

Adjuvant therapy applies to radiation therapy when it is used postoperatively, more commonly seen in patients with early-stage cancers or used in order to avoid the recurrence of tumors (Rutqvist et al. 2003). In many cases where a surgery is performed, an adjuvant therapy with radiation therapy can greatly reduce the chances of recurrence as well as decrease the chances of death following breast cancer and treatment (Rutqvist et al. 2003). Adjuvant therapy is also known as curative therapy when radiation is the sole mode of treatment. However, radiation treatment is rarely used as the primary modality; rather, a multimodal strategy is favored, with the choice of modalities determined by the kind of cancer, stage, and, if feasible, patient desire.

Neoadjuvant therapy with radiation therapy is when the radiation is applied before the primary treatment or modality is used. This can shrink the tumor size so that it is more manageable or in order for smaller resection to be made; this allows more healthy tissues to be spared and preserved as well as lowering the chances of death following the treatment (Papademetriou et al. 2010). Older techniques used in breast cancer did not accommodate for patient choices such as preserving the function or aesthetics of breasts; neoadjuvant therapy can also assist with that, as less cells have to be removed if surgery is conducted.

As opposed to the curative goal seen in adjuvant and neoadjuvant therapies, palliative radiation therapy does not aim to cure cancer. Rather it aims to shrink tumor cells, slow the rate of growth, and to alleviate or to reduce symptoms. Palliative radiation therapy particularly in breast cancer is effective in reducing pain, bleeding, ulceration, discharge, and general discomfort (Jacobson et al. 2021).

Different techniques allow the delivery of radiation directed toward the cancer cells and its surrounding tissues. This can be provided from the outside, known as external beam radiation therapy (EBRT) or the radiation that can be provided internally, known as brachytherapy. EBRT is delivered from the outside of the tumor with different techniques allowing for varied results regarding healthy tissue sparing. Broadly speaking the radiation is created through the use of Linear Accelerators or Linac which can create X-rays, gamma rays, proton beams, etc., which can be targeted and conformed to the boundaries of the cancer tissue (Huh and Kim 2020; Wang et al. 2017). The machines delivering the radiation are able to position the beams to be able to enter from almost any angles; better manipulation toward targeting the tumor tissue and avoiding healthy tissue. Combining these machines along with imaging techniques such as computer tomography (CT) and magnetic resonance imaging (MRI), thus creating detailed 3D models of the tumor, can allow for greater accuracy in guiding the beams, thus allowing for higher doses of radiation therapy to be used; the 3D images allow the healthy cells to be avoided (Chandarana et al. 2018; Commandeur et al. 2017).

Radiation therapy delivered as brachytherapy (also known as internal radiation therapy) relies on delivering the radiation to the tumor tissue by placing the radiation source within the tissue, rather than as a beam to the tumor tissue as seen in EBRT. By placing the source within the tumor tissue, the ability to increase the radiation dose is gained. In 1915, the first use of brachytherapy took place, where a radium implant was used to treat cervical cancer (Banerjee and Kamrava 2014). Brachytherapy in breast cancer often follows straight after surgery. The dosage is delivered as intracavitary, intraluminal, or multicatheter interstitial (Skowronek and Chicheł 2014).

## 5 Uses of Radiotherapy in Breast Cancer

Radiotherapy is generally divided into 2 broad categories of External Beam Radiation Therapy and Internal Radiation Therapy (more commonly referred to as Brachytherapy), both named so due to the method delivery; from the use of external beams directed toward the tumor or through the placement of a solid radiation source internally, within the tumor tissue or in the cavity following the removal of tumor tissue. Large randomized studies have demonstrated the benefits of conducting radiation therapy following a wide excision of tumor; studies have also shown that radiation therapy is more beneficial in ductal carcinoma in situ (Rutqvist et al. 2003).

External Beam Radiation Therapy utilizes beams that are produced by machines such as a linear accelerator. The beams themselves are produced by colliding high-speed particles such as electrons to a metal target such as tungsten resulting in X-ray or other high-energy ionizing rays (Baskar et al. 2012). As opposed to Brachytherapy, External Beam Therapy is painless with most patients not having to stay in the hospital following a treatment (Waddle et al. 2015). EBRT can be delivered to patients using many different techniques and the commonly used techniques and sources of radiation are given below.

3D conformal radiotherapy (3D CRT) is a type of EBRT treatment that can target the tumor and to match the shape of the tumor itself. Compared to earlier methods, which could only target the tumor's height and width and significantly damaged the surrounding healthy tissue (Gupta et al. 2012). The beams can be manipulated in which angles they enter to optimize the target area (Hansen 2010). The technique relies on the use of special imaging techniques to visualize the size, shape, margins, and location of the tumor such as computer tomography (CT), magnetic resonance imaging (MRI), positron emission tomography (PET scan) or PET-CT scan (Jaffray and Gospodarowicz 2015). These images will then be used to fabricate a plan for the radiation to be applied to the patient. Due to the nature of detailed imaging and healthy tissue sparing, recovery will be quicker as surrounding healthy tissues receive less radiation. This also enables for larger radiation doses to be used, resulting in greater tumor shrinking efficiency (Jaffray and Gospodarowicz 2015; Pereira et al. 2014).

Intensity-modulated radiotherapy (IMRT) is a more specialized form of 3D CRT which allows more precise shaping of the radiation applied to fit the tumor (Jaffray and Gospodarowicz 2015). The radiation beam is divided into "beamlets", the intensity of which may be adjusted over the radiation field (Gupta et al. 2012). This level of control allows the healthy tissue that surrounds the tumor to be spared even more than it can be through 3D CRT, as well as increasing the dose delivered. Randomized studies have shown that the use of IMRT is able to reduce toxicity in breast cancer patients (Buwenge et al. 2017).

Volumetric Modulated Arc Therapy (VMAT) is a novel radiation therapy extension to IMRT, which allows for an optimized 3D dose distribution delivery in a single gantry rotation (Popescu et al. 2010). It works similarly to IMRT where the radiation dose is varied through the treatment with intensity differences to allow healthy tissue sparing. Comparing VMAT to 3D CRT, VMAT offers more efficiency with the same or higher level of dose conformity while delivering a lower dose to the ipsilateral lung in breast cancer treatment (Qiu et al. 2010). As with 3D CRT and IMRT, a high level of accuracy is maintained in targeting the tumor.

Image Guided Radiation Therapy (IGRT) uses frequent imaging techniques in order to visualize and aid radiation therapy. This is especially useful in areas that may move and is rarely used in breast cancer treatment.

Proton Beam Therapy (PBT) is a specialized form of radiation therapy that utilizes protons being delivered to the target rather than X-rays as seen in conventional radiation therapy (Jaffray and Gospodarowicz 2015). The use of this technique for breast cancer is conflicted with countries such as the United Kingdom not approving the use whereas countries such as Switzerland allow the use (Thorpe et al. 2019). As a relatively novel technique, more studies will be required in order to identify if the technique can provide a better solution compared to other radiation techniques. The advantage of PBT is that minimal dose is deposited past the treatment target sparing many of the organs that otherwise would have been affected in traditional Radiation Therapy techniques, specifically the heart, lungs, bones, and muscles (Chowdhary et al. 2019).

Neutron Beam Therapy, like PBT, is a specialized form of EBRT that uses neutrons. The use of NBT in breast cancer is less frequent than the past levels of use, with only a handful of countries that are still using this technique such as the United States of America, Russia, and South Africa (Loap and Kirova 2020). NBT can be of great benefit in patients where the tumors are radioresistant to conventional radiation therapy techniques; they have a higher impact on cells compared to other radiation types (Jaffray and Gospodarowicz 2015).

## 6 Brachytherapy

Brachytherapy involves placement of a radiation or radiation therapy placed close to or within tumor tissue (Skowronek 2017). The dosage is delivered as intracavitary, intraluminal, or multicatheter interstitial. Brachytherapy can be classed by the delivery method or by the dose given; low dose rate implant, high-dose rate implants and permanent implants or Interstitial, Intracavitary, and Intraluminal. The main advantage of brachytherapy lies within the method of delivery where the radiation is placed within the tumor tissue or the site where the tumor tissue is removed, allowing a higher dose to be applied. In the case of Breast Cancer, the radiation is almost always given following surgical removal of the tumor (Czajka and Pfeifer 2021).

Brachytherapy requires easy access to the site of the tumor; here lies the advantage of brachytherapy for breast cancer as it can be routinely used following surgical removal of the advances in 3D scanning that allows visualization to make access to the tumor site (Skowronek 2017). American Brachytherapy Society guidelines for brachytherapy suggest that women 45 or above diagnosed with DCIS or early-stage invasive breast cancer with lumpectomy are appropriate candidates for brachytherapy (Shah et al. 2018). Additional guidelines from American Brachytherapy Society also include that the patient's cancer is:

- Hormone-receptor positive or negative
- Node-negative
- Have not invaded the lymphovascular system
- Has a negative or a clear margin

Interstitial Brachytherapy is a type of brachytherapy which involves implanting hollow tubes, needles, or wires that are radioactive to the tumor site or the area where the tumor was removed. This is one of the main types of brachytherapy used in breast cancer compared to other types. Interstitial brachytherapy can be delivered using a single or multi-catheter device, in which a remote afterload is performed following the puncture (Gurram et al. 2016).

Intracavitary Brachytherapy is when the radiation is placed within the body cavity. It is of great significance in breast cancer as it can be carried out, following mastectomy or lumpectomy.

Intraluminal Brachytherapy is most commonly used in hollow organs where the radiation is delivered through the means of a tube or an applicator of some sort to the lumen of the target organ.

The main advantage of Brachytherapy is the fact that it can spare a lot of the healthy tissue surrounding the tumor and reduce the side effects that may come with External Beam Therapy (Baskar et al. 2012).

## 7 Combination Therapy and Management of Breast Cancer

Combination therapy is a treatment strategy that combines two or more therapeutic components that help in cancer treatment. The combination of anti-cancer medications improves efficacy as compared to monotherapy because it targets important pathways in a synergistic or additive manner. Separate types of therapies can be used in sequence, at the same time, or at different times (Mokhtari et al. 2017). This technique has the potential to minimize medication resistance while also providing anti-cancer effects such as lowering tumor formation and metastatic potential, trapping mitotically active cells, decreasing cancer stem cell populations, and activating apoptosis. The integration of two or more types of therapy is referred to as a combination treatment (Mokhtari et al. 2017). Surgery, radiation therapy, and chemotherapy are the most often used cancer treatments. The term "combined-modality treatment" refers to the use of two or more of these treatment modalities on a single patient (Mendenhall 2020).

### 7.1 Chemotherapy

Chemotherapy is a method of cancer treatment that involves the use of chemicals to kill cancer cells. It usually acts by inhibiting cancer cell growth, division, and replication. Chemotherapy has a greater impact on cancer cells because they grow and divide faster than normal cells. In any event, chemotherapy medications are effective, yet they can still cause harm to healthy cells. This damage is responsible for the adverse effects associated with chemotherapy (Lebaron et al. 1988; Natale 2015). Anticancer medications are often classed by either their mode of action or their origin. Alkylating agents bind to DNA and cellular proteins, and their major method of action is cross-linking of DNA strands, repression of DNA replication, and transcription of RNA (Ogawa 1997). Cancer chemotherapeutic medications are categorized chemically and function as alkylating agents, cytotoxic pharmaceuticals for targeted treatment, antimicrobial antimetabolites, topoisomerase inhibitors, and corticosteroids. A targeted cancer therapy is one that use medications or other substances to precisely identify and attack certain types of cancer cells (Tsimberidou 2015). mTOR inhibitors, angiogenesis inhibitors, apoptosis-inducing medicines, tyrosine kinase inhibitors, monoclonal antibodies, and hormonal therapies are examples of targeted therapy (Dray 2011).

There are many classifications of drugs used in chemotherapy that can be used as a treatment modality in breast cancer.

***Cytotoxic Drugs*** Cytotoxic pharmaceuticals, often known as antineoplastics, are a class of medications that include toxic and harmful compounds to cells. This class of medications inhibits their replication or growth and is used to treat cancer (Ribeiro et al. 2012). Examples include: cyclophosphamide (cytoxan), 5-fluorouracil (5-FU) or capecitabine, taxanes such as paclitaxel (taxol), epirubicin (ellence), doxorubicin (adriamycin), anthracyclines, and docetaxel (taxotere), and carboplatin (paraplatin). Combinations of two or three of these medicines are most commonly employed (Conte et al. 2001; Ribeiro et al. 2012).

***Monoclonal Antibodies*** These are synthesized in a lab and target proteins on the surface of cancer cells. On the surface of some malignant cells are unique antigens known as tumor markers. Monoclonal antibodies can be administered to specifically bind with these antigens. When these medications are administered to a person, the monoclonal antibodies attach to the cancer cells and clump them together (Henricks et al. 2015). Trastuzumab inhibits HER2-mediated signaling via the phosphatidylinositol 3-kinase (PI3K) and mitogen-activated protein kinase (MAPK) pathways. Reduced downstream signaling via these pathways activates the cyclin-dependent kinase inhibitor p27kip1, leading to cell-cycle arrest and death in breast cancer (Nahta and Esteva 2006).

## 7.2 Targeted Therapy

Hormone therapy is a treatment option that helps regulate cancer cells from proliferating by using hormones. About two-thirds of all cases of breast cancer are hormone receptor-positive. Their cells include receptors for the hormone's estrogen-positive malignancies and/or progesterone-positive cancers, which aid in the development and spread of cancer cells. Hormone therapy for breast cancer comes in a variety of forms (Abdulkareem and Zurmi 2012). The large percentage of hormone therapies work by lowering estrogen levels or blocking estrogen's action on breast cancer cells.

Adjuvant therapy is recommended for early-stage breast cancer. Tamoxifen has been approved by the FDA for adjuvant hormone therapy of premenopausal and postmenopausal women with ER-positive early-stage breast cancer, and the aromatase inhibitors exemestane, letrozole, and anastrozole have been approved for postmenopausal women (Eisen et al. 2008; Krásenská 2016; de Cremoux et al. 2004). Aromatase inhibitors reduce estrogen levels by inhibiting aromatase, a molecule in adipose tissue that converts other hormones into estrogen (Mouridsen

2004). Letrozole is now the most effective aromatase inhibitor in the first-line situation of advanced breast cancer (Mouridsen 2004).

Currently, neoadjuvant chemotherapy is widely utilized in early-stage breast cancer (EBC) and locally advanced breast cancer (LABC) to improve the odds of breast-conserving surgery (BCS) and to detect tumor response to therapy (Ashbeck et al. 2007; Tai et al. 2007). Breast cancer surgery is often accomplished by a lumpectomy followed by whole-breast irradiation or mastectomy (Loibl and Gianni 2017). Local irradiation, systemic treatment with cytotoxic medicines, molecular targeted therapies, or any combination of these may be used as adjuvant therapy for breast cancer (Anderson et al. 2014; Kumar and Aggarwal 2016; Sullivan et al. 2016).

## 7.3 Management of Non-invasive (In Situ) Breast Cancer

Ductal carcinoma in situ (DCIS) and lobular carcinoma in situ (LCIS) are not often symptomatic, although the broad pre-invasive disease may display with a mass or thickening of breast tissue. DCIS is often diagnosed during screening.

There are many management methods for non-invasive breast carcinoma.

***Simple Mastectomy*** In situ breast cancer was rarely diagnosed before the approach of mammographic screening. The standard treatment until the 1980s was mastectomy. After the introduction of mastectomy, relapse rates have reduced exceptionally low (2–3%). Loco-regional recurrence and metastatic cancers are credited to undiagnosed micro-invasive cancer. Mastectomy remains the standard treatment for large in situ cancers and for multifocal illness with breast reconstruction if desired.

***Wide Excision Alone*** With the expanded diagnosis of small localized non-invasive cancers by mammographic screening, breast-conserving surgery has been adopted predominantly. Since the lesion is, usually screen-detected and impalpable, the lesion which is often a microcalcification, is localized first by the radiologist. An example X-ray is fundamental to ensure total excision of the abnormality. Cavity wall shavings that prohibit residual disease are increasingly routine.

In any case, 20–30% create local recurrence within 5 years. The most elevated chance of recurrence is with high-grade DCIS and excision margin involved or <1 mm clear—excision of margin of at least 2 mm is now recommended.

Almost half of the recurrences are non-invasive, but the rest contains invasive cancer.

***Wide Excision and Postoperative Radiotherapy*** Three large studies have now affirmed the advantage of radiotherapy in this setting.

- The entire breast is irradiated.
- Chance of recurrence of <10% at 5 years.

- The major portion of patients is treated with radiation following breast-conserving surgery for DCIS, with the exemption of patients with minor foci of low-grade DCIS removed with broad resection margins and adjuvant hormone treatment.
- Tamoxifen 20 mg every day, taken for 5 years, diminishes the recurrence of DCIS by approximately 30%, but there's instability with respect to the recurrence of invasive cancer.
- Clinical trials with aromatase inhibitors are also ongoing.

## 7.4 Management of Early Invasive Breast Cancer

Early breast cancer is characterized as a disease that can be totally removed by surgery, i.e., T1–3, N0–1 tumor. The management of this infection comprises the following:

- Endocrine therapy
- Radiotherapy
- Chemotherapy
- Adjuvant therapy
- Pathological evaluation and staging for adjuvant therapy coordination
- Surgical treatment of the axilla and breast
- Follow-up

Again, multidisciplinary discussion and arranging of treatment are basic to optimize treatment outcomes. Consensus guidelines don't recommend routine staging investigations such as CXR, liver ultrasound, bone scan, or CT imaging of the chest and abdomen. Although, people at high risk of metastatic cancer should be identified by the MDT and imaged accordingly. Essentially, breast MRI can enhance loco-regional staging in chosen cases (Dray 2011).

## 7.5 Loco-Regional Radiotherapy

Breast illumination diminishes the chance of local repeat after breast-conserving surgery from around 30% to <10% at 10 years. In expansion to the 75% reduction within the local recurrence rate, it is estimated that, for every four local recurrences avoided, one cancer-related death at 10 years is avoided. Breast irradiation is suggested for all patients after wide local excision for invasive breast cancer. The entire breast is treated with tangential areas to a dosage of 40 Gy in 15 fractions. Care should be taken to reduce the volume of the lung and heart irradiated which is why most centers now use CT planning. Boost of 10–15 Gy is commonly conveyed to the tumor bed, using electrons or 192Ir implant.

Postoperative irradiation of the breast presently accounts for a large proportion of the work of most radiotherapy departments, and efforts are being made to diminish this. Risk-adjusted procedures are beneath examination, e.g., no radiotherapy in ladies above 70 years who have experienced wide local excision for ER-positive T1, N0 breast cancer. Accelerated partial breast irradiation, e.g., high-dose-rate brachytherapy, IMRT, or IORT are moreover being surveyed in clinical trials. Post-mastectomy radiotherapy to the chest wall decreases the hazard of loco-regional recurrence. There are trials progressing to determine the benefits of post-mastectomy radiotherapy in patients with 1–3 positive lymph nodes.

### 7.6 Axillary Radiotherapy

This method is certainly indicated after a lymph node testing that's positive. In general, this should be avoided after axillary clearance because of the high chance of lymphoedema and brachial plexopathy. The SCF may be irradiated after axillary clearance if four or more nodes are positive. In common, radiotherapy should start as soon as conceivable after surgery. However, enhanced normal tissue damage can result when radiotherapy and adjuvant chemotherapy are given together, and radiotherapy is often postponed until the completion of chemotherapy (Dray 2011).

## 8 Adverse Effects of Radiation Therapy

Adverse effects of Radiation Therapy can be different from one case to another. As with any medication or treatment plan, some of those may affect the patients more frequently than others; it may depend on the patient, the type of cancer, the aggressive nature of cancer, the location of therapy (systemic/local), type/strength/duration of doses used, etc. The most common side effects seen in breast cancer are fatigue, skin abnormalities (hyperpigmentation, dry desquamation, or erythema), myelosuppression as well as lymphedema (Shapiro and Recht 2001). Older techniques, lack of control over delivery of radiation and dosage of radiation, resulted in more healthy tissue damage. With advances, using other technology to aid direct delivery, use of higher-energy sources and lowered daily doses, the exposure to normal healthy tissue can be limited and reduce a lot of symptoms (Shapiro and Recht 2001).

Sensitivity to radiation therapy varies depending on the tissue types; lymphocytes and germ cells show high sensitivity, epithelial cells are moderately sensitive, whereas, the central nervous system and connective tissues tend to show resistance to radiation (Cassidy et al. 2015). This can have an impact on the patient as some hematopoietic cells may also undergo apoptosis following the radiation along with the tumor cells; these lost cells tend to be replaced by proliferation of the surviving progenitor cells. Generally adverse effects occurring following radiation therapy can be divided as acute and late. The acute effects are seen with a high proliferative rate which can replace the dead cells quicker and these tend to resolve within 8 weeks; the replacement turnaround time depends on the level of depletion of the progenitor cells. On the other hand, the late effects predominantly affect tissues of lower proliferation rates.

Acute effects are seen up to 8 weeks of radiation treatment, with the severity depending on the total radiation and length of time the radiation therapy delivered (Cassidy et al. 2015). It tends to manifest more with skin GI, tract and the bone marrow with a high tendency for complete recovery with the right dose selection. Late effects are more complicated to define but tend to affect cells of lower proliferation rates such as cells of the central nervous system, heart, liver, kidney, and lung. These effects may not be apparent straight after the treatment; effects can develop several years following the radiation therapy.

When administering radiation therapy, giving rise to a second malignancy is of high concern. An increased chance of malignancy following has been seen in patients given radiation therapy, mostly arising on the contralateral breast (Shapiro and Recht 2001). Risk of acquiring leukemia is also increased with most occurring 6–8 years after therapy and an increased chance of solid cancer is mostly seen in 10–30 years (Cassidy et al. 2015). Radiation therapy can impact the lymphatic system of the body. This can manifest as lymphedema with an incident rate of around 5–25% (Warren et al. 2014). However, early detection can allow successful treatment: complete decongestive physiotherapy and elastic pressure sleeves.

Skin abnormalities are very common in patients receiving radiation therapy. It can present itself as both acute and late effects. Manifestation may occur after 10–14 days following the start of radiation therapy. Acute effects on skin can include:

- Erythema – "sun-burnt appearance" (with or without pruritus, may feel hot or sore)
- Desquamation
  - Initially starts as dry desquamation with limited peeling seen on the epidermis
  - Later painful, wet/moist desquamation can be seen (Image A)
  - Ulceration and bleeding can occur if the above symptoms doesn't resolve
  - Late effects on skin can include
- Atrophy
- Fibrosis
- Telangiectasia
- Hair loss specifically seen under armpit area – it may or may not grow back
- Hyper- or hypo-pigmentation

Radiation pneumonitis is another common side effect of radiation therapy. Occurring as a late effect, it can be characterized by cough, shortness of breath, and fever, often self-resolving a few months following the therapy (Shapiro and Recht 2001). The incident rate increases when radiation therapy is conducted with chemotherapy.

Effects on the gastrointestinal tract and the oral mucosa can cause discomfort to the patient to continue their day-to-day life, specifically sore mouth, nausea, vomiting, loss of appetite, discomfort swallowing, diarrhea, bloody stool, etc. are commonly seen (Cassidy et al. 2015; Shapiro and Recht 2001). Most effects can be seen as acute. Acute effects on the oral mucosa and GI tract may include:

- Erythema and painful ulceration in the oral mucosa
- Dry, irritated, or sore mouth
- Nausea and vomiting
- Diarrhea
- Bloody or mucus discharge

Late effects on the oral mucosa and GI tract may include:

- Mucosal ulceration in GI tract
- Obstruction or fibrosis of GI tract

Musculoskeletal effects can be seen as stiffness, swelling, and general uncomfortableness in joints and muscles. Fracture incidents are also increased with more chances of bones around the breast tissue such as the ribs being the main affected bone. Bone marrow suppression can be seen; as stated above, radiation therapy can have a significant effect on hematopoietic cells.

The heart, kidney, and the central nervous system do not show responses that may be attributed as acute. It was long believed that the regeneration of late affected tissues was not possible however, studies have shown vast improvement in long-term regeneration specifically within many of these tissues (Cassidy et al. 2015). Fatigue is generally seen in patients receiving therapy. Late effects on the heart may include:

- Pericarditis and myocarditis
- Conduction blocks
- Cardiomyopathy
- Cardiac failure

Late effects on the kidneys may include:

- Radiation nephropathy
- Proteinuria
- Hypertension
- Renal failure

Late on the central nervous system may include:

- Necrosis due to radiation (irreversible)
- Demyelination – drowsiness, Lhermitte's syndrome (electric shock sensation which occurs with neck flexion that radiates down the spine)

Cardiac toxicity-related issues used to be one of the main reported problems with radiotherapy in general (CollaborativeGroup 1995). A small increase in long-term

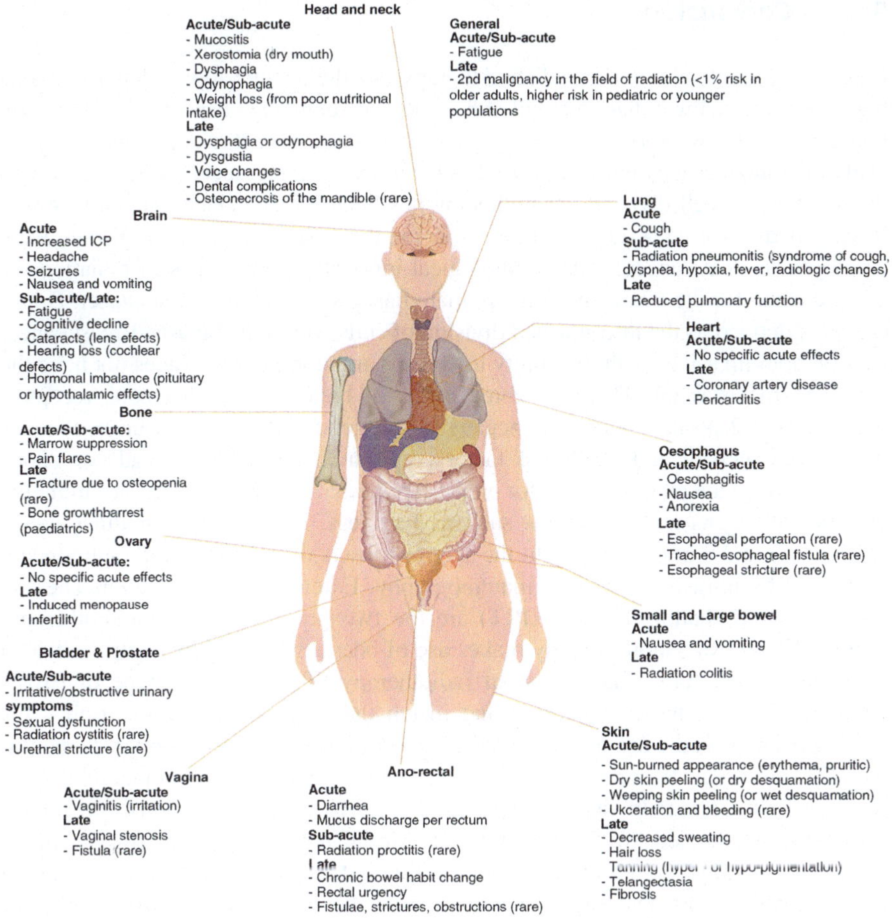


**Fig. 2** General side effects seen in patients undergoing radiation therapy

mortality was reported from cardiac causes, however, much of the data and information regarding this used outdated techniques delivering unsafe amounts of radiation to cardiac tissues (Nixon et al. 1998). Similarly, there is an increase in the risk of other cardiac complications such as pericarditis, and cardiac failure is also rarely seen but can include pericarditis, myocarditis, and cardiac failure, if a higher dose is given (Shapiro and Recht 2001). The risk can be greatly reduced with planning such as optimizing the angles of the beam, using intensity-modulated radiotherapy (IMRT), treating the patient in prone position rather than supine and treating patient in deep inspiration (Taylor and Kirby 2015) (Fig. 2).

## 9 Conclusion

Radiotherapy, often known as radiation therapy, is a therapeutic approach that employs high-energy ionizing radiation to kill or halt the development of cancer cells. Radiation treatment works by producing DNA damage within the cell. Radiation therapy, especially the ionizing radiation employed, has the greatest effect on cells when they are in the M phase of cell division. Radiotherapy has been utilized since Wilhelm Conrad Roentgen discovered X-rays in 1895, and for the first time in 1896, X-rays were employed to treat cancer. With technological innovation, new types of radiation are being created for the treatment of many malignancies, including breast cancer. Breast cancer is defined as the uncontrolled division of cells within the breasts. Breast cancer may be both invasive (with the ability to spread) and non-invasive (does not have the ability to metastasize). There are three forms of breast cancer: hormone receptor-positive, HER2-positive, and triple-negative. Age, gender, prior breast cancer, changes in histology, genetic and family risk factors, reproductive risk factors, and exogenous hormone usage are all risk factors for breast carcinoma. Patients with breast carcinoma present with signs and symptoms such as hardness, inconsistency, nodular focus, asymmetry, and a lump attached to the skin or muscle of the breast; these tumors are diagnosed by ultrasonography or mammography. External Beam Radiation Therapy and Internal Radiation Therapy (IRT) are the two primary kinds of radiotherapy. External Beam Radiation Therapy makes use of beams generated by equipment such as a linear accelerator. 3D conformal radiotherapy (3D CRT) is a type of EBRT treatment that can target the tumor and match the shape of the tumor itself. This technique relies on the use of special imaging techniques such as Computer Tomography (CT), Magnetic Resonance Imaging (MRI), Positron Emission Tomography (PET scan), or PET-CT scan to visualize the size, shape, margins, and location of the tumor. Volumetric Modulated Arc Therapy, Intensity Modulated Radiotherapy, Image Guided Radiation Therapy, and Proton and Neutron Beam therapy are some of the additional EBRT treatment options. Brachytherapy entails the implantation of a radiation or radiation therapy close or into tumor tissue. Aside from these, combination therapy, which involves chemotherapy (the use of specific medications to treat cancer), targeted therapy (hormone treatment that moderates or stops the development of cancer that employs hormones to grow), is also utilized in the treatment of breast cancer. Certain management criteria, such as mastectomy, extensive excision, surgical treatment, radiation, and so on, are used to limit the risk of occurrence of breast cancer and to minimize symptoms. These techniques of treating cancer with radiations have certain side effects on the body, the most prevalent of which are weariness, skin abnormalities (hyperpigmentation, dry desquamation, or erythema), myelosuppression, and lymphedema. Despite substantial advancements in breast radiation therapy in recent years, treatment paradigms continue to evolve, and treatment deintensification is being re-evaluated.

## References

Abdulkareem IH, Zurmi IB (2012) Review of hormonal treatment of breast cancer. Niger J Clin Pract 15(1):9–14. https://doi.org/10.4103/1119-3077.94088
Alkabban FM (2021) Breast cancer. StatPearls Publishing. Retrieved from https://pubmed.ncbi.nlm.nih.gov/29493913/
Anderson KN, Schwab RB, Martinez ME (2014) Reproductive risk factors and breast cancer subtypes: a review of the literature. Breast Cancer Res Treat 144(1):1–10. https://doi.org/10.1007/S10549-014-2852-7
Arps DP, Healy P, Zhao L, Kleer CG, Pang JC (2013) Invasive ductal carcinoma with lobular features: a comparison study to invasive ductal and invasive lobular carcinomas of the breast. Breast Cancer Res Treat 138(3):719–726. https://doi.org/10.1007/S10549-013-2493-2
Ashbeck EL, Rosenberg RD, Stauber PM, Key CR (2007) Benign breast biopsy diagnosis and subsequent risk of breast cancer. Cancer Epidemiol Biomark Prevent 16(3):467–472. https://doi.org/10.1158/1055-9965.EPI-06-0394
Banerjee R, Kamrava M (2014) Brachytherapy in the treatment of cervical cancer: a review. Int J Women's Health 6(1):555. https://doi.org/10.2147/IJWH.S46247
Baskar R, Lee KA, Yeo R, Yeoh KW (2012) Cancer and radiation therapy: current advances and future directions. Int J Med Sci 9(3):193. https://doi.org/10.7150/IJMS.3635
Baskar R, Dai J, Wenlong N, Yeo R, Yeoh KW (2014) Biological response of cancer cells to radiation treatment. Front Mol Biosci 1(NOV). https://doi.org/10.3389/FMOLB.2014.00024
Bilous M (2010) Breast core needle biopsy: issues and controversies. Mod Pathol 23(2):S36–S45. https://doi.org/10.1038/modpathol.2010.34
Buwenge M, Cammelli S, Ammendolia I, Tolento G, Zamagni A, Arcelli A et al (2017) Intensity modulated radiation therapy for breast cancer: current perspectives. Breast Cancer 9:121. https://doi.org/10.2147/BCTT.S113025
Cassidy J, Bissett D, Spence RAJ, Payne M, Morris-Stiff G (2015) Oxford handbook of oncology, 4th edn. Edition by Donald Bissett, pp 261–299. https://doi.org/10.1093/med/9780199689842.001.0001
Chandarana H, Wang H, Tijssen RHN, Das IJ (2018) Emerging role of MRI in radiation therapy. J Magn Resonan Imag 48(6):1468. https://doi.org/10.1002/JMRI.26271
Chowdhary M, Lee A, Gao S, Wang D, Barry PN, Diaz R et al (2019) Is proton therapy a "pro" for breast cancer? A comparison of proton vs. non-proton radiotherapy using the national cancer database. Front Oncol 9(JAN). https://doi.org/10.3389/FONC.2018.00678/PDF
CollaborativeGroup E (1995) Effects of radiotherapy and surgery in early breast cancer. An overview of the randomized trials. N Engl J Med 333(22):1444–1456. https://doi.org/10.1056/NEJM199511303332202
Commandeur F, Simon A, Mathieu R, Nassef M, Arango JDO, Rolland Y et al (2017) MRI to CT prostate registration for improved targeting in cancer external beam radiotherapy. IEEE J Biomed Health Inf 21(4):1015–1026. https://doi.org/10.1109/JBHI.2016.2581881
Conte P, Gennari A, Guarneri V, Landucci E, Donati S, Salvadori B, Bengala C, Orlandini C (2001) Epirubicin/taxane combinations in breast cancer: experience from several Italian trials. Oncology (Williston Park) 15(5):21–23. Retrieved from https://pubmed.ncbi.nlm.nih.gov/11396360/
Cserni G, Chmielik E, Cserni B, Tot T (2018) The new TNM-based staging of breast cancer. Virchows Archiv 472(5):697–703. https://doi.org/10.1007/S00428-018-2301-9
Czajka ML, Pfeifer C (2021) Breast cancer surgery. Br J Med Pract 13(1):1. https://doi.org/10.1093/med/9780198839248.003.0013
de Cremoux P, Diéras V, Poupon M-F Magdelénat H, Sigal-Zafrani B, Alain Fourquet J-YP (2004) [Tamoxifen and aromatase inhibitors in the treatment of breast cancer in menopausal women: pharmacological and clinical aspects]. Bull Cancer 91(12):917–927. Retrieved from https://pubmed.ncbi.nlm.nih.gov/15634633/
Dray N (2011) Harrison's hematology and oncology. Yale J Biol Med 84(1):64. Retrieved from /pmc/articles/PMC3064251/

Eisen A, Trudeau M, Shelley W, Messersmith H, Pritchard KI (2008) Aromatase inhibitors in adjuvant therapy for hormone receptor positive breast cancer: a systematic review. Cancer Treat Rev 34(2):157–174. https://doi.org/10.1016/J.CTRV.2007.11.001

Evans A, Trimboli RM, Athanasiou A, Balleyguier C, Baltzer PA, Bick U et al (2018) Breast ultrasound: recommendations for information to women and referring physicians by the European Society of Breast Imaging. Insights into Imaging 9(4):449. https://doi.org/10.1007/S13244-018-0636-Z

Feng Y, Spezia M, Huang S, Yuan C, Zeng Z, Zhang L et al (2018) Breast cancer development and progression: Risk factors, cancer stem cells, signaling pathways, genomics, and molecular pathogenesis. Genes Dis 5(2):77–106. https://doi.org/10.1016/J.GENDIS.2018.05.001

Gajdos C, Ian Tartter P, Bleiweiss IJ, Hermann G, De Csepel J, Estabrook A, Rademaker AW (2002) Mammographic appearance of nonpalpable breast cancer reflects pathologic characteristics. Ann Surg 235(2):246. https://doi.org/10.1097/00000658-200202000-00013

Gianfaldoni S, Gianfaldoni R, Wollina U, Lotti J, Tchernev G, Lotti T (2017) An overview on radiotherapy: from its history to its current applications in dermatology. Open Access Macedonian J Med Sci 5(4):521. https://doi.org/10.3889/OAMJMS.2017.122

Goh CW, Wu J, Ding S, Lin C, Chen X, Huang O et al (2019) Invasive ductal carcinoma with coexisting ductal carcinoma in situ (IDC/DCIS) versus pure invasive ductal carcinoma (IDC): a comparison of clinicopathological characteristics, molecular subtypes, and clinical outcomes. J Cancer Res Clin Oncol 145(7):1877–1886. https://doi.org/10.1007/S00432-019-02930-2

Gokhale S (2009) Ultrasound characterization of breast masses. Indian J Radiol Imag 19(3):242. https://doi.org/10.4103/0971-3026.54878

Gupta T, Agarwal J, Jain S, Phurailatpam R, Kannan S, Ghosh-Laskar S et al (2012) Three-dimensional conformal radiotherapy (3D-CRT) versus intensity modulated radiation therapy (IMRT) in squamous cell carcinoma of the head and neck: a randomized controlled trial. Radiother Oncol 104(3):343–348. https://doi.org/10.1016/J.RADONC.2012.07.001

Gurram L, Wadasadawala T, Joshi K, Phurailatpam R, Paul S, Sarin R (2016) Multi-catheter interstitial brachytherapy for partial breast irradiation: an audit of implant quality based on dosimetric evaluation comparing intra-operative versus post-operative placement. J Contemp Brachyther 8(2):116–121. https://doi.org/10.5114/JCB.2016.59195

Hansen EK (2010) Handbook of evidence-based radiation oncology. https://doi.org/10.1007/978-0-387-92988-0

Healy BJ, van der Merwe D, Christaki KE, Meghzifene A (2017) Cobalt-60 machines and medical linear accelerators: competing technologies for external beam radiotherapy. Clin Oncol (Royal College of Radiologists (Great Britain)) 29(2):110–115. https://doi.org/10.1016/J.CLON.2016.11.002

Henricks LM, Schellens JHM, Huitema ADR, Beijnen JH (2015) The use of combinations of monoclonal antibodies in clinical oncology. Cancer Treat Rev 41(10):859–867. https://doi.org/10.1016/J.CTRV.2015.10.008

Huang CS, Wu CY, Chu JS, Lin JH, Hsu SM, Chang KJ (1999) Microcalcifications of non-palpable breast lesions detected by ultrasonography: correlation with mammography and histopathology. Ultrasound Obstetr Gynecol 13(6):431–436. https://doi.org/10.1046/J.1469-0705.1999.13060431.X

Hubenak JR, Zhang Q, Branch CD, Kronowitz SJ (2014) Mechanisms of injury to normal tissue after radiotherapy: a review. Plast Reconstr Surg 133(1):49e. https://doi.org/10.1097/01.PRS.0000440818.23647.0B

Huh HD, Kim S (2020) History of radiation therapy technology. Korean Soc Med Phys 31(3):124–134. https://doi.org/10.14316/PMP.2020.31.3.124

Ingledew PA, Kong T (2021). Side effects. Radiation oncology basics. Retrieved from Learn Oncology website: https://www.learnoncology.ca/modules/radiation-oncology-basics#section-6

Jacobson G, Kaidar-Person O, Haisraely O, Galper S, Rabin T, Dromi Shahadi I et al (2021) Palliative radiation therapy for symptomatic advance breast cancer. Sci Rep 11(1):5282. https://doi.org/10.1038/S41598-021-84872-9

Jaffray DA, Gospodarowicz MK (2015) Radiation therapy for cancer. J Korean Med Assoc 6:142–146. https://doi.org/10.1596/978-1-4648-0349-9_CH14

Krásenská M (2016) [Treatment with aromatase inhibitors in postmenopausal women with breast cancer and the possibility of influencing side effects]. Klinicka Onkologie: Casopis Ceske a Slovenske Onkologicke Spolecnosti 29(Suppl 3):S39–S49. https://doi.org/10.14735/AMKO20163S39

Kumar P, Aggarwal R (2016) An overview of triple-negative breast cancer. Arch Gynecol Obstet 293(2):247–269. https://doi.org/10.1007/S00404-015-3859-Y

Lebaron S, Zeltzer LK, Lebaron C, Scott SE, Zeltzer PM (1988) Chemotherapy side effects in pediatric oncology patients: drugs, age, and sex as risk factors. Med Pediatr Oncol 16(4): 263–268. https://doi.org/10.1002/MPO.2950160408

Lee E, McKean-Cowdin R, Ma H, Spicer DV, Van Den Berg D, Bernstein L, Ursin G (2011) Characteristics of triple-negative breast cancer in patients with a BRCA1 mutation: results from a population-based study of young women. J Clin Oncol 29(33):4373. https://doi.org/10.1200/JCO.2010.33.6446

Loap P, Kirova Y (2020) Fast neutron therapy for breast cancer treatment: an effective technique sinking into oblivion. Int J Particle Ther 7(3):61–64. https://doi.org/10.14338/IJPT-19-00080.1

Loibl S, Gianni L (2017) HER2-positive breast cancer. Lancet (London, England) 389(10087): 2415–2429. https://doi.org/10.1016/S0140-6736(16)32417-5

Mendenhall WM (2020) Radiation oncology. combined-modality therapy. Retrieved February 17, 2022, from https://radonc.med.ufl.edu/patient-care/information-for-patients/cancer-treatment-results/combined-modality-therapy/

Menezes GLG, Knuttel FM, Stehouwer BL, Pijnappel RM, Van Den Bosch MAAJ (2014) Magnetic resonance imaging in breast cancer: a literature review and future perspectives. World J Clin Oncol 5(2):61. https://doi.org/10.5306/WJCO.V5.I2.61

Mokhtari RB, Homayouni TS, Baluch N, Morgatskaya E, Kumar S, Das B, Yeger H (2017) Combination therapy in combating cancer. Oncotarget 8(23):38022–38043. https://doi.org/10.18632/ONCOTARGET.16723

Mouridsen HT (2004) Aromatase inhibitors in advanced breast cancer. Semin Oncol 31(6 Suppl 12):3–8. https://doi.org/10.1053/J.SEMINONCOL.2004.09.020

Nahta R, Esteva FJ (2006) HER2 therapy: molecular mechanisms of trastuzumab resistance. Breast Cancer Res 8(6):1–8. https://doi.org/10.1186/BCR1612/TABLES/1

Natale JJ (2015) Reviewing current and emerging antiemetics for chemotherapy-induced nausea and vomiting prophylaxis. Hosp Practice (1995) 43(4):226–234. https://doi.org/10.1080/21548331.2015.1077095

Nixon AJ, Manola J, Gelman R, Bornstein B, Abner A, Hetelekidis S et al (1998) No long-term increase in cardiac-related mortality after breast-conserving surgery and radiation therapy using modern techniques. J Clin Oncol Off J Am Soc Clin Oncol 16(4):1374–1379. https://doi.org/10.1200/JCO.1998.16.4.1374

Ogawa M (1997) [Anticancer drugs and pharmacologic actions]. Retrieved February 17, 2022, from Japanese website: https://pubmed.ncbi.nlm.nih.gov/9155146/

Papademetriou K, Ardavanis A, Kountourakis P (2010) Neoadjuvant therapy for locally advanced breast cancer: Focus on chemotherapy and biological targeted treatments' armamentarium. J Thoracic Dis 2(3):160. https://doi.org/10.3978/J.ISSN.2072-1439.2010.02.03.8

Peddi PF (2018) Hormone receptor positive breast cancer: state of the art. Curr Opin Obstet Gynecol 30(1):51–54. https://doi.org/10.1097/GCO.0000000000000424

Pereira GC, Traughber M, Muzic RF (2014) The role of imaging in radiation therapy planning: past, present, and future. Biomed Res Int 2014. https://doi.org/10.1155/2014/231090

Popescu CC, Olivotto IA, Beckham WA, Ansbacher W, Zavgorodni S, Shaffer R et al (2010) Volumetric modulated arc therapy improves dosimetry and reduces treatment time compared to conventional intensity-modulated radiotherapy for locoregional radiotherapy of left-sided breast cancer and internal mammary nodes. Int J Radiat Oncol Biol Phys 76(1):287–295. https://doi.org/10.1016/J.IJROBP.2009.05.038

Qiu JJ, Chang Z, Wu QJ, Yoo S, Horton J, Yin FF (2010) Impact of volumetric modulated arc therapy technique on treatment with partial breast irradiation. Int J Radiat Oncol Biol Phys 78(1):288–296. https://doi.org/10.1016/J.IJROBP.2009.10.036

Raina V, Bhutani M, Bedi R, Sharma A, Deo SVS, Shukla NK et al (2005) Clinical features and prognostic factors of early breast cancer at a major cancer center in North India. Indian J Cancer 42(1):36–41. https://doi.org/10.4103/0019-509X.15099

Ribeiro JT, Macedo LT, Curigliano G, Fumagalli L, Locatelli M, Dalton M et al (2012) Cytotoxic drugs for patients with breast cancer in the era of targeted treatment: back to the future? Ann Oncol 23(3):547–555. https://doi.org/10.1093/ANNONC/MDR382

Russo J, Russo IH (2006) The role of estrogen in the initiation of breast cancer. J Steroid Biochem Mol Biol 102(1–5):89–96. https://doi.org/10.1016/J.JSBMB.2006.09.004

Rutqvist LE, Rose C, Cavallin-Ståhl E (2003) A systematic overview of radiation therapy effects in breast cancer. Acta Oncol (Stockholm, Sweden) 42(5–6):532–545. https://doi.org/10.1080/02841860310014444

Shah C, Vicini F, Shaitelman SF, Hepel J, Keisch M, Arthur D et al (2018) The American Brachytherapy Society consensus statement for accelerated partial-breast irradiation. Brachytherapy 17(1):154–170. https://doi.org/10.1016/J.BRACHY.2017.09.004

Shapiro CL, Recht A (2001) Side effects of adjuvant treatment of breast cancer. N Engl J Med 344(26):1997–2008. https://doi.org/10.1056/NEJM200106283442607

Sharma GN, Dave R, Sanadya J, Sharma P, Sharma KK (2010) Various types and management of breast cancer: an overview. J Adv Pharma Technol Res 1(2):109. Retrieved from /pmc/articles/PMC3255438/

Skowronek J (2017) Current status of brachytherapy in cancer treatment – short overview. J Contemp Brachyther 9(6):581. https://doi.org/10.5114/JCB.2017.72607

Skowronek J, Chicheł A (2014) Brachytherapy in breast cancer: an effective alternative. Przeglad Menopauzalny = Menopause Rev 13(1):48–55. https://doi.org/10.5114/pm.2014.41090

Soerjomataram I, Louwman MWJ, Ribot JG, Roukema JA, Coebergh JWW (2008) An overview of prognostic factors for long-term survivors of breast cancer. Breast Cancer Res Treat 107(3):309. https://doi.org/10.1007/S10549-007-9556-1

Sullivan SD, Sarrel PM, Nelson LM (2016) Hormone replacement therapy in young women with primary ovarian insufficiency and early menopause. Fertil Steril 106(7):1588. https://doi.org/10.1016/J.FERTNSTERT.2016.09.046

Sun J, Lee MC (2020) Clinical presentation, diagnosis and prognosis of pregnancy-associated breast cancer. Adv Exp Med Biol 1252:87–93. https://doi.org/10.1007/978-3-030-41596-9_11

Sung H, Ferlay J, Siegel RL, Laversanne M, Soerjomataram I, Jemal A, Bray F (2021) Global cancer statistics 2020: GLOBOCAN estimates of incidence and mortality worldwide for 36 cancers in 185 countries. CA Cancer J Clin 71(3):209–249. https://doi.org/10.3322/CAAC.21660

Tai YC, Domchek S, Parmigiani G, Chen S (2007) Breast cancer risk among male BRCA1 and BRCA2 mutation carriers. J Natl Cancer Inst 99(23):1811–1814. https://doi.org/10.1093/JNCI/DJM203

Taylor CW, Kirby AM (2015) Cardiac side-effects from breast cancer radiotherapy. Clin Oncol (Royal College of Radiologists (Great Britain)) 27(11):621–629. https://doi.org/10.1016/J.CLON.2015.06.007

Thorpe CS, Niska JR, Girardo ME, Kosiorek HE, McGee LA, Hartsell WF et al (2019) Proton beam therapy reirradiation for breast cancer: multi-institutional prospective PCG registry analysis. Breast J 25(6):1160–1170. https://doi.org/10.1111/TBJ.13423

Tsimberidou AM (2015) Targeted therapy in cancer. Cancer Chemother Pharmacol 76(6):1113–1132. https://doi.org/10.1007/S00280-015-2861-1

Waddle MR, Chen RC, Arastu NH, Green RL, Jackson M, Qaqish BF et al (2015) Unanticipated hospital admissions during or soon after radiation therapy: incidence and predictive factors. Pract Radiat Oncol 5(3):e245–e253. https://doi.org/10.1016/J.PRRO.2014.08.004

Wang J, Trovati S, Borchard PM, Loo BW, Maxim PG, Fahrig R (2017) Thermal limits on MV x-ray production by bremsstrahlung targets in the context of novel linear accelerators. Med Phys 44(12):6610–6620. https://doi.org/10.1002/MP.12615

Warren LEG, Miller CL, Horick N, Skolny MN, Jammallo LS, Sadek BT et al (2014) The impact of radiation therapy on the risk of lymphedema after treatment for breast cancer: a prospective cohort study. Int J Radiat Oncol Biol Phys 88(3):565–571. https://doi.org/10.1016/J.IJROBP.2013.11.232

White RR, Halperin TJ, Olson JA, Soo MS, Bentley RC, Seigler HF (2001) Impact of core-needle breast biopsy on the surgical management of mammographic abnormalities. Ann Surg 233(6):769. https://doi.org/10.1097/00000658-200106000-00006

Wu M, Burstein DE (2004) Fine needle aspiration. Cancer Investig 22(4):620–628. https://doi.org/10.1081/CNV-200027160

Yersal O, Barutca S (2014) Biological subtypes of breast cancer: prognostic and therapeutic implications. World J Clin Oncol 5(3):412. https://doi.org/10.5306/WJCO.V5.I3.412

# Percutaneous Breast Cancer Treatment

Seda Aladag Kurt and Inci Kizildag Yirgin

**Abstract**

The minimally invasive approach remains popular in the treatment of breast cancer. Breast surgery has evolved from radical mastectomy to breast-conserving surgery. Axillary surgery adopts the same goal as sentinel lymph node biopsy. Nowadays, surgery is replaced by ablative treatments in selected cases and for specific indications. Ablation techniques are based on the principle of creating acute cellular necrosis in tissue using thermal or electrical energy under local or general anesthesia. The most popular of these are cryoablation and radiofrequency ablation. Cryoablation is the most preferred method among ablative therapies because of causing less pain and having better cosmetic results. In the literature, data on the short- and medium-term results of ablative treatments is increasing daily. Success rates close to surgery have been reported in these minimally invasive methods applied percutaneously. However, long-term results are needed with larger multicenter randomized controlled studies.

**Keywords**

Ablative therapy · Breast cancer · Cryoablation · High-intensity focused ultrasound · Laser ablation · Microwave ablation · Radiofrequency ablation

S. Aladag Kurt
Cerrahpasa Faculty of Medicine, Department of Radiology, Istanbul University-Cerrahpasa, Istanbul, Türkiye

I. Kizildag Yirgin (✉)
Department of Radiology, The Institute of Oncology, Istanbul University, Istanbul, Türkiye
e-mail: inci.kizildagyirgin@istanbul.edu.tr

 
Interdisciplinary Cancer Research, https://doi.org/10.1007/16833_2024_204
Published online: 8 March 2024

# 1 Introduction

Breast cancer can be detected at an earlier stage by means of the increasingly effective use of screening programs and the developing imaging technology (Kaufman 2005). As a result, survival rates are increasing (Kaufman 2005).

The gold standard in breast cancer treatment is surgical removal, and treatment methods continue to evolve. Its history dates back to 3,000 BC, and according to Egyptian tablets, the first known breast cancer in history was treated by cauterization with fire. With the development of the science of anesthesia, radical mastectomy performed by William Halsted in 1884 was the first breast surgery described (Halsted 1907). The first attempts at non-surgical methods date back to the beginning of the twentieth century. Although chemotherapy gained momentum in the mid-twentieth century, it was also brought to the fore that radiotherapy could be an alternative to radical mastectomy. Toward the end of the twentieth century, the treatment trend with lumpectomy was adopted, and immunotherapy-based methods became popular in the last 10 years.

In parallel with modern medicine, breast cancer treatment, which started with radical mastectomy, has evolved to less invasive approaches over the years (Roknsharifi et al. 2021). In addition to breast-conserving surgery, axillary dissection, which caused high morbidity toward the 2000s, was replaced by sentinel lymph node biopsy in selected cases. The approach that has become popular with the aim of a minimally invasive approach is ablation therapy. Today, it is used as an alternative to surgery in many different systems, such as abdominal solid organ tumors, bone tumors, prostate cancer, and neurogenic tumors.

Any new treatment modality should adhere to three criteria, as per the American College of Surgeons: it should demonstrate its safety and efficacy; it should achieve the same or better effectiveness than pre-existing therapies, with fewer adverse effects; and it should be cost-effective (ACS 1995). Studies in the literature over time support the efficacy of ablative treatments on breast tumors (Beji et al. 2018; Pfleiderer et al. 2002; Simmons et al. 2016; Fine et al. 2021).

Indications for ablative treatments continue to develop and change with each passing year. The literature shows results regarding its use in benign and malignant breast lesions and its mid-term results over time.

Ablation is based on the direct destruction of the tumor by chemical or thermal effects. Ablative treatments are minimally invasive, require less anesthesia, are easily repeatable, allow outpatient treatment, and have low morbidity, short recovery time, and high cosmetic satisfaction (Roknsharifi et al. 2021). However, like every interventional method, there are major and minor complications. The most frequently observed complications are local minor complications such as skin burn or skin necrosis, hematoma, infection, seroma, and nipple retraction. The most frequently reported major complications are pectoral muscle necrosis or pneumothorax (Roknsharifi et al. 2021). One of the most critical risks that may be encountered is inadequate ablation (Roknsharifi et al. 2021).

Nowadays, ablative treatments are used more frequently in benign lesions of the breast. The most commonly used ablative techniques today are cryoablation and radiofrequency ablation (RFA). Other methods defined in the treatment of benign and malignant breast tumors are microwave ablation, high-intensity focused ultrasound (HIFU), laser ablation, vacuum-assisted excision (VAE), and BLES (breast lesion excision system). Cryoablation attracts more attention than other percutaneous methods because of being less painful (Manenti et al. 2013).

In malignant breast lesions, the indication must be determined correctly. Ablation treatment may be ideal, especially in single-focus luminal A tumors smaller than 2 cm, have no accompanying obvious in situ component (<25%), and do not have axillary involvement. As in other tumor groups, the leading indication group for ablative treatment in breast cancer is patients with advanced age, various comorbidity factors, and patients who cannot tolerate the surgical approach. The patient's preference for non-surgical treatment is another indication. The only absolute contraindication to ablative therapy is incurable coagulopathy. Relative contraindications for breast cancer can be listed as in situ ductal carcinoma, invasive lobular carcinoma, multifocal or multicentric tumors, and tumors closer than 1 cm to the skin or chest wall.

Methods such as immunohistochemical staining, vacuum-assisted biopsy, and core biopsy have been used for complete ablation and tumor margin evaluation in various studies (Nguyen et al. 2014). Although surgical excision and immunohistochemistry can confirm the adequate treatment, core biopsy alone may be misleading. In this process, where we prioritize the minimally invasive approach, radiological evaluation with contrast-enhanced magnetic resonance imaging (MRI) is the most frequently used and proven reliable method (Nguyen et al. 2014).

A detailed article was published by Ahmed et al. in 2014 for reporting and terminological standardization in ablative therapies (Ahmed et al. 2014). Technical details regarding the procedure, radiological terms, and follow-up recommendations are discussed in detail. Accordingly, the detection of tumor enhancement at the first control after the procedure is defined as "residual unablated tumor", the detection of tumor foci in the ablation zone after at least one normal control is defined as "local tumor progression", and the new tumor foci that emerge after local eradication is achieved is defined as "local tumor recurrence" (Ahmed et al. 2014).

No standard algorithm exists for the follow-up interval and imaging method after ablative treatment. In the literature, imaging every 6 months is recommended for follow-up for the first 2 years, and the most appropriate modality is dynamic breast MRI with high sensitivity. Some centers perform early imaging (6 weeks after ablation) with PET-CT or PET-MRI (Kwong et al. 2023). With breast MRI, the ablation cavity and the typical rim-shaped circular enhancement surrounding the cavity can be easily observed. Thickening, nodularity in this annular enhancement, or enhancement within or adjacent to the cavity are considered suspicious regarding residue or recurrence.

The ablation methods used in malignant lesions of breast are briefly mentioned below.

## 2 Cryoablation

Cryoablation, also known as cryotherapy, cryosurgery, or cryo destruction, was first applied to superficially located cancers like breast, cervical, and skin in the mid-nineteenth century (Arnott 1850).

The direct mechanism of action of cryoablation is cold-induced cellular damage. Its indirect effect is in the form of tissue ischemia by disrupting the microvascular structure. The ultimate response in the tissue is coagulation necrosis. Cryoablation causes a general tumor-specific immune response in the body with the release of tumor-specific antigens, and it is thought that it may reduce secondary systemic micrometastases.

In the cryoablation procedure, the tissue is frozen suddenly to prevent the proliferation of the tumoral cells, followed by active or passive thawing processes. Then, this cycle is repeated for the second time. The cell membrane ruptures due to the effect of ice crystals formed during the first rapid freezing process. The inflammatory reaction and endothelial damage that develops at the cellular level due to cytotoxic oxygen radicals and nitric oxide emergent in the tissue cause hypoxia. The second cycle allows the area of necrosis to expand. Necrosis and irreversible damage caused by repeated periods of freezing and thawing cause tumor destruction. Months after ablation, necrosis is replaced by fibrous collagen bands (Weber and Lee 2005).

The cooling agents are liquid nitrogen or argon used in cryoprobes placed inside the mass to freeze the tumor. While 10–12 Gauge (G) needles are used in devices operating with liquid nitrogen, argon gas systems have needles of 12–17 G thickness and varying shapes. Helium gas is used during the thawing period, which creates destruction. The procedure can be performed with multiple needles depending on the lesion size. For irreversible damage, the temperature must be between −20 and −40 °C. Tissue damage is related to the rate and duration of freezing and the speed of the thawing process.

An ice ball is formed around the cryoprobes during the process, forming thermal isotherms (Fig. 1). The formation of this ice ball can be simultaneously followed by imaging methods. Ultrasonography (US) is the most commonly used method for this purpose. The expanding hyperechoic ice ball is sharply circumscribed and produces posterior acoustic shadowing. The shape and size of the ice ball varies depending on the cryoprobe used. Generally, every 2 mm probe creates a necrosis area of approximately 2 cm in diameter. Depending on the lesion diameter, it may be necessary to use more than one cryoprobe (Fig. 2). The depth at which the cryoprobe is placed is significant for a safe ablation margin. The tip of the cryoprobe should be placed at the deepest border of the lesion. For a safe ablation margin, the ice ball formed is expected to cover an area at least 1 cm larger than all tumor edges. Because ablation is performed under US guidance, localization of the needle to the desired area and ice ball formation are visualized simultaneously. Procedure time varies depending on the nature and size of the lesion, the number of probes, and the device used. After the second melting, the remaining ice ball passively melts and disappears within 20–30 min. Procedure time is shorter in fibroadenomas. For a fibroadenoma with a

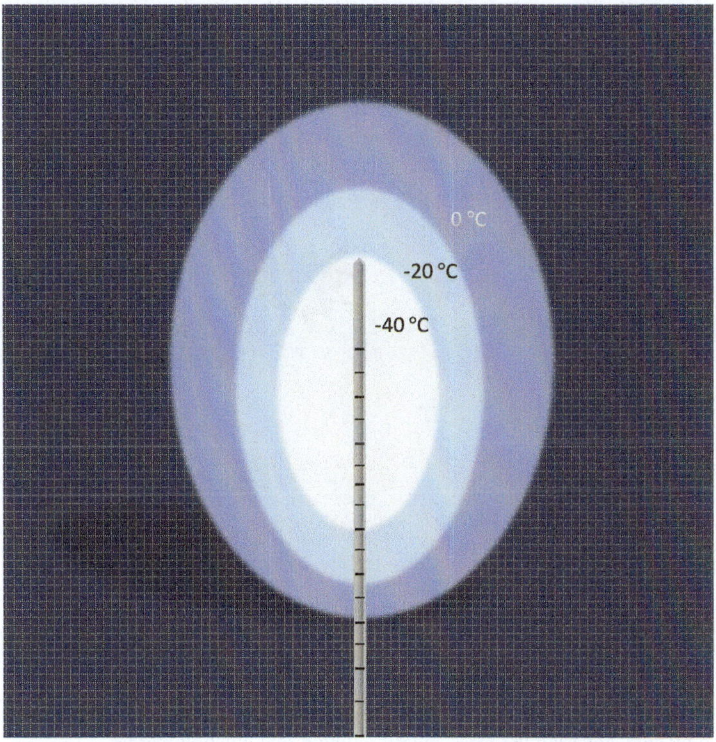


**Fig. 1** Ice ball formation with thermal isotherms formed around the cryoprobe

diameter of 1 cm, a freezing time of 2 min is sufficient for each cycle, while for an invasive carcinoma of the same size, this period is approximately 6 min.

Indications and recommendations for cryoablation are defined in the current guideline published by the American Society of Breast Surgeons (ASBrS) (ASBrS 2018). In this guideline, it is reported that the effectiveness and reliability of cryoablation in fibroadenomas are similar to surgery. It can be used in fibroadenomas with a widest diameter of less than 4 cm (ASBrS 2018). Cryoablation may be an alternative treatment method in cases with a previous history of fibroadenoma excision and those who do not want surgical treatment, especially in patients with multiple fibroadenomas. It is ineligible for use in ductal carcinoma in situ (DCIS), invasive lobular carcinoma (ILC), or invasive ductal carcinomas accompanied by an extended ductal component, as the size of the area observed with US may be misleading (Ward et al. 2019). Estrogen receptor-positive and human epidermal growth factor receptor 2 (HER2)-negative tumors are more proper to cryotherapy because they tend to be less aggressive, unifocal, and unilateral. The diameter, location of the mass, and breast size are factors that will affect the treatment decision. The distance of the mass from the skin must be at least 5 mm. Ideally, it should be at least 1 cm away (Ward et al. 2020; Littrup et al. 2009). This distance is the required

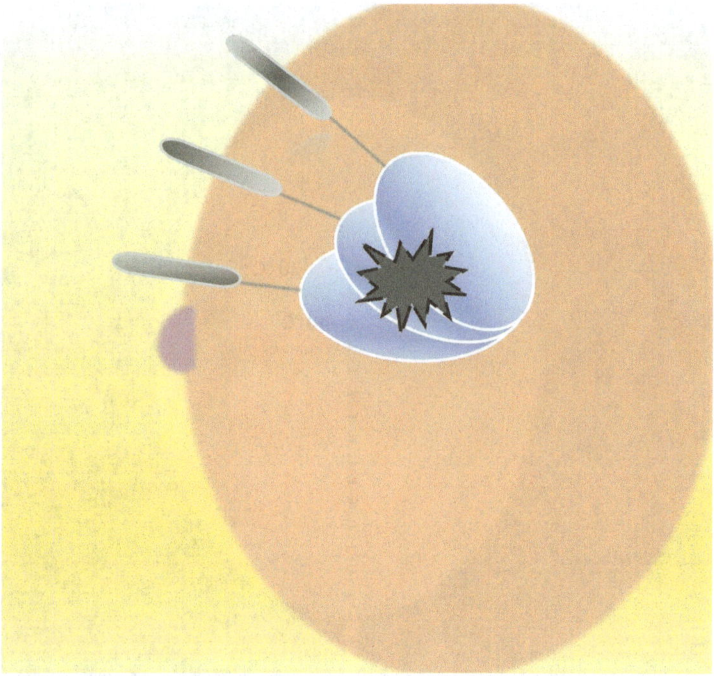

**Fig. 2** The ablation zone formed by the coalescing of ice balls formed around the cryoprobes placed side by side, covering the mass and providing a safe ablation margin

safety threshold to prevent the resulting ice ball from damaging the skin. In lesions close to the skin or chest wall, hydro-dissection is performed by injecting sterile saline between the mass and the tissue to prevent damage to the surrounding tissue (Littrup et al. 2009). Cryoablation, performed under local anesthesia, can be completed in less than 45 min. Its advantage over other ablative treatment methods is that it freezes the tissue, causing less pain during and after the procedure. Disadvantages are technical reasons such as placing more than one cryoprobe depending on the tumor size to create a sufficiently large ice ball, the cost of the argon gas used, and the difficulty of storage. Complications that may occur include skin burns, bleeding, infection, skin or chest wall damage, and inadequate treatment.

One of the most important clinical studies investigating the effectiveness of cryoablation in early-stage breast cancer is the ACOSOG (Alliance) Z1072 phase II study published in 2016 (Simmons et al. 2016). In this multicenter study conducted with 87 cases diagnosed with invasive ductal carcinoma, lumpectomy specimens were evaluated within 28 days after ablation. When multifocal foci outside the lesion are excluded, the success rate for lesions smaller than 1 cm is 100%, while complete ablation has been reported in 92% (Simmons et al. 2016). In a prospective study by Poplack et al. (2015), which included 20 patients, a 15% rate of residual disease was reported.

The FROST (Freezing Instead of Removal of Small Tumors) Trial, which investigated the effectiveness of cryoablation on early-stage breast cancer, was completed in 2021 (FROST 2019). It was studied on patients aged 50 and over, ≤ 1.5 cm in maximum tumor size, clinically node-negative, hormone receptor-positive with <25% intraductal component. Final data is expected to be announced in December 2023. ICE3 Trial, expected to be completed in 2024, was recorded as 96.91% of patients recurrence-free, according to the interim analysis reported in 2022 (Ice3 Trial 2019; Fine et al. 2021).

The first control after cryoablation is ultrasonography, performed immediately after the procedure. The ablation zone is observed as an echogenic area. In the following months, this area shrinks, and the findings evolve into fat necrosis (Ward et al. 2019). This fat necrosis can be observed with mammography (MG) within 1–2 months. However, the most effective method for follow-up is MRI. No contrast enhancement in the ablation zone means complete ablation success (Machida et al. 2019). After ablation of malignancy, control is recommended at 6-month intervals within the first 1 or 2 years. Thick nodular enhancement in the ablation zone is considered suspicious and should be sampled. There is no indication for an MRI after fibroadenoma ablation.

## 3 Radiofrequency Ablation

Radiofrequency (RF) waves were first used by Beer in the treatment of bladder cancer in 1910 (Beer 1910). Radiofrequency ablation (RFA), a minimally invasive technique, has found wide use over the years, especially in cases where surgical treatment is not possible or as an alternative method to surgery in organ neoplasms such as liver, kidney, lung, and thyroid.

In radiofrequency ablation, an alternating electric field is created in the tissue by transmitting low-voltage, high-frequency alternating current from uninsulated electrode tips to the target tissue. Significant ion agitation occurs around the electrode since there is more resistance in the tissue than in the electrode. The movements of the ions cause friction, and friction causes heat to increase. The heat concentrates around the needle electrode. Exposure to high heat over a period of time initiates a chain of events that will result in coagulation necrosis in the tissue. This process begins with cell membrane damage, followed by protein denaturation and cell destruction (Burak et al. 2003; Sinletary 2003). Cell death occurs at temperatures above 60 °C. Therapeutic temperatures are maintained throughout the ablation without exceeding 100 °C to prevent tissue evaporation.

Technically, it is a closed circuit system that creates alternating current, including an RF generator, active electrode, and a grounding auxiliary return electrode. The RF needle to be used in the process consists of thin electrodes placed close to each other, opened as a bundle, and connected to the generator. Opening the electrodes in this way provides a larger ablation volume.

Since RFA is a slightly more painful procedure than cryoablation, it is performed under sedation or local/general anesthesia in most centers. Radiofrequency ablation of breast lesions can be performed under US or MRI guidance. To prevent complications such as burns and necrosis, the tumor should be at least 1 cm away from the skin or chest wall. If necessary, hydro dissection can be performed before starting ablation. Correct placement of the RF electrode is critical for ablation success (Singletary 2003). The electrode is placed in the central center of the target lesion with the guidance of imaging methods. Air bubbles formed by ablation can be monitored with US during the procedure. Tissue temperature is monitored with a simultaneously placed fiberoptic probe. It is aimed to ensure a safe tumor margin by destroying the 1 cm thick healthy tissue surrounding the lesion by ablation. For this purpose, applying more than one ablation session may be necessary. The procedure is terminated by performing tract ablation to prevent potential tumor transplantation and hemorrhage (Rhim et al. 2001). Procedure time varies between 5 and 20 min depending on the number of ablations. After the procedure, the patient is monitored for a few hours and can be discharged on the same day if no complications develop.

In addition to the general benefits of ablative treatments, additional advantages include short ablation time compared to cryotherapy and not requiring gas tanks. The most significant disadvantage is that the boundaries of the ablation zone cannot be monitored during the procedure. Another issue is that in tumors with high vascularization, the blood flow passing through the ablation area absorbs heat and has a cooling effect named the heat sink effect. This situation may reduce the effectiveness of RFA.

The hard, painless swelling that develops immediately after the ablation may grow in the early period but subsides within 2–6 months. As with cryotherapy, the most common complications are bleeding, ecchymosis, skin burn, and fat necrosis. Also, breast abscess, nipple retraction, and fistula formation may rarely be observed (Izzo et al. 2001; Palussière et al. 2012).

One of the factors determining the indication for ablation for breast tumors is size. RFA is not recommended for tumors larger than 5 cm due to the risk of residual disease (Leylek et al. 2013; Fornage et al. 2004). Apart from this, RFA is not an appropriate option in the presence of large ILC, >2 cm DCIS, multifocal or multicentric breast cancer. Additionally, it is recommended that the lesion be at least 1 cm away from the skin or pectoral muscle planes to avoid burns and tissue damage (Leylek et al. 2013).

In most of the studies in the literature, surgical excision was performed after RFA, and the effectiveness of the procedure was investigated. Accordingly, complete treatment response rates of up to 80–100% following ablation and satisfactory cosmetic results compared to lumpectomy have been reported (Leylek et al. 2013). In another large series study including 386 patients published in 2018, the 5-year recurrence-free survival rates in the ipsilateral breast were reported as 97%, 94%, and 87%, respectively, according to tumor diameter of 1 cm, 1–2 cm, and > 2 cm (Ito et al. 2018).

Although MRI is considered the most successful method for follow-up after ablation, secondary changes in procedure localization are also observed with conventional imaging methods such as US and MG (Fig. 3). Like in other ablative methods, MRI correlates well with histopathological results in RFA and is preferred

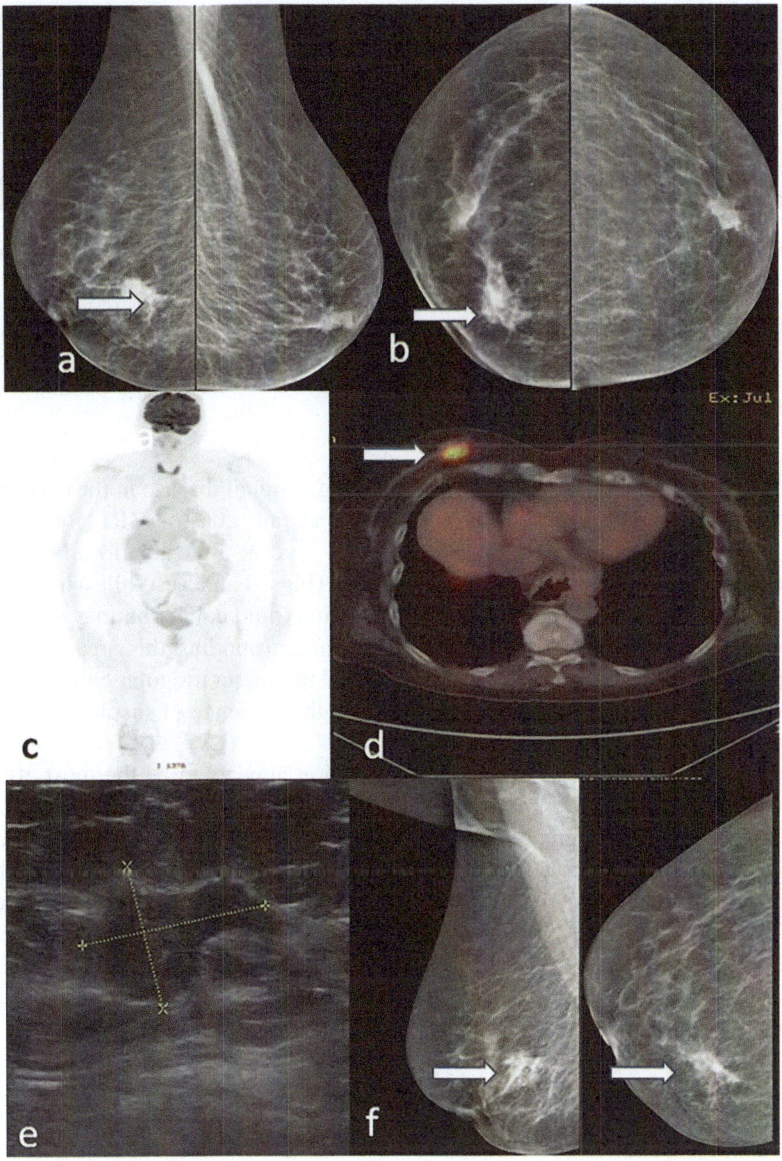


**Fig. 3** An 84-year-old female patient presented with a palpable mass in her right breast. (**a** and **b**) In MLO and CC views, a lesion with a diameter of 28 mm, highly suspicious for malignancy, is observed in the inner quadrant of the right breast (white arrows). As a result of the tru-cut biopsy, the diagnosis of ER (+), PR (+), HER2 (−), Grade-2, and invasive ductal carcinoma was confirmed. (**c** and **d**) Axillary and distant metastases were not detected in PET-CT examinations performed due to systemic evaluation. FDG uptake was detected at the primary lesion level (white arrow) (**e**) The patient was not found suitable for the operation due to her advanced age and comorbidities. First, hormonotherapy treatment was started, and the lesion diameter was measured as 18 mm at the ultrasonography examination performed 3 months later. (**f**) Radiofrequency ablation treatment was performed, and early changes and regression due to treatment were observed in the mammographic examination obtained in the first month after the treatment. The patient will continue to be followed up

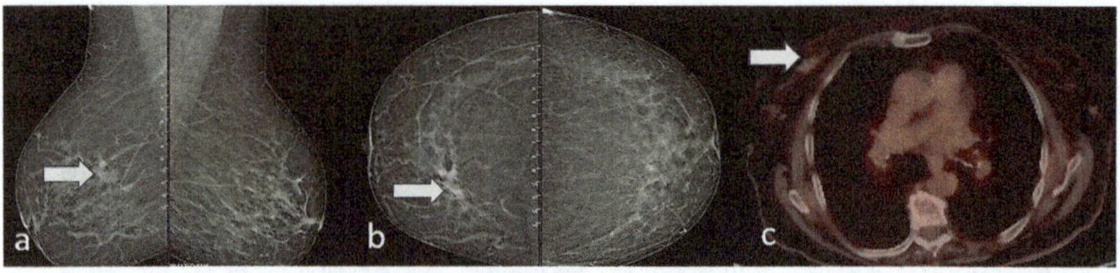


**Fig. 4** A 91-year-old female patient presented with a palpable mass in her right breast. (**a** and **b**) In MLO and CC views, a lesion with a diameter of 28 mm, highly suspicious for malignancy, is observed in the upper inner quadrant of the right breast (white arrows). As a result of the tru-cut biopsy, the diagnosis of ER (+), PR (+), HER2 (−), Grade-2, and invasive lobular carcinoma was confirmed. (**c**) Axillary and distant metastases were not detected in PET-CT examinations performed due to systemic evaluation. Low FDG uptake was detected at the primary lesion level (white arrow)

in monitoring response to treatment and imaging residual disease in the early period after ablation (Vilar et al. 2012; Manenti et al. 2009). MRI follow-up is recommended every 6 months during the first 2 years after ablation. No residual enhancement is considered as complete ablation (Figs. 4 and 5). While a hypointense ablation zone is observed in the procedure location, fluid collection may accompany it. The appearance of uniform ring enhancement surrounding this area may persist for up to 12 months. In the presence of a residual tumor, an irregular enhancement or mass image is observed at the ablation edge. In the presence of suspicious findings, sampling should be performed.

Performing the surgical resection after ablation appears to be the most effective method to achieve local control. However, in the meta-analysis, Xia et al. (2021) found no significant difference in local recurrence between groups with and without surgical resection after ablation. According to the same meta-analysis, RF ablation supported by postoperative RT and systemic adjuvant therapy is considered to provide good local control and is a reliable and effective method for breast tumors smaller than 2 cm (Xia et al. 2021). Despite all this, accurate determination of the indication for RFA is critical due to challenges, such as reported poor cosmetic results and the restriction of the use of radiotherapy after RFA (Brem 2018).

## 4 Microwave Ablation

Microwave ablation (MWA) is a minimally invasive thermal therapy method that can be an alternative to surgery in selected cases. This method destroys tissue by causing coagulation necrosis, similar to RFA. Compared to worldwide RFA technology, MWA has many advantages, such as creating higher intratumoral temperature, larger ablation volume, shorter ablation time, improved heat distribution, less pain, and optimal heating in cystic masses (Martin et al. 2007).

In MWA, the electromagnetic field created in the tissue by high-frequency microwave energy causes water molecules to oscillate at high frequencies. The

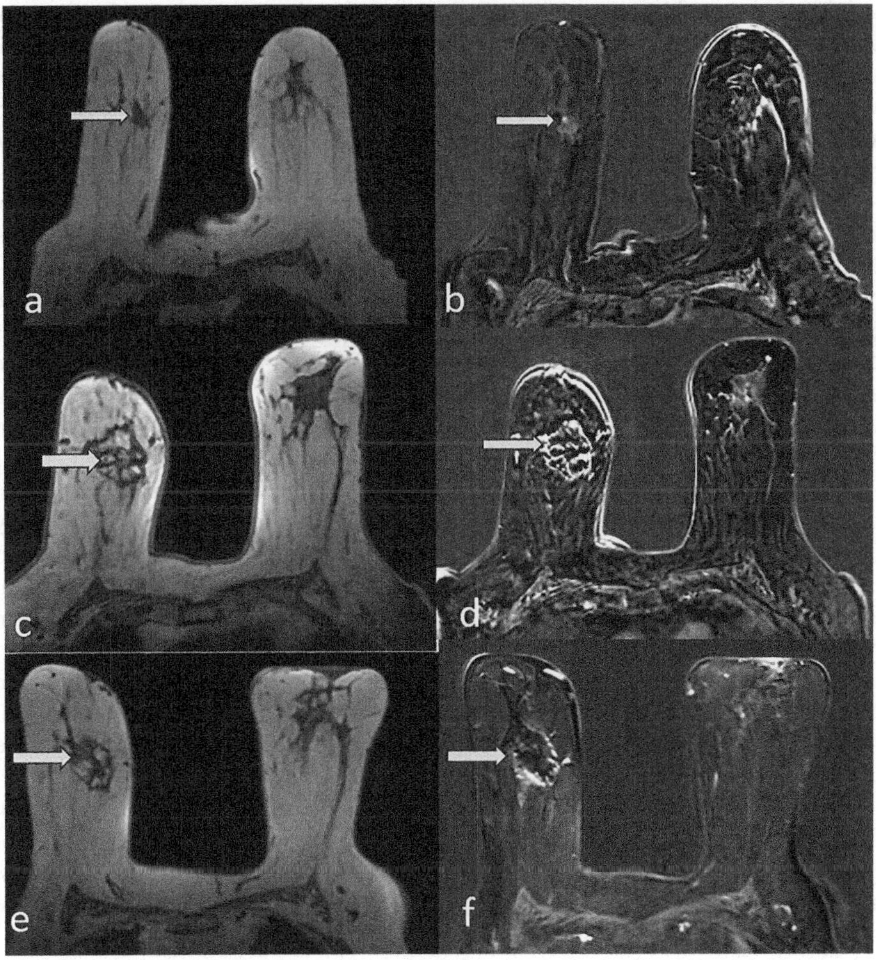


**Fig. 5** Figure 4 continued MRI examinations of the same patient. (**a**) T1W pre-contrast, non-fat-sat image and (**b**) T1W post-contrast fat-sat image showing the primary lesion in the upper inner quadrant of the right breast before radiofrequency ablation (white arrows). (**c**) T1W pre-contrast, non-fat-sat image and (**d**) T1W post-contrast fat-sat image showing the fat necrosis in the treated area at the third month control after RFA (white arrows). (**d**) T1W pre-contrast, non-fat-sat image and (**e**) T1W post-contrast fat-sat image show the fat necrosis in the treated area, with no progression findings at the sixth-month control after RFA (white arrows). The patient will continue to be followed up

resulting kinetic energy turns into heat due to friction and increases the temperature of the tissue. The resulting temperature increase causes coagulation necrosis, similar to the mechanism described in RFA (Simon et al. 2005; Sindram et al. 2010). On the other hand, RF requires electrical and thermal conductivity, while microwaves can also propagate through tissues with zero or low conductivity. While RF current cannot be transmitted in tissue above 100 °C, microwaves can be transmitted at any temperature and with any water density. A larger and deeper ablation volume can be achieved with MWA than RFA (Feng and Liang 2012). In addition to providing all the benefits of RFA, MWA is more effective, especially in lesions adjacent to vessels, as it can reach higher temperatures in a shorter time (Brace 2009). However, since it is a more painful procedure than RFA and cryoablation, it may need to be performed under general anesthesia in some cases.

After anesthesia, the procedure starts by placing a microwave antenna of favorable thickness and length along the long axis of the lesion under US guidance. Hydro-dissection can be performed for lesions close to the skin or chest wall if necessary. Proper power and frequency output are determined according to the lesion's size, structure, and surrounding tissue's characteristics. The process is completed within minutes since lethal temperatures are reached in microwave ablation in a very short time. Fiberoptic probes are used to detect temperature in tissue. Besides that, cooling systems can be used to reduce the risk of skin burns. For large lesions, multiple entries and multiple ablation zones can be performed.

Although MWA has been reported to be used in both benign and malignant breast lesions, knowledge of its use in the literature is still limited. In 2012, a study including 41 patients who underwent completion mastectomy after ablation reported that 90% complete tumor necrosis was achieved with microwave ablation (Zhou et al. 2012). However, there is not enough literature data regarding the long-term results of the method.

Although there are centers that use US for follow-up after MWA in benign lesions, the best modality that can be used to evaluate the ablation zone is MRI. Irregular nodular enhancement in the ablation zone or its periphery indicates residue in both benign and malignant lesions.

In addition to benign lesions of the breast, it may be a promising treatment option that can be combined with adjuvant endocrine therapy, especially in cases of hormone receptor-positive, HER2-negative, advanced age, and early-stage breast cancer (Yu et al. 2021). This statement needs to be supported by prospective multicenter studies in the future.

## 5 High-Intensity Focused Ultrasound (HIFU)

High-intensity focused ultrasound (HIFU) is a non-invasive procedure that does not require probe insertion and can be used in the treatment of benign and malignant lesions of the breast. Focused on a small tumor via a transducer, US waves have the potential to propagate through tissue layers toward the target area without causing direct damage to adjacent soft tissues (Peek and Wu 2018). Low-frequency and high-energy sound waves between the frequencies of 0.8–3.5 MHz are used for ablation (Brenin 2011).

HIFU causes tissue ablation through hyperthermia, acoustic cavitation, and immune system modulation. With this energy, which can be applied from outside the body, necrosis is provided in the desired area due to temperature increase. As ultrasonographic sound waves propagate through tissue layers, some energy is lost in rapidly spreading heat. When US waves carrying sufficient high energy are applied to the tumor focus on the targeted depth, the energy gathered in this area causes coagulation necrosis with mechanical stress and temperature increase in the tissue. Additionally, HIFU is thought to cause tumor antigen exposure in dendritic cells, strengthening the immune mechanism against tumor cells (van den Bijgaart et al. 2022).

HIFU can be applied under US and MRI guidance. US guidance is cheaper and more accessible and provides excellent positioning. However, the increase in echogenicity in the targeted area due to the procedure limits the visualization of the lesion and the surrounding tissue. MRI guidance allows both simultaneous monitoring and following the temperature. The most crucial disadvantage of the procedure is that it takes 45–180 min. Accordingly, sedation or general anesthesia is required to reduce patient movement.

Although specific indication criteria for HIFU have not been defined, it has been applied to various breast malignancies such as IDC, ILC, and DCIS (Peek et al. 2015). The tumor must be clearly visible on US or MRI. It is recommended that the target lesion be at least 1 cm away from the skin and chest wall and at least 2 cm away from the nipple (Brenin 2011). It is not applied in implant cases due to the risk of thermal damage or burns (Brenin 2011).

The procedure begins after positioning the patient and following local anesthesia or sedation. The target lesion volume, borders, and surrounding organs are evaluated, and the number of sonications to cover sufficient ablation area is planned. A sonication causes temperature increase and coagulation in a cylindrical area measuring 1–3 × 10–15 mm. For large lesions, sonications arranged side by side are used. Cooling periods are allowed between sonications to prevent heat accumulation in adjacent tissues.

Ablation by HIFU is used to reduce the volume of fibroadenomas and the associated pain, anxiety, and palpable mass effect (Kovatcheva et al. 2015). A limited number of studies in the literature investigating the effectiveness of HIFU on breast cancer were designed to include mastectomy after ablation, and complete ablation rates vary (Peek et al. 2015; Furusawa et al. 2006; Merckel et al. 2016). In addition, Chen et al.'s study with mice reported that HIFU applied during the chemotherapy process, in addition to paclitaxel, increased the distribution and effectiveness of the drug, which is quite remarkable (Chen et al. 2022).

MRI, as in other methods, is the modality that most effectively demonstrates the presence of a residual tumor. Rapid enhancement in the early phase and peripheral nodular or irregular enhancement are significant for residual disease (Peek et al. 2015). In the follow-up of fibroadenomas, US can be used alone to evaluate treatment response 6, 12, and 24 months after HIFU (Hahn et al. 2018).

Since HIFU provides less anesthesia requirement, lower risk of infection, minimal pain, and better cosmesis, it can be recommended as an alternative treatment method in selected cases who do not want to undergo surgery (Brenin 2011). Its most important advantages are that it is non-invasive and non-ionizing, as well as being reproducible and having a lower risk of complications compared to other methods.

## 6 Laser Ablation

Laser ablation is a minimally invasive method that causes irreversible tissue damage with minimal side effects. The method is also terminologically called laser interstitial tumor therapy (LITT), laser coagulation therapy, and laser interstitial photocoagulation (Ahmed et al. 2014). Laser ablation aims at thermal destruction by electively creating a sudden increase in temperature in abnormal tissue while preserving normal tissue. It can be performed under local anesthesia or conscious sedation under US and MRI guidance. Its advantage over other ablative methods is that the procedure can be performed in a short time, approximately 15 min. A laser fiber containing a thermal sensor at its tip and passing electromagnetic radiation through it is placed on the target tumor. The thermal probe parallel to the laser probe monitors the tissue temperature. An infusion pump that continuously infuses saline is added to the laser probe to prevent tissue overheating. In this method, light energy transmitted through the fiberoptic probe causes a rapid increase of heat and evaporation in the tissue. As a result, direct and indirect tissue damage occurs. Although it is recommended that the lesion size be less than 2 cm and be more than 1 cm away from the skin or chest wall, the use of the method in breast cancer is limited to experimental clinical studies (Schwartzberg et al. 2018). Although the clinical trial LASERBREAST1, which started in 2012 and included 16 patients, was last updated in 2018, no relevant data has been published (LASERBREAST1 2018).

## 7 Conclusion

The current status of ablative treatments is controversial but promising. Although ablative treatments are less invasive methods compared to surgery, they are still not widely used in breast cancer treatment for various reasons. The surgically removed tissue allows pathological evaluation, and the most appropriate oncological treatment and follow-up decision can be made for the patient. Axillary intervention can be performed as a sentinel lymph node biopsy or dissection in the same session as breast surgery. This conventional approach, and especially the condition of the axilla, limits the use of local ablative treatments. In addition, the patient's desire to be saved from the mass in her breast and the option of cosmetic surgery along with cancer surgery may cause the patient to turn to surgery instead of ablative treatment. Although the literature has proven that ablative treatments are almost as effective as surgery, reasons such as the risk of leaving residual tumor behind with ablation or the possibility of fibrosis and fat necrosis after ablation may interfere with the residual tumor and reduce the effectiveness of imaging. Therefore, medium- and long-term follow-up results are needed.

Immunotherapy studies combined with ablative therapies on stage 4 breast cancer, where surgical treatment is not possible, are promising. In the coming years, the breast cancer treatment algorithm will continue to change and develop as minimally invasive approaches such as targeted endocrine therapies, immunotherapy, and ablative treatments become widespread.

## References

Ahmed M, Solbiati L, Brace CL et al (2014) International working group on image-guided tumor ablation; Interventional oncology sans frontières expert panel; Technology assessment committee of the society of interventional radiology; Standard of practice committee of the cardiovascular and interventional radiological society of Europe. Image-guided tumor ablation: standardization of terminology and reporting criteria – a 10-year update. Radiology 273(1): 241–260

American College of Surgeons (1995) Statement on issues to be considered before new surgical technology is applied to the care of patients. Committee on emerging surgical technology and education. Bull Am Coll Surg 80:46–47

Arnott J (1850) Practical illustrations of the remedial efficacy of a very low or anesthetic temperature in cancer. Lancet 2:257–259

Beer E (1910) Removal of neoplasms of the urinary bladder: a new method employing high frequency (oudin) currents through a cauterizing cystoscope. JAMA 54:1768–1769

Beji H, Pilleul F, Picard R et al (2018) Percutaneous cryoablation of breast tumours in patients with stable metastatic breast cancer: safety, feasibility and efficacy. Br J Radiol 91(1083):20170500

Brace CL (2009) Radiofrequency and microwave ablation of the liver, lung, kidney and bone: what are the differences. Curr Probl Diagn Radiol 38(3):135–143

Brem RF (2018) Radiofrequency ablation of breast cancer: a step forward. Radiology 289:325–326

Brenin DR (2011) Focused ultrasound ablation for the treatment of breast cancer. Ann Surg Oncol 18(11):3088–3094

Burak WE Jr, Agnese DM, Povoski SP et al (2003) Radiofrequency ablation of invasive breast carcinoma followed by delayed surgical excision. Cancer 98(7):1369–1376

Chen S, Bian H, Duan J (2022) High-intensity focused ultrasound enhanced anti-tumor activities of paclitaxel in breast cancer in vitro and in vivo. Cancer Manag Res 14:1303–1312

Cryoablation of Low Risk Small Breast Cancer- Ice3 Trial (2019). Available from: https://clinicaltrials.gov/ct2/show/NCT02200705. Accessed 20 July 2019

Cryoablation of Small Breast Tumors in Early Stage Breast Cancer (FROST) (2019). Available from: https://clinicaltrials.gov/ct2/show/NCT01992250. Accessed 20 July 2019

Feng B, Liang P (2012) Local thermal ablation of renal cell carcinoma. Eur J Radiol 81(3):437–440

Fine RE, Gilmore RC, Dietz JR et al (2021) Cryoablation without excision for low-risk early-stage breast cancer: 3-year interim analysis of ipsilateral breast tumor recurrence in the ICE3 trial. Ann Surg Oncol 28(10):5525–5534

Fornage BD, Sneige N, Ross MI et al (2004) Small (< or = 2-cm) breast cancer treated with US-guided radiofrequency ablation: feasibility study. Radiology 231(1):215–224

Furusawa H, Namba K, Thomsen S et al (2006) Magnetic resonance-guided focused ultrasound surgery of breast cancer: reliability and effectiveness. J Am Coll Surg 203(1):54–63

Hahn M, Fugunt R, Schoenfisch B et al (2018) High intensity focused ultrasound (HIFU) for the treatment of symptomatic breast fibroadenoma. Int J Hyperth 35(1):463–470

Halsted WS (1907) The results of radical operations for the cure of cancer of the breast. Trans Am Surg Assoc 25:61–79

https://www.breastsurgeons.org/docs/statements/Consensus-Guideline-on-the-Use-of-Transcutaneous-and-Percutaneous-Methods-for-the-Treatment-of-Benign-and-Malignant-Tumors-of-the-Breast.pdf

https://www.clinicaltrials.gov/study/NCT01706016

Ito T, Oura S, Nagamine S et al (2018) Radiofrequency ablation of breast cancer: a retrospective study. Clin Breast Cancer 18(4):e495–e500

Izzo F, Thomas R, Delrio P et al (2001) Radiofrequency ablation in patients with primary breast carcinoma: a pilot study in 26 patients. Cancer 92(8):2036–2044

Kaufman CS (2005) Cryoablation for breast cancer: is it time? Breast Dis Year Book Q 16(3): 227–229

Kovatcheva R, Guglielmina JN, Abehsera M, Boulanger L, Laurent N, Poncelet E (2015) Ultrasound-guided high-intensity focused ultrasound treatment of breast fibroadenoma-a multicenter experience. J Ther Ultrasound 3(1):1

Kwong A, Co M, Fukuma E (2023) Prospective clinical trial on expanding indications for cryosurgery for early breat cancers. Clin Breast Cancer 23(4):363–368

Leylek AM, Whitman GJ, Vilar VS, Kisilevzky N, Faintuch S (2013) Radiofrequency ablation for breast cancer. Tech Vasc Interv Radiol 16(4):269–276

Littrup PJ, Jallad B, Chandiwala-Mody P, D'Agostini M, Adam BA, Bouwman D (2009) Cryotherapy for breast cancer: a feasibility study without excision. J Vasc Interv Radiol 20(10): 1329–1341

Machida Y, Shimauchi A, Igarashi T, Fukuma E (2019) MRI findings after cryoablation of primary breast cancer without surgical resection. Acad Radiol 26(6):744–751

Manenti G, Bolacchi F, Perretta T et al (2009) Small breast cancers: in vivo percutaneous US-guided radiofrequency ablation with dedicated cool-tip radiofrequency system. Radiology 251(2):339–346

Manenti G, Scarano AL, Pistolese CA et al (2013) Subclinical breast cancer: minimally invasive approaches. Our experience with percutaneous radiofrequency ablation vs. Crytotherapy Breast Care (Basel) 8:356–360

Martin RCG, Scoggins CR, McMasters KM (2007) Microwave hepatic ablation: initial experience of safety and efficacy. J Surg Oncol 96(6):481–486

Merckel LG, Knuttel FM, Deckers R et al (2016) First clinical experience with a dedicated MRI-guided high-intensity focused ultrasound system for breast cancer ablation. Eur Radiol 26(11):4037–4046

Nguyen T, Hattery E, Khatri VP (2014) Radiofrequency ablation and breast cancer: a review. Gland Surg 3(2):128–135

Palussière J, Henriques C, Mauriac L et al (2012) Radiofrequency ablation as a substitute for surgery in elderly patients with nonresected breast cancer: pilot study with long-term outcomes. Radiology 264(2):597–605

Peek MCL, Wu F (2018) High-intensity focused ultrasound in the treatment of breast tumours. Ecancermedicalscience 12:794

Peek MC, Ahmed M, Napoli A et al (2015) Systematic review of high-intensity focused ultrasound ablation in the treatment of breast cancer. Br J Surg 102(8):873–882

Pfleiderer SO, Freesmeyer MG, Marx C, Kühne-Heid R, Schneider A, Kaiser WA (2002) Cryotherapy of breast cancer under ultrasound guidance: initial results and limitations. Eur Radiol 12: 3009–3014

Poplack SP, Levine GM, Henry L et al (2015) A pilot study of ultrasound-guided cryoablation of invasive ductal carcinomas up to 15 mm with MRI follow-up and subsequent surgical resection. AJR Am J Roentgenol 204(5):1100–1108

Rhim H, Goldberg SN, Dodd GD 3rd, Solbiati L, Lim HK, Tonolini M, Cho OK (2001) Essential techniques for successful radiofrequency thermal ablation of malignant hepatic tumors. Radiographics 21:S17–S35

Roknsharifi S, Wattamwar K, Fishman M, Ward RC et al (2021) Image-guided microinvasive percutaneous treatment of breast lesions: where do we stand? Radiographics 41(4):945–966

Schwartzberg B, Lewin J, Abdelatif O et al (2018) Phase 2 open-label trial investigating percutaneous laser ablation for treatment of early-stage breast cancer: MRI, pathology, and outcome correlations. Ann Surg Oncol 25(10):2958–2964

Simmons RM, Ballman KV, Cox C et al (2016) ACOSOG investigators. A phase II trial exploring the success of cryoablation therapy in the treatment of invasive breast carcinoma: results from ACOSOG (Alliance) Z1072. Ann Surg Oncol 23(8):2438–2445

Simon CJ, Dupuy DE, Mayo-Smith WW (2005) Microwave ablation: principles and applications. Radiographics 25(Suppl 1):S69–S83

Sindram D, Lau KN, Martinie JB, Iannitti DA (2010) Hepatic tumor ablation. Surg Clin N Am 90: 863–876

Singletary ES (2003) Feasibility of radiofrequency ablation for primary breast cancer. Breast Cancer 10(1):4–9

van den Bijgaart RJE, Mekers VE, Schuurmans F, Raaijmakers TK et al (2022) Mechanical high-intensity focused ultrasound creates unique tumor debris enhancing dendritic cell-induced T cell activation. Front Immunol 13:1038347

Vilar VS, Goldman SM, Ricci MD et al (2012) Analysis by MRI of residual tumor after radiofrequency ablation for early stage breast cancer. AJR Am J Roentgenol 198(3):W285–W291

Ward RC, Lourenco AP, Mainiero MB (2019) Ultrasound-guided breast cancer cryoablation. AJR Am J Roentgenol 213(3):716–722

Ward RC, Lourenco AP, Mainiero MB (2020) Implementing breast cryoablation in practice. J Breast Imaging 2(1):61–66

Weber SM, Lee FT (2005) Cryoablation: history, mechanism of action and guidance modalities. In: Tumor ablation: principles and practice, p 250–265

Xia LY, Hu QL, Xu WY (2021) Efficacy and safety of radiofrequency ablation for reast cancer smaller than 2 cm: a systematic review and meta-analysis. Front Oncol 11:651646

Yu M, Pan H, Che N et al (2021) Microwave ablation of primary breast cancer inhibits metastatic progression in model mice via activation of natural killer cells. Cell Mol Immunol 18:2153–2164

Zhou W, Zha X, Liu X et al (2012) US-guided percutaneous microwave coagulation of small breast cancers: a clinical study. Radiology 263(2):364–373

# Revolutionizing Breast Cancer Care: Cutting-Edge Breakthroughs and Future Frontiers in Precision Medicine

Muhammad Ali Qureshi, Muhammad Younas Khan, Ahmed Imran, Quratulain Maqsood, Nazim Hussain, and Shinawar Waseem Ali

**Abstract**

This chapter aims to provide current research and future perspectives on treatment strategies for breast cancer. New technologies that are being developed and used include targeted therapies and immunotherapies. Targeted therapies and immunotherapies target specific checkpoints in cell growth and metabolism. These include hormone receptor–targeted therapies, HER2-targeted therapies, cell cycle–targeted therapies, and immune checkpoint inhibitors. Oncolytic viruses and cancer vaccines are still under development. New technologies that are paving their way in the treatment of breast cancer include nanotechnology, artificial intelligence (AI), and clustered regularly interspaced short palindromic repeats (CRISPR) gene editing. Nanotechnology is proving beneficial by improving drug delivery and targeting specific tumour sites, thus providing precision control. Artificial intelligence has allowed us to make more accurate predictions based on the data provided, significantly improving decision systems by reducing false positives and false negatives. CRISPR gene editing allows altering the genome of the cancer cells or immune cells, thus preventing their proliferation. Along with these, personalized treatments and modern bioinformatics are also making their way into the treatment of this disease. Although curing this disease has seen significant breakthroughs in the last 30 years and the areas discussed are still in their infancy, further research is necessary to reduce the burden of this malignancy in the female population, especially triple-negative type of breast cancer.

M. A. Qureshi · M. Y. Khan · A. Imran · S. W. Ali
Faculty of Agricultural Sciences, Department of Food Science, University of the Punjab, Lahore, Pakistan

Q. Maqsood (✉) · N. Hussain
Centre for Applied Molecular Biology, University of the Punjab, Lahore, Pakistan
e-mail: qmabc160403473@gmail.com; nazim.camb@pu.edu.pk

 
Interdisciplinary Cancer Research, https://doi.org/10.1007/16833_2024_254
Published online: 21 May 2024

**Keywords**

Artificial intelligence · Bioinformatics · Breast cancer treatment · CRISPR · Immunotherapies · Nanotechnology · Targeted therapies

# 1 Introduction

Globally, cancer is the primary cause of death (Mitra et al. 2022). Breast cancer is still the second most common type of cancer worldwide and the most common cancer in women; 1.4 million cases are reported annually, with an age-standardized incidence of 39.0 cases per 100,000 people. Men are almost never impacted, but women are almost always impacted; this usually happens to women in their forties or fifties. Breast cancer is responsible for 6.1% of all cancer-related deaths in the population as a whole and 13.7% of all cancer-associated deaths in women (Jemal et al. 2011; Siegel et al. 2021). Breast tissue can grow breast tumours in a variety of locations. The majority of them originate from cells lining the ducts (ductal cancers). Some develop from other tissues, while others (lobular cancers) grow inside the lobules. MiRNA expression proliferation can be utilized to determine tumour origin. Although invasive BC is defined as malignant cells spreading from the ductal and lobular surface into the breast's fatty and connective tissues (Sharma et al. 2010; Tang et al. 2017). By analysing the expression profiles of the ER, PR, and/or HER2 receptors, five distinct molecular subtypes of breast cancer (BC) are discovered (Tong et al. 2018). The most deadly phenotypic category of breast carcinoma is known as triple-negative breast cancer (TNBC) (Koedoot et al. 2019; Grief et al. 2022; Kvokačková et al. 2023). Because of its diversity, TNBC is difficult to address; therefore, the molecular subtyping of TNBC would aid in the management of this fatal illness (Qin et al. 2019; da Silva et al. 2020; Selmin et al. 2020; Hua et al. 2022).

As a result of medical breakthroughs, the worldwide mortality rate from BC has decreased. The prevalence of this disease was higher in postmenopausal women (78%), with only 22% of patients being under 50. Breast cancer in females under the age of 35 is extremely rare, accounting for only 3% of all cases, and it rarely affects women under the age of 25 (Jassem et al. 2003; Krzakowski et al. 2009; Ferlay et al. 2010; Didkowska et al. 2013; Jeziorski et al. 2018). Poland's data on breast cancer suggests a downward trend, which could point to advancements in the disease's diagnosis and treatment (0.64 in 1963, 0.38 in 2005) (Didkowska et al. 2013; Wojciechowska et al. 2016). Despite the pandemic, the cancer death rate in the United States decreased by 1.5% in 2019 and 2020 compared to other major causes of death. This decline contributed to a 33% overall decline in cancer-related deaths since 1991, and an estimated 3.8 million deaths averted. This progress is gradually reflecting improvements in therapy, as seen by the substantial decreases in mortality (approximately 2% per year from 2016 to 2020) for kidney, leukaemia, and skin cancer, as well as the rapid falls for lung cancer. Even if cancer death rates continue to fall, the incidence of breast, prostate, and uterine corpus cancers is rising,

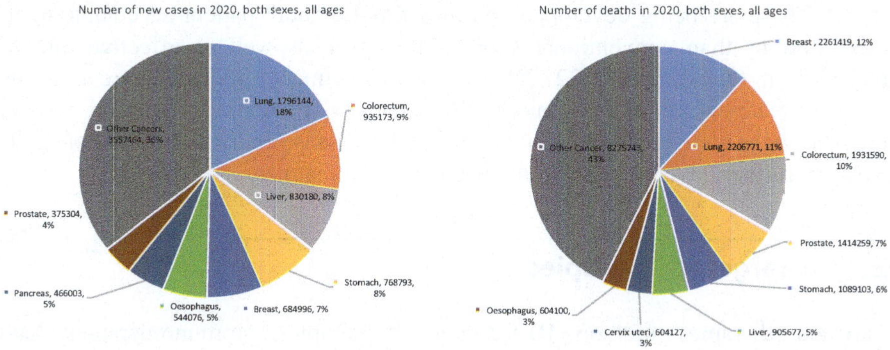


**Fig. 1** Data of new cases and deaths for various cancer types in 2020 (Source: GLOBOCAN Data for Cancer 2020 (https://gco.iarc.fr/today))

which might stymie future improvement. These tumours are also associated with the greatest racial differences in fatality (Siegel et al. 2023). Figure 1 shows the GLOBOCAN data of the new cases and deaths for various cancer types reported in 2020.

Despite indications that the prevalence of breast cancer has dropped in the last two decades, it made up 25% of all female cancer cases in 2020 and was by far the most often diagnosed malignancy. Breast cancer is becoming more common worldwide, particularly in underdeveloped countries (Heer et al. 2020; Arnold et al. 2022; Berdzuli 2023; Choi et al. 2023). The year 2020 saw 685,000 breast cancer deaths and over 2.3 million new cases. A great deal of variations in these trends exist with respect to the geographical region and area. Despite having less regional variation in death rates, developing countries are generally showing a radical increase in deaths related to BC than the developed countries. It is estimated that by 2040, there will be over 3 million new cases of breast cancer and 1 million annual deaths from the disease (Arnold et al. 2022; Hong and Xu 2022; Sun et al. 2022; Berdzuli 2023; Choi et al. 2023). These treatments have been the cause of the declining cancer death rate. Patients' lives have been considerably prolonged by improved early screening, high-resolution imaging, and development of efficient chemotherapy, radiation, targeted, and immunotherapy regimens (Gupta et al. 2020). It is necessary to find novel antineoplastic agents with fewer side effects because of the risks and toxicity problems associated with chemotherapeutic agents (Sak 2012; Gupta et al. 2020). Early detection and treatment of breast cancer with minimally invasive methods are proving to be beneficial. Open surgical biopsies have been superseded by percutaneous image-guided biopsy techniques, which are now the accepted method for diagnosing breast cancer (Vlastos and Verkooijen 2007). A novel technique called sentinel lymph node biopsy eliminates the need for axillary lymph node dissection by predicting the status of the lymph nodes (Li et al. 2022). The treatment of primary tumours without surgery presents the next challenge. Traditional surgical therapies may be replaced by minimally invasive techniques like cryotherapy, focused ultrasound ablation, radiofrequency ablation, and interstitial laser ablation (Kwak et al.

2022). These are being developed and may provide more patient-acceptable treatment options than conventional surgical therapies, as well as effective tumour management (Kwak et al. 2022). We review the treatment options that are available for the treatment of breast cancer. We will also be looking at what current breakthroughs were made in the treatment of this disease and what future potentials exist for further developments in its diagnosis and treatment.

## 2 Targeted Therapies

Targeted therapies include HER2-targeted therapies, immunotherapies, and therapies against critical checkpoints and pathways in cell growth. The area is becoming noteworthy for research on triple-negative breast cancer (TNBC), which is an aggressive and lethal type of breast cancer (Lau et al. 2022).

### 2.1 Hormone Receptor–Targeted Therapies

Adjuvant therapy for hormone receptor–positive breast cancers frequently involves endocrine therapy, which has been demonstrated to be effective in these cases. Many studies have been conducted on the PI3K/Akt/mTOR pathway (PAM pathway) in the treatment of endocrine-resistant breast cancers. p85 and p110 are the two subunits that make up PI3K. The p85 subunit docks at the phosphorylated tyrosine residues upon the activation of receptor tyrosine kinase (RTK). Phosphatidylinositol-4,5-biphosphate (PIP2) is subsequently converted to phosphatidylinositol-3,4,5-triphosphate (PIP3) when the p110 subunit is attracted and activated by Ras proteins. Encouraging protein synthesis to support cell growth and proliferation is the main way that mTOR activation promotes anabolic cell growth (Chang et al. 2003; Paplomata and O'Regan 2014; Glaviano et al. 2023). Even in the context of endocrine treatment, activating PI3K mutations and/or aberrant signalling in the absence of growth factors may contribute to the progression of breast cancer. There are several therapies available that target the PAM pathway (Chang et al. 2003; Arena 2014; Paplomata and O'Regan 2014; Carbognin et al. 2019; Glaviano et al. 2023).

mTOR inhibitors, such as everolimus, temsirolimus, and sirolimus, target the mTOR pathway to regulate cell cycle progression, growth, and proliferation. By forming complexes with specific proteins, these inhibitors disrupt mTOR function, inhibiting key processes in cancer cells (Chan et al. 2005; Kwitkowski et al. 2010; Fleming et al. 2012; Arena 2014; Steelman et al. 2016; Riccardi et al. 2018; Glaviano et al. 2023). Everolimus is approved for renal cell carcinoma and breast cancer (Royce and Osman 2015; Riccardi et al. 2018). Temsirolimus for renal cell carcinoma and sirolimus for organ rejection prevention. Their mechanisms involve interfering with the mTOR pathway, offering potential for cancer treatment (Chan et al. 2005; Kwitkowski et al. 2010; Fleming et al. 2012; Arena 2014).

PI3K inhibitors, including alpelisib (BYL719), taselisib (GDC-0032), pictilisib (GDC-0941), and buparlisib (BKM120), target the PI3K pathway to regulate cell growth and proliferation (Janku 2017; Schöffski et al. 2018; André et al. 2019; Garrido-Castro et al. 2020; Zhang et al. 2020; Dent et al. 2021; Jhaveri et al. 2021; Rugo et al. 2021; Turner et al. 2021). Alpelisib, FDA-approved for HR-positive, HER2-negative breast cancer, disrupts the activated PI3K p110 subunit alpha isoform, offering effective therapy for PIK3CA-mutant cancers (André et al. 2019; Rugo et al. 2021; Turner et al. 2021). Taselisib is being studied for advanced breast cancer, with ongoing clinical trials assessing its efficacy and safety (Schöffski et al. 2018; Zhang et al. 2020). Pictilisib demonstrates effectiveness against PIK3CA-mutated, HER2-positive, and HER2-negative cancers, albeit with safety concerns. Buparlisib shows potential in clinical trials in the early stages for endocrine-resistant metastatic breast cancers but has dose-limited toxicities (Schöffski et al. 2018; Garrido-Castro et al. 2020; Zhang et al. 2020).

PTEN, which is also an inhibitor of PI3K pathway, is often diminished in aggressive cancers, contributing to uncontrolled cell growth. Alpelisib-fulvestrant combination therapy proves effective in treating cancers with endocrine resistance. Additionally, studies explore gene editing and natural compounds like bergapten and oridonin to upregulate the activity of PTEN in breast cancer cells, indicating potential directions for future therapeutics (De Amicis et al. 2015; Wu et al. 2016; Moses et al. 2019). The tumour suppressor AMP-activated protein kinase (AMPK), which is activated by liver kinase B1 (LKB1), stops cell growth and metabolism. Honokiol, a natural polyphenol, activates AMPK through LKB1 phosphorylation, inhibiting breast cancer tumourigenesis and demonstrating potential synergy with rapamycin for apoptosis induction. The LKB1-AMPK pathway presents a promising therapeutic target for cancer treatment, particularly in inhibiting TGF-β signalling and metastasis (Arora et al. 2012; Nagalingam et al. 2012; Li et al. 2015, 2016).

## 2.2 HER2-Targeted Therapies

HER2 is an essential therapeutic target in some types of breast cancer. Approved drugs for HER2-positive breast cancer include trastuzumab, pertuzumab, lapatinib, neratinib, and T-DM1. HER2-targeted treatments are also utilized to treat malignancies of the stomach, biliary system, colon, non-small cell lung, and bladder, in addition to BC (Klapper et al. 2000; Oh and Bang 2020). However, we will just be talking about breast cancer here. Trastuzumab, the first HER2-directed drug licensed for the treatment of breast cancer, was approved in the late 1990s (Kreutzfeldt et al. 2020; Zubair and Bandyopadhyay 2023). By inhibiting the MAPK and PIK3 pathways, lapatinib prevents the phosphorylation of tyrosine kinase residues, which inhibits cell proliferation26 (Xia et al. 2002; Zubair and Bandyopadhyay 2023). Neratinib is an irreversible inhibitor that targets EGFR, HER2, and HER437. It shows potential for treating HER2-positive breast cancer. Notably, neratinib achieves higher pathologic complete response rates and improved invasive-free survival, particularly beneficial in HR+ disease. Approved as extended adjuvant therapy in the United States, its application post-trastuzumab is constrained by a notable incidence of diarrhoea, mitigated through prophylaxis and dose escalation

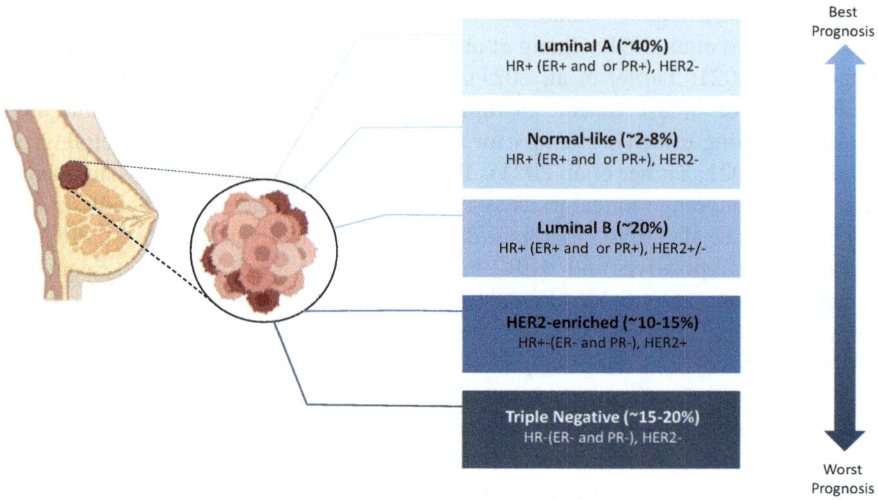


**Fig. 2** Molecular subtype of breast cancer

(Saura et al. 2020; Chan et al. 2021; Schlam and Swain 2021). Clinical outcomes for HER2-positive breast cancer have significantly changed since the advent of monoclonal HER2-targeting antibodies and antibody-drug conjugates. A greater percentage of patients in the non-metastatic setting are cured as a result of this change. However, despite these advancements, resistance to anti-HER2 treatments remains a significant hurdle in the metastatic setting as shown in Fig. 2. This underscores the critical need for the development of innovative HER2-directed therapies (Kunte et al. 2020).

## 2.3 Cell Cycle–Targeted Therapies

Cyclin-dependent kinase 4 (CDK4) and CDK6 are two cell cycle proteins that are frequently dysregulated in cancerous cells and have attracted a lot of interest as possible targets for cancer treatment. There are currently three small molecules—palbociclib, abemaciclib, and ribociclib—approved for the treatment of breast cancer patients among the several inhibitors that have been developed that target CDK4 and CDK6. These inhibitors are undergoing extensive testing in individuals with various solid and haematological malignancies, highlighting their potential broader applications in cancer treatment (Petroni et al. 2020). Palbociclib is an orally administered drug with a low enzymatic inhibition for CDK4 and CDK6. It stands as an inaugural member of CDK4 and CDK6 inhibitors, showcasing its efficacy in breast cancer cells. Studies led by Finn and colleagues underscored its synergistic effects with tamoxifen in ER-positive breast cancer cells. Its effectiveness in inhibiting cancer propagation, even in tamoxifen-resistant tumours, was validated in vivo using patient-derived xenograft mouse models (PDX) (Finn et al. 2009;

Wardell et al. 2015; Sobhani et al. 2019). Ribociclib is also an orally administered CDK4/6 inhibitor featuring a low enzymatic like palbociclib. It is permitted to be used at a beginning dose of 600 mg daily for 3 weeks, followed by 1 week off (Malorni et al. 2012; Kim et al. 2013; Prat et al. 2013; Asghar et al. 2017). Additionally, targeted treatments that interfere with the cyclin D–CDK4/6 pathway are beginning to yield encouraging outcomes. Clinical activity of these inhibitors in TNBC can be enhanced by combinations with other targeted therapies or immunotherapies (Malorni et al. 2012; Prat et al. 2013; Asghar et al. 2017; Liu et al. 2017a, b; Teo et al. 2017; O'Brien et al. 2018; Foidart et al. 2019; Yamamoto et al. 2019).

## 3 Immunotherapy

Traditionally considered immunogenically 'cold', breast cancer has seen a shift with the approval of two immune checkpoint inhibitors in combination with chemotherapy for PD-L1 positive advanced triple-negative breast cancer (TNBC). For example, pembrolizumab has received recent approval for early-stage TNBC. These approvals have now paved the way for immunotherapies (Howard et al. 2022).

### 3.1 Immune Checkpoint Inhibitors

Immune checkpoint inhibitors, like atezolizumab (ATE), increase T cell–mediated cytotoxicity and apoptosis in PD-L1+ triple-negative breast cancer (TNBC) cells, demonstrating promising anti-tumour activity (Mohan et al. 2019). The difficult prognosis of triple-negative breast cancer (TNBC) and the lack of targeted therapies have prompted research into new molecular targets and combination tactics. TNBCs exhibit high biological heterogeneity, making not all patients responsive to immunotherapeutic approaches like atezolizumab plus nab-paclitaxel or pembrolizumab plus nab-paclitaxel, paclitaxel, or gemcitabine/carboplatin. Patients with PD-L1+ disease are currently the only ones eligible for treatment, which highlights the need for an adaptive and mutation-based therapeutic strategy (Gumusay et al. 2021). These studies suggest that immune checkpoint inhibitors possess the potential for treatment for an aggressive breast cancer subtype such as TNBC (Bareche et al. 2019).

### 3.2 Cancer Vaccines

Many cancer vaccines target tumour-associated antigens (TAAs), which are proteins overexpressed in various cancers and are categorized into five types (Chiang et al. 2015; Nicolás-Morales et al. 2022). Cancer vaccines come in various types, such as proteins, peptides, whole cells, and nucleic acids (Emens 2012; Morse et al. 2021). Identifying the antigen of interest is crucial for developing effective breast cancer vaccines. HER2, MUC-1, and carbohydrate antigens, for example, have been used in a variety of techniques to boost immune responses, stimulating T cells and antibody formation (Emens 2012). Whole cell vaccines utilize the quantity of antigens on cancer cells to elicit a powerful immune response, profiting from their aptitude to digest and deliver many tumour-associated antigens (TAAs). This includes antigen coupling to MHC class I and II, recycling of pre-existing MHC-peptide complexes shed by cancer cells, and exogenous cell transfer of TAA peptides to dendritic cells (DC)(Mittendorf et al. 2007; Perez and De Palma 2019). DNA vaccines for breast cancer involve direct injection of plasmids encoding antigens, facilitating transient expression in the host's nucleus. An example is MVA-BN®-HER2, a vaccinia virus-based vector encoding a modified form of HER2. In preclinical breast cancer models, it induced a protective immune response, suggesting potential in clinical phase I, possibly in combination with CTLA-4 checkpoint inhibitors (Mandl et al. 2012; Yang et al. 2014; Foy et al. 2016; Li and Petrovsky 2016; Lee et al. 2018). Protein or peptide-based vaccines for breast cancer focus on specific antigens, promoting immune responses to selected peptides. In contrast, whole cell vaccines offer a broader immune response with multiple tumour-associated antigens (TAAs). Nucleic acid–based vaccines are designed to express proteins that are identical to the target protein, allowing for the inclusion of several targets to boost immune responses (Burke et al. 2019; Crosby et al. 2019; Maeng and Berzofsky 2019). Table 1 shows the targeted therapies for the four molecular subtypes of BC.

## 4 Precision Medicine

Recent advancements in breast cancer care emphasize precision medicine, targeting specific molecular defects associated with tumours. This approach, marked by multigene assays and molecular profiling, enables more precise and effective treatments tailored to individual patients, minimizing side effects. The future of cancer therapy holds promise for increased precision through the integration of "big data" and high-throughput analysis, allowing for personalized and highly effective interventions (Greenwalt et al. 2020). In oncology, precision medicine aims to adapt each patient's therapy and management based on a precise evaluation of their diagnosis, risk of recurrence, or rate of progression (Muzahir 2020). Precision medicine has risen to the forefront of drug research and cancer patient care due to the introduction of cutting-edge technologies, data-sharing systems, and novel clinical trial designs. In 2017, the FDA accepted 16 new applications, five of which were for breast cancer treatments, highlighting the rapid increase of accessible medicines. However, challenges such as identifying functionally significant targets from vast genomic and proteomic data, identifying optimal patients for targeted therapies, ensuring therapy access, and addressing resistance and toxicities associated with new treatments such as checkpoint inhibitors entering breast cancer protocols remain. Collaborative research, multidisciplinary information exchange, and raising the bar for clinically relevant trial goals are critical to achieving precision medicine's goal of improving breast cancer therapies and outcomes (Sachdev et al. 2019).

**Table 1** Drugs used for targeted therapies of the four breast cancer subtypes

| Drug name | Drug category | Breast cancer subtype | Patient population | Therapy given | References |
|---|---|---|---|---|---|
| Talazoparib | PARP inhibitors | HER2 negative | HER2− deleterious gBRCA mutant | Single-agent talazoparib | Bellare and Patro (2022), Eskiler and Ozturk (2022) |
| Olaparib | PARP inhibitors | HER2 negative | HER2− deleterious gBRCA mutant | Single-agent Olaparib | Loap et al. (2022), Tung and Garber (2022) |
| Abemaciclib | CDK4/CDK6 inhibitors | HER2 negative | HR+, HER2−<br>HR+, HER2 | Abemaciclib + AI<br>Single-agent abemaciclib | Royce et al. (2022), Johnston et al. (2023) |
| Ribociclib | CDK4/CDK6 inhibitors | HER2-negative | HR+, HER2− post/premenopausal<br>HR+, HER2− postmenopausal | Ribociclib + letrozole | Al-Ziftawi et al. (2023) |
| E75 | Cancer vaccines | TNBC | HER2+ | E75 + GM-CSF | Zhu and Yu (2022) |
| Atezolizumab | Immune checkpoint inhibitors | TNBC | TNBC | Atezolizumab + chemotherapy | Huober et al. (2022), Nakamoto et al. (2023) |
| Pembrolizumab | Immune checkpoint inhibitors | TNBC | TNBC | Pembrolizumab+ chemotherapy | Schmid et al. (2022)<br>Tarekegn et al. (2023) |
| GP2 | Cancer vaccines | TNBC | HER2+ | Single-agent GP2 | Zhou (2023) |

# 5 Other Emerging Therapies

## 5.1 Oncolytic Viruses

Mary E. Carter and colleagues have provided an in-depth systematic review of all published clinical trials with oncolytic viruses that included breast cancer patients. Three distinct types of oncolytic viruses are employed in clinical trials: (i) those with inherent anti-neoplastic properties; (ii) viruses crafted for tumour-selective replication; and (iii) viruses engineered to stimulate the immune system. Presently, 18 published clinical trials have explored 14 diverse oncolytic viruses. These trials affirm the safety and tolerability of oncolytic viruses in patients, showcasing clinical activity. However, the focus has predominantly been on a limited number of patients with various advanced tumours, including some with breast cancer (Carter et al. 2021). Oncolytic viruses (OVs) are effective in cancer treatment because they combine targeted cancer cell killing with the development of a long-lasting anti-tumour immune response. Recent studies with triple-negative breast cancer murine models show that administering oncolytic Maraba virus (MRB) before surgical tumour resection effectively reduces the metastatic burden, provides protection against tumour recurrence, achieves cures in some mice, and improves the responsiveness of refractory tumours to immune checkpoint blockade. Extending this surgical model to other OVs—vesicular stomatitis virus (VSV), adenovirus (Ad), reovirus (Reo), and herpes simplex virus (HSV)—shows favourable results across many OVs, with varying effects on primary and secondary tumour formation. The study proves that OV therapy, when delivered during the pre-operative window, provides long-term protection without the need for further treatment after rechallenge (Martin et al. 2019; Monaco et al. 2023; Sherlock and Coster 2023; Yuan et al. 2023).

## 5.2 PARP Inhibitors

Certain mutations, notably deleterious alterations in the BRCA gene, are linked to a higher susceptibility to triple-negative breast cancer (TNBC). Mutations in BRCA 1 and BRCA 2 genes heighten the risk of breast cancer by compromising DNA strand break repair, resulting in chromosomal instability. Because poly-ADP ribose polymerase (PARP) inhibitors have improved response and progression-free survival than standard chemotherapy, the FDA has approved them for the treatment of BRCA-associated HER2-negative metastatic breast cancer (O'Donovan and Livingston 2010; Tung et al. 2016; Robson et al. 2017, Robson et al. 2019; Litton et al. 2018). PARP inhibitors (PARPis) have become crucial in curing BC with BRCA1 and BRCA2 mutations, showing benefits in survival and quality of life, particularly in metastatic cases. Trials like OlympiAD and EMBRACA highlight their advantages over chemotherapy, despite notable side effects. Talazoparib and olaparib, registered for metastatic or locally advanced cases, require prior anthracycline and taxane treatment. While no direct comparison with platinum

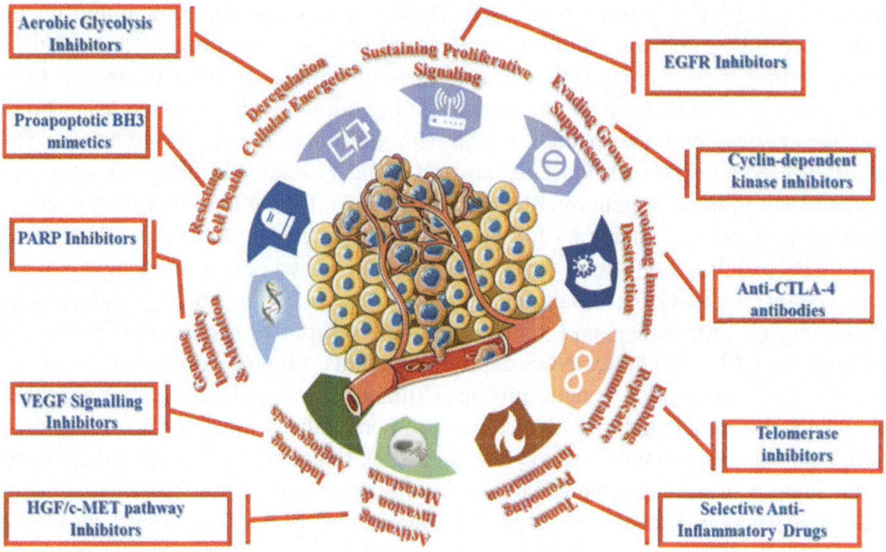


**Fig. 3** Therapeutic approaches for breast cancer (El-Tanani et al. 2023). (https://www.sciencedirect.com/science/article/pii/S0898656822002546)

salts exists, PARPis demonstrate a reduction in progression risk and improved progression-free survival in breast cancer. Their impact extends beyond the metastatic setting, showcasing potential in adjuvant and neoadjuvant contexts (Palleschi et al. 2021; Schettini et al. 2021). Ongoing combinatorial clinical trials explore the synergy between PARP inhibitors and immunotherapies, particularly in triple-negative subtype of breast cancer. Constantia Pantelidou demonstrated that the PARP inhibitor olaparib has immunomodulatory effects in BRCA1-deficient TNBC. Their findings demonstrated that olaparib stimulates in vivo CD8+ T cell infiltration and activation, an essential process for antitumour efficaciousness. Through the activation of the cGAS/STING pathway in tumour cells, dendritic cells are paracrinely activated in this recruitment process. Interestingly, STING knockout reverses this effect, which is more noticeable in HR-deficient TNBC cells and provides insight into a possible method for treating TNBC by combining PARP inhibition with immunotherapies (Pantelidou et al. 2019). Figure 3 shows the common therapeutic targeting approaches in the treatment of BC.

# 6 Personalized Treatment Approaches

## 6.1 Integration of Genomic and Clinical Data

The search for more potent treatments and a deeper comprehension of mechanisms is still ongoing in the field of personalized medicine. Sanger sequencing was used in a study involving 50 patients with breast cancer in Qatar to find new harmful mutations in the estrogen receptor 1 gene (ESR1). Rapid and accurate protein-ligand binding affinity interaction studies were carried out for a few drugs and the natural ligand estrogen to evaluate the effect of these mutations on drug treatment. Four nonsynonymous mutations in the ligand-binding domain were assessed using binding free energy techniques in conjunction with molecular dynamics simulations. The study offered a framework for a customized clinical decision system that integrates molecularly individual genomic data and provides information about the effectiveness of particular medications for patients with particular mutations (Wan et al. 2021). By enabling high-throughput sequencing of varied genetic data, next generation sequencing (NGS) is changing genomics research. The study, conducted by A. Kosvyra, C. Maramis, and I. Chouvarda and exemplified in chronic lymphocytic leukaemia (CLL), integrates NGS data with patients' clinical information to create a comprehensive clinico-genetic profile. Utilizing machine learning, the approach correlates different data types, contributing to personalized medicine and prognostic insights in cancer treatment (Kosvyra et al. 2019). Integrating genomic and radiomic data, known as radiogenomics, holds significant potential for cancer diagnosis, prognosis, and treatment. The study, conducted by Mario Zanfardino, proposed using the multiassay experiment (MAE), an R package, as a versatile tool for managing radiogenomic experiment data. By successfully integrating radiomic and genomic data from 91 patients into a single MAE object, the approach demonstrated its suitability for handling and analysing radiogenomic information (Zanfardino et al. 2019).

## 6.2 Development of Decision Support Tools

Clinical decision support systems (CDSS) are computer-based data collection technologies that assist clinicians in promoting cardiovascular disease prevention and healthy weight control (Papandreou et al. 2021). The effectiveness of CDSSs in breast cancer treatment varies depending on how they are developed, what data they contain, and how they are used in clinical settings. However, CDSSs aid in decision-making, enhance patient care, and lower costs (Mazo et al. 2020). Using patient data and the electronic health record (EHR), the decision support system for making personalized assessments and recommendations concerning breast cancer patients (DPAC) seeks to develop a clinical decision support system (CDSS). DPAC is designed to accurately classify patients and provide personalized treatment recommendations based on individual features. The study conducted by Xia Jiang and colleagues showed that patients who followed DPAC's advice collectively experienced significantly better outcomes compared to those who did not. The effectiveness of DPAC in gathering and analysing data for treatment recommendations is evident. However, certain findings, such as DPAC's assertion that chemotherapy may elevate metastasis risk for many node-negative patients, are controversial. This underscores the need to refine DPAC to offer conclusive and optimized patient-specific treatment recommendations (Jiang et al. 2019). CDSS also holds promise as a valuable tool to aid breast cancer (BC) patients in making lifestyle modifications in situations such as the COVID-19 pandemic (Papandreou et al. 2021).

# 7 Nanotechnology in Breast Cancer Treatment

Different types of nanoparticles, including liposomes, micelles, polymeric nanoparticles, solid lipid nanoparticles, and gold nanoparticles, have been made available for use in the treatment of breast cancer due to recent advances in nanotechnology. Nowadays, nanotechnology is essential to targeted drug delivery because it allows precise targeting of tumours and precise control of drug release at specific sites, improving therapeutic efficacy. This method reduces the amount of toxicity to healthy tissues or organs. Furthermore, nanoparticles have demonstrated the ability to stimulate immune cells in opposition to tumours. As a result, nanoparticles are a possible future option for cancer research and treatment (Tran et al. 2020). Figure 4 shows different nanoparticle types that are in use for breast cancer treatment. We shall look at various nanoparticles and their involvement in breast cancer treatment:

## 7.1 Folic Acid–Engineered Nanocarriers

Research has shown the effect of anti-cancer nanomedicines, particularly those with folic acid conjugation, on mitigating the side effects of chemotherapy. These nanocarriers exhibit dual targeting abilities, effectively inducing apoptosis and inhibiting tumour growth. Surface modifications enable sustained drug release, enhancing cytotoxicity in folate-overexpressed cell lines. Both hydrophilic and hydrophobic medications can be released continuously thanks to the amphiphilic characteristics of nanocarriers. Drug loading and release efficiency can be improved

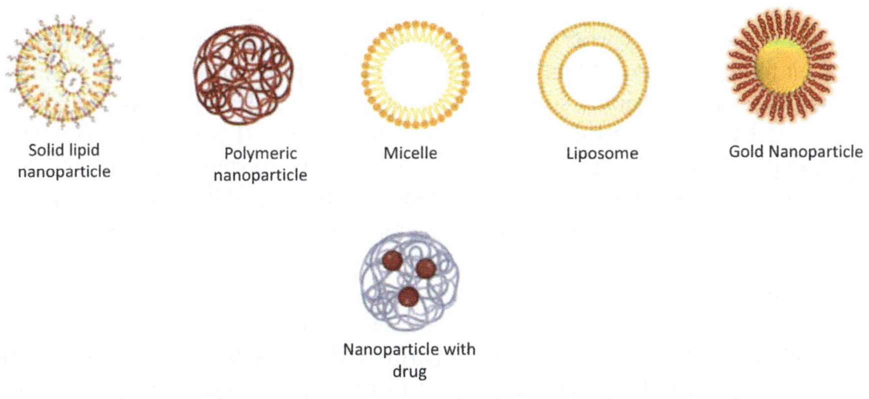


**Fig. 4** Nanoparticle types that are in use for the treatment of breast cancer

by modifying the size, shape, and characteristics of synthesized nanoparticles. Certain medications are released quicker from nanocarriers in acidic pH settings, accumulating at distinct tumour locations dependent on their metabolic pathway. When targeted to cancer cells, folic acid–conjugated nanosystems promote tumour formation while also acting as co-delivery vehicles for imaging agents and anti-cancer medications. Hurdles in this research include retaining the biochemical activity of nanoparticles, addressing issues like short half-life and production costs and overcoming obstacles for commercialization. While folate receptor–targeted drug delivery has shown promise in preclinical studies, its translation to clinical settings requires considering the more moderate folate receptor expression observed in clinical situations. Recent advancements in phase III trials suggest progress in folate receptor targeting, but continued research is essential (Tagde et al. 2020).

## 7.2 Antineoplastic Biogenic Gold Nanoparticles

Biologically synthesized gold nanoparticles (AuNPs) have shown promising effects against breast cancer cells through in vitro studies. Various naturally derived materials serve as reducing and capping agents for the eco-friendly and cost-effective synthesis of AuNPs with diverse morphologies and size distributions. Despite reported heterogeneities, biogenic AuNPs exhibit higher cytotoxicity against breast cancer cells compared to healthy cells, with significant synergy when combined with other anti-cancer drugs in vitro. Proposed anti-cancer mechanisms include overproducing reactive oxygen species (ROS) and activating caspase cascades to induce apoptosis. While initial evidence suggests the anti-cancer potential of eco-friendly AuNPs, further in vivo studies are needed for validation through animal models. Future research should address challenges such as genotoxicity, safety profiles, therapeutic windows, and pharmacokinetics/pharmacodynamics (Tagde et al. 2020).

## 7.3 Lipid-Based Nanoparticles (LNPs)

The use of LNPs as a carrier system for drug delivery is well established, leveraging non-toxic, biocompatible, and biodegradable lipids such as those found in liposomes, nanoemulsions, solid lipid nanoparticles (SLNs), nanostructured lipid carriers (NLCs), and lipid-polymer hybrid nanoparticles (LPH-NPs). These lipids offer opportunities to customize the drug delivery system's physicochemical properties, including particle size, morphology, drug loading, and release profile. LNPs demonstrate passive and active targeting capabilities, enhancing therapeutic efficacy and cancer site specificity, especially when surface functionalized. Concerns about LNPs' safety and efficacy are alleviated by the use of generally recognized as safe (GRAS)-regulated lipids, cGMP-grade manufacturing processes, and supportive preclinical data on their absorption, distribution, metabolism, excretion (ADME) status, and toxicity. LNPs are anticipated to have a significant impact on cancer management and treatment, with examples like Doxil paving the way for transformative anti-cancer therapeutics (Chaudhuri et al. 2022; Akanda et al. 2023; Aziz et al. 2023; Fatima Qizilbash et al. 2023; Lu et al. 2023a, b; Luiz et al. 2023) in the treatment of breast cancer and has successfully shown to suppress tumour growth while avoiding toxicity (Tang et al. 2020). The research led by Wanxing Lu and colleagues, as outlined by Tiwari et al. (2023), introduced a novel approach based on polydopamine-coated liposomes for accurate and effective combinatorial cancer treatment, which includes chemotherapy and laser-induced PDT/PTT. The nanoplatform, co-incorporating indocyanine green (ICG) and doxorubicin (DOX), demonstrated enhanced toxicity on breast cancer cells. Upon endocytosis, PDA@Lipo/DOX/ICG generated significant ROS for photodynamic therapy (PDT) upon 808 nm laser irradiation, achieving a notable cell inhibition rate of 80.4%. In an in vivo mouse model, the nano platform efficiently accumulated at the tumour site and, upon laser irradiation, effectively suppressed tumour proliferation without observed side effects. PDA@Lipo/DOX/ICG presents a promising multifunctional strategy for accurate and efficient cancer therapy (Lu et al. 2023a, b).

## 7.4 Curcumin Nanoparticles

A systematic review conducted by Alicia S. Ombredane and colleagues highlights the potential of nanoparticles (NPs) delivering curcumin for breast cancer treatment. Results demonstrate significant tumour volume reduction, increased apoptosis, decreased cell proliferation, and angiogenesis inhibition. Customizing NP structures and ensuring cost-effective manufacturing are essential for clinical translation. Future research should explore combined therapies. Overall, the evidence supports further investigation of Cur-NPs in breast cancer clinical trials (Ombredane et al. 2021).

## 7.5 Antibody-Conjugated Polymeric Nanoparticles

In the realm of breast cancer, particularly the HER2+ subtype, antibodies have been crucial in targeted therapy. However, natural and learned resistance offers difficulties, particularly in the later phases. To solve these issues, antibody conjugation to polymeric NPs emerges as a tempting technique. This approach offers several advantages, including enhanced biocompatibility, precise control over drug delivery, and increased efficiency in functionalization. By incorporating antibodies into polymeric NPs, researchers aim to optimize targeted therapies, overcome resistance, and improve treatment outcomes (Juan et al. 2020a, b). Antibody-conjugated nanoparticles (ACNPs) hold promise for treating resistant breast cancers by preserving drug structure, enabling controlled delivery, and reducing toxicity. This approach enhances treatment effectiveness and safety, offering potential solutions for challenging cases in breast cancer therapy (Juan et al. 2020a, b).

# 8 Other Treatment Strategies

## 8.1 Liquid Biopsies

Liquid biopsy, involving circulating analytes like tumour DNA and extracellular vesicles, offers non-invasive, real-time monitoring of tumour dynamics, and treatment response. Recent studies demonstrate the utility of liquid biopsy in predicting disease progression and monitoring treatment efficacy in breast cancer. With ongoing clinical trials and technological advancements, liquid biopsy holds promise for personalized breast cancer medicine, providing valuable insights more conveniently than traditional tissue biopsies (Cayrefourcq and Alix-Panabières 2020). Liquid biopsy procedures provide useful information. In breast cancer (Bca) patients, ctDNA offers information on somatic mutations and copy number changes (CNAs), and it is used to monitor tumour burden, screen, analyse medication response, determine prognosis, and discover minimum residual illness. ctDNA levels are directly related to clinical variables such as tumour size, lymph node metastases, stage, and grade. Liquid biopsy can detect BRCA1 gene mutations and HER2 receptor status, as well as tumour-derived genomic changes. ctDNA and circulating tumour cells (CTCs) have been found in BCa patients, and liquid biopsy has the potential to diagnose resistance to therapy and identify tumour recurrence (Cohen et al. 2008; Zanetti-Dällenbach et al. 2008; Aceto et al. 2014; Madhavan et al. 2014; Janku et al. 2015; Mayor et al. 2017; Wang et al. 2017; Tzeng et al. 2018; Mathai et al. 2019; Stergiopoulou et al. 2023; Venetis et al. 2023).

## 8.2 Artificial Intelligence

Recent developments in digital analysis of images with artificial intelligence (AI) show potential value in improving multimodal breast imaging for the diagnosis and treatment of breast cancer (Negrão de Figueiredo et al. 2019). An artificial intelligence (AI) system designed for breast cancer prediction outperforms human experts, demonstrating notable reductions in false positives and false negatives. Using datasets from the United Kingdom and the United States, the AI system achieves an absolute reduction of 5.7% and 1.2% in false positives and 9.4% and 2.7% in false negatives. Simulation results indicate non-inferior performance in the double-reading process, reducing the workload of the second reader by 88%. These statistics highlight the AI system's potential to enhance breast cancer screening accuracy and efficiency (McKinney et al. 2020; Kuhl 2023; Zheng et al. 2023).

# 9 Role of Modern Bioinformatics in the Treatment of Breast Cancer

Bioinformatics is an interdisciplinary field that combines computer science, mathematics, statistics, and biological and biomedical sciences. Its purpose is to facilitate the management, analysis, and interpretation of data obtained from biological experiments and observational studies (Bilotta et al. 2018). Bioinformatics is showing its potential in the treatment of breast cancer, especially in the identification of novel prognostic biomarkers in metastatic breast cancer, potentially improving treatment and predicting clinical results (Kim 2021). Research conducted by Jian Qi and coworkers showed that FPR3, a G protein–coupled receptor essential for neutrophil activation, is the sole factor that predicts poor prognosis. Gene set enrichment analysis showed FPR3 upregulation synergizes with the activation of many pathways involved in carcinogenesis. In this study, bioinformatics played a key role in the identification of FPR3 as a key immune-related biomarker predicting a poor prognosis of breast cancer (Qi et al. 2021). Bioinformatics also helps in the identification of specific alterations at the molecular level, such as homologous recombination deficiencies (HRDs), which can help in BC treatment decisions (Galland et al. 2023). In another study, long, non-coding RNA (lncRNAs) with altered expression levels were identified using bioinformatics in breast cancer cells that can potentially play a role in cancer studies (Zaheed et al. 2020). These studies suggest that bioinformatics is a key tool in the roles related to genetic study on cancer cells, gene regulation, and clinical decision systems (Zaheed et al. 2020; Qi et al. 2021; Galland et al. 2023).

## 10 Role of CRISPR Gene Editing in Treating Breast Cancer

Gene editing is a technique that manipulates gene function by modifying the sequence in the target organism's DNA. Current gene editing methods rely on targeting specific sites in an organism's genome using nucleases. There are three major engineered nuclease tools for gene editing: zinc finger nucleases, transcription activator-like effector nucleases, and clustered regularly interspaced short palindromic repeats (CRISPR)/CRISPR-associated protein 9 (Cas9) (Boch et al. 2009, Carroll 2011). CRISPR-Cas systems have now become the most widely used genome editing technology in molecular biology laboratories. This specifically due to their simple design, low cost, high efficiency, good repeatability and short cycle (Xu and Li 2020). CRISPR gene editing has shown its potential in the treatment of cancer due to its ability to alter the genome of cancer cells (Song et al. 2021). It can significantly inhibit tumour cell growth in BC by targeting oncogenes, tumour-suppressive genes, and genes associated with therapies (Hazafa et al. 2020). This technology holds a great promise at the gene-level treatment of cancer, thus showing its potential for personalized and targeted therapy of BC (Jiang et al. 2020).

## 11 Conclusion

Current data and statistics show that breast cancer has a prominent impact on the global morbidity and mortality rates related to breast cancer. Although the declining mortality rates over the past three decades are a witness of the ongoing advancements in the treatment strategies, the rising global burden necessitates ongoing research and innovative treatment strategies, including minimally invasive approaches. Targeted therapies such as hormone receptor–targeted therapies, HER2-targeted and cell cycle–targeted therapies are showing promising results. Immunotherapies along with chemotherapy are also paving the way for the treatment of breast cancer treatment. Cancer vaccines, oncolytic viruses, and PARP inhibitors are still in their infancy, but these treatments have great potential to deal with TNBC, which is a heterogeneous and aggressive type of breast cancer. Personalized treatment approaches developed by integrating clinical and genomic data, and by developing decision support tools, are providing treatment options for a larger group of population. Nanoparticles have enabled the delivery of anti-cancer drugs that can help target the desired tissues and minimize toxicity problems. Combining artificial intelligence and liquid biopsies would help provide more accurate results for treatment strategies. While all these technologies are still in development, further research in the mentioned areas is necessary to deal with the rising burden of this disease in the female population.

## References


Aceto N, Bardia A, Miyamoto DT, Donaldson MC, Wittner BS, Spencer JA, Yu M, Pely A, Engstrom A, Zhu H (2014) Circulating tumor cell clusters are oligoclonal precursors of breast cancer metastasis. Cell 158:1110–1122

Akanda M, Mithu MSH, Douroumis D (2023) Solid lipid nanoparticles: an effective lipid-based technology for cancer treatment. J Drug Deliv Sci Technol 104709

Al-Ziftawi NH, Alam MF, Elazzazy S, Shafie AA, Hamad A, Mohamed Ibrahim MI (2023) Cost-effectiveness and cost-utility of Palbociclib versus Ribociclib in women with stage IV breast cancer: a real-world data evaluation. Int J Environ Res Public Health 20:512

André F, Ciruelos E, Rubovszky G, Campone M, Loibl S, Rugo HS, Iwata H, Conte P, Mayer IA, Kaufman B (2019) Alpelisib for PIK3CA-mutated, hormone receptor–positive advanced breast cancer. N Engl J Med 380:1929–1940

Arena F (2014) Clinical implications of recent studies using mTOR inhibitors to treat advanced hormone receptor-positive breast cancer. Cancer Manag Res 389–395

Arnold M, Morgan E, Rumgay H, Mafra A, Singh D, Laversanne M, Vignat J, Gralow JR, Cardoso F, Siesling S (2022) Current and future burden of breast cancer: global statistics for 2020 and 2040. Breast 66:15–23

Arora S, Singh S, Piazza GA, Contreras CM, Panyam J, Singh AP (2012) Honokiol: a novel natural agent for cancer prevention and therapy. Curr Mol Med 12:1244–1252

Asghar US, Barr AR, Cutts R, Beaney M, Babina I, Sampath D, Giltnane J, Lacap JA, Crocker L, Young A (2017) Single-cell dynamics determines response to CDK4/6 inhibition in triple-negative breast cancer. Clin Cancer Res 23:5561–5572

Aziz A, Rehman U, Sheikh A, Abourehab MA, Kesharwani P (2023) Lipid-based nanocarrier mediated CRISPR/Cas9 delivery for cancer therapy. J Biomater Sci Polym Ed 34:398–418

Bareche Y, Buisseret L, Gruosso T, Girard E, Venet D, Dupont F, Desmedt C, Larsimont D, Park M, Rothé F, Stagg J, Sotiriou C (2019) Unraveling triple-negative breast cancer tumor microenvironment heterogeneity: towards an optimized treatment approach. JNCI J Natl Cancer Inst 112:708–719

Bellare GP, Patro BS (2022) Resveratrol sensitizes breast cancer to PARP inhibitor, talazoparib through dual inhibition of AKT and autophagy flux. Biochem Pharmacol 199:115024

Berdzuli N (2023). Breast cancer: from awareness to access. British Medical Journal Publishing Group, p 380

Bilotta M, Tradigo G, Veltri P (2018) Bioinformatics data models, representation and storage. In: Encyclopedia of bioinformatics and computational biology: ABC of Bioinformatics, p 110
Boch J, Scholze H, Schornack S, Landgraf A, Hahn S, Kay S, Lahaye T, Nickstadt A, Bonas U (2009) Breaking the code of DNA binding specificity of TAL-type III effectors. Science 326: 1509–1512
Burke EE, Kodumudi K, Ramamoorthi G, Czerniecki BJ (2019) Vaccine therapies for breast cancer. Surg Oncol Clin N Am 28:353–367
Carbognin L, Miglietta F, Paris I, Dieci MV (2019) Prognostic and predictive implications of PTEN in breast cancer: unfulfilled promises but intriguing perspectives. Cancers 11:1401
Carroll D (2011) Genome engineering with zinc-finger nucleases. Genetics 188:773–782
Carter ME, Koch A, Lauer UM, Hartkopf AD (2021) Clinical trials of oncolytic viruses in breast cancer. Front Oncol 11
Cayrefourcq L, Alix-Panabières C (2020) Clinical relevance of liquid biopsy in breast cancer: update in 2020. Expert Rev Mol Diagn 20:913–919
Chan S, Scheulen ME, Johnston S, Mross K, Cardoso F, Dittrich C, Eiermann W, Hess D, Morant R, Semiglazov V (2005) Phase II study of temsirolimus (CCI-779), a novel inhibitor of mTOR, in heavily pretreated patients with locally advanced or metastatic breast cancer. J Clin Oncol 23:5314–5322
Chan A, Moy B, Mansi J, Ejlertsen B, Holmes FA, Chia S, Iwata H, Gnant M, Loibl S, Barrios CH (2021) Final efficacy results of neratinib in HER2-positive hormone receptor-positive early-stage breast cancer from the phase III ExteNET trial. Clin Breast Cancer 21(80–91):e87
Chang F, Lee J, Navolanic P, Steelman L, Shelton J, Blalock W, Franklin R, McCubrey J (2003) Involvement of PI3K/Akt pathway in cell cycle progression, apoptosis, and neoplastic transformation: a target for cancer chemotherapy. Leukemia 17:590–603
Chaudhuri A, Kumar DN, Shaik RA, Eid BG, Abdel-Naim AB, Md S, Ahmad A, Agrawal AK (2022) Lipid-based nanoparticles as a pivotal delivery approach in triple negative breast cancer (TNBC) therapy. Int J Mol Sci 23:10068
Chiang CL-L, Coukos G, Kandalaft LE (2015) Whole tumor antigen vaccines: where are we? Vaccine 3:344–372
Choi JE, Kim Z, Park CS, Park EH, Lee SB, Lee SK, Choi YJ, Han J, Jung K-W, Kim HJ (2023) Breast cancer statistics in Korea, 2019. J Breast Cancer 26:207
Cohen SJ, Punt C, Iannotti N, Saidman BH, Sabbath KD, Gabrail NY, Picus J, Morse M, Mitchell E, Miller MC (2008) Relationship of circulating tumor cells to tumor response, progression-free survival, and overall survival in patients with metastatic colorectal cancer
Crosby EJ, Gwin W, Blackwell K, Marcom PK, Chang S, Maecker HT, Broadwater G, Hyslop T, Kim S, Rogatko A (2019) Vaccine-induced memory CD8+ T cells provide clinical benefit in HER2 expressing breast cancer: a mouse to human translational study. Clin Cancer Res 25: 2725–2736
da Silva JL, Nunes NCC, Izetti P, de Mesquita GG, de Melo AC (2020) Triple negative breast cancer: a thorough review of biomarkers. Crit Rev Oncol Hematol 145:102855
De Amicis F, Aquila S, Morelli C, Guido C, Santoro M, Perrotta I, Mauro L, Giordano F, Nigro A, Andò S (2015) Bergapten drives autophagy through the up-regulation of PTEN expression in breast cancer cells. Mol Cancer 14:1–12
Dent S, Cortés J, Im Y-H, Diéras V, Harbeck N, Krop I, Wilson T, Cui N, Schimmoller F, Hsu J (2021) Phase III randomized study of taselisib or placebo with fulvestrant in estrogen receptor-positive, PIK3CA-mutant, HER2-negative, advanced breast cancer: the SANDPIPER trial. Ann Oncol 32:197–207
Didkowska J, Wojciechowska U, Zatoński W (2013) Nowotwory złośliwe w Polsce w 2011 roku. Krajowy Rejestr Nowotworów, Warszawa
El-Tanani M, Al Khatib AO, Al-Najjar BO, Shakya AK, El-Tanani Y, Lee YF, Serrano-Aroca Á, Mishra V, Mishra Y, Aljabali AA, Goyal R (2023) Cellular and molecular basis of therapeutic approaches to breast cancer. Cell Signal 101:110492

Emens LA (2012) Breast cancer immunobiology driving immunotherapy: vaccines and immune checkpoint blockade. Expert Rev Anticancer Ther 12:1597–1611
Eskiler GG, Ozturk M (2022) Therapeutic potential of the PI3K inhibitor LY294002 and PARP inhibitor Talazoparib combination in BRCA-deficient triple negative breast cancer cells. Cell Signal 91:110229
Fatima Qizilbash F, Sartaj A, Qamar Z, Kumar S, Imran M, Mohammed Y, Ali J, Baboota S, Ali A (2023) Nanotechnology revolutionises breast cancer treatment: harnessing lipid-based nanocarriers to combat cancer cells. J Drug Target 31:794–816
Ferlay J, Shin H-R, Bray F, Forman D, Mathers C, Parkin DM (2010) Estimates of worldwide burden of cancer in 2008: GLOBOCAN 2008. Int J Cancer 127:2893–2917
Finn RS, Dering J, Conklin D, Kalous O, Cohen DJ, Desai AJ, Ginther C, Atefi M, Chen I, Fowst C (2009) PD 0332991, a selective cyclin D kinase 4/6 inhibitor, preferentially inhibits proliferation of luminal estrogen receptor-positive human breast cancer cell lines in vitro. Breast Cancer Res 11:1–13
Fleming GF, Ma CX, Huo D, Sattar H, Tretiakova M, Lin L, Hahn OM, Olopade F, Nanda R, Hoffman PC (2012) Phase II trial of temsirolimus in patients with metastatic breast cancer. Breast Cancer Res Treat 136:355–363
Foidart P, Yip C, Radermacher J, Blacher S, Lienard M, Montero-Ruiz L, Maquoi E, Montaudon E, Château-Joubert S, Collignon J (2019) Expression of MT4-MMP, EGFR, and RB in triple-negative breast cancer strongly sensitizes tumors to erlotinib and palbociclib combination therapy. Clin Cancer Res 25:1838–1850
Foy SP, Mandl SJ, dela Cruz T, Cote JJ, Gordon EJ, Trent E, Delcayre A, Breitmeyer J, Franzusoff A, Rountree RB (2016) Poxvirus-based active immunotherapy synergizes with CTLA-4 blockade to increase survival in a murine tumor model by improving the magnitude and quality of cytotoxic T cells. Cancer Immunol Immunother 65:537–549
Galland L, Roussot N, Desmoulins I, Mayeur D, Kaderbhai C, Ilie S, Hennequin A, Reda M, Albuisson J, Arnould L (2023) Clinical utility of genomic tests evaluating homologous recombination repair deficiency (HRD) for treatment decisions in early and metastatic breast cancer. Cancers 15:1299
Garrido-Castro AC, Saura C, Barroso-Sousa R, Guo H, Ciruelos E, Bermejo B, Gavilá J, Serra V, Prat A, Paré L (2020) Phase 2 study of buparlisib (BKM120), a pan-class I PI3K inhibitor, in patients with metastatic triple-negative breast cancer. Breast Cancer Res 22:1–13
Glaviano A, Foo AS, Lam HY, Yap KC, Jacot W, Jones RH, Eng H, Nair MG, Makvandi P, Geoerger B (2023) PI3K/AKT/mTOR signaling transduction pathway and targeted therapies in cancer. Mol Cancer 22:138
Greenwalt I, Zaza N, Das S, Li BD (2020) Precision medicine and targeted therapies in breast cancer. Surg Oncol Clin N Am 29:51–62
Grief D, Pal A, Gonzalez-Malerva L, Eaton S, Xu C, Christensen G, Blain J, Mellor N, Steel J, Gupta C (2022) Abstract P1-02-05: a genome-wide functional genomics screen reveals unique co-driver mutations of mutant TP53 promoting cellular heterogeneity during breast cancer progression. Cancer Res 82:P1–02-05-P01–02-05
Gumusay O, Wabl CA, Rugo HS (2021) Trials of immunotherapy in triple negative breast cancer. Curr Breast Cancer Rep 13:171–185
Gupta GK, Collier AL, Lee D, Hoefer RA, Zheleva V, Siewertsz van Reesema LL, Tang-Tan AM, Guye ML, Chang DZ, Winston JS (2020) Perspectives on triple-negative breast cancer: current treatment strategies, unmet needs, and potential targets for future therapies. Cancers 12:2392
Hazafa A, Mumtaz M, Farooq MF, Bilal S, Chaudhry SN, Firdous M, Naeem H, Ullah MO, Yameen M, Mukhtiar MS (2020) CRISPR/Cas9: a powerful genome editing technique for the treatment of cancer cells with present challenges and future directions. Life Sci 263:118525
Heer E, Harper A, Escandor N, Sung H, McCormack V, Fidler-Benaoudia MM (2020) Global burden and trends in premenopausal and postmenopausal breast cancer: a population-based study. Lancet Glob Health 8:e1027–e1037

Hong R, Xu B (2022) Breast cancer: an up-to-date review and future perspectives. Cancer Commun 42:913–936

Howard FM, Villamar D, He G, Pearson AT, Nanda R (2022) The emerging role of immune checkpoint inhibitors for the treatment of breast cancer. Expert Opin Investig Drugs 31:531–548

Hua Z, White J, Zhou J (2022) Cancer stem cells in TNBC. Elsevier, Seminars in cancer biology

Huober J, Barrios CH, Niikura N, Jarząb M, Chang Y-C, Huggins-Puhalla SL, Pedrini J, Zhukova L, Graupner V, Eiger D (2022) Atezolizumab with neoadjuvant anti–human epidermal growth factor receptor 2 therapy and chemotherapy in human epidermal growth factor receptor 2–positive early breast cancer: primary results of the randomized phase III IMpassion050 trial. J Clin Oncol 40:2946

Janku F (2017) Phosphoinositide 3-kinase (PI3K) pathway inhibitors in solid tumors: from laboratory to patients. Cancer Treat Rev 59:93–101

Janku F, Angenendt P, Tsimberidou AM, Fu S, Naing A, Falchook GS, Hong DS, Holley VR, Cabrilo G, Wheler JJ (2015) Actionable mutations in plasma cell-free DNA in patients with advanced cancers referred for experimental targeted therapies. Oncotarget 6:12809

Jassem J, Krzakowski M, Olszewski W (2003) Rak piersi. Zalecenia diagnostyczno-terapeutyczne Polskiej Unii Onkologii. Nowotwory 53:300–324

Jemal A, Bray F, Center MM, Ferlay J, Ward E, Forman D (2011) Global cancer statistics. CA Cancer J Clin 61:69–90

Jeziorski A, Rutkowski P, Wysocki WM (2018). Chirurgia onkologiczna 1, Wydawnictwo Lekarskie PZWL

Jhaveri K, Chang MT, Juric D, Saura C, Gambardella V, Melnyk A, Patel MR, Ribrag V, Ma CX, Aljumaily R (2021) Phase I basket study of taselisib, an isoform-selective PI3K inhibitor, in patients with PIK3CA-mutant cancers. Clin Cancer Res 27:447–459

Jiang X, Wells A, Brufsky A, Neapolitan R (2019) A clinical decision support system learned from data to personalize treatment recommendations towards preventing breast cancer metastasis. PLoS One 14:e0213292

Jiang C, Meng L, Yang B, Luo X (2020) Application of CRISPR/Cas9 gene editing technique in the study of cancer treatment. Clin Genet 97:73–88

Johnston SR, Toi M, O'Shaughnessy J, Rastogi P, Campone M, Neven P, Huang C-S, Huober J, Jaliffe GG, Cicin I (2023) Abemaciclib plus endocrine therapy for hormone receptor-positive, HER2-negative, node-positive, high-risk early breast cancer (monarchE): results from a preplanned interim analysis of a randomised, open-label, phase 3 trial. Lancet Oncol 24:77–90

Juan A, Cimas FJ, Bravo I, Pandiella A, Ocaña A, Alonso-Moreno C (2020a) Antibody conjugation of nanoparticles as therapeutics for breast cancer treatment. Int J Mol Sci 21:6018

Juan A, Cimas FJ, Bravo I, Pandiella A, Ocaña A, Alonso-Moreno C (2020b) An overview of antibody conjugated polymeric nanoparticles for breast cancer therapy. Pharmaceutics 12:802

Kim J (2021) In silico analysis of differentially expressed genesets in metastatic breast cancer identifies potential prognostic biomarkers. World J Surg Oncol 19:188

Kim S, Loo A, Chopra R, Caponigro G, Huang A, Vora S, Parasuraman S, Howard S, Keen N, Sellers W (2013) Abstract PR02: LEE011: an orally bioavailable, selective small molecule inhibitor of CDK4/6–reactivating Rb in cancer. Mol Cancer Ther 12:PR02

Klapper LN, Waterman H, Sela M, Yarden Y (2000) Tumor-inhibitory antibodies to HER-2/ErbB-2 may act by recruiting c-Cbl and enhancing ubiquitination of HER-2. Cancer Res 60:3384–3388

Koedoot E, Fokkelman M, Rogkoti V-M, Smid M, van de Sandt I, de Bont H, Pont C, Klip JE, Wink S, Timmermans MA (2019) Uncovering the signaling landscape controlling breast cancer cell migration identifies novel metastasis driver genes. Nat Commun 10:2983

Kosvyra A, Maramis C, Chouvarda I (2019) Developing an integrated genomic profile for cancer patients with the use of NGS data. Emerg Sci J 3:157–167

Kreutzfeldt J, Rozeboom B, Dey N, De P (2020) The trastuzumab era: current and upcoming targeted HER2+ breast cancer therapies. Am J Cancer Res 10:1045

Krzakowski M, Herman K, Jassem J, Jędrzejczak W, Kowalczyk R, Podolak-Dawidziak M (2009). Zalecenia postępowania diagnostyczno-terapeutycznego w nowotworach złośliwych, Wydawnictwo Via Medica

Kuhl CK (2023) What the future holds for the screening, diagnosis, and treatment of breast cancer. Radiology 306:e223338

Kunte S, Abraham J, Montero AJ (2020) Novel HER2–targeted therapies for HER2–positive metastatic breast cancer. Cancer 126:4278–4288

Kvokačková B, Fedr R, Kužílková D, Stuchlý J, Vávrová A, Navrátil J, Fabian P, Ondrušček R, Ovesná P, Remšík J (2023) Single-cell protein profiling defines cell populations associated with triple-negative breast cancer aggressiveness. Mol Oncol

Kwak K, Yu B, Lewandowski RJ, Kim D-H (2022) Recent progress in cryoablation cancer therapy and nanoparticles mediated cryoablation. Theranostics 12:2175

Kwitkowski VE, Prowell TM, Ibrahim A, Farrell AT, Justice R, Mitchell SS, Sridhara R, Pazdur R (2010) FDA approval summary: temsirolimus as treatment for advanced renal cell carcinoma. Oncologist 15:428–435

Lau KH, Tan AM, Shi Y (2022) New and emerging targeted therapies for advanced breast cancer. Int J Mol Sci 23:2288

Lee J, Kumar SA, Jhan YY, Bishop CJ (2018) Engineering DNA vaccines against infectious diseases. Acta Biomater 80:31–47

Li L, Petrovsky N (2016) Molecular mechanisms for enhanced DNA vaccine immunogenicity. Expert Rev Vaccines 15:313–329

Li N, Huang D, Lu N, Luo L (2015) Role of the LKB1/AMPK pathway in tumor invasion and metastasis of cancer cells. Oncol Rep 34:2821–2826

Li N-S, Zou J-R, Lin H, Ke R, He X-L, Xiao L, Huang D, Luo L, Lv N, Luo Z (2016) LKB1/AMPK inhibits TGF-β1 production and the TGF-β signaling pathway in breast cancer cells. Tumour Biol 37:8249–8258

Li F-M, Xu D-Y, Xu Q, Yuan Y (2022) Breast-conserving surgery and sentinel lymph node biopsy for breast cancer and their correlation with the expression of polyligand proteoglycan-1. World J Clin Cases 10:3113

Litton JK, Rugo HS, Ettl J, Hurvitz SA, Gonçalves A, Lee K-H, Fehrenbacher L, Yerushalmi R, Mina LA, Martin M (2018) Talazoparib in patients with advanced breast cancer and a germline BRCA mutation. N Engl J Med 379:753–763

Liu C-Y, Lau K-Y, Hsu C-C, Chen J-L, Lee C-H, Huang T-T, Chen Y-T, Huang C-T, Lin P-H, Tseng L-M (2017a) Combination of palbociclib with enzalutamide shows in vitro activity in RB proficient and androgen receptor positive triple negative breast cancer cells. PLoS One 12: e0189007

Liu T, Yu J, Deng M, Yin Y, Zhang H, Luo K, Qin B, Li Y, Wu C, Ren T (2017b) CDK4/6-dependent activation of DUB3 regulates cancer metastasis through SNAIL1. Nat Commun 8:1–12

Loap P, Loirat D, Berger F, Rodrigues M, Bazire L, Pierga J-Y, Vincent-Salomon A, Laki F, Boudali L, Raizonville L (2022) Concurrent olaparib and radiotherapy in patients with triple-negative breast cancer: the phase 1 olaparib and radiation therapy for triple-negative breast cancer trial. JAMA Oncol 8:1802–1808

Lu W, Liu W, Hu A, Shen J, Yi H, Cheng Z (2023a) Combinatorial Polydopamine-liposome Nanoformulation as an effective anti-breast cancer therapy. Int J Nanomed 861–879

Lu Y, Pan X, Nie Q, Zhou Z, Dai X, Liu O (2023b) Administration methods of lipid-based nanoparticle delivery systems for cancer treatment. Biomater Sci 11:3800–3812

Luiz MT, Dutra JAP, Viegas JSR, de Araújo JTC, Tavares Junior AG, Chorilli M (2023) Hybrid magnetic lipid-based nanoparticles for cancer therapy. Pharmaceutics 15:751

Madhavan D, Wallwiener M, Bents K, Zucknick M, Nees J, Schott S, Cuk K, Riethdorf S, Trumpp A, Pantel K (2014) Plasma DNA integrity as a biomarker for primary and metastatic breast cancer and potential marker for early diagnosis. Breast Cancer Res Treat 146:163–174

Maeng HM, Berzofsky JA (2019) Strategies for developing and optimizing cancer vaccines. F1000Res 8
Malorni L, Shetty P, De Angelis C, Hilsenbeck S, Rimawi M, Elledge R, Osborne C, De Placido S, Arpino G (2012) Clinical and biologic features of triple-negative breast cancers in a large cohort of patients with long-term follow-up. Breast Cancer Res Treat 136:795–804
Mandl SJ, Rountree RB, Dalpozzo K, Do L, Lombardo JR, Schoonmaker PL, Dirmeier U, Steigerwald R, Giffon T, Laus R (2012) Immunotherapy with MVA-BN®-HER2 induces HER-2-specific Th1 immunity and alters the intratumoral balance of effector and regulatory T cells. Cancer Immunol Immunother 61:19–29
Martin NT, Roy DG, Workenhe ST, van den Wollenberg DJM, Hoeben RC, Mossman KL, Bell JC, Bourgeois-Daigneault M-C (2019) Pre-surgical neoadjuvant oncolytic virotherapy confers protection against rechallenge in a murine model of breast cancer. Sci Rep 9:1865
Mathai RA, Vidya RVS, Reddy BS, Thomas L, Udupa K, Kolesar J, Rao M (2019) Potential utility of liquid biopsy as a diagnostic and prognostic tool for the assessment of solid tumors: implications in the precision oncology. J Clin Med 8:373
Mayor P, Gay LM, Lele S, Elvin JA (2017) BRCA1 reversion mutation acquired after treatment identified by liquid biopsy. Gynecol Oncol Rep 21:57
Mazo C, Kearns C, Mooney C, Gallagher WM (2020) Clinical decision support Systems in Breast Cancer: a systematic review. Cancers 12:369
McKinney SM, Sieniek M, Godbole V, Godwin J, Antropova N, Ashrafian H, Back T, Chesus M, Corrado GS, Darzi A, Etemadi M, Garcia-Vicente F, Gilbert FJ, Halling-Brown M, Hassabis D, Jansen S, Karthikesalingam A, Kelly CJ, King D, Ledsam JR, Melnick D, Mostofi H, Peng L, Reicher JJ, Romera-Paredes B, Sidebottom R, Suleyman M, Tse D, Young KC, De Fauw J, Shetty S (2020) International evaluation of an AI system for breast cancer screening. Nature 577:89–94
Mitra S, Lami MS, Ghosh A, Das R, Tallei TE, Fatimawali IF, Dhama K, Begum MY, Aldahish A, Chidambaram K, Emran TB (2022) Hormonal therapy for gynecological cancers: how far has science progressed toward clinical applications? Cancers 14:759
Mittendorf EA, Peoples GE, Singletary SE (2007) Breast cancer vaccines: promise for the future or pipe dream? Cancer 110:1677–1686
Mohan N, Hosain S, Zhao J, Shen Y, Luo X, Jiang J, Endo Y, Wu WJ (2019) Atezolizumab potentiates Tcell-mediated cytotoxicity and coordinates with FAK to suppress cell invasion and motility in PD-L1+ triple negative breast cancer cells. Onco Targets Ther 8:e1624128
Monaco ML, Idris OA, Essani K (2023) Triple-negative breast cancer: basic biology and Immuno-oncolytic viruses. Cancers 15:2393
Morse MA, Gwin WR III, Mitchell DA (2021) Vaccine therapies for cancer: then and now. Target Oncol 16:121–152
Moses C, Nugent F, Waryah CB, Garcia-Bloj B, Harvey AR, Blancafort P (2019) Activating PTEN tumor suppressor expression with the CRISPR/dCas9 system. Mol Therapy-Nucleic Acids 14: 287–300
Muzahir S (2020) Molecular breast cancer imaging in the era of precision medicine. Am J Roentgenol 215:1512–1519
Nagalingam A, Arbiser JL, Bonner MY, Saxena NK, Sharma D (2012) Honokiol activates AMP-activated protein kinase in breast cancer cells via an LKB1-dependent pathway and inhibits breast carcinogenesis. Breast Cancer Res 14:1–16
Nakamoto S, Shien T, Itoh M, Yamamoto Y, Ohsumi S, Yoshitomi S, Hikino H, Miyoshi K, Notsu A, Taira N (2023) Systemic immunity markers are associated with clinical outcomes of atezolizumab treatment in patients with triple-negative advanced breast cancer: a retrospective multicenter observational study. Clin Exp Med 1–10
Negrão de Figueiredo G, Ingrisch M, Fallenberg Eva M (2019) Digital analysis in breast imaging. Breast Care 14:142–150

Nicolás-Morales ML, Luisa-Sanjuan A, Gutiérrez-Torres M, Vences-Velázquez A, Ortuño-Pineda C, Espinoza-Rojo M, Navarro-Tito N, Cortés-Sarabia K (2022) Peptide-based vaccines in clinical phases and new potential therapeutic targets as a new approach for breast cancer: a review. Vaccine 10:1249

O'Brien N, Conklin D, Beckmann R, Luo T, Chau K, Thomas J, Mc Nulty A, Marchal C, Kalous O, von Euw E (2018) Preclinical activity of abemaciclib alone or in combination with antimitotic and targeted therapies in breast cancer. Mol Cancer Ther 17:897–907

O'Donovan PJ, Livingston DM (2010) BRCA1 and BRCA2: breast/ovarian cancer susceptibility gene products and participants in DNA double-strand break repair. Carcinogenesis 31:961–967

Oh D-Y, Bang Y-J (2020) HER2-targeted therapies—a role beyond breast cancer. Nat Rev Clin Oncol 17:33–48

Ombredane AS, Silva VRP, Andrade LR, Pinheiro WO, Simonelly M, Oliveira JV, Pinheiro AC, Gonçalves GF, Felice GJ, Garcia MP, Campos PM, Luz GVS, Joanitti GA (2021) In Vivo efficacy and toxicity of curcumin nanoparticles in breast cancer treatment: a systematic review. Front Oncol 11

Palleschi M, Tedaldi G, Sirico M, Virga A, Ulivi P, De Giorgi U (2021) Moving beyond PARP inhibition: current state and future perspectives in breast cancer. Int J Mol Sci 22:7884

Pantelidou C, Sonzogni O, De Oliveria TM, Mehta AK, Kothari A, Wang D, Visal T, Li MK, Pinto J, Castrillon JA, Cheney EM, Bouwman P, Jonkers J, Rottenberg S, Guerriero JL, Wulf GM, Shapiro GI (2019) PARP inhibitor efficacy depends on CD8+ T-cell recruitment via Intratumoral STING pathway activation in BRCA-deficient models of triple-negative breast cancer. Cancer Discov 9:722–737

Papandreou P, Gioxari A, Nimee F, Skouroliakou M (2021) Application of clinical decision support system to assist breast cancer patients with lifestyle modifications during the COVID-19 pandemic: a randomised controlled trial. Nutrients 13:2115

Paplomata E, O'Regan R (2014) The PI3K/AKT/mTOR pathway in breast cancer: targets, trials and biomarkers. Therap Adv Med Oncol 6:154–166

Perez CR, De Palma M (2019) Engineering dendritic cell vaccines to improve cancer immunotherapy. Nat Commun 10:5408

Petroni G, Formenti SC, Chen-Kiang S, Galluzzi L (2020) Immunomodulation by anticancer cell cycle inhibitors. Nat Rev Immunol 20:669–679

Prat A, Adamo B, Cheang MC, Anders CK, Carey LA, Perou CM (2013) Molecular characterization of basal-like and non-basal-like triple-negative breast cancer. Oncologist 18:123–133

Qi J, Liu Y, Hu J, Lu L, Dou Z, Dai H, Wang H, Yang W (2021) Identification of FPR3 as a unique biomarker for targeted therapy in the immune microenvironment of breast cancer. Front Pharmacol 11

Qin J-J, Yan L, Zhang J, Zhang W-D (2019) STAT3 as a potential therapeutic target in triple negative breast cancer: a systematic review. J Exp Clin Cancer Res 38:1–16

Riccardi F, Colantuoni G, Diana A, Mocerino C, Cartenì G, Lauria R, Febbraro A, Nuzzo F, Addeo R, Marano O (2018) Exemestane and Everolimus combination treatment of hormone receptor positive, HER2 negative metastatic breast cancer: a retrospective study of 9 cancer centers in the Campania region (southern Italy) focused on activity, efficacy and safety. Mol Clin Oncol 9:255–263

Robson M, Im S-A, Senkus E, Xu B, Domchek SM, Masuda N, Delaloge S, Li W, Tung N, Armstrong A (2017) Olaparib for metastatic breast cancer in patients with a germline BRCA mutation. N Engl J Med 377:523–533

Robson M, Tung N, Conte P, Im S-A, Senkus E, Xu B, Masuda N, Delaloge S, Li W, Armstrong A (2019) OlympiAD final overall survival and tolerability results: Olaparib versus chemotherapy treatment of physician's choice in patients with a germline BRCA mutation and HER2-negative metastatic breast cancer. Ann Oncol 30:558–566

Royce ME, Osman D (2015) Everolimus in the treatment of metastatic breast cancer. Breast Cancer 9:BCBCR.S29268

Royce M, Osgood C, Mulkey F, Bloomquist E, Pierce WF, Roy A, Kalavar S, Ghosh S, Philip R, Rizvi F (2022) FDA approval summary: abemaciclib with endocrine therapy for high-risk early breast cancer. J Clin Oncol 40:1155–1162

Rugo HS, Lerebours F, Ciruelos E, Drullinsky P, Ruiz-Borrego M, Neven P, Park YH, Prat A, Bachelot T, Juric D (2021) Alpelisib plus fulvestrant in PIK3CA-mutated, hormone receptor-positive advanced breast cancer after a CDK4/6 inhibitor (BYLieve): one cohort of a phase 2, multicentre, open-label, non-comparative study. Lancet Oncol 22:489–498

Sachdev JC, Sandoval AC, Jahanzeb M (2019) Update on precision medicine in breast cancer. In: Precision medicine in cancer therapy. Springer, Cham, pp 45–80

Sak K (2012) Chemotherapy and dietary phytochemical agents. Chemother Res Pract 2012

Saura C, Oliveira M, Feng Y-H, Dai M-S, Chen S-W, Hurvitz SA, Kim S-B, Moy B, Delaloge S, Gradishar W (2020) Neratinib plus capecitabine versus lapatinib plus capecitabine in HER2-positive metastatic breast cancer previously treated with≥ 2 HER2-directed regimens: phase III NALA trial. J Clin Oncol 38:3138

Schettini F, Giudici F, Bernocchi O, Sirico M, Corona SP, Giuliano M, Locci M, Paris I, Scambia G, De Placido S (2021) Poly (ADP-ribose) polymerase inhibitors in solid tumours: systematic review and meta-analysis. Eur J Cancer 149:134–152

Schlam I, Swain SM (2021) HER2-positive breast cancer and tyrosine kinase inhibitors: the time is now. NPJ Breast Cancer 7:56

Schmid P, Cortes J, Dent R, Pusztai L, McArthur H, Kümmel S, Bergh J, Denkert C, Park YH, Hui R (2022) Event-free survival with pembrolizumab in early triple-negative breast cancer. N Engl J Med 386:556–567

Schöffski P, Cresta S, Mayer IA, Wildiers H, Damian S, Gendreau S, Rooney I, Morrissey KM, Spoerke JM, Ng VW (2018) A phase Ib study of pictilisib (GDC-0941) in combination with paclitaxel, with and without bevacizumab or trastuzumab, and with letrozole in advanced breast cancer. Breast Cancer Res 20:1–12

Selmin OI, Donovan MG, Stillwater BJ, Neumayer L, Romagnolo DF (2020) Epigenetic regulation and dietary control of triple negative breast cancer. Front Nutr 7:159

Sharma GN, Dave R, Sanadya J, Sharma P, Sharma K (2010) Various types and management of breast cancer: an overview. J Adv Pharm Technol Res 1:109

Sherlock BD, Coster AC (2023) Oncolytic virus treatment of human breast cancer cells: modelling therapy efficacy. J Theor Biol 560:111394

Siegel RL, Miller KD, Fuchs HE, Jemal A (2021) Cancer statistics, 2021. CA Cancer J Clin 71:7–33

Siegel RL, Miller KD, Wagle NS, Jemal A (2023) Cancer statistics, 2023. CA Cancer J Clin 73:17–48

Sobhani N, D'Angelo A, Pittacolo M, Roviello G, Miccoli A, Corona SP, Bernocchi O, Generali D, Otto T (2019) Updates on the CDK4/6 inhibitory strategy and combinations in breast cancer. Cells 8:321

Song X, Liu C, Wang N, Huang H, He S, Gong C, Wei Y (2021) Delivery of CRISPR/Cas systems for cancer gene therapy and immunotherapy. Adv Drug Deliv Rev 168:158–180

Steelman LS, Martelli AM, Cocco L, Libra M, Nicoletti F, Abrams SL, McCubrey JA (2016) The therapeutic potential of mTOR inhibitors in breast cancer. Br J Clin Pharmacol 82:1189–1212

Stergiopoulou D, Markou A, Strati A, Zavridou M, Tzanikou E, Mastoraki S, Kallergi G, Georgoulias V, Lianidou E (2023) Comprehensive liquid biopsy analysis as a tool for the early detection of minimal residual disease in breast cancer. Sci Rep 13:1258

Sun X, Liu K, Lu S, He W, Du Z (2022) Targeted therapy and immunotherapy for heterogeneous breast cancer. Cancers 14:5456

Tagde P, Kulkarni GT, Mishra DK, Kesharwani P (2020) Recent advances in folic acid engineered nanocarriers for treatment of breast cancer. J Drug Deliv Sci Technol 56:101613

Tang W, Wan S, Yang Z, Teschendorff AE, Zou Q (2017) Tumor origin detection with tissue-specific miRNA and DNA methylation markers. Bioinformatics 34:398–406

Tang H, Chen J, Wang L, Li Q, Yang Y, Lv Z, Bao H, Li Y, Luan X, Li Y, Ren Z, Zhou X, Cong D, Liu Z, Jia J, Chen H, Zhao W, Meng Q, Sun F, Pei J (2020) Co-delivery of epirubicin and paclitaxel using an estrone-targeted PEGylated liposomal nanoparticle for breast cancer. Int J Pharm 573:118806

Tarekegn K, Keskinkilic M, Kristoff TJ, Evans ST, Kalinsky K (2023) The role of immune checkpoint inhibition in triple negative breast cancer. Expert Rev Anticancer Ther 23: 1095–1106

Teo ZL, Versaci S, Dushyanthen S, Caramia F, Savas P, Mintoff CP, Zethoven M, Virassamy B, Luen SJ, McArthur GA (2017) Combined CDK4/6 and PI3Kα inhibition is synergistic and immunogenic in triple-negative breast cancer. Cancer Res 77:6340–6352

Tiwari H, Rai N, Kajal, Salvi P, Singh SK, Gautam V (2023) Recent advances in nanomaterials-based targeted drug delivery for preclinical cancer diagnosis and therapeutics. Bioengineering 10:760

Tong CWS, Wu M, Cho WCS, To KKW (2018) Recent advances in the treatment of breast cancer. Front Oncol 8

Tran P, Lee S-E, Kim D-H, Pyo Y-C, Park J-S (2020) Recent advances of nanotechnology for the delivery of anticancer drugs for breast cancer treatment. J Pharm Investig 50:261–270

Tung N, Garber JE (2022) PARP inhibition in breast cancer: Progress made and future hopes. NPJ Breast Cancer 8:47

Tung N, Lin NU, Kidd J, Allen BA, Singh N, Wenstrup RJ, Hartman A-R, Winer EP, Garber JE (2016) Frequency of germline mutations in 25 cancer susceptibility genes in a sequential series of patients with breast cancer. J Clin Oncol 34:1460

Turner S, Chia S, Kanakamedala H, Hsu W-C, Park J, Chandiwana D, Ridolfi A, Yu C-L, Zarate JP, Rugo HS (2021) Effectiveness of alpelisib+ fulvestrant compared with real-world standard treatment among patients with HR+, HER2–, PIK3CA-mutated breast cancer. Oncologist 26: e1133–e1142

Tzeng Y-DT, Chang S-E, Mei R, Javey M (2018) Liquid biopsy prevents inaccurate Her2 status determination by in situ hybridization in a patient with invasive ductal adenocarcinoma of the breast: case report. Case Rep Oncol 10:857–862

Venetis K, Cursano G, Pescia C, D'Ercole M, Porta FM, Blanco MC, Frascarelli C, Ivanova M, Rocco EG, Fusco N (2023) Liquid biopsy: cell-free DNA based analysis in breast cancer. J Liquid Biopsy : 100002

Vlastos G, Verkooijen HM (2007) Minimally invasive approaches for diagnosis and treatment of early-stage breast cancer. Oncologist 12:1–10

Wan S, Kumar D, Ilyin V, Al Homsi U, Sher G, Knuth A, Coveney PV (2021) The effect of protein mutations on drug binding suggests ensuing personalised drug selection. Sci Rep 11:13452

Wang R, Li X, Zhang H, Wang K, He J (2017) Cell-free circulating tumor DNA analysis for breast cancer and its clinical utilization as a biomarker. Oncotarget 8:75742

Wardell SE, Ellis MJ, Alley HM, Eisele K, VanArsdale T, Dann SG, Arndt KT, Primeau T, Griffin E, Shao J (2015) Efficacy of SERD/SERM hybrid-CDK4/6 inhibitor combinations in models of endocrine therapy–resistant breast cancer. Clin Cancer Res 21:5121–5130

Wojciechowska U, Olasek P, Czaderny K, Didkowska J (2016). Nowotwory złośliwe w Polsce w 2014 roku, Centrum Onkologii-Instytut im. Marii Sklodowskiej-Curie, Krajowy Rejestr …

Wu Q-X, Yuan S-X, Ren C-M, Yu Y, Sun W-J, He B-C, Wu K (2016) Oridonin upregulates PTEN through activating p38 MAPK and inhibits proliferation in human colon cancer cells. Oncol Rep 35:3341–3348

Xia W, Mullin RJ, Keith BR, Liu L-H, Ma H, Rusnak DW, Owens G, Alligood KJ, Spector NL (2002) Anti-tumor activity of GW572016: a dual tyrosine kinase inhibitor blocks EGF activation of EGFR/erbB2 and downstream Erk1/2 and AKT pathways. Oncogene 21:6255–6263

Xu Y, Li Z (2020) CRISPR-Cas systems: overview, innovations and applications in human disease research and gene therapy. Comput Struct Biotechnol J 18:2401–2415

Yamamoto T, Kanaya N, Somlo G, Chen S (2019) Synergistic anti-cancer activity of CDK4/6 inhibitor palbociclib and dual mTOR kinase inhibitor MLN0128 in pRb-expressing ER-negative breast cancer. Breast Cancer Res Treat 174:615–625

Yang B, Jeang J, Yang A, Wu T, Hung C-F (2014) DNA vaccine for cancer immunotherapy. Hum Vaccin Immunother 10:3153–3164

Yuan Y, Egelston C, Colunga Flores O, Chaurasiya S, Lin D, Chang H, Chong LMO, Seiz A, Shah M, Meisen WH (2023) CF33-hNIS-anti-PD-L1 oncolytic virus followed by trastuzumab-deruxtecan in a patient with metastatic triple negative breast cancer: a case study. Therc Adv Med Oncol 15:17588359231210675

Zaheed O, Samson J, Dean K (2020) A bioinformatics approach to identify novel long, non-coding RNAs in breast cancer cell lines from an existing RNA-sequencing dataset. Non-coding RNA Res 5:48–59

Zanetti-Dällenbach R, Wight E, Fan AX-C, Lapaire O, Hahn S, Holzgreve W, Zhong XY (2008) Positive correlation of cell-free DNA in plasma/serum in patients with malignant and benign breast disease. Anticancer Res 28:921–925

Zanfardino M, Franzese M, Pane K, Cavaliere C, Monti S, Esposito G, Salvatore M, Aiello M (2019) Bringing radiomics into a multi-omics framework for a comprehensive genotype–phenotype characterization of oncological diseases. J Transl Med 17:337

Zhang M, Jang H, Nussinov R (2020) PI3K inhibitors: review and new strategies. Chem Sci 11: 5855–5865

Zheng D, He X, Jing J (2023) Overview of artificial intelligence in breast cancer medical imaging. J Clin Med 12:419

Zhou Y (2023) HER2/neu-based vaccination with li-key hybrid, GM-CSF immunoadjuvant and trastuzumab as a potent triple-negative breast cancer treatment. J Cancer Res Clin Oncol: 1–8

Zhu S-Y, Yu K-D (2022) Breast cancer vaccines: disappointing or promising? Front Immunol 13: 828386

Zubair T, Bandyopadhyay D (2023) Small molecule EGFR inhibitors as anti-cancer agents: discovery, mechanisms of action, and opportunities. Int J Mol Sci 24:2651

# Updates on the Management of Ductal Carcinoma In Situ of the Breasts

Michael Tiong-Hong Co

**Abstract**

The current gold standard treatment for ductal carcinoma in situ (DCIS) is by complete excision with a general consensus of 2 mm surgical margin requirement. Adjuvant radiotherapy is usually advised in patients who received lumpectomy. However, over the last decades, de-escalating DCIS treatment has been a debatable topic. Ongoing randomized controlled trials (RCTs) are now in progress to evaluate whether or not we could safely omit surgery in selected low-risk DCIS patients.

This chapter aims to summarize the updated evidence on de-escalating DCIS treatment.

Retrospective analysis of cancer registry data and systematic review have suggested that selected low-risk DCIS (low-grade and older age patients) can be safely treated conservatively. Ongoing randomized controlled trials (e.g. LORIS trial) are underway to evaluate the long-term clinical outcomes of conservative treatment of selected low-risk DCIS patients. However, apart from oncologic outcomes, the patient's acceptance and the psychosocial impact of conservative treatment for DCIS should be taken into consideration in the treatment planning for DCIS.

The importance of oncologic safety is obvious in decision-making, but to patients, all other aforementioned factors should be equally important. After all, as clinicians, we treat patients, not the DCIS.

M. T.-H. Co (✉)
Division of Breast Surgery, Department of Surgery, Queen Mary Hospital, Hong Kong SAR, China

Department of Surgery, School of Clinical Medicine, LKS Faculty of Medicine, The University of Hong Kong, Hong Kong SAR, China
e-mail: mcth@hku.hk



Interdisciplinary Cancer Research, https://doi.org/10.1007/16833_2023_188
Published online: 5 October 2023

**Keywords**

Ductal carcinoma in situ · Over-diagnosis · Over-treatment

## 1 Introduction

Ductal carcinoma in situ (DCIS) is the earliest detectable form of breast cancer; its incidence has increased significantly from 1.87 in 100,000 in the 1970s to 32.5 in 100,000 in the early 2000s in America. It is postulated that both increased breast cancer awareness and wide introduction of breast screening programmes are the reasons for the increased incidence (Virnig et al. 2010; Allegra et al. 2010).

Treatment of DCIS has not changed much over the decades, of which the standard of care for treatment is complete surgical excision. The extent of surgical margin requirement has, however, been controversial. Currently, the generally accepted surgical margin is more than or equal to 2 mm (Burstein et al. 2004); this surgical principle was agreed and endorsed by a group of expert panel at the 15th St. Gallen International Breast Cancer Consensus Conference held in March 2017 in Vienna, Austria, although some members from the panel suggested narrower margins (St. Gallen 2017).

## 2 Breast Cancer Screening and Diagnosis of DCIS

One intrinsic drawback of screening is over-diagnosis/over-treatment. An independent review by the Department of Health of the United Kingdom in 2013 has concluded that screening saves lives but also results in over-diagnosis (Marmot et al. 2013). Screening is a common strategy to pick up early or pre-malignant conditions and to improve treatment outcomes of cancers. Commonly screened malignancies include colonic cancers, cervical cancers and breast cancers. Let us take colonic cancer as an example, where colonic polyp is known to be the true precursor of colorectal malignancy. The introduction of screening colonoscopy and polypectomy results in the drop in colorectal cancer incidence (Siegel et al. 2017). On the contrary, increased detection and early treatment and excision of DCIS over the last decades have not decreased the incidence of breast cancers (Bleyer and Welch 2012). This finding is truly intriguing.

In fact, long-term survival analyses of DCIS have demonstrated that the disease-specific survival rates were constantly higher than 95% in patients with DCIS regardless of the type of surgical treatments received (Narod et al. 2015; Sagara et al. 2016; Worni et al. 2015). Studies have also evaluated the natural course of DCIS without surgical excision – retrospective studies have found that not all DCIS eventually become invasive. Progression to invasive disease was reported in approximately 25–50% of cases over one or two decades (Page et al. 1982, 1995). With these figures in mind, international researchers have been studying the possibility of over-diagnosis and over-treatment of DCIS.

Nonetheless, some other studies have also suggested that nearly half of the recurrences of DCIS were invasive despite 'adequate' surgical treatments with satisfactory surgical margins (Early Breast Cancer Trialists' Collaborative G 2010; Wapnir et al. 2011a, b). Under-treatment would become a real concern if surgical treatment is completely omitted from the management of DCIS.

## 3 Classification Systems of DCIS

To understand over-diagnosis and over-treatment of DCIS, one has to understand the risk stratification system of DCIS. Conventional classification of DCIS includes the Van Nuy classification which is based mainly on the Bloom and Richardson nuclear grading (Bloom and Richardson 1957; Silverstein et al. 1995). Van Nuys Prognostic Index (VNPI) was developed to predict the risk of local recurrence after DCIS excision. It is a straightforward scoring system based on histopathological features such as tumour size, surgical margin, nuclear grade and presence of comedo necrosis. VPNI has been shown to be reproducible in real-life clinical practice (Poller et al. 1995). Like many scoring systems, some studies also failed to demonstrate the reliable predictability of recurrence by VPNI (Macausland et al. 2007). Apart from VPNI, the Armed Forces Institute of Pathology (AFIP) classification system is another DCIS classification system which classifies DCIS based on the presence of atypia or necrosis (Tavassoli 1998). At this moment, none of the aforementioned classifications have been universally accepted to guide DCIS treatment.

## 4 Challenges in Histopathological Diagnosis

Instead of being considered as one single disease entity, DCIS is now regarded as a spectrum of disease, ranging from pre-malignant atypical ductal hyperplasia to high-grade DCIS with or without microinvasion (Fig. 1) (Co 2020).

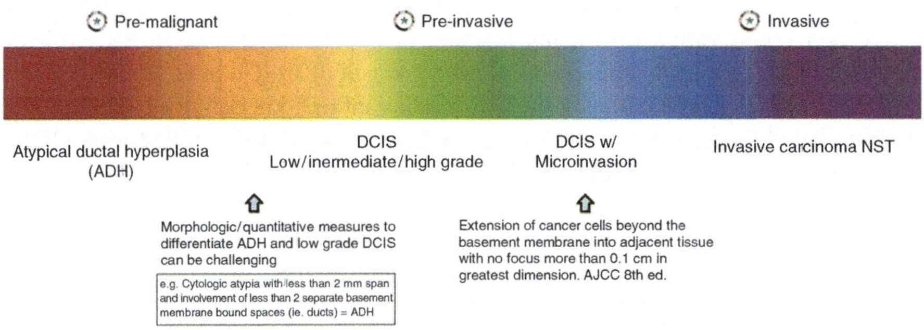


**Fig. 1** Spectrum of pre-malignant, pre-invasive and invasive breast neoplasms. (Co 2020)

Despite being classified under the same group, clinical behaviour and natural history vary significantly between different pre-invasive or pre-malignant breast neoplasms. What complicates the matter further is that there is no single reproducible pathognomonic pathological feature that could distinguish atypical ductal hyperplasia (ADH) from low-grade DCIS, ADH and low-grade DCIS are closely related at the molecular level. Over the decades, pathologists have tried to look for criteria to distinguish ADH from DCIS; one simple way was to distinguish them by size – involvement of two separate ducts and size of more than 2 mm have been proposed as arbitrary cut-off numbers for DCIS (Tavassoli and Norris 1990). With this vague pathological definition and classification, variability in the pathological diagnosis of DCIS becomes inevitable. A study on 6900 slides of breast biopsies with histopathological diagnosis of DCIS revealed significant inter-observer (pathologist) variability (Elmore et al. 2016). With the help of modern technology including artificial intelligence (AI), scientists have been trying to look for ways to improve the diagnosis accuracy of DCIS in biopsy specimens. A recent systematic review has evaluated 23 studies on the use of AI in diagnosing DCIS in pathology specimens. A mean accuracy of 82.52% and a mean sensitivity of 83.00% were achieved with the use of AI (Co et al. 2023a, b).

## 5 Paradigm Change of the Current Standard of Care

As mentioned earlier, the current standard of care for DCIS is complete surgical excision – either by lumpectomy followed by adjuvant radiotherapy or by mastectomy. Factors affecting the choice of surgical options include 1. breast size, 2. extent of DCIS (i.e. tumour size), 3. multi-focality or multi-centricity of the tumour, 4. location of the tumour, and 5. patient preference.

However, retrospective studies have shown that with excellent survival of DCIS in the long run, there was no significant difference in breast cancer–specific survival rates among patients with low-grade DCIS regardless of the treatment received (surgery or active surveillance) (Sagara et al. 2017). A recent systematic review has summarized the updated evidence of breast cancer–specific survival after conservative treatment of DCIS. A total of 34,007 women with low-risk DCIS were included in the analysis in this systematic review – pooled analysis found that both active and conservative treatments resulted in excellent 10-year BCSS, with no statistically insignificant difference (98.6% versus 96.0%) (Co et al. 2023a, b).

Apart from surgical treatment, the use of adjuvant radiotherapy for DCIS has also been revisited by a recent meta-analysis and a randomized controlled trial (RCT). Meta-analysis of four RCTs on more than 3700 patients found that radiotherapy after lumpectomy for DCIS resulted in a decreased risk of local recurrence (hazard ratio of 0.46) but did not improve survival (Correa et al. 2010). RCT by Radiation Therapy Oncology Group (RTOG) revealed similar findings that administering adjuvant radiotherapy after lumpectomy only resulted in a slight decrease in the risk of local recurrence. More importantly, the survival data was lacking (McCormick et al. 2015).

It is not until more recently that gene expression analysis (e.g. Oncotype DX DCIS recurrence score) has been studied as a predictor as to which patient with DCIS could be safely spared from adjuvant radiotherapy after lumpectomy. The Oncotype DX DCIS recurrence score was validated on patients with DCIS who participated in the Eastern Cooperative Oncology Group E5194 trial (Hughes et al. 2009). Patients with a high-risk DCIS recurrence score were found to have higher rates of local recurrence and experienced a higher absolute benefit by adding radiotherapy (RT) to surgery when compared to those with a lower recurrence risk score.

Evidence on adjuvant hormonal therapy for DCIS has been controversial. The best evidence comes from a systematic review and a large randomized controlled trial that concluded that additional hormonal treatment in DCIS patients might reduce the risk of contralateral invasive breast cancer in DCIS patients, but there was no reduction in invasive breast cancer events if the ipsilateral breast was treated by surgery and/or radiotherapy (Staley et al. 2014; Forbes et al. 2016).

With the emergence of new evidence suggesting the possibility of over-treatment of DCIS, several ongoing prospective studies, including randomized controlled trials, are currently on the way to evaluate the long-term oncological outcomes of conservative treatment for selected low-risk DCIS, i.e. de-escalating DCIS treatment. These include the Low-risk DCIS (LORIS) trial from the United Kingdom, the Management of low-risk DCIS (LORD) trial from Europe and the Comparing an Operation to Monitoring, With or Without Endocrine Therapy (COMET) study from the United States (Francis et al. 2014). A similar study design was adopted across all three trials, comparing standard surgical treatment (+/− radiotherapy and endocrine therapy) with active surveillance. Patients on the surveillance arm were closely monitored with serial mammogram for 5–10 years. Please refer to Table 1 for the summary of the three ongoing clinical trials on de-escalating DCIS treatments.

Results of these ongoing randomized controlled trials are yet to be published, but some researchers have already applied the "low-risk" criteria to their local territory-wide cancer registry data, aiming to evaluate the clinical outcomes of conservative treatment of "low-risk" DCIS. One study applied the LORIS criteria to stratify its cancer registry database into low-risk versus high-risk DCIS. This study found that the 10-year breast cancer–specific survival in the non-operated low-risk DCIS group was 97.8%, which was comparable to the operative counterpart (Co et al. 2021a, b), concluding that LORIS-low-risk DCIS could be safely managed conservatively.

## 6 Treating DCIS or Treating Patients with DCIS?

It is very obvious that the future direction of research for DCIS is to assess the feasibility to further de-escalate its treatment in selected low-risk patients. While the three major randomized controlled trials (LORD, LORIS and COMET) and a number of other ongoing prospective studies provide good evidence in terms of the long-term survival outcomes of conservative treatment for DCIS, it is important that clinicians never forget that it is the patient that they are treating, not the disease itself.

**Table 1** Trial criteria in the three DCIS clinical trials

| Trial | COMET | LORIS | LORD |
|---|---|---|---|
| *Inclusion criteria* | | | |
| Age | >=40 | >=46 | >=45 |
| Nuclear grade | Low and intermediate | Low and intermediate | Low |
| Morphology | Calcifications only | Calcifications only | Calcification only |
| Hormonal receptor | ER PR +ve and HER-2 −ve | Not applicable | Not applicable |
| *Exclusion criteria* | | | |
| Symptoms | Exclude if symptomatic | Exclude if symptomatic | Exclude if symptomatic |
| Comedo necrosis | Exclude | Exclude | Not applicable |
| Synchronous invasive cancer | Exclude | Exclude | Exclude |
| Bilaterality | Exclude bilateral | Include bilateral | Exclude bilateral |
| High risk | Include | Exclude if high risk (National Institute for Health and Care Excellence (NICE) guidelines) | Exclude for family history of BRCA carrier |
| History of chemoprevention | Exclude | Not applicable | Not applicable |

*ER* oestrogen receptor, *PR* progestrogen receptor

Retrospective cancer registry studies have already shown promising oncological outcomes of conservative treatment for DCIS. However, oncological outcomes should never be taken as the only factor to consider in the treatment decision of DCIS. Acceptance of de-escalating DCIS treatment from the patient's perspective is equally important. The psychological burden of 'living with' DCIS should be addressed; costs of intensive clinical and radiological surveillance should also be taken into account. Being oncologically safe for de-escalating DCIS treatment does not necessarily equate to the standard of care. In one recent study on patients' perspectives towards conservative treatment for DCIS, 85% of patients decided on operative treatment with the hypothetical scenario of having a 'low-risk' DCIS.

## 7 Clinical Factors to Be Considered When Treating Low-Risk DCIS

One suggested holistic approach to manage low-risk DCIS is to follow the OSCAR criteria i.e. oncologic safety, symptom, costs of treatment, anxiety of patient, and risks of the treatment (Co 2021). The importance of oncologic safety is obvious in decision-making, but to patients, all other aforementioned factors should be equally important. After all, as clinicians, we treat patients, not the DCIS.

# 8 Conclusion

The long-term survival of DCIS is excellent. The current trend of DCIS research is along the lines of de-escalating treatment. Retrospective analysis of cancer registry data and systematic review have suggested that selected low-risk DCIS (low-grade and older age patients) can be safely treated conservatively. Ongoing randomized controlled trials are underway to evaluate the long-term clinical outcomes of conservative treatment of selected low-risk DCIS patients. However, apart from oncologic outcomes, the patient's acceptance and psychosocial impact of conservative treatment for DCIS should also be taken into consideration in the treatment planning for DCIS.

**Compliance with Ethical Standards** The author of this chapter does not have a conflict of interest to report.

This chapter does not involve unpublished studies on human subjects or animals.

The chapter is not funded by external sources of sponsorships.

# References

Allegra CJ, Aberle DR, Ganschow P, Hahn SM, Lee CN, Millon-Underwood S et al (2010) National Institutes of Health State-of-the-Science Conference statement: diagnosis and management of ductal carcinoma in situ September 22–24, 2009. J Natl Cancer Inst 102(3):161–169. https://doi.org/10.1093/jnci/djp485

Bleyer A, Welch HG (2012) Effect of three decades of screening mammography on breast-cancer incidence. N Engl J Med 367:1998–2005

Bloom HJG, Richardson WW (1957) Histologic grading and prognosis in breast cancer: a study of 1409 cases of which 359 have been followed for 15 years. Br J Cancer 11:359–377

Burstein HJ, Polyak K, Wong JS et al (2004) Ductal carcinoma in situ of the breast. N Engl J Med 350:1430–1441

Co M (2020) Ductal carcinoma in situ of the breasts: over-diagnosis, over-treatment and a decade of lost direction. Precis Med Sci 9:4–8

Co M (2021) De-escalating treatment for ductal carcinoma in situ – has the pendulum swung too far? (Invited opinion). Cancer Treat Res Commun 28:100438

Co M, Ngan RKC, Mang OWK, Tam AHP, Wong KH, Kwong A (2021a) Long-term survival outcomes of 'low risk' ductal carcinoma in situ from a territory-wide cancer registry. Clin Oncol (R Coll Radiol) 33(1):40–45

Co M, Lee A, Kwong A (2021b) Non-surgical treatment for ductal carcinoma in situ of the breasts – a prospective study on patient's perspective. Cancer Treat Res Commun 26:100241. https://doi.org/10.1016/j.ctarc.2020.100241

Co M, Lau YCC, Qian YXY et al (2023a) Artificial intelligence in histologic diagnosis of ductal carcinoma in situ. Mayo Clin Proc Digit Health 1(3):267–275

Co M, Cheng KCK, Yeung YH, Lau KC, Qian Z, Wong CM, Wong BY, Sin ELK, Wong HYS, Ma CH (2023b) Clinical outcomes of conservative treatment for low-risk ductal carcinoma in situ: a systematic review and pooled analysis. Clin Oncol (R Coll Radiol) 35(4):255–261

Correa C, McGale P, Taylor C, Wang Y, Clarke M, Davies C et al (2010) Overview of the randomized trials of radiotherapy in ductal carcinoma in situ of the breast. J Natl Cancer Inst Monogr 41:162–177. https://doi.org/10.1093/jncimonographs/lgq039

Early Breast Cancer Trialists' Collaborative G, Correa C, McGale P et al (2010) Overview of the randomized trials of radiotherapy in ductal carcinoma in situ of the breast. J Natl Cancer Inst Monogr 2010(41):162–177
Elmore JG, Nelson HD, Pepe MS et al (2016) Variability in pathologists' interpretations of individual breast biopsy slides: a population perspective. Ann Intern Med 164:649–655. https://doi.org/10.7326/M15-0964
Forbes JF, Sestak I, Howell A et al (2016) Anastrozole versus tamoxifen for the prevention of locoregional and contralateral breast cancer in postmenopausal women with locally excised ductal carcinoma in situ (IBIS-II DCIS): a double-blind, randomised controlled trial. Lancet 387(10021):866–873
Francis A, Thomas J, Fallowfield L, Wallis M, Bartlett JMS, Brookes C et al (2014) Addressing overtreatment of screen detected DCIS; the LORIS trial. Eur J Cancer 51(16):2296
Hughes LL, Wang M, Page DL et al (2009) Local excision alone without irradiation for ductal carcinoma in situ of the breast: a trial of the Eastern Cooperative Oncology Group. J Clin Oncol 27(32):5319–5324. https://doi.org/10.1200/JCO.2009.21.8560
Macausland SG, Hepel JT, Chong FK et al (2007) An attempt to independently verify the utility of the Van Nuys Prognostic Index for ductal carcinoma in situ. Cancer 110:2648–2653
Marmot MG, Altman DG, Cameron DA et al (2013) The benefits and harms of breast cancer screening: an independent review. Br J Cancer 108:2205–2240
McCormick B, Winter K, Hudis C, Kuerer HM, Rakovitch E, Smith BL et al (2015) RTOG 9804: a prospective randomized trial for good-risk ductal carcinoma in situ comparing radiotherapy with observation. J Clin Oncol 33(7):709–716. https://doi.org/10.1200/JCO.2014.57.9029
Narod SA, Iqbal J, Giannakeas V et al (2015) Breast cancer mortality after a diagnosis of ductal carcinoma in situ. JAMA Oncol 1:888–896
Page DL, Dupont WD, Rogers LW et al (1982) Intraductal carcinoma of the breast: follow-up after biopsy only. Cancer 49(4):751–758
Page DL, Dupont WD, Rogers LW et al (1995) Continued local recurrence of carcinoma 15–25 years after a diagnosis of low grade ductal carcinoma in situ of the breast treated only by biopsy. Cancer 76(7):1197–1200
Poller DN, Barth A, Slamon DJ et al (1995) Prognostic classification of breast ductal carcinoma-in-situ. Lancet 345(8958):1154–1157. https://doi.org/10.1016/S0140-6736(95)90982-6
Sagara Y, Freedman RA, Vaz-Luis I et al (2016) Patient prognostic score and associations with survival improvement offered by radiotherapy after breast-conserving surgery for ductal carcinoma in situ: a population-based longitudinal cohort study. J Clin Oncol 34:1190–1196
Sagara Y, Julia W, Golshan M, Toi M (2017) Paradigm shift toward reducing overtreatment of ductal carcinoma in situ of breast. Front Oncol 7:192. Published 2017 Aug 28. https://doi.org/10.3389/fonc.2017.00192
Siegel RL, Miller KD, Fedewa SA et al (2017) Colorectal cancer statistics 2017. CA Cancer J Clin 67:177–193
Silverstein MJ, Poller DN, Waisman JR, Colburn WJ, Barth A, Gierson ED, Blamey RW et al (1995) Prognostic classification of breast ductal carcinoma-in-situ. Lancet 345:1154–1157
St. Gallen International Breast Cancer Consensus Conference, March 2017, Vienna
Staley H, McCallum I, Bruce J (2014) Postoperative Tamoxifen for ductal carcinoma in situ: cochrane systematic review and meta-analysis. Breast 23(5):546–551
Tavassoli FA (1998) Ductal carcinoma in situ: introduction of the concept of ductal intraepithelial neoplasia. Mod Pathol 11:140–154
Tavassoli FA, Norris HJ (1990) A comparison of the results of long-term follow-up for atypical intraductal hyperplasia and intraductal hyperplasia of the breast. Cancer 65:518–529
Virnig BA, Tuttle TM, Shamliyan T, Kane RL (2010) Ductal carcinoma in situ of the breast: a systematic review of incidence, treatment, and outcomes. J Natl Cancer Inst 102:170–178. https://doi.org/10.1093/jnci/djp482

Wapnir IL, Dignam JJ, Fisher B et al (2011a) Long-term outcome of invasive ipsilateral breast tumor recurrences after lumpectomy in NSABP-17 and B-24 randomized clinical trials for DCIS. J Natl Cancer Inst 103:478–488
Wapnir IL, Dignam JJ, Fisher B et al (2011b) Long-term outcomes of invasive ipsilateral breast tumor recurrences after lumpectomy in NSABP B-17 and B-24 randomized clinical trials for DCIS. J Natl Cancer Inst 103(6):478–488
Worni M, Akushevich I, Greenup R et al (2015) Trends in treatment patterns and outcomes for ductal carcinoma in situ. J Natl Cancer Inst 107

# Nanostructured Lipid Carrier as a Strategy for the Treatment of Breast Cancer

Kammila Martins Nicolau Costa, Camila Beatriz Barros Araújo, Ana Lígia Santiago Barros, Mariana Rillo Sato, and João Augusto Oshiro-Júnior

**Abstract**

Breast cancer is a malignant neoplasm of epithelial cells that affects both sexes and represents a critical public health issue that causes thousands of deaths annually worldwide. Conventional treatment, although clinically effective, is associated with the development of resistance in more advanced stages of the disease and high toxicity in healthy cells, leading to a poor prognosis. Furthermore, the active compounds that represent these therapeutic classes possess inadequate biopharmaceutical classifications, which hinders their solubility in a physiological environment. Aiming to optimize clinical practice, the nanoencapsulation of these active substances in nanostructured lipid carriers is an attractive and viable alternative for the pharmaceutical industry, benefiting patients. Therefore, this chapter will approach an overview regarding breast cancer and the present NLC developed and used in the preclinical studies against the disease in focus, addressing their physicochemical characteristics, advantages, and disadvantages, among other important aspects.

K. M. N. Costa · C. B. B. Araújo
Pharmaceutical Sciences Postgraduate Program, Center for Biological and Health Sciences, State University of Paraíba - UEPB, Campina Grande, PB, Brazil

A. L. S. Barros
Centro Universitário UNIFACISA, Campina Grande, PB, Brazil

M. R. Sato
Department of Drugs and Medicines, Faculdade de Ciências Farmacêuticas, UNESP - Universidade Estadual Paulista, Campus Araraquara, Araraquara, SP, Brazil

J. A. Oshiro-Júnior (✉)
Pharmaceutical Sciences Postgraduate Program, Center for Biological and Health Sciences, State University of Paraíba - UEPB, Campina Grande, PB, Brazil

Centro Universitário UNIFACISA, Campina Grande, PB, Brazil
e-mail: joaooshiro@yahoo.com.br; joao.oshiro@pq.cnpq.br

 
Interdisciplinary Cancer Research, https://doi.org/10.1007/16833_2022_13
Published online: 28 October 2022

**Keywords**

Drug delivery system · Nanodrugs · Oncology · Target

## 1 Introduction

Breast cancer is an epithelial cell malignant neoplasm that affects both sexes, more commonly diagnosed in the female population (25%) and rarely among the male population (1%). Additionally, the estimated worldwide incidence is approximately 1.6 million new cases per year. It is a heterogeneous disease at the molecular level, and its causes are not fully understood. At its initial stage (non-metastatic), it has a chance of cure in approximately 70–80% of patients. At the metastatic stage, despite the advances in clinical therapy, it has the ability to transfer and spread within other distant organs, such as the bones, liver, lungs, and brain, which contributes to its high mortality rate (Sun et al. 2017; Harbeck et al. 2019; Liang et al. 2020).

In this context, the conventional treatment for the advanced cancer stage is performed with chemotherapy regimens involving anthracycline antibiotics, mitotic spindle inhibitors, alkylating agents, platinum compounds, and antimetabolites, among others, and, in many cases, with an unfavorable prognosis. Approximately 18% of patients are free of the disease for more than 5 years and 10% of patients for periods longer than 20 years. In addition, these chemotherapeutics have problems related to physicochemical and pharmacokinetic characteristics, resulting in low selectivity and, consequently, low drug concentration in the tumor mass and intense cytotoxic activity against healthy cells, generating numerous side effects for the patient (Abotaleb et al. 2018; Tong et al. 2018; Zhang et al. 2020).

Furthermore, these active substances have inadequate biopharmaceutical classification, which limits their solubility in a physiological environment, requiring chemical structural modifications, such as salt and prodrug formation, or physical modifications, such as particle size reduction, use of surfactants, complexation, or nanoencapsulation in nanostructured drug delivery systems (NDDS) (Dolatkhah et al. 2020; Liu et al. 2020). Among these strategies, the NDDS is an attractive alternative, well accepted by the scientific community, the pharmaceutical industry, and patients, due to results obtained through more than 20 years of use in clinical practice (Szebeni et al. 2016; Gondim et al. 2019; Alves et al. 2020; de Assis et al. 2020; de Oliveira et al. 2020).

NDDS exhibit advantages over conventional delivery systems, such as (a) ability to maintain the therapeutic range with small variations; (b) prevention of local and systemic toxic effects; (c) increased plasma concentrations of substances with a short half-life; (d) better target specificity; (e) ability to carry hydrophilic and hydrophobic molecules; and (f) reduction in the number of daily doses, which can increase patient adherence to treatment (Oshiro-Júnior et al. 2018; Barros et al. 2020; Oshiro-Junior et al. 2020; Costa et al. 2021).

In the literature, numerous NDDS are used for encapsulation of chemotherapeutics; however, nanostructured lipid carriers (CLNs), which are characterized by having a matrix composed of a mixture of liquid lipid and solid lipid at room temperature, have multifunctionalities such as higher incorporation of active substances and low toxicity (Sato et al. 2017, 2019). Advantages produced by the excipients that constitute its matrix (lipophilic) have a greater vectorization for the cells of the neoplastic tissue (Jaiswal et al. 2016; Salvi and Pawar 2019; Haider et al. 2020), in addition to being reproduced on a large scale and presenting higher stability in relation to other NDDS (Beloqui et al. 2016; Gordillo-Galeano and Mora-Huertas 2018; Khosa et al. 2018).

Therefore, based on the above panorama, in this chapter, we will present an overview of the application of different CLNs in breast cancer therapy. Accordingly, papers previously published in renowned journals have been used, exploring the earliest and the most recent publications about the subject.

Thus, the authors expect that this chapter helps researchers and others interested in the matter better understand and update their knowledge about the use of nanostructured lipid carriers as a strategy for the treatment of breast cancer.

## 2 Conventional Treatment of Breast Cancer

The treatment for breast cancer comprises a process with several steps and numerous protocols, including surgery, radiotherapy, systemic treatment (chemotherapy, hormone therapy, immunotherapy, among others), and rehabilitation. The treatment aims to promote the local removal of the tumor, interrupt its drainage pathways through lymph nodes in the region, and carry out the prognostic characterization and treatment orientation through chemotherapy or hormonal drugs, as they aim at adjuvant therapy, that is, protection against the development of metastases or palliative therapy in case of already established metastases. The association of surgical, chemotherapy, and radiotherapy treatments is more effective in the probabilistic reduction of disease recurrence (Akram et al. 2017; Sun et al. 2017; Waks and Winer 2019).

In most cases, the practice of treatment consists of a combination of different techniques, such as surgery and chemotherapy. Until the 1950s, tumor removal surgeries were practically the only form of cancer therapy. Surgeries, however, cannot treat metastatic cancers. For that, treatments that affect essentially every organ in the body are needed. Consequently, chemotherapeutics, immune-mediated therapies, and radiotherapy have become the focus of cancer research (Akram et al. 2017; Liyanage et al. 2019).

In the 1940s, during the combats of the Second World War, with the use of mustard gas and the discovery of its activity against lymphomas, chemotherapy began, one of the most used systemic treatments to treat cancer, and has been carried out in combination therapy, with one or more drugs. Since the 1940s, several drugs have been obtained through extraction from natural organisms or chemical synthesis and proved to be very useful to increase the survival of patients (Agarwal 2016; Smith 2017; ElBagoury and Kotb 2018).

Over the past 50 years, research into new anticancer drugs has introduced approximately 70 drugs into the therapy. However, the relative similarity between cancer cells and normal cells represents to cancer treatment the great challenge of distinguishing between these cells (Agarwal 2016; Tong et al. 2018). Currently, the most effective drug treatment for advanced cases should be performed with anthracycline (doxorubicin or epirubicin) and taxane (paclitaxel, docetaxel) chemotherapeutics. Other agents (capecitabine, gemcitabine, methotrexate, 5-fluorouracil, vinorelbine, vinblastine, mitomycin, etoposide, cyclophosphamide, cisplatin, irinotecan) are also used in some protocols (Brasil 2018).

The indication for cytotoxic therapy arises when the breast cancer is considered advanced, metastatic, or relapsed. Chemotherapy for breast cancer can be given as a single agent or in combination. The latter is aimed at those patients who need an immediate pharmacological response. Studies have shown that the association of chemotherapy agents has increased response rate, time to disease progression, and a small overall survival benefit. However, it raised the toxicity (Brasil 2018).

Nevertheless, the numerous side effects lead to problems that hinder treatment adherence and, consequently, the recovery of the patient's well-being (Machado et al. 2019). Chemotherapy is based on the ability to destroy neoplastic cells, which feature rapid cell division when compared to normal cells. However, important side effects can occur on normal cells that also show rapid division, such as those in the gastrointestinal system, capillaries, and immune system, causing diarrhea, nausea, vomiting, alopecia, and higher susceptibility to infections. In addition, several tumors show resistance to chemotherapies and other therapies applied concomitantly. This phenomenon happens in several types of breast cancer, although the mechanisms leading to this are not yet fully understood. Nanotechnology offers several features that have proven to circumvent these disadvantages (Pearce et al. 2017; Li et al. 2020).

## 2.1 Lipidic Nanoparticle as a Strategy for the Treatment of Breast Cancer

Nanotechnology has been advancing over the years in the most diverse areas, being a multidisciplinary science. In the pharmaceutical field, one of the research targets is its employment in the treatment of cancer, as the latter is a disease that has been progressively increasing over the years, with limited therapy and major side effects. In order to overcome these disadvantages, the drug's specific targeting strategy toward tumor cells is used, without affecting healthy cells. In this way, it is possible to reduce side effects using nanotechnology (Avitabile et al. 2018; Chiari-Andréo et al. 2019; de Oliveira et al. 2020; Jain et al. 2020; dos Santos et al. 2021).

In the early 1990s, the first nanodrug CaelyxTM/Doxil was approved by the FDA, the first liposome system containing doxorubicin (DOX) measured approximately 100 nm and was able to encapsulate a high concentration of DOX (10,000–15,000) (Pandey et al. 2016; Wibroe et al. 2016; Russell et al. 2018) for the treatment of breast cancer.

Most of the FDA-approved nanodrugs heretofore are composed of liposomes, which consist of colloidal structures and vesicles composed of lipid bilayers, usually phospholipids and cholesterol, surrounded by an aqueous core, thus allowing the incorporation of hydrophilic or hydrophobic active pharmaceutical ingredients (APIs) (Pandey et al. 2016; Peretz Damari et al. 2018; Robson et al. 2018; Van Tran et al. 2019; Ghasemiyeh and Mohammadi-Samani 2020).

Nevertheless, Doxil® should be used with caution so as not to induce fatal congestive heart failure. The half-life is 55 h, and the recommended dose is 20–50 mg/m$^2$, given every 3–4 weeks. Elimination is slower when compared to conventional DOX, and side effects occur when the formulation is dispensed and administered inappropriately (Barenholz 2012; Russell et al. 2018).

Despite the great euphoria at the time, considered as the "magic bullet," the clinical results did not follow expectations. Thus, the intrinsic limitations related to liposomes such as low solubility in the physiological environment and inadequate shelf life, high production cost, factors resulting from the excipients used, and suffering from oxidation of phospholipids (Pandey et al. 2016; Kanásová and Nesměrák 2017; Maja et al. 2020) associated with instability in administration, in which after reconstitution they show low stability (Shao et al. 2017; Doskocz et al. 2020), led researchers to start looking for NDDS to circumvent these disadvantages.

Solid lipid nanoparticles (NLS) appear in this context as they comprise the first generation of solid lipid transport systems. NLS emerged as an alternative to liposomes around 1990 by the research group of Müller, Gasco, and Westesen, presenting advantages such as controlled release, increased stability, biocompatibility, low production cost, and a wide spectrum of applications (Shidhaye et al. 2008; Karamchedu et al. 2020).

Solid lipid nanoparticles (NLS) are colloidal systems with a structure similar to nanoemulsions and size that usually ranges from 50 to 1000 nm. The lipid core in nanoemulsions is liquid, opposed to nanoparticles, in which the lipid core is in the solid state and has the advantage of increasing control over the release kinetics of encapsulated compounds and improving the stability of chemically sensitive lipophilic ingredients (Yadav et al. 2013).

The disadvantages of NLS are their low encapsulation capacity, since the use of very similar lipids generates perfect crystals; while the drug is located between the lipid chains and in crystal imperfections, the high organization of the crystals decreases the encapsulation efficiency. In order to improve the encapsulation efficiency and minimize the removal of the active compound from the particles during storage, in the late 1990s, nanostructured lipid carriers emerged (NLC) (Yadav et al. 2013).

Considered the second generation of solid lipid particles, it emerged in 1999, as an improvement of NLS, due to the partial replacement of the solid lipid by a liquid lipid at room temperature. The association of these lipids promotes a less organized lipid matrix with more imperfections, which can accommodate a greater amount of the active compound and provide more flexibility for the modulation of release. In addition, NLCs possess no toxicity; are protected against chemical degradation of labile substances; have increased drug stability, easy sterilization and transposition to large scale, and good biocompatibility; and are also capable of increasing the chemical stability of compounds sensitive to light, oxidation, and hydrolysis

(Bahari and Hamishehkar 2016; Salvi and Pawar 2019; Haider et al. 2020; Medeiros et al. 2020), arousing interest through the scientific and commercial community for their use in anticancer drugs (Iqbal et al. 2012; Selvamuthukumar and Velmurugan 2012; Fang et al. 2013; Naseri et al. 2015; Zhou et al. 2016).

## 2.2 Nanostructured Lipid Carriers as a Strategy for the Treatment of Breast Cancer

Generally, the main components present in the formulation of NLCs are solid and liquid lipids, surfactants, water, and the active compound to be incorporated, varying the proportion of these components according to the methodological technique and type of carrier in development. When used for specific purposes, as in the case of chemotherapy, some measures must be applied more strictly, such as the biocompatibility between the components, route of administration, and toxicity (Beloqui et al. 2016; Gordillo-Galeano and Mora-Huertas 2018; Khosa et al. 2018; Haider et al. 2020).

Among the most applied components for the development of CLN, solid lipids glyceryl behenate, carnauba wax, cetyl palmitate, glyceryl monostearate, and solid lipid stearic acid are included. As for liquid lipids, among the most cited are oleic acid, squalene, medium-chain triglycerides (MCT)/caprylic, and capric triglycerides. As for surfactants, Pluronic®, Tween®, poloxamer, polyvinyl alcohol (PVA), and soy lecithin may be used (Khosa et al. 2018; Haider et al. 2020) in order to prevent particle aggregation and stabilize them (Beloqui et al. 2016; Ganesan and Narayanasamy 2017).

As the main techniques for obtaining these various NLCs are listed as effective, they can be separated according to their methods of organic solvents, high energy, and low energy. The preparation methods used are high-pressure homogenization (HPH), ultrasonically assisted emulsification, emulsification solvent diffusion method, multiple emulsion, phase inversion, and microemulsion, being the first cited preferred by the industry (Gordillo-Galeano and Mora-Huertas 2018; Khosa et al. 2018; Salvi and Pawar 2019).

The lipid matrix may be influenced by the nature of the lipids used in the preparation of NLC, resulting in three known models. The "imperfect NLC" (Fig. 1a) consists of a mixture of glycerides composed of different fatty acids. This mixture can increase the distance between the fatty acid chains of the glycerides, causing crystal imperfections (Marcato 2009; Müller et al. 2002; Junior 2016).

These imperfections allow the formation of more space to accommodate the active compound, thus increasing encapsulation efficiency. The second model, called "amorphous NLC" (Fig. 1b), occurs due to a mixture of solid and liquid lipids (oils), leading to solidification of the lipid matrix, resulting in an amorphous structure, without crystallization, a process responsible for the removal of the drug that is always undesirable for drug transport systems (Beloqui et al. 2017; Ganesan and Narayanasamy 2017). The third model is called "multiple NLC" (Fig. 1c), and

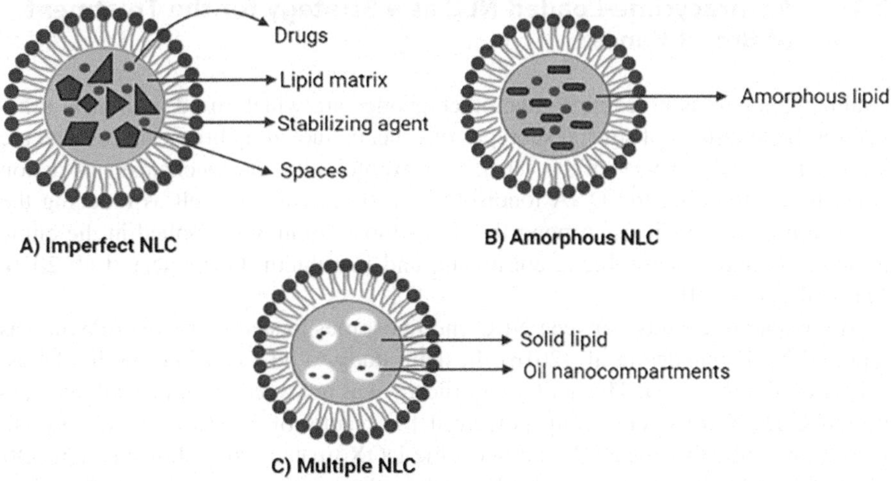


**Fig. 1** Structural diagram of nanoparticles: (**a**) imperfect NLC, (**b**) amorphous NLC, (**c**) multiple NLC. Elaborated with the assistance of BioRender

they are formed by several gaps due to the large amount of liquid lipid, which favors the solubility of the drug inside the droplets. The "multiple" type can incorporate a higher amount of actives and control their release, due to the encapsulation of the matrix of solid lipid particles attributed to the use of liquid lipids in high concentrations well above their solubility in solid lipid at room temperature, resulting in the development of nano-compartments (Ganesan and Narayanasamy 2017; Khosa et al. 2018; Salvi and Pawar 2019).

Regardless of the type, NLCs take advantage of the enhanced permeability and retention (EPR). It is possible to develop a vectorization strategy that aims to improve the therapeutic efficacy of chemotherapeutics and reduce their adverse effects with the use of nanostructured systems (Oshiro-Junior et al. 2020; Silvestre et al. 2020). This strategy aims for the space among healthy cells to be only between 15 and 30 nm and the drugs to be encapsulated within 50 to 800 nm nanoparticles, making them unable to cross the vessel wall of healthy regions of the body. However, in inflamed regions or even those where tumors are located, the endothelial cells are less joined together, resulting in an accumulation of nanoparticles in the tumor tissue adjacent to the blood vessel (Vieira and Gamarra 2016).

## 3 Cancer Drug-Loaded NLC

In this section, we introduce the main drugs carried in nanostructured lipid carriers as strategy for the treatment of breast cancer.

### 3.1 Anthracycline-Loaded NLC as a Strategy for the Treatment of Breast Cancer

Anthracyclics or anthracycline chemotherapeutics are widely used in clinical practice for the treatment of several subtypes of cancers, including breast cancer, and are known for their adverse effects, such as cardiotoxicity. Its mechanism of action consists of inhibiting the DNA topoisomerase II enzyme, as well as inducing the formation of free radicals. Among the drugs most commonly applied in the clinic are doxorubicin, daunorubicin, epirubicin, and idarubicin (Guenancia et al. 2016; Marinello et al. 2018).

A comparative study between NLC and liposomes loaded with doxorubicin was reported by Fernandes et al. (2016). In their study, a 4T1 cell line murine breast cancer model was used. The antitumor efficacy was analyzed by tumor volume, and the NLC-DOX treatment group portrayed the results of 73.5% of tumor growth inhibition, while the free DOX and liposome-DOX treated group showed the results of 48.8% and 68.0%, respectively. Regarding the scintigraphic images, the calculated ratio for the animals treated with NLC-DOX showed tumor-muscle compliance 3 and 2 times lower than the control and liposome-DOX. Therefore, the NLC system can be a strategy to obtain an effective antitumor activity.

Assuming that the compounds docosahexaenoic acid (DHA) and α-tocopherol succinate (TS) can increase the antitumor activity of chemotherapy agents, Lages et al. (2020) studied the incorporation of doxorubicin, DHA, and TS into NLCs with the hypothesis of increasing the efficacy of DOX and reducing its toxicity. The results of the in vitro studies revealed synergistic effects (IC ranging from 0.59 to 0.78) against 4T1 tumor cells, while in the in vivo study, the NLC-DHA-DOX-TS showed a higher rate of tumor inhibition (76.6%) compared to the groups treated with NLC-DOX (66.5%) and free DOX (64.6%) and decreased mice mortality. The death of mice was observed from the eighth day after the start of the study in the groups treated with free DOX and NLC-DOX, while in the groups treated with the blank-NLC and NLC-DHA-DOX-TS systems, no deaths and no clinical changes were observed (Lages et al. 2020).

Fernandes et al. (2018) also incorporated TS with DOX into NLC, and in their evaluation, the NLC-TS-DOX system showed higher tumor absorption, along with long blood clearance, blood levels decreased so that T1/2 α was 72.17 min, T ½ β was 1107.71 min, and area under the curve (AUC) was 2750.32% ID $min^{-1}$. The T ½ β of NLC-TS-DOX was fourfold higher compared to free Tc-DOX. Tumor growth was significantly controlled, and animals receiving control treatment (blank-NLC) had tumor volume of 1,324 ± 156 $mm^3$. In contrast, the NLC-DOX and DOX groups showed volumes of 721 ± 82 and 642 ± 27 $mm^3$, respectively. Notably, the NLC-TS-DOX group had the smallest tumor volume (293 ± 91 $mm^3$). Therefore, the authors have concluded that it would be an alternative to increase the effectiveness of DOX.

Deng et al. (2017) studied the antineoplastic drug pirarubicin (THP), a specific agent in the S phase of the cell cycle. Incorporation of THP into NLC was proposed, and it was observed that the system (THP-NLC) reduced the toxicity compared to the free drug and improved the immune system of mice with breast cancer. THP-NLC was also co-injected with the peptide iRGD responsible for increasing the penetration of drugs into the tumor, thus achieving the control of tumor growth in addition to an increase in cytotoxic T lymphocyte mediators and cytokines IFN-γ and INF- α.

## 3.2 Taxane-Loaded NLC as a Strategy for the Treatment of Breast Cancer

Another pharmacological class highly employed in clinical practice is taxane. They are often applied in the treatment of metastatic breast cancer. Taxanes act by intervening in cell division processes during mitosis by stabilizing actin microtubules and, consequently, leading to tumor growth delay. Among their derivatives, the most used drugs are paclitaxel and docetaxel (Willson et al. 2019; Kanda et al. 2020).

Paclitaxel (PTX) is a widespread drug in cancer therapy. However, its low solubility limits intravenous administration. The inclusion of this drug in NLC has been promising to overcome these limitations. Pedro et al. (2019) incorporated PTX into NLC with high encapsulation efficiency. The in vitro results demonstrate cytotoxic and anti-clonogenic effects of NLC-PTX against larger MDA-MB-231 cells in comparison to free PTX. At the highest concentration tested (2500 nM), cell viability was 56.0 ± 3.2% (free PTX) and 38.0 ± 5.0% (NLC-PTX) after 24 h. NLC-PTX was able to reduce the viability of cancer cells given the greater uptake of the carrier system by these, in addition to observing the greater accumulation of NLC-PTX in the tumor region in mice with Ehrlich tumor.

Cabazitaxel (CBZ) is a semi-synthetic antineoplastic drug that works by disrupting the microtubule network in cells. Widely widespread in prostate cancer therapy, Chand et al. (2021) developed CBZ-loaded NLCs against breast cancer cell lines. The results showed a homogeneous system with uniformity capable of incorporating the drug without chemical interference. In in vitro studies, NLC-CBZ showed 6 and 2.5 times greater cytotoxicity compared to the free drug, in addition to greater absorption and apoptosis of MDA-MB-468 and MCF-7 cells. The percentage of apoptotic cells was (control group) 3.756 ± 0.568%, (CBZ) 7.84 ± 1.24%, and (NLCs loaded with CBZ) 14.53 ± 2.76% in 24 h.

Varshosaz et al. (2018) used a drug extensively used in the treatment of cancers for the functionalization of NLC incorporated with docetaxel. Herceptin®, or trastuzumab, is an injectable drug based on monoclonal antibodies. The formulation's toxicity was tested in two breast cancer cell lines: MDA-MB-468 and BT-474, HER-2 negative and positive, respectively. The results obtained showed that the functionalization with trastuzumab negatively influenced the trapping of docetaxel inside the NLC matrix, decreasing its incorporation capacity in the matrix, but increasing the cell death rate, with higher mortality for the BT-474 lineage when compared to pure NLC, in addition to increasing cellular uptake. When comparing the cell response of each lineage, the docetaxel incorporated into the functionalized NLC became more toxic to cells of the BT-474 lineage than those of the MDA-MB-468 lineage.

### 3.3 Antagonist of Estrogen Receptor-Loaded NLC as a Strategy for the Treatment of Breast Cancer

Estrogen is a term used for substances analogous to endogenous hormones. Natural human estrogens are estrone, estradiol, and estriol. Selective estrogen receptor modulators (SERMs) are variations of the 17-β estradiol molecule. Widely used in hormone therapy for the treatment of breast cancer, hormonal antagonists prevent estrogens from binding to their receptors, blocking their action as growth factors for neoplastic cells. The main drugs used are triphenylethylene derivatives, such as raloxifene, tamoxifen, benzothiophenes, and aromatase inhibitors (Diniz et al. 2017; Gabriel et al. 2017).

Shete et al. (2013) aimed to evaluate the in vitro efficacy of nanostructured lipid carriers using glyceryl monostearate (GMS) as the solid lipid and Labrafil WL 2609 BS as oil, loaded with tamoxifen (Tmx-NLC) against breast cancer cell lines MCF-7 and ZR-75-1. The study demonstrated better activity for the Tmx-NLC system in comparison to the free drug against MCF-7 cells, drug retention in the NLC nucleus over 80% after 6 h, and an increase in bioavailability, as plasma concentration was 2.71 times higher and extended the T1/2 by 7.10 times when compared to tamoxifen suspension (Tmx-susp).

Sabzichi et al. (2016a) investigated the combination of tamoxifen and melatonin applied into NLCs to increase cytotoxicity and apoptosis in MCF-7 breast cancer cells. The results indicated that the formulation remained stable during 4 months in terms of size and drug release. The mean zeta potential was $-19 \pm 07$ mV, and the polydispersity index (PDI) of 0.29 suggests a narrow particle size distribution. IC50 values for melatonin and tamoxifen were $1.3 \pm 0.4$ mM and $30.7 \pm 5.2$ μM, respectively. Hence, the inhibition of MCF-7 cell growth occurred more effectively in the NLC-melatonin system in comparison to the free substance. However, the percentage of apoptotic cells increased from $35 \pm 2\%$ to $40 \pm 4\%$ in cells treated with melatonin-loaded NLCs in combination with tamoxifen. Free melatonin and tamoxifen indicated the percentage of apoptosis of $13 \pm 0.4\%$ and $4 \pm 0.2\%$, respectively. The maximum trapping efficiency and drug load of the NLCs were calculated as 84.3% and 18.7%.

Another example of tamoxifen-loaded NLC was investigated by How et al. (2013), which reveals the reduction in cytotoxicity and side effects compared to the free drug due to the more selective targeting provided by the system. Among the results, it was possible to observe that the NLC system showed a constant release rate of 96.5% after 24 h. IC50 values of TAM and TAM-NLC in MCF-7 cell lines were 2,72 and 5,56 μg/mL, respectively; for 4T1 cell lines, they were 5.13 μg/mL−1 and 5.19 μg/mL−1, respectively. NLCs exhibited greater drug loading capability, in addition to increasing drug solubility and stability.

Raloxifene is a benzothiophene derivative that possesses preventive activity against breast cancer. Given that fact, Soni et al. (2020) developed nanostructured lipid carriers incorporated with raloxifene using the simple ultrasound method, Compritol 888 ATO as a solid lipid and Transcutol® HP as a liquid lipid. The results of the in vitro study revealed an initial rapid release for 4 h and sustained release for up to 24 h. The RLN-NLCs system was in the 121 nm nanometer range with a trapping efficiency of 81%. The in vivo study demonstrated a 4.79-fold increase in bioavailability for RLN-NLCs, as well as an increase in cytotoxicity against MCF-7 cells.

Letrozole (LTZ) is an antineoplastic agent of choice against breast tumors. It is a nonsteroidal aromatase inhibitor, acting by inhibiting estrogen biosynthesis. Based on this knowledge, Sabzichi et al. (2016b) investigated the ability of folate-targeted nanostructured lipid carriers to enhance the drug's (LTZ) efficacy against the MCF-7 lineage. The results confirmed activation of apoptosis and arrest in Sub-G1 of the cell cycle through folate-LTZ NLCs. In the study of scanning electron microscopy (SEM), the uniformity in the size and homogeneity of the particles were observed, in addition to reaching an IC50 of 2.2 μM. There was also a significant increase in the percentage of apoptotic cells to 42.2% when compared to free LTZ.

## 3.4 HIT, LEAD, or Drug Candidate-Loaded NLC as a Strategy for the Treatment of Breast Cancer

The development of new drugs comprises a complex and extensive process that includes basic research steps, target identification, compound optimization, and preclinical and clinical studies. According to the terms used in medicinal chemistry, bioactive molecules (hits) are identified through screenings of natural and synthetic products or combinatorial collections. The initially identified molecules are hardly commercialized due to low potency and affinity, consequently requiring optimization in terms of their pharmacodynamic and pharmacokinetic properties. The optimized compounds are called lead compounds or prototypes and have confirmed activity in a cell assay; however, they are remodeled through structure-activity relationships (SAR) in order to increase their potency and selectivity (Guido et al. 2010; Ferreira 2018).

Quercetin (Q) is a type of flavonoid frequently used because of its antioxidant properties. In order to increase the uptake of quercetin by cancer cells, Sun et al. (2014) synthesized nanostructured lipid carriers (Q-NLC). Q-NLC showed good chemical and physical stability with diameter and zeta potential without significant changes at 4 °C and polydispersion index below 0.3. The aqueous solubility of quercetin increased 1,000-fold, and the Q-NLC system containing 20 mg of Q was fully dissolved in 10 mL of 1×PBS (equivalent to 2 g/L). The release pattern was sustained, and the 1:1 ratio of lipids and surfactant resulted in a size of 32 nm and greater encapsulation efficiency (95%). After 24 and 48 h, the system decreased the viability of cancer cells to about 80% and 68%, respectively, and induced apoptosis by 20 μm in MCF-7 and MDA-MB-231 cells.

Godugu et al. (2016) developed NLC formulations based on diindolylmethane (DIM) derivatives, a phytochemical formed from the metabolization of indole-3-carbinol, derived from cruciferous vegetables. DIM has low solubility and high lipophilicity, which makes the compound with limited bioavailability. This study was focused on the improvement in oral bioavailability and anti-breast cancer effects. The results demonstrated that the DIM-10 NLC formulation significantly reduced the tumor by 2.56-fold when compared to animals treated with the free drug. In vitro studies showed a sustained release profile of the drug. The oral bioavailability of DIM-10 increased from 27% to 80% in NLC formulations, in comparison to the free drug. The system produced a two- to threefold increase in anticancer effects compared to the respective free drug-treated groups.

Another phytochemical compound that exhibits antitumor properties is thymoquinone, found in the *Nigella sativa* plant. In the study performed by Ng et al. (2015), a thymoquinone-loaded nanostructured lipid carrier (TQ-NLC) was developed. The formulations have an average diameter of less than 50 nm, a polydispersity index (PDI) of less than 0.25, and negative zeta potential. MDA-MB-231 breast cancer cells treated with TQ-NLC showed apoptosis and formation of apoptotic bodies, as well as induced cell cycle arrest in G2/M and S phases and increased cell population in the sub-G1 phase in the concentrations 1.56 and 3.125 μM of TQ-NLC. IC50 values from the MTT assay showed that TQ-NLC was more potent against the MDA-MB-231 cell line. The formulation was relatively nontoxic to normal cell lines.

Nordin et al. (2020) carried out two studies using the natural product citral incorporated in the nanostructured lipid carrier (NLC-Citral) and later evaluated its antitumor effects. In response to in vitro and in vivo studies, it was possible to verify the antitumor effect on MDA-MB 231 cells, antimetastatic effect, and induction of apoptosis, with a higher number of cells in tumors treated with NLC-Citral (3.5-fold) and citral (2.1-fold) compared to blank-NLC. NLC-Citral treated tumors demonstrated a more prominent size reduction, in addition to decreasing the expression of cytokines, specifically IL-6 and IL-1α and the tumor pro-growth cytokine IL-10, which was reduced to 291 ± 1, 5 pg/mL in NLC-Citral at 360 ± 0.8 pg/mL. In the cell viability study, NLC-Citral and free citral reduced the viability of 4T1 cells by 20.3 ± 1.76 μg/mL and 21.4 ± 2.82 μg/mL, respectively, suggestive that NLC is a potential and effective delivery system.

Gambogic acid (GA) is a compound of natural origin and is extensively studied due to its pharmacological activities, which include an antitumor action. However, the respective molecule has low selectivity and delivery to tumor cells, as well as a low solubility. In this context, Kebebe et al. (2019) made use of dimeric c-peptides, aiming to increase the targeting of the AG incorporated into the NLC, proposing to circumvent its disadvantages. The results of the in vitro cytotoxicity studies revealed that the functionalized nanosystem showed higher cytotoxicity (0.25 and 0.5 μg/mL) in comparison to the nanosystem without the presence of the peptide, in addition to presenting a better rate of cell accumulation in the tumoral area and, consequently, the inhibition of its development.

Resveratrol (RSV) is a polyphenol commonly found in black grape seeds. It has been reported to have anticancer activities in numerous studies. Poonia et al. (2019) resolved to incorporate resveratrol into folic acid-functionalized carriers to generate a parenteral formulation for effective, optimized, and selective delivery to breast tumors. With a particle size of 88.3 ± 3.1 nm and high trapping efficiency of 88.0 ± 2.6%, the formulation presented a significant targeting potential when compared to NLC that were not functionalized, in addition to having ninefold more

(57.92 ± 4.15 μg h/mLh) Cmax compared to free resveratrol (6.37 ± 1.16 μg h/mLh). Regarding the release study, RSV-NLCs and RSV-FA-NLCs showed high drug release up to 35.12 and 28.17%, in the first 2 h, followed by sustained release up to 99.95 and 97.65% until 72 h, respectively. The systems were shown to be significantly more cytotoxic ($p < 0.05$) against MCF-7 and A549 cells compared to free RSV.

Reinforcing their study, Gadag et al. (2021) also used resveratrol (RSV) incorporated into NLCs, using microneedle (MN) arrays as an alternative for a treatment possibly less invasive. As in the previous study, the RVT-NLCs administered by microneedles showed an increase in Cmax, and the IC50 of the RVT-NLCs was lower (27.50 ± 3.43) than that of the pure RVT (33.93 ± 7.34), highlighting that RVT-NLCs are more potent. The developed system showed greater anticancer activity in MDA-MB-231, in addition to an improved pharmacokinetic profile via MN array and increased cytotoxicity.

## 3.5 Multidrug-Loaded NLC as a Strategy for the Treatment of Breast Cancer

Multidrug resistance is the main factor responsible for the unsatisfactory result in breast cancer therapy. The actions of multidrug NLCs can be optimized concerning pharmacological properties and system stabilization. In order to suppress this multidrug resistance in breast cancer therapy, Li et al. (2018) developed a nanostructured lipid carrier co-administering β-lapachone (Lapa) and doxorubicin (DOX) (LDNLC). The release behavior was sustained compared to free drugs, and both DOX and Lapa were efficiently loaded into NLC and uniformly distributed in the nanoparticle matrix. The size distribution of LDNLC was close to 100 nm, and the drug loading reached 6.4% for Lapa and 5.1% for DOX, with an efficiency of 83.8% and 92.3%, respectively. The in vitro drug uptake was studied to observe the subcellular distribution and release of the drug in ADR MCF-7 cells. It was observed that the cells treated with LDNLC showed intense fluorescence clustered in the nucleus. In addition, LDNLC revealed a half-life of 5.3-fold longer than free DOX.

NLC containing vitamin D was studied by Sabzichi et al. (2017) relying on a concomitant administration of doxorubicin to observe whether there was an increase in the efficacy of DOX in the treatment of breast cancer. The synergy of the treatment caused a significant increase in the cancer cell apoptosis index (MCF-7) since vitamin D had a percentage of apoptosis of 14 and DOX of 16, but when incorporated into NLC, an increase of twice in the apoptosis index (33%) was observed, in addition to the decrease in cell proliferation.

Further aiming at the concomitant use of agents attempting better efficacy, Di et al. (2016) proposed to produce nanocarriers based on the co-delivery of doxorubicin (DOX) and cisplatin (CDDP) for the treatment of breast cancer. As a result of their in vitro study, the D-C-NLC system exhibited greater cytotoxicity and synergistic effect of the two drugs on tumor cells. Regarding the in vivo study, D-C-NLCs showed greater antitumor activity and efficacy, establishing the synergism of the drugs and presenting it as promising in nanomedicine.

Paclitaxel (PTX), a synthetic drug, as aforementioned, is known to be widely used in the treatment of breast cancer, but it has several disadvantages, from low solubility to relevant adverse effects. Attempting to reduce these disadvantages, the group composed by Zhang et al. (2019) decided to incorporate PTX with chlorin e6 (Ce6) into FA-functionalized NLC. As a photosensitizing agent, Ce6 acts synergistically with PTX, generating reactive oxygen species that are harmful to target cells. It was observed that, in addition to increasing the solubility of both and the intracellular absorption as they are dispersed in NLC, the AF present on the carrier surface helped in the delivery of the compounds in a targeted manner to the tumor.

## 3.6 Developing Theory

Atorvastatin is a drug belonging to the class of statins, acknowledged and widespread for reducing cholesterol. Nevertheless, studies have emerged encompassing its effects in modulating cell growth and apoptosis. Gambhire et al. (2018) proposed to improve tumor targeting of atorvastatin by integrating it into NLCs. At first, the lot of NLCs had standard sizes <200 nm and encapsulation efficiency of 90.42% and remained stable both in acidic pH and in buffer solution. Due to the presence of Span 80, the dispersion was stable with a zeta potential of 51.01 mV. In the in vitro study using human breast adenocarcinoma cell lines, total growth inhibition (TGI) and 50% growth inhibition (GI50) results were observed at 27.4 μg/mL and <10 μg/mL, respectively.

In the article designed by Lanna et al. (2021), artemether (ART) was combined with docosahexaenoic acid (DHA) triglyceride and incorporated into NLC. ART is derived from artemisinin, a semi-synthetic sesquiterpene lactone used as an antimalarial. In his study, he showed a reduction in tumor viability in addition to apoptosis in MCF-7 cells, as well as higher selectivity for tumor cells and reduced clonogenicity. Cytotoxic activity was 4.5-fold improved for the 50/30 μM ART/DHA combination compared to NLC loaded with ART alone, and the improvement in nanocarrier-mediated activity was evidenced in MDA-MB-231 cells. The potency of the ART/DHA combination was >400 times for NLC in MCF-7, as well as reduced colony formation by 59%.

Tran et al. (2016) developed NLCs through hot homogenization to incorporate artesunate (ART), another semi-synthetic derivative of artemisinin. The ART-NLC system showed good stability, with high encapsulation efficiency (over 75%) and good uptake by breast cancer cells, in addition to stronger cytotoxicity compared to the free drug. The viability of MCF-7 after 24 h was 60%, after treatment with ART at a concentration of 25 μg/mL, and 20% with ART-NLCs, while after 48 h of incubation, cell viability was 46% and 29% for ART and ART-NLCs, respectively.

Zerumbone is a natural compound, extracted from ginger, which is popularly known for its anti-inflammatory, antifungal, and antitumor activities. Similar to other natural compounds, one of its biggest pharmacokinetic disadvantages, when referring to administration, is its solubility. Mohamad et al. (2015) suggested incorporating zerumbone into NLC systems as an alternative to enable its use. The antitumor activity of this active compound is not well defined in the literature, but the authors reported that they noticed controlled tumor growth and metastases through in vivo tests with the use of mice and 4T1 lineage cells. The results in the level of nitric oxide and malondialdehyde in the lipid peroxidation of the tumors reduced 1.6 and 1.16 times, respectively, compared to the control rats. In parallel, the levels of COX-2 and VEGF proteins were significantly reduced, as well as a 35-fold downregulation in tumors harvested from mice treated with NLC-zerumbone. When encapsulated in NLC, the effects were more promising, in addition to an increase in the anti-inflammatory activity of the compound.

## 3.7 Stimuli-Responsive NLC as a Strategy for the Treatment of Breast Cancer

Drug delivery co-administration systems have also been extensively studied as a result of their synergistic effects associated with selective delivery and improvement in the bioavailability of the active incorporated therein. Often, this co-administration can also positively influence the characteristics of the nanosystem used, causing it to be activated through a pre-established stimulus, such as pH or temperature variation (Almeida et al. 2017; Deirram et al. 2019).

Vohra et al. (2013) performed an experiment in which nanostructured lipid carriers were loaded with docetaxel. Furthermore, this entire system was incorporated into a thermoreversible polymeric gel system. The outcome of the process resulted in 85% encapsulation efficiency of docetaxel and 96% active release in the in vitro release profile analysis. In addition, the gel showed a good rate of mucoadhesion. An in vitro cytotoxicity assay was performed against two cell lines, B-16 and HeLa, revealing lower cytotoxicity of the system when compared to the free active, although with a high concentration rate of the active compound in the target tissue that remained during nearly 60 h.

Ong et al. (2020) employed the co-administration principle and developed superparamagnetic iron oxide nanoparticles containing methotrexate, which was later loaded into nanostructured lipid carriers. This co-loading system displayed high storage stability for 3 months at room temperature. It presented a time-dependent cytotoxicity in MDA-MB-231 breast cancer cells, presenting an IC50 rate of 137 μg/mL in 48 h, and later, following 72 h, it presented a value of 12 μg/mL. The presence of nanoparticles also influenced the local temperature increase, leading to a higher percentage of apoptosis to 50% and 65% after exposure to SPIONs-NLC and MTX-SPIONs NLC, respectively.

Aiming at new approaches to disable the several mechanisms of cancer, Li et al. (2017) developed a NLC targeting the chemokine receptor CXCR4, which, collectively with its cognate ligand SDF-1, is part of the process metabolism of breast cancer and may also be present in metastasis traces. The NLCs were targeted with plerixafor, a receptor antagonist that prevents the binding of the receptor and its ligand, and were loaded with IR780 iodide, a dye used in photothermal therapy characterized as an alternative therapy that uses the combination of a specific wavelength activated by light to obtain the effect (113). The analyses determined that the formulation was able to reduce cancer cell invasion and metastases both in vitro and in vivo, in addition to being able to accumulate in tumors. The results showed that IR780-AMD-NLCs exhibited greater photothermal efficacy in comparison to AMD-NLCs. The proportion of 4T1-luc cells in apoptosis and late necrosis was 54.7%, while the IR780 solution corresponded to the value of 14.1%. Encapsulated IR780 portrayed higher cytotoxic potential than free IR780, with and without laser irradiation.

## 4 New Trends: Functionalized CLN

The modification of nanoparticle properties by modifying their surface is a possibility by the coating with polymers that provide attributes, such as the ability to escape from the reticuloendothelial system (RES), thus delaying elimination by the liver and prolonging its circulation in the bloodstream. Therefore, these may be designated stealth nanoparticles through pegylated nanocarriers. The use of pegylated polymers can also minimize opsonization during blood circulation, contribute to the biocompatibility and reduction of the immunogenicity of the coated nanoparticles, and, being hydrophilic, allow greater solubilization and stability of polymeric nanoparticles in aqueous solutions (Jafari et al. 2016; Thiruppathi et al. 2017). These strategies aim to improve the characteristics of the system.

Polyethylene glycol (PEG), also known as macrogol, is currently considered the polymer of choice for several pharmaceutical applications, including in the drug delivery system. It is approved by the FDA and displays several advantages, such as being biocompatible, reducing the interaction of foreign molecules with the distribution system that is coupled, and allowing greater stability for the active compound (Anari et al. 2016; D'souza and Shegokar 2016). PEG-containing functionalizations can easily be combined with other polymers, such as poly(glycolic acid) (PGA), poly (lactic acid-co-glycolic acid) (PLGA), and even folic acid itself.

The use of polymers by the pharmaceutical industry, such as Pluronic F127, has been increasingly employed to enhance the solubility of hydrophobic drugs and improve the biocompatibility of biomaterials. Furthermore, poloxamer (Pluronic F127) is a copolymer that has a triblock structure, PEO-PPO-PEO, composed of two PEO chains, poly(ethylene oxide), and a PPO chain, poly(propylene oxide), with great potential to achieve targeted drug delivery. It is also considered one of the most promising biological transport systems for presenting the ability to self-organize in the aqueous phase and carry out the transport of poorly soluble substances in physiological environments and for controlled release (Chiappetta and Sosnik 2007; Gonzales and Krishnan 2007; Basak and Bandyopadhyay 2013).

Li et al. (2013) developed nanocarriers using Pluronic F127 functionalized with a peptide of seven amino acids PV7, produced for intracellular administration of the hydrophobic anticancer drug doxorubicin (DOX). Presenting an encapsulation efficiency result of 72.68%, the functionalized nanocarriers were more potent in DOX release and with more considerable cellular absorption.

Another technique used to improve NLCs is the functionalization strategy of nanocarriers with a ligand specific to the chosen target. This prospect is of great interest to researchers, considering that it aims to increase the selectivity of the system. When the NLCs target tumor cells, folic acid (FA) is more frequently used as a signaling molecule on the surface of these systems, due to overexpression of folate receptors on tumor cells (ovarian, prostate, and breast) and absence in most normal tissues (Nakamura et al. 2014).

In the research developed by Oshiro-Junior et al. (2020), a new carrier was developed with the incorporation of phthalocyanine, which also played its role as a photosensitizing agent, having its disadvantages mitigated. Once again, AF was present to assist the tumor-directed NLC delivery. This time, AF was coupled to the polymer used in the NLC formulation, causing the formation of a new functionalized molecule. The final formulation showed a controlled and sustained release, in addition to a reduction in cell viability. Cytotoxicity assays showed that NLC, at concentrations of 09.1 μM of photosensitizer, is nontoxic with $>80 \pm 6.8\%$ viable, and after 90s of exposure to light, the results showed a significant decrease in cell viability ($57 \pm 4\%$). Ultimately, the results obtained demonstrate that the functionalized carrier incorporated into the photosensitizer associated with the PDT technique has characteristics that make it a potential alternative treatment for breast cancer.

An NLC encapsulating PTX and gambogic acid (GA) was formulated by Ma et al. (2021), functionalized with cleavable AF in conjunction with a cell-penetrating peptide called dNP2. The developed formulation showed an increased rate of penetration into tumor tissues, in addition to a more selective targeting attributed to the presence of the combination between cleavable FA and the peptide.

Two different NLC formulations functionalized with folic acid were developed by Moraes et al. (2021) to obtain an encapsulation efficiency of doxorubicin higher than 65%. Using the MDA-MB-231 breast cancer cell line, an in vitro cytotoxic study was performed, revealing an increase in cell uptake. Functionalized NLCs exhibited more toxicity in cancer cells compared to non-functionalized ones. The cell viability study demonstrated a result higher than 98%, and the FA-PEG-NLC responsive system for DOX delivery was able to increase bioavailability and improve therapeutic efficacy in situ. Furthermore, it was observed that through oral administration of the carrier, 50% of the encapsulated DOX reached the tumor target and showed stability for more than 40 days, demonstrating that the DOX incorporated into the functionalized carrier is a promising alternative for the treatment of the disease.

An example of a conjugate composed of AF and PEG was proposed in the article developed by Lin et al. (2016) for the delivery of curcumin (CUR) to target cells (MCF-7), aiming at increasing targeting and maximizing cell cytotoxicity, resulting in tumor regression. In vitro and in vivo analyses were performed, confirming the inhibition of the growth of MCF-7 cells, in addition to showing an improved antitumor activity in mice, showing results of 83% for functionalized NLC-CUR, versus 31% obtained by the solutions of CUR. A different study group used PEG containing different combinations of lipids to formulate an NLC containing simvastatin, where the PEG-containing portions were responsible for specific activities such as prolonging simvastatin release, optimizing the zeta potential and particle size (Safwat et al. 2017).

17-Allylamino-17-demethoxygeldanamycin (17AAG) is an Hsp90 inhibitor, used as a treatment for several types of cancer, including breast cancer. However, it has numerous adversities such as low solubility, hepatotoxicity, low stability, and short half-life. With 17AAG incorporated into PEGylated NLC, the intention was to reduce or even correct these disadvantages. In addition to presenting an easy and fast production methodology, a sustained release of the drug was observed after the first moments. It was also possible to observe a significantly improved tumor activity when compared to the free drug and a prolongation in its half-life (Wang et al. 2016).

Wang et al. (2015) presented a paper intending to develop a PEGylated NLC containing biochanin A (BCA), an isoflavone characterized by its estrogenic activity, exhibiting a significant affinity activity for estrogen receptors in humans. The study identified a slower release rate of BCA-PEG-NLC when compared to the suspension containing the free molecule, as well as Cmax values approximately 16 times higher when compared to a suspension of BCA. In addition to these factors, NLC also showed a higher rate of cytotoxicity in MCF-7 lineage cells when compared to free molecule.

Adopting the same strategies as Kebebe et al. (2019), Huang et al. (2018) used functionalized NLC to increase the efficacy and selectivity of gambogic acid, distinguishing due to the aid of two cell penetration peptides. In vitro cytotoxicity analyses were performed on MDA-MB-231 breast cancer cells, in addition to in vivo analyses performed on mice, resulting in satisfactory results for functionalized NLC containing gambogic acid, demonstrating a significant increase in cytotoxicity, more elevated accumulation where the tumor was located, and an increase in antitumor activity, all of these factors being attributed to cell targeting induced by the functionalization of the NLC.

Vincristine (VCR) is an alkaloid-derived chemotherapeutic agent widely used in breast cancer treatment. Given its properties, Gao et al. (2017) developed NLCs modified with hyaluronic acid (HA-NLC) for vincristine sulfate (VCR) targeting. HA-NLCs exhibited the most potent effect in increasing cell apoptosis, and both HA-NLCs and VCR-NLCs showed a 1.8-fold improvement in bioavailability. The in vitro cytotoxicity of HA-NLCs was 2 and 7 times higher than VCR-NLCs and VCR solution, respectively. Similarly, the Cmax values of VCR-NLCs and HA-NLCs were 1.49 and 1.65 times higher than the VCR solution. Pharmacokinetic studies also displayed prolonged Tmax, and according to the results, the nanocarrier proved to be effective for VCR delivery, with an increase in oral absorption.

## 5 Conclusion

Despite being a pathology extensively discussed in the worldwide media, there are still several concepts that have not been approached regarding what breast cancer encompasses. Obstacles related to the treatments are the points most explored by researchers. With the help of nanotechnology, the numerous disadvantages presented both by conventional products found on the market and by formulations still under study, including active pharmaceutical ingredients of natural origin, were gradually being overcome.

Liposomes, as the first lineage of lipid nanoparticles, showed an effective ability to protect the API, consequently leading to a decrease in adverse effects and improvement in the bioavailability of the active compound. Nevertheless, their instability and high production cost led researchers to develop two more distinct types of lipid nanoparticles that subsequently became promising alternatives for the insertion of nanotechnology in formulations.

Solid lipid nanoparticles have a broad similarity in comparison to nanostructured lipid carriers, such as the methods of obtainment, particle size, and even some of the essential materials for matrix formation. However, the importance of their divergences is highlighted, such as the use of liquid lipids by the carriers, generating a disordered matrix and producing a larger space for the deposit of inputs, making it difficult to expel them from the matrix, as usually occurs with solid lipid nanoparticles when they have a considerable amount of input dispersed in their matrix.

Numerous studies have been performed over the years intending to investigate the effect of nanosystems directly on breast cancer and on the drugs used to treat it. The standard characteristics inherent to these nanosystems need to be preserved since the inputs are incorporated in their matrices and have been adopted throughout the vast majority of the studies addressed here. Among the main advantages presented, the significant increase in cytotoxicity against breast cancer lineage cells, improved bioavailability and half-life of the active compound, and decrease of adverse effects are the most cited, consequently leading to better patient adherence to the treatment and presenting themselves as promising alternatives to be used as a treatment for breast cancer.

**Acknowledgments** None

**Compliance with Ethical Standards** The authors declare that there is no conflict of interest.

## References

Abotaleb M, Kubatka P, Caprnda M, Varghese E, Zolakova B, Zubor P et al (2018) Chemotherapeutic agents for the treatment of metastatic breast cancer: an update. Biomed Pharmacother 101:458–477. https://doi.org/10.1016/j.biopha.2018.02.108

Agarwal M (2016) Is cancer chemotherapy dying? Asian J Transfus Sci 10(Suppl 1):S1–S7. https://doi.org/10.4103/0973-6247.182735

Akram M, Iqbal M, Daniyal M, Khan AU (2017) Awareness and current knowledge of breast cancer. Biol Res 50(1):1–23. https://doi.org/10.1186/s40659-017-0140-9

Almeida H, Lobão P, Frigerio C, Fonseca J, Silva R, Sousa Lobo JM et al (2017) Preparation, characterization and biocompatibility studies of thermoresponsive eyedrops based on the combination of nanostructured lipid carriers (NLC) and the polymer Pluronic F-127 for controlled delivery of ibuprofen. Pharm Dev Technol 22(3):336–349. https://doi.org/10.3109/10837450.2015.1125922

Alves LP, da Silva Oliveira K, da Paixão Santos JA, da Silva Leite JM, Rocha BP, de Lucena NP et al (2020) A review on developments and prospects of anti-inflammatory in microemulsions. J Drug Deliv Sci Technol 60:102008. https://doi.org/10.1016/j.jddst.2020.102008

Anari E, Akbarzadeh A, Zarghami N (2016) Chrysin-loaded PLGA-PEG nanoparticles designed for enhanced effect on the breast cancer cell line. Artif Cells Nanomed Biotechnol 44(6): 1410–1416. https://doi.org/10.3109/21691401.2015.1029633

Avitabile E, Bedognetti D, Ciofani G, Bianco A, Delogu LG (2018) How can nanotechnology help the fight against breast cancer? Nanoscale 10(25):11719–11731. https://doi.org/10.1039/C8NR02796J

Bahari LAS, Hamishehkar H (2016) The impact of variables on particle size of solid lipid nanoparticles and nanostructured lipid carriers; a comparative literature review. Adv Pharm Bull 6(2):143–151. https://doi.org/10.15171/apb.2016.021

Barenholz Y (2012) Doxil® – the first FDA-approved nano-drug: lessons learned. J Control Release 160(2):117–134. https://doi.org/10.1016/j.jconrel.2012.03.020

Barros RM, de Oliveira MS, Costa KMN, Sato MR, Santos KLM, de L Damasceno BPG et al (2020) Physicochemical characterization of bioactive compounds in nanocarriers. Curr Pharm Des 26(33):4163–4173. https://doi.org/10.2174/1381612826666200310144533

Basak R, Bandyopadhyay R (2013) Encapsulation of hydrophobic drugs in pluronic F127 micelles: effects of drug hydrophobicity, solution temperature, and pH. Langmuir 29(13):4350–4356. https://doi.org/10.1021/la304836e

Beloqui A, Solinís MÁ, Rodríguez-Gascón A, Almeida AJ, Préat V (2016) Nanostructured lipid carriers: promising drug delivery systems for future clinics. Nanomed Nanotechnol Biol Med 12(1):143–161. https://doi.org/10.1016/j.nano.2015.09.004

Beloqui A, del Pozo-Rodríguez A, Isla A, Rodríguez-Gascón A, Solinís MÁ (2017) Nanostructured lipid carriers as oral delivery systems for poorly soluble drugs. J Drug Deliv Sci Technol 42:144–154. https://doi.org/10.1016/j.jddst.2017.06.013

BRASIL (2018) Diretrizes Diagnósticas e Terapêuticas do Carcinoma de Mama. Ministério da Saúde. Portaria Conjunta N° 04, de 23 de Janeiro de 2018

Chand P, Kumar H, Badduri N, Gupta NV, Bettada VG, Madhunapantula SRV et al (2021) Design and evaluation of cabazitaxel loaded NLCs against breast cancer cell lines. Colloids Surf B Biointerfaces 199:111535. https://doi.org/10.1016/j.colsurfb.2020.111535

Chiappetta DA, Sosnik A (2007) Poly(ethylene oxide)-poly(propylene oxide) block copolymer micelles as drug delivery agents: improved hydrosolubility, stability and bioavailability of drugs. Eur J Pharm Biopharm 66(3):303–317. https://doi.org/10.1016/j.ejpb.2007.03.022

Chiari-Andréo BG, Abuçafy MP, Manaia EB, da Silva BL, Rissi NC, Oshiro-Júnior JA et al (2019) Drug delivery using theranostics: an overview of its use, advantages and safety assessment. Curr Nanosci 16(1):3–14. https://doi.org/10.2174/1573413715666190618162321

Costa KMN, Sato MR, Barbosa TLA, Rodrigues MGF, Medeiros ACD, de Lima Damasceno BPG et al (2021) Curcumin-loaded micelles dispersed in ureasil-polyether materials for a novel sustained-release formulation. Pharmaceutics 13(5):1–10. https://doi.org/10.3390/pharmaceutics13050675

D'souza AA, Shegokar R (2016) Polyethylene glycol (PEG): a versatile polymer for pharmaceutical applications. Expert Opin Drug Deliv 13(9):1257–1275. https://doi.org/10.1080/17425247.2016.1182485

de Assis KMA, de A Rêgo RI, de Melo DF, da Silva LM, Oshiro-Júnior JA, Formiga FR et al (2020) Therapeutic potential of Melaleuca alternifolia essential oil in new drug delivery systems. Curr Pharm Des 26(33):4048–4055. https://doi.org/10.2174/1381612826666200305124041
de Oliveira MS, Oshiro-Junior JA, Dantas MM, da Fonsêca NF, Ramos HA, da Silva JVB et al (2020) An overview of the antimicrobial activity of polymeric nanoparticles against enterobacteriaceae. Curr Pharm Des 27(10):1311–1322. https://doi.org/10.2174/1381612826666201029095327
Deirram N, Zhang C, Kermaniyan SS, Johnston APR, Such GK (2019) pH-responsive polymer nanoparticles for drug delivery. Macromol Rapid Commun 40(10):1–23. https://doi.org/10.1002/marc.201800917
Deng C, Jia M, Wei G, Tan T, Fu Y, Gao H et al (2017) Inducing optimal antitumor immune response through coadministering iRGD with pirarubicin loaded nanostructured lipid carriers for breast cancer therapy. Mol Pharm 14(1):296–309. https://doi.org/10.1021/acs.molpharmaceut.6b00932
Di H, Wu H, Gao Y, Li W, Zou D, Dong C (2016) Doxorubicin- and cisplatin-loaded nanostructured lipid carriers for breast cancer combination chemotherapy. Drug Dev Ind Pharm 42(12):2038–2043. https://doi.org/10.1080/03639045.2016.1190743
Diniz RNT, Sousa IJO, Negreiros Santos PS, Negreiros FS, Gonçalves RLG (2017) Pharmacological strategies in breast cancer therapy: a review on estrogenic modulators. Rev UNINGÁ Rev 30(2):61–63. Available from: https://www.mastereditora.com.br/periodico/20170503_210913.pdf
Dolatkhah M, Hashemzadeh N, Barar J, Adibkia K, Aghanejad A, Barzegar-Jalali M et al (2020) Graphene-based multifunctional nanosystems for simultaneous detection and treatment of breast cancer. Colloids Surf B Biointerfaces 193:111104. https://doi.org/10.1016/j.colsurfb.2020.111104
dos Santos DDL, Besegato JF, de Melo PBG, Oshiro Junior JA, Chorilli M, Deng D et al (2021) Curcumin-loaded Pluronic® F-127 micelles as a drug delivery system for curcumin-mediated photodynamic therapy for oral application. Photochem Photobiol 97(5):1072–1088. https://doi.org/10.1111/php.13433
Doskocz J, Dałek P, Foryś A, Trzebicka B, Przybyło M, Mesarec L et al (2020) The effect of lipid phase on liposome stability upon exposure to the mechanical stress. Biochim Biophys Acta Biomembr 1862(9):183361. https://doi.org/10.1016/j.bbamem.2020.183361
ElBagoury M, Kotb M (2018) Chemotherapy over the years. J Pharm Sci Res 10(2):316–318
Fang C-L, Al-Suwayeh SA, Fang J-Y (2013) Nanostructured lipid carriers (NLCs) for drug delivery and targeting. Recent Pat Nanotechnol 7(1):41–55. https://doi.org/10.2174/1872210511307010041
Fernandes RS, Silva JO, Monteiro LOF, Leite EA, Cassali GD, Rubello D et al (2016) Doxorubicin-loaded nanocarriers: a comparative study of liposome and nanostructured lipid carrier as alternatives for cancer therapy. Biomed Pharmacother 84:252–257. https://doi.org/10.1016/j.biopha.2016.09.032
Fernandes RS, Silva JO, Seabra HA, Oliveira MS, Carregal VM, Vilela JMC et al (2018) α-Tocopherol succinate loaded nano-structured lipid carriers improves antitumor activity of doxorubicin in breast cancer models *in vivo*. Biomed Pharmacother 103:1348–1354. https://doi.org/10.1016/j.biopha.2018.04.139
Ferreira DD (2018) Estudos Farmacológicos de Novos Compostos Semissintéticos com Potencial Terapêutico na Doença de Chagas. Programa de Pós-Graduação em Ciências. Coordenadoria de Controle de Doenças da Secretaria de Estado da Saúde-SP
Gabriel GH, Nepomuceno LL, de Sousa Cruz Pimenta V, Araújo EG (2017) Chemotherapy, hormoniotherapy and new alternatives for treatment of mammary adenocarcinoma. Enciclopédia Biosf 14(26):584. https://doi.org/10.18677/EnciBio_2017B56
Gadag S, Narayan R, Nayak AS, Catalina Ardila D, Sant S, Nayak Y et al (2021) Development and preclinical evaluation of microneedle-assisted resveratrol loaded nanostructured lipid carriers for localized delivery to breast cancer therapy. Int J Pharm 606:120877. https://doi.org/10.1016/j.ijpharm.2021.120877

Gambhire VM, Salunkhe SM, Gambhire MS (2018) Atorvastatin-loaded lipid nanoparticles: antitumor activity studies on MCF-7 breast cancer cells. Drug Dev Ind Pharm 44:1685–1692. https://doi.org/10.1080/03639045.2018.1492605

Ganesan P, Narayanasamy D (2017) Lipid nanoparticles: different preparation techniques, characterization, hurdles, and strategies for the production of solid lipid nanoparticles and nanostructured lipid carriers for oral drug delivery. Sustain Chem Pharm 6:37–56. https://doi.org/10.1016/j.scp.2017.07.002

Gao X, Zhang J, Xu Q, Huang Z, Wang Y, Shen Q (2017) Hyaluronic acid-coated cationic nanostructured lipid carriers for oral vincristine sulfate delivery. Drug Dev Ind Pharm 43(4): 661–667. https://doi.org/10.1080/03639045.2016.1275671

Ghasemiyeh P, Mohammadi-Samani S (2020) Potential of nanoparticles as permeation enhancers and targeted delivery options for skin: advantages and disadvantages. Drug Des Devel Ther 14: 3271–3289. https://doi.org/10.2147/DDDT.S264648

Godugu C, Doddapaneni R, Safe SH, Singh M (2016) Novel diindolylmethane derivatives based NLC formulations to improve the oral bioavailability and anticancer effects in triple negative breast cancer. Eur J Pharm Biopharm 108:168–179. https://doi.org/10.1016/j.ejpb.2016.08.006

Gondim BLC, Oshiro-Júnior JA, Fernanandes FHA, Nóbrega FP, Castellano LRC, Medeiros ACD (2019) Plant extracts loaded in nanostructured drug delivery systems for treating parasitic and antimicrobial diseases. Curr Pharm Des 25(14):1604–1615. https://doi.org/10.2174/1381612825666190628153755

Gonzales M, Krishnan KM (2007) Phase transfer of highly monodisperse iron oxide nanocrystals with Pluronic F127 for biomedical applications. J Magn Magn Mater 311(1):59–62. https://doi.org/10.1016/j.jmmm.2006.10.1150

Gordillo-Galeano A, Mora-Huertas CE (2018) Solid lipid nanoparticles and nanostructured lipid carriers: a review emphasizing on particle structure and drug release. Eur J Pharm Biopharm 133:285–308. https://doi.org/10.1016/j.ejpb.2018.10.017

Guenancia C, Lefebvre A, Cardinale D, Yu AF, Ladoire S, Ghiringhelli F et al (2016) Obesity as a risk factor for anthracyclines and trastuzumab cardiotoxicity in breast cancer: a systematic review and meta-analysis. J Clin Oncol 34(26):3157–3165. https://doi.org/10.1200/JCO.2016.67.4846

Guido RVC, Andricopulo AD, Oliva G (2010) Drug design, biotechnology and medicinal chemistry: applications to infectious diseases. Estud Avançados 24(70):81–98

Haider M, Abdin SM, Kamal L, Orive G (2020) Nanostructured lipid carriers for delivery of chemotherapeutics: a review. Pharmaceutics 12(3):288. https://doi.org/10.3390/pharmaceutics12030288

Harbeck N, Penault-Llorca F, Cortes J, Gnant M, Houssami N, Poortmans P et al (2019) Breast cancer. Nat Rev Dis Primers 5(1):66. https://doi.org/10.1038/s41572-019-0111-2

How CW, Rasedee A, Manickam S, Rosli R (2013) Tamoxifen-loaded nanostructured lipid carrier as a drug delivery system: characterization, stability assessment and cytotoxicity. Colloids Surf B Biointerfaces 112:393–399. https://doi.org/10.1016/j.colsurfb.2013.08.009

Huang R, Li J, Kebebe D, Wu Y, Zhang B, Liu Z (2018) Cell penetrating peptides functionalized gambogic acid-nanostructured lipid carrier for cancer treatment. Drug Deliv 25(1):757–765. https://doi.org/10.1080/10717544.2018.1446474

Iqbal MA, Md S, Sahni JK, Baboota S, Dang S, Ali J (2012) Nanostructured lipid carriers system: recent advances in drug delivery. J Drug Target 20(10):813–830. https://doi.org/10.3109/1061186X.2012.716845

Jafari M, Heidari D, Ebrahimnejad P (2016) Synthesizing and characterizing functionalized short multiwall carbon nanotubes with folate, magnetite and polyethylene glycol as multitargeted nanocarrier of anti-cancer drugs. Iran J Pharm Res 15(2):449–456

Jain V, Kumar H, Anod HV, Chand P, Gupta NV, Dey S et al (2020) A review of nanotechnology-based approaches for breast cancer and triple-negative breast cancer. J Control Release 326: 628–647. https://doi.org/10.1016/j.jconrel.2020.07.003

Jaiswal P, Gidwani B, Vyas A (2016) Nanostructured lipid carriers and their current application in targeted drug delivery. Artif Cells Nanomed Biotechnol 44(1):27–40. https://doi.org/10.3109/21691401.2014.909822

Junior APD (2016) Preparação de carreadores lipídicos nanoestruturados a partir de cera de carnaúba e óleo de pracaxi contendo dexametasona para tratamento tópico de inflamações cutâneas 1–191. Available from: https://repositorio.ufpe.br/handle/123456789/18883

Kanásová M, Nesměrák K (2017) Systematic review of liposomes' characterization methods. Monatshefte fur Chemie 148(9):1581–1593. https://doi.org/10.1007/s00706-017-1994-9

Kanda Y, Ishihara Y, Wilde NC, Baran PS (2020) Two-phase Total synthesis of taxanes: tactics and strategies. J Org Chem 85(16):10293–10320. https://doi.org/10.1021/acs.joc.0c01287

Karamchedu S, Tunki L, Kulhari H, Pooja D (2020) Morin hydrate loaded solid lipid nanoparticles: characterization, stability, anticancer activity, and bioavailability. Chem Phys Lipids 233: 104988. https://doi.org/10.1016/j.chemphyslip.2020.104988

Kebebe D, Wu Y, Zhang B, Yang J, Liu Y, Li X et al (2019) Dimeric c(RGD) peptide conjugated nanostructured lipid carriers for efficient delivery of Gambogic acid to breast cancer. Int J Nanomedicine 14:6179–6195. https://doi.org/10.2147/IJN.S202424

Khosa A, Reddi S, Saha RN (2018) Nanostructured lipid carriers for site-specific drug delivery. Biomed Pharmacother 103:598–613. https://doi.org/10.1016/j.biopha.2018.04.055

Lages EB, Fernandes RS, de Oliveira Silva J, de Souza ÂM, Cassali GD, de Barros ALB et al (2020) Co-delivery of doxorubicin, docosahexaenoic acid, and α-tocopherol succinate by nanostructured lipid carriers has a synergistic effect to enhance antitumor activity and reduce toxicity. Biomed Pharmacother 132:110876. https://doi.org/10.1016/j.biopha.2020.110876

Lanna EG, Siqueira RP, Machado MGC, de Souza A, Trindade IC, Branquinho RT et al (2021) Lipid-based nanocarriers co-loaded with artemether and triglycerides of docosahexaenoic acid: effects on human breast cancer cells. Biomed Pharmacother 134:111114. https://doi.org/10.1016/j.biopha.2020.111114

Li YY, Li L, Dong HQ, Cai XJ, Bin RT (2013) Pluronic F127 nanomicelles engineered with nuclear localized functionality for targeted drug delivery. Mater Sci Eng C 33(5):2698–2707. https://doi.org/10.1016/j.msec.2013.02.036

Li H, Wang K, Yang X, Zhou Y, Ping Q, Oupicky D et al (2017) Dual-function nanostructured lipid carriers to deliver IR780 for breast cancer treatment: anti-metastatic and photothermal antitumor therapy. Acta Biomater 53:399–413. https://doi.org/10.1016/j.actbio.2017.01.070

Li X, Jia X, Niu H (2018) Nanostructured lipid carriers co-delivering lapachone and doxorubicin for overcoming multidrug resistance in breast cancer therapy. Int J Nanomedicine 13:4107–4119. https://doi.org/10.2147/IJN.S163929

Li S, So TH, Tang G, Tan HY, Wang N, Ng BFL et al (2020) Chinese herbal medicine for reducing chemotherapy-associated side-effects in breast cancer patients: a systematic review and meta-analysis. Front Oncol 10:1–20. https://doi.org/10.3389/fonc.2020.599073

Liang Y, Zhang H, Song X, Yang Q (2020) Metastatic heterogeneity of breast cancer: molecular mechanism and potential therapeutic targets. Semin Cancer Biol 60:14–27. https://doi.org/10.1016/j.semcancer.2019.08.012

Lin M, Teng L, Wang Y, Zhang J, Sun X (2016) Curcumin-guided nanotherapy: a lipid-based nanomedicine for targeted drug delivery in breast cancer therapy. Drug Deliv 23(4):1420–1425. https://doi.org/10.3109/10717544.2015.1066902

Liu P, Liu X, Cheng Y, Zhong S, Shi X, Wang S et al (2020) Core-shell nanosystems for self-activated drug-gene combinations against triple-negative breast cancer. ACS Appl Mater Interfaces 12(48):53654–53664. https://doi.org/10.1021/acsami.0c15089

Liyanage PY, Hettiarachchi SD, Zhou Y, Ouhtit A, Seven ES, Oztan CY et al (2019) Nanoparticle-mediated targeted drug delivery for breast cancer treatment. Biochim Biophys Acta Rev Cancer 1871(2):419–433. https://doi.org/10.1016/j.bbcan.2019.04.006

Ma Z, Pi J, Zhang Y, Qin H, Zhang B, Li N et al (2021) Enhanced anticancer efficacy of dual drug-loaded self-assembled nanostructured lipid carriers mediated by ph-responsive folic acid and human-derived cell penetrating peptide dnp2. Pharmaceutics 13(5):600. https://doi.org/10.3390/pharmaceutics13050600

Machado FC, Adum de Matos RP, Primo FL, Tedesco AC, Rahal P, Calmon MF (2019) Effect of curcumin-nanoemulsion associated with photodynamic therapy in breast adenocarcinoma cell line. Bioorg Med Chem 27(9):1882–1890. https://doi.org/10.1016/j.bmc.2019.03.044

Maja L, Željko K, Mateja P (2020) Sustainable technologies for liposome preparation. J Supercrit Fluids 165:104984. https://doi.org/10.1016/j.supflu.2020.104984

Marcato PD (2009) Preparation, characterization and application in drugs and cosmetics of solid lipid nanoparticles. Rev Eletrônica Farmácia 6(2):1–37. ISSN 1808-0804

Marinello J, Delcuratolo M, Capranico G (2018) Anthracyclines as topoisomerase II poisons: from early studies to new perspectives. Int J Mol Sci 19(11):3480. https://doi.org/10.3390/ijms19113480

Medeiros TS, Moreira LMCC, Oliveira TMT, Melo DF, Azevedo EP, Gadelha AEG et al (2020) Bemotrizinol-loaded carnauba wax-based nanostructured lipid carriers for sunscreen: optimization, characterization, and *in vitro* evaluation. AAPS PharmSciTech 21(8):1–13. https://doi.org/10.1208/s12249-020-01821-x

Mohamad NE, Abu N, Rahman HS, Ky H, Ho WY, Lim KL et al (2015) Nanostructured lipid carrier improved *in vivo* anti-tumor and immunomodulatory effect of Zerumbone in 4T1 challenged mice. RSC Adv 5(28):22066–22074. https://doi.org/10.1039/C5RA00144G

Moraes S, Marinho A, Lima S, Granja A, Araújo JP, Reis S et al (2021) Targeted nanostructured lipid carriers for doxorubicin oral delivery. Int J Pharm 592:120029. https://doi.org/10.1016/j.ijpharm.2020.120029

Müller RH, Radtke M, Wissing SA (2002) Solid lipid nanoparticles (SLN) and nanostructured lipid carriers (NLC) in cosmetic and dermatological preparations. Adv Drug Deliv Rev 54(SUPPL 1): S131–S155. https://doi.org/10.1016/s0169-409x(02)00118-7

Nakamura T, Kawano K, Shiraishi K, Yokoyama M, Maitani Y (2014) Folate-targeted gadolinium-lipid-based nanoparticles as a bimodal contrast agent for tumor fluorescent and magnetic resonance imaging. Biol Pharm Bull 37(4):521–527. https://doi.org/10.1248/bpb.b13-00484

Naseri N, Valizadeh H, Zakeri-Milani P (2015) Solid lipid nanoparticles and nanostructured lipid carriers: structure preparation and application. Adv Pharm Bull 5(3):305–313. https://doi.org/10.15171/apb.2015.043

Ng WK, Yazan LS, Yap LH, Hafiza WAGN, How CW, Abdullah R (2015) Thymoquinone-loaded nanostructured lipid carrier exhibited cytotoxicity towards breast cancer cell lines (MDA-MB-231 and MCF-7) and cervical cancer cell lines (HeLa and SiHa). Biomed Res Int 28(1):1–10. https://doi.org/10.1155/2015/263131

Nordin N, Yeap SK, Rahman HS, Zamberi NR (2020) Antitumor and anti-metastatic effects of citral-loaded nanostructured lipid carrier in 4T1-induced breast cancer mouse model. Molecules 25(11):2670. https://doi.org/10.3390/molecules25112670

Ong YS, Bañobre-López M, Costa Lima SA, Reis S (2020) A multifunctional nanomedicine platform for co-delivery of methotrexate and mild hyperthermia towards breast cancer therapy. Mater Sci Eng C Mater Biol Appl 116:111255. https://doi.org/10.1016/j.msec.2020.111255

Oshiro-Júnior JA, Rodero C, Hanck-Silva G, Sato MR, Alves RC, Eloy JO et al (2018) Stimuli-responsive drug delivery nanocarriers in the treatment of breast cancer. Curr Med Chem 27(15): 2494–2513. https://doi.org/10.2174/0929867325666181009120610

Oshiro-Junior JA, Sato MR, Boni FI, Santos KLM, de Oliveira KT, de Freitas LM et al (2020) Phthalocyanine-loaded nanostructured lipid carriers functionalized with folic acid for photodynamic therapy. Mater Sci Eng C 108:110462. https://doi.org/10.1016/j.msec.2019.110462

Pandey H, Rani R, Agarwal V (2016) Liposome and their applications in cancer therapy. Braz Arch Biol Technol 59:150477. https://doi.org/10.1590/1678-4324-2016150477

Pearce A, Haas M, Viney R, Pearson S, Haywood P, Brown C et al (2017) Incidence and severity of self-reported chemotherapy side effects in routine care: a prospective cohort study. PLoS One 12(10):1–12. https://doi.org/10.1371/journal.pone.0184360

Pedro IDR, Almeida OP, Martins HR, Lemos J d A, Branco de Barros AL, Leite EA et al (2019) Optimization and *in vitro*/*in vivo* performance of paclitaxel-loaded nanostructured lipid carriers for breast cancer treatment. J Drug Deliv Sci Technol 54:101370. https://doi.org/10.1016/j.jddst.2019.101370

Peretz Damari S, Shamrakov D, Varenik M, Koren E, Nativ-Roth E, Barenholz Y et al (2018) Practical aspects in size and morphology characterization of drug-loaded nano-liposomes. Int J Pharm 547(1–2):648–655. https://doi.org/10.1016/j.ijpharm.2018.06.037

Poonia N, Kaur Narang J, Lather V, Beg S, Sharma T, Singh B et al (2019) Resveratrol loaded functionalized nanostructured lipid carriers for breast cancer targeting: systematic development, characterization and pharmacokinetic evaluation. Colloids Surf B Biointerfaces 181:756–766. https://doi.org/10.1016/j.colsurfb.2019.06.004

Robson AL, Dastoor PC, Flynn J, Palmer W, Martin A, Smith DW et al (2018) Advantages and limitations of current imaging techniques for characterizing liposome morphology. Front Pharmacol 9:1–8. https://doi.org/10.3389/fphar.2018.00080

Russell LM, Hultz M, Searson PC (2018) Leakage kinetics of the liposomal chemotherapeutic agent doxil: the role of dissolution, protonation, and passive transport, and implications for mechanism of action. J Control Release 176(1):139–148. https://doi.org/10.1016/j.jconrel.2017.11.007

Sabzichi M, Samadi N, Mohammadian J, Hamishehkar H, Akbarzadeh M, Molavi O (2016a) Sustained release of melatonin: a novel approach in elevating efficacy of tamoxifen in breast cancer treatment. Colloids Surf B Biointerfaces 145:64–71. https://doi.org/10.1016/j.colsurfb.2016.04.042

Sabzichi M, Mohammadian J, Khosroushahi AY, Bazzaz R, Hamishehkar H (2016b) Folate-targeted nanostructured lipid carriers (NLCs) enhance (Letrozol) efficacy in MCF-7 breast cancer cells. Asian Pac J Cancer Prev 17(12):6085–6088. https://doi.org/10.22034/APJCP.2016.17.12.5185

Sabzichi M, Mohammadian J, Mohammadi M, Jahanfar F, Movassagh Pour AA, Hamishehkar H et al (2017) Vitamin D-loaded nanostructured lipid carrier (NLC): a new strategy for enhancing efficacy of doxorubicin in breast cancer treatment. Nutr Cancer 69(6):840–848. https://doi.org/10.1080/01635581.2017.1339820

Safwat S, Ishak RAH, Hathout RM, Mortada ND (2017) Nanostructured lipid carriers loaded with simvastatin: effect of PEG/glycerides on characterization, stability, cellular uptake efficiency and *in vitro* cytotoxicity. Drug Dev Ind Pharm 43(7):1112–1125. https://doi.org/10.1080/03639045.2017.1293681

Salvi VR, Pawar P (2019) Nanostructured lipid carriers (NLC) system: a novel drug targeting carrier. J Drug Deliv Sci Technol 51:255–267. https://doi.org/10.1016/j.jddst.2019.02.017

Sato MR, Oshiro Junior JA, Machado RTA, de Souza PC, Campos DL, Pavan FR et al (2017) Nanostructured lipid carriers for incorporation of copper(II) complexes to be used against Mycobacterium tuberculosis. Drug Des Devel Ther 11:909–921. https://doi.org/10.2147/DDDT.S127048

Sato MR, Oshiro-Junior JA, Souza PC, Campos DL, Pereira-Da-Silva MA, Pavan FR et al (2019) Copper(II) complex-loaded castor oil-based nanostructured lipid carriers used against mycobacterium tuberculosis : development, characterisation, *in vitro* and *in vivo* biological assays. Pharmazie 74(12):715–720. https://doi.org/10.1691/ph.2019.9110

Selvamuthukumar S, Velmurugan R (2012) Nanostructured lipid carriers: a potential drug carrier for cancer chemotherapy. Lipids Health Dis 11:159. https://doi.org/10.1186/1476-511X-11-159

Shao XR, Wei XQ, Zhang S, Fu N, Lin YF, Cai XX et al (2017) Effects of micro-environmental pH of liposome on chemical stability of loaded drug. Nanoscale Res Lett 12:504. https://doi.org/10.1186/s11671-017-2256-9

Shete H, Chatterjee S, De A, Patravale V (2013) Long chain lipid based tamoxifen NLC. Part II: pharmacokinetic, biodistribution and *in vitro* anticancer efficacy studies. Int J Pharm 454(1):584–592. https://doi.org/10.1016/j.ijpharm.2013.03.036

Shidhaye S, Vaidya R, Sutar S, Patwardhan A, Kadam V (2008) Solid lipid nanoparticles and nanostructured lipid carriers – innovative generations of solid lipid carriers. Curr Drug Deliv 5(4):324–331. https://doi.org/10.2174/156720108785915087

Silvestre ALP, Oshiro-Júnior JA, Garcia C, Turco BO, da Silva Leite JM, de Lima Damasceno BPG et al (2020) Monoclonal antibodies carried in drug delivery nanosystems as a strategy for cancer treatment. Curr Med Chem 28(2):401–418. https://doi.org/10.2174/0929867327666200121121409

Smith SL (2017) War! What is it good for? Mustard gas medicine. CMAJ 189(8):E321–E322. https://doi.org/10.1503/cmaj.161032

Soni NK, Sonali LJ, Singh A, Mangla B, Neupane YR, Kohli K (2020) Nanostructured lipid carrier potentiated oral delivery of raloxifene for breast cancer treatment. Nanotechnology 31(47): 475101. https://doi.org/10.1088/1361-6528/abaf81

Sun M, Nie S, Pan X, Zhang R, Fan Z, Wang S (2014) Quercetin-nanostructured lipid carriers: characteristics and anti-breast cancer activities *in vitro*. Colloids Surf B Biointerfaces 113:15–24. https://doi.org/10.1016/j.colsurfb.2013.08.032

Sun YS, Zhao Z, Yang ZN, Xu F, Lu HJ, Zhu ZY et al (2017) Risk factors and preventions of breast cancer. Int J Biol Sci 13(11):1387–1397. https://doi.org/10.7150/ijbs.21635

Szebeni J, Fülöp T, Dézsi L, Metselaar B, Storm G (2016) Liposomal doxorubicin: the good, the bad and the not-so-ugly. J Drug Target 24(9):765–767. https://doi.org/10.3109/1061186X.2016.1172591

Thiruppathi R, Mishra S, Ganapathy M, Padmanabhan P, Gulyás B (2017) Nanoparticle functionalization and its potentials for molecular imaging. Adv Sci 4(3):1600279. https://doi.org/10.1002/advs.201600279

Tong CWS, Wu M, Cho WCS, To KKW (2018) Recent advances in the treatment of breast cancer. Front Oncol 8:227. https://doi.org/10.3389/fonc.2018.00227

Tran TH, Nguyen AN, Kim JO, Yong CS, Nguyen CN (2016) Enhancing activity of artesunate against breast cancer cells via induced-apoptosis pathway by loading into lipid carriers. Artif Cells Nanomed Biotechnol 44(8):1979–1987. https://doi.org/10.3109/21691401.2015.1129616

Van Tran V, Moon JY, Lee YC (2019) Liposomes for delivery of antioxidants in cosmeceuticals: challenges and development strategies. J Control Release 300:114–140. https://doi.org/10.1016/j.jconrel.2019.03.003

Varshosaz J, Davoudi MA, Rasoul-Amini S (2018) Docetaxel-loaded nanostructured lipid carriers functionalized with trastuzumab (Herceptin) for HER2-positive breast cancer cells. J Liposome Res 28(4):285–295. https://doi.org/10.1080/08982104.2017.1370471

Vieira DB, Gamarra LF (2016) Advances in the use of nanocarriers for cancer diagnosis and treatment. Einstein 14(1):99–103. https://doi.org/10.1590/S1679-45082016RB3475

Vohra T, Kaur I, Heer H, Murthy RR (2013) Nanolipid carrier-based thermoreversible gel for localized delivery of docetaxel to breast cancer. Cancer Nanotechnol 4(1–3):1–12. https://doi.org/10.1007/s12645-013-0032-9

Waks AG, Winer EP (2019) Breast cancer treatment: a review. JAMA 321(3):288–300. https://doi.org/10.1001/jama.2018.19323

Wang L, Luo Q, Lin T, Li R, Zhu T, Zhou K et al (2015) PEGylated nanostructured lipid carriers (PEG-NLC) as a novel drug delivery system for biochanin A. Drug Dev Ind Pharm 41(7): 1204–1212. https://doi.org/10.3109/03639045.2014.938082

Wang Z, Wang J, Yang S, Hou S (2016) Construction and *in vitro*/*in vivo* evaluation of 17-allylamino-17-demethoxygeldanamycin (17AAG)-loaded PEGylated nanostructured lipid carriers. Drug Dev Ind Pharm 42(1):91–98. https://doi.org/10.3109/03639045.2015.1031138

Wibroe PP, Ahmadvand D, Oghabian MA, Yaghmur A, Moghimi SM (2016) An integrated assessment of morphology, size, and complement activation of the PEGylated liposomal doxorubicin products Doxil®, Caelyx®, Doxorubicin, and SinaDoxosome. J Control Release 221:1–8. https://doi.org/10.1016/j.jconrel.2015.11.021

Willson ML, Burke L, Ferguson T, Ghersi D, Nowak AK, Wilcken N (2019) Taxanes for adjuvant treatment of early breast cancer. Cochrane Database Syst Rev 9(9):CD004421. https://doi.org/10.1002/14651858.CD004421.pub3

Yadav N, Khatak S, Sara UVS (2013) Solid lipid nanoparticles – a review. Int J Appl Pharm 5(2): 8–18

Zhang Q, Zhao J, Hu H, Yan Y, Hu X, Zhou K et al (2019) Construction and *in vitro* and *in vivo* evaluation of folic acid-modified nanostructured lipid carriers loaded with paclitaxel and chlorin e6. Int J Pharm 569:118595. https://doi.org/10.1016/j.ijpharm.2019.118595

Zhang Y, Li H, Zhang J, Zhao C, Lu S, Qiao J et al (2020) The combinatory effects of natural products and chemotherapy drugs and their mechanisms in breast cancer treatment. Phytochem Rev 19(5):1179–1197. https://doi.org/10.1007/s11101-019-09628-w

Zhou J, Guo B, Zhu W, Sui X et al (2016) Novel biomimetic nanostructured lipid carriers for cancer therapy: preparation, characterization, and *in vitro*/*in vivo* evaluation. Pharm Dev Technol 26(1):81–91. https://doi.org/10.1080/10837450.2020.1835957

# Managing Breast Cancer Using the Cell-Surface GRP78

Abdo A. Elfiky

**Abstract**

Breast cancer remains one of the leading causes of cancer-related deaths in women worldwide. While significant advancements have been made in the treatment of breast cancer, there is a need for more effective management strategies.

Different avenues emerged recently use the glucose-regulated protein 78 (GRP78) in cancer patients both in diagnosis and in treatment. In addition to managing the unfolded protein response under the stress of the unfolded proteins in cancer cells, GRP78 plays critical roles in cellular survival and response to chemotherapeutics or radiotherapy treatment.

In this chapter, we will explore how cell-surface GRP78 can be utilized to manage breast cancer and potentially improve patient outcomes.

This paves the way for understanding the molecular origin that links cell stress to cancer propagation and resistance to therapy.

**Keywords**

Breast cancer · Chemoresistance · GRP78 · Proteostasis · Unfolded protein response

## 1 Introduction

Breast cancer is a complex disease characterized by the uncontrolled growth of malignant cells in the breast tissue (Waks and Winer 2019). These include lobular carcinoma in situ, ductal carcinoma in situ, medullary carcinoma, mucinous carcinoma, and tubular carcinoma (Sharma et al. 2010). GRP78, also known as binding

A. A. Elfiky (✉)
Biophysics Department, Faculty of Science, Cairo University, Giza, Egypt
e-mail: abdo@sci.cu.edu.eg



Interdisciplinary Cancer Research, https://doi.org/10.1007/16833_2024_234
Published online: 4 April 2024

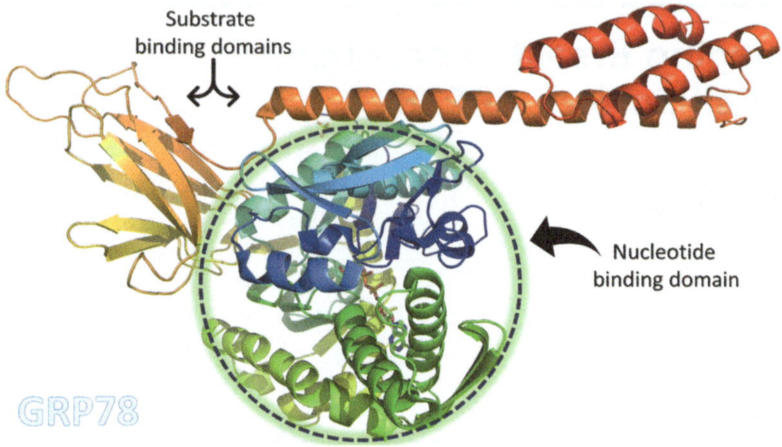


**Fig. 1** The structure of GRP78 represented by PyMOL software for the solved structure (PDB ID: 5E84, Chain A) retrieved from the protein data bank website (https://www.rcsb.org/). Two distinct domains are defined as the nucleotide-binding domain (NBD) and the substrate-binding domain (SBD)

immunoglobulin protein (BiP), is an endoplasmic reticulum (ER) chaperone protein that plays a critical role in maintaining homeostasis (Li et al. 2023). It regulates proteins' proper folding and assembly and controls cellular stress responses (Ibrahim et al. 2019; Elfiky et al. 2020). GRP78 has two coordinated domains; the nucleotide-binding domain (NBD) that binds either adenosine triphosphate (ATP) (activated) or adenosine diphosphate (ADP) and the substrate-binding domain (SBD) that binds to the misfolded proteins or peptides (Fig. 1). Recent research has revealed that cell-surface GRP78 is closely associated with breast cancer progression and resistance to therapy, making it an attractive target for management strategies (Gifford and Hill 2018; Amaresan and Gopal 2023). In a recent study, we identified GRP78 as a potential target for recently approved anticancer agents, such as abemaciclib, ribociclib, and tunicamycin (Abu-Mahfouz et al. 2022). This paves the way for a better understanding of the mechanism of action of these therapeutic regimens. Additionally, several studies have reported the role of cell-surface GRP78 in recognizing viral and fungal proteins (Elfiky and Ibrahim 2021; Elgohary et al. 2021; Elshemey et al. 2023; Elfiky et al. 2023).

## 1.1 Targeting GRP78 for Early Detection and Diagnosis

GRP78 expression level is elevated in many cancer types including breast cancer. This makes its expression level detection important in many different cancer types including brain lower-grade glioma, cholangiocarcinoma, colon adenocarcinoma, esophageal carcinoma, lymphoid neoplasm diffuse large B-cell lymphoma, glioblastoma multiforme, pancreatic adenocarcinoma, rectum adenocarcinoma, prostate adenocarcinoma, stomach adenocarcinoma, thymoma, uterine corpus endometrial carcinoma, skin cutaneous melanoma, and uterine carcinosarcoma. Early detection is crucial for successful breast cancer management (Corbex et al. 2012; Coleman 2017). Current diagnostic techniques, such as mammography, have proven to be effective, but there is still room for improvement. Studies have shown that GRP78 is overexpressed in breast cancer cells, even at the early stages of the disease (Li and Lee 2006; Lee 2007). This suggests that GRP78 could serve as a potential biomarker for early detection. By developing sensitive and specific assays to measure GRP78 levels, physicians could identify breast cancer cases earlier, leading to timely intervention and improved patient outcomes (López-Muñoz et al. 2019; Raiter et al. 2020). It was reported that the blood, serum, pneumocytes, and leukocytes level of GRP78 is elevated in viral infection (SARS-CoV-2) and could be a biomarker in the coronavirus disease of 2019 (COVID-19) infection (Puzyrenko et al. 2021; Muthu et al. 2022; Angeles-Floriano et al. 2023).

## 1.2 Modulating GRP78 Expression for Therapeutic Intervention

In addition to its role as a diagnostic biomarker, GRP78 has also been implicated in breast cancer treatment (Farshbaf et al. 2020). Research has shown that high levels of GRP78 are associated with resistance to chemotherapy and radiation therapy. Treatment sensitivity could be restored by targeting GRP78 and reducing its expression (Lee 2007; Lu et al. 2020; Farshbaf et al. 2020). Various approaches are being explored, including small molecule inhibitors, gene therapy, and immunotherapy. These strategies aim to disrupt GRP78 function and resensitize breast cancer cells to conventional therapies, ultimately improving treatment outcomes (Sato et al. 2010; Katanasaka et al. 2010; Farshbaf et al. 2020; Hernandez and Cohen 2022). We have identified the nucleotide-binding domain (NBD) of GRP78 as a potential target for abemaciclib, ribociclib, and tunicamycin in silico. This is yet to be verified in the wet lab.

## 1.3 GRP78 as a Prognostic Indicator and Monitoring Tool

Beyond its role in detection and treatment, GRP78 also holds promise as a prognostic indicator and monitoring tool for cancer patients. It has been found that high levels of GRP78 are associated with poor prognosis and risk of disease recurrence (Uramoto et al. 2005; Zhang et al. 2006; Niu et al. 2015). Monitoring GRP78 expression levels during and after treatment could provide valuable insights into patient response and the likelihood of disease recurrence (Uramoto et al. 2005; Zhang and Zhang 2010; Yerushalmi et al. 2015; Zielinska et al. 2020). This information can then be used to tailor personalized treatment plans and optimize patient care.

## 1.4 GRP78 Mediates Cell Survival and Apoptosis

GRP78 has been shown to play a critical role in cell survival and apoptosis, which are crucial processes in cancer progression and treatment response (Zhang and Zhang 2010; Zheng et al. 2014; Casas 2017). The overexpression of GRP78 in breast cancer cells enhances cell survival pathways, allowing them to evade cell death signals. This increases the cancer cells' resistance to therapy and hence enhances tumor progression. By targeting GRP78, researchers aim to disrupt these survival pathways and induce apoptosis in breast cancer cells, ultimately improving treatment outcomes (Daneshmand et al. 2007; Zhang and Zhang 2010; Casas 2017). Figure 2 summarizes the possible roles of CS-GRP78 in breast cancer research.

## 1.5 Challenges and Future Prospects

While the potential of GRP78 in breast cancer management is promising, several challenges still need to be addressed. Research is necessary to fully understand the mechanisms underlying GRP78's functions in breast cancer cells. Additionally, developing targeted therapies that modulate GRP78 expression without affecting normal physiological processes is crucial since GRP78 has a master role in the proteostasis inside the ER (Pfaffenbach and Lee 2011). Despite these challenges, the utilization of GRP78 in breast cancer management holds great promise and could significantly improve patient outcomes in the future. This is especially critical for the most resistant types of breast cancer, like the triple negative (estrogen receptor, progesterone receptor, and the human epidermal growth factor-2) breast cancer cases. GRP78 was reported to be associated with programmed death ligand-1

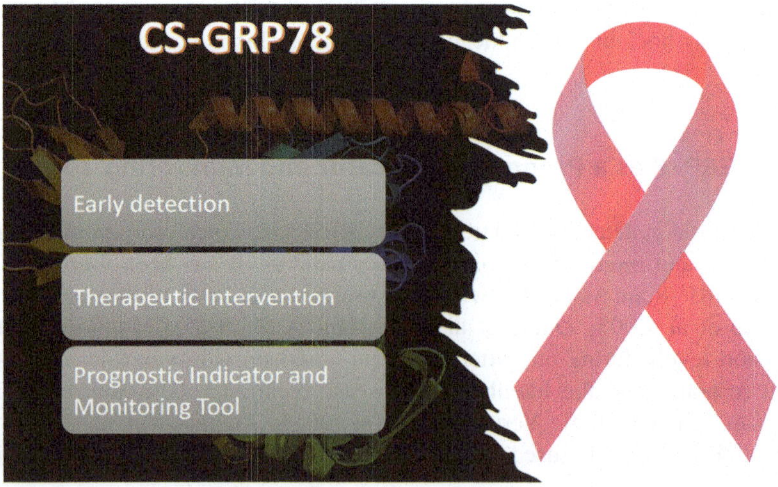


**Fig. 2** The possible roles of CS-GRP78 in the breast cancer research field. The figure was generated using PyMOL and MS Power Point

(PDL-1), increasing its level and subsequently the GRP78 level is elevated as well. These elevated PDL-1 levels trigger immune responses that suppress apoptosis (Chou et al. 2020).

## 2 Conclusion

Managing breast cancer using GRP78 as a diagnostic, therapeutic, prognostic, and monitoring tool represents a promising avenue for cancer research. Its role in early detection, treatment, and patient care highlights its potential to revolutionize breast cancer management. Continued research and development in this area are crucial to fully harness the potential benefits of GRP78 and improve the lives of breast cancer patients worldwide.

**Acknowledgments** The author is thankful to Alaa Elgohary, Alaa Mohamed, Albraa Mohamed, and Yousuf Mohamed for their help and support during the writing and revisions of this chapter.

**Funding** N/A.

**Compliance with Ethical Standards** The author declares no conflict of interest in this work. No humans or animals are involved in this chapter.

## References

Abu-Mahfouz A, Ali M, Elfiky A (2022) Anti-breast cancer drugs targeting cell-surface glucose-regulated protein 78: a drug repositioning in silico study. J Biomol Struct Dyn 41(16): 7794–7808. https://doi.org/10.1080/07391102.2022.2125076

Amaresan R, Gopal U (2023) Cell surface GRP78: a potential mechanism of therapeutic resistant tumors. Cancer Cell Int 23:1–13. https://doi.org/10.1186/S12935-023-02931-9

Angeles-Floriano T, Sanjuan-Méndez A, Rivera-Torruco G et al (2023) Leukocyte surface expression of the endoplasmic reticulum chaperone GRP78 is increased in severe COVID-19. J Leukoc Biol 113:1–10. https://doi.org/10.1093/JLEUKO/QIAC017

Casas C (2017) GRP78 at the centre of the stage in cancer and neuroprotection. Front Neurosci 11: 254955. https://doi.org/10.3389/FNINS.2017.00177/BIBTEX

Chou C-W, Yang R-Y, Chan L-C et al (2020) The stabilization of PD-L1 by the endoplasmic reticulum stress protein GRP78 in triple-negative breast cancer. Am J Cancer Res 10:2621

Coleman C (2017) Early detection and screening for breast cancer. Semin Oncol Nurs 33:141–155. https://doi.org/10.1016/J.SONCN.2017.02.009

Corbex M, Burton R, Sancho-Garnier H (2012) Breast cancer early detection methods for low and middle income countries, a review of the evidence. Breast 21:428–434. https://doi.org/10.1016/J.BREAST.2012.01.002

Daneshmand S, Quek ML, Lin E et al (2007) Glucose-regulated protein GRP78 is up-regulated in prostate cancer and correlates with recurrence and survival. Hum Pathol 38:1547–1552. https://doi.org/10.1016/J.HUMPATH.2007.03.014

Elfiky AA, Ibrahim IM (2021) Zika virus envelope–heat shock protein A5 (GRP78) binding site prediction. J Biomol Struct Dyn 39:5248–5260. https://doi.org/10.1080/07391102.2020.1784794/SUPPL_FILE/TBSD_A_1784794_SM7299.TIF

Elfiky AA, Baghdady AM, Ali SA, Ahmed MI (2020) GRP78 targeting: hitting two birds with a stone. Life Sci 260:118317. https://doi.org/10.1016/J.LFS.2020.118317

Elfiky AA, Amr A, Mosaad A et al (2023) Cs-GRP78 recognition site on dengue virus envelope protein: in silico perspective. Future Virol 18:411–420. https://doi.org/10.2217/FVL-2022-0192

Elgohary AM, Elfiky AA, Barakat K (2021) GRP78: a possible relationship of COVID-19 and the mucormycosis; in silico perspective. Comput Biol Med 139:104956. https://doi.org/10.1016/J.COMPBIOMED.2021.104956

Elshemey W, Ibrahim IM, Elfiky AA, Elgohary AM (2023) Prediction of HCV E2 association with the host-cell chaperone, GRP78. Inf Med Unlocked 39:101257. https://doi.org/10.1016/J.IMU.2023.101257

Farshbaf M, Khosroushahi AY, Mojarad-Jabali S et al (2020) Cell surface GRP78: an emerging imaging marker and therapeutic target for cancer. J Control Release 328:932–941. https://doi.org/10.1016/J.JCONREL.2020.10.055

Gifford JB, Hill R (2018) GRP78 influences chemoresistance and prognosis in cancer. Curr Drug Targets 19:701–708. https://doi.org/10.2174/1389450118666170615100918

Hernandez I, Cohen M (2022) Linking cell-surface GRP78 to cancer: from basic research to clinical value of GRP78 antibodies. Cancer Lett 524:1–14. https://doi.org/10.1016/J.CANLET.2021.10.004

Ibrahim IM, Abdelmalek DH, Elfiky AA (2019) GRP78: a cell's response to stress. Life Sci 226:156–163. https://doi.org/10.1016/J.LFS.2019.04.022

Katanasaka Y, Ishii T, Asai T et al (2010) Cancer antineovascular therapy with liposome drug delivery systems targeted to BiP/GRP78. Int J Cancer 127:2685–2698. https://doi.org/10.1002/IJC.25276

Lee AS (2007) GRP78 induction in cancer: therapeutic and prognostic implications. Cancer Res 67:3496–3499. https://doi.org/10.1158/0008-5472.CAN-07-0325

Li J, Lee AS (2006) Stress induction of GRP78/BiP and its role in cancer. Curr Mol Med 6:45–54

Li T, Fu J, Cheng J et al (2023) New progresses on cell surface protein HSPA5/BiP/GRP78 in cancers and COVID-19. Front Immunol 14:1166680. https://doi.org/10.3389/FIMMU.2023.1166680/BIBTEX

López-Muñoz E, Corres-Molina M, García-Hernández N (2019) Correlation of the protein expression of GRP78 and BIK/NBK with prognostic markers in patients with breast cancer and neoadjuvant chemotherapy. J Obstet Gynaecol 40:419–426. https://doi.org/10.1080/01443615.2019.1652886

Lu G, Luo H, Zhu X (2020) Targeting the GRP78 pathway for cancer therapy. Front Med (Lausanne) 7:540458. https://doi.org/10.3389/FMED.2020.00351/BIBTEX

Muthu V, Dhaliwal M, Sharma A et al (2022) Serum glucose-regulated protein 78 (GRP78) levels in COVID-19-associated mucormycosis: results of a case–control study. Mycopathologia 187:355–362. https://doi.org/10.1007/S11046-022-00645-6/TABLES/3

Niu Z, Wang M, Zhou L et al (2015) Elevated GRP78 expression is associated with poor prognosis in patients with pancreatic cancer. Sci Rep 5:1–12. https://doi.org/10.1038/srep16067

Pfaffenbach KT, Lee AS (2011) The critical role of GRP78 in physiologic and pathologic stress. Curr Opin Cell Biol 23:150–156. https://doi.org/10.1016/J.CEB.2010.09.007

Puzyrenko A, Jacobs ER, Sun Y et al (2021) Pneumocytes are distinguished by highly elevated expression of the ER stress biomarker GRP78, a co-receptor for SARS-CoV-2, in COVID-19 autopsies. Cell Stress Chaperones 26:859–868. https://doi.org/10.1007/S12192-021-01230-4/FIGURES/3

Raiter A, Lipovetzki J, Lubin I, Yerushalmi R (2020) GRP78 expression in peripheral blood mononuclear cells is a new predictive marker for the benefit of taxanes in breast cancer neoadjuvant treatment. BMC Cancer 20:1–10. https://doi.org/10.1186/S12885-020-06835-Z/FIGURES/4

Sato M, Yao VJ, Arap W, Pasqualini R (2010) GRP78 signaling hub: a receptor for targeted tumor therapy. Adv Genet 69:97–114. https://doi.org/10.1016/S0065-2660(10)69006-2

Sharma GN, Dave R, Sanadya J et al (2010) Various types and management of breast cancer: an overview. J Adv Pharm Technol Res 1:109

Uramoto H, Sugio K, Oyama T et al (2005) Expression of endoplasmic reticulum molecular chaperone Grp78 in human lung cancer and its clinical significance. Lung Cancer 49:55–62. https://doi.org/10.1016/J.LUNGCAN.2004.12.011

Waks AG, Winer EP (2019) Breast cancer treatment: a review. JAMA 321:288–300. https://doi.org/10.1001/JAMA.2018.19323

Yerushalmi R, Raiter A, Nalbandyan K, Hardy B (2015) Cell surface GRP78: a potential marker of good prognosis and response to chemotherapy in breast cancer. Oncol Lett 10:2149–2155. https://doi.org/10.3892/OL.2015.3579/HTML

Zhang LH, Zhang X (2010) Roles of GRP78 in physiology and cancer. J Cell Biochem 110:1299–1305. https://doi.org/10.1002/JCB.22679

Zhang J, Jiang Y, Jia Z et al (2006) Association of elevated GRP78 expression with increased lymph node metastasis and poor prognosis in patients with gastric cancer. Clin Exp Metastasis 23:401–410. https://doi.org/10.1007/S10585-006-9051-9/FIGURES/3

Zheng YZ, Cao ZG, Hu X, Shao ZM (2014) The endoplasmic reticulum stress markers GRP78 and CHOP predict disease-free survival and responsiveness to chemotherapy in breast cancer. Breast Cancer Res Treat 145:349–358. https://doi.org/10.1007/S10549-014-2967-X/FIGURES/3

Zielinska HA, Daly CS, Alghamdi A et al (2020) Interaction between GRP78 and IGFBP-3 affects tumourigenesis and prognosis in breast cancer patients. Cancers 3821(12):3821. https://doi.org/10.3390/CANCERS12123821

# A Hormone Immunotherapy (HIT) Combination in Advanced Breast Cancer

Andrea Nicolini and Paola Ferrari

**Abstract**

The NCCN guidelines on the use of hormone therapy (HT), immunotherapy (IT), and hormone immunotherapy (HIT) in advanced breast cancer are briefly reported. Then, investigational and clinical findings supporting the relationships between tumor growth and immune evasion and its impact on both HT and IT are examined. Finally, the rationale for an efficacious combination of a novel IT with conventional HT (HIT) in the largest subset of ER+ breast cancer patients is introduced and discussed. ER+ breast cancer is considered not amenable for IT as being immunologically "cold," and currently, aromatase inhibitors (AIs) and fulvestrant, a selective estrogen receptor downregulator (SERD), are recommended as first-line hormone therapy (HT) in association with cyclin kinase inhibitors (CKIs) for ER+ HER2− breast cancer patients. However, a pilot clinical study and a successive larger, observational, retrospective 2:1 ratio control-case study where ER+ breast cancer patients who benefited clinically during conventional first-line antiestrogen HT additionally received cyclic interferon beta interleukin-2 sequence, showed superior results to those who received HT alone, and results also were comparable to those receiving HT-CKIs. In the observational 2:1 ratio control-case study, the median PFS and OS were significantly longer in the 42 cases than in the 95 controls (median time 33 vs. 18 months, $p = 0.002$ and 81 vs. 62 months, $p = 0.019$, respectively). In the 42 cases, unlike the HT-CKI combination where grade 3–4 adverse events (AEs) occurred, grade 1–2 flu-like syndrome and injection site reaction were the only

A. Nicolini (retired) (✉)
Department of Oncology, Transplantations and New Technologies in Medicine, University of Pisa, Pisa, Italy
e-mail: andrea.nicolini@med.unipi.it

P. Ferrari
Unit of Oncology 1, Azienda Ospedaliera Universitaria Pisana, Pisa, Italy

 
Interdisciplinary Cancer Research, https://doi.org/10.1007/16833_2022_38
Published online: 21 October 2022

two more commonly observed serious AEs. Moreover, the proposed IT is 8–18 times cheaper than HT combined with CKIs. Despite the promising findings, no prospective multicenter validation trial has been sponsored and launched by pharmaceutical industries or government agencies. Overall, this and other considerations raise a major concern for the ongoing promotion of research in oncology.

**Keywords**

Breast cancer · Hormone immunotherapy (HIT) · Immunotherapy

## 1 Introduction

Breast cancer is the most common cancer in women worldwide, and there were 2,260,000 estimated new cases in 2020 and will be 2,500,000 new cases in 2025 (Globocan n.d., https://gco.iarc.fr). ER+ and/or PR+ endocrine-dependent breast cancer represents up to 60–80% of cases among the different molecular subtypes. Therefore, it includes the majority of all breast cancer patients, and the incidence increases with age (Senkus et al. 2014; Carey et al. 2006). In these patients, from the middle to almost the end of the twentieth century, hormone therapy (HT), including surgical oophorectomy and successively tamoxifen (TAM), was seminal to curing advanced disease. Tamoxifen was discovered more than 50 years ago, and it was the first therapy to enter clinical practice among a class of new drugs called selective estrogen receptor modulators (SERMs). In 1977, it was approved for use in patients with advanced breast cancer, and for decades, it has been uniformly administered to millions of patients with great benefit. As a therapeutic class, SERMs have estrogenic and antiestrogenic activity depending on the target tissue and the hormonal environment. In breast cancer cells, SERMS and tamoxifen in particular exert antiestrogenic action, as they prevent ER signaling by binding to the ER and causing an inactive complex (Shagufta 2018; Patel and Bihani 2018). Therefore, tamoxifen was an *ante-litteram* "target therapy" that was followed by another SERM drug, toremifene citrate, which was developed and entered clinical practice more than 20 years ago. Toremifene (TOR) resulted from investigational research aimed at discovering a SERM maintaining the helpful ER antagonist actions of TAM without the concomitant ER agonist ones that were supposed to be responsible for important harmful side effects, mainly cervical cancer promotion (Hayes et al. 1995). Clinical data supported the efficacy and safety of TOR for the treatment of breast cancer in postmenopausal patients. However, overall, these studies did not confirm a clear safety advantage or disadvantage for TOR compared to TAM (Vogel et al. 2014). Over a prolonged period of SERM use, first- and second-generation aromatase inhibitors (AIs) and progestins were the other available hormone therapies. Between the end of the twentieth and the beginning of the twenty-first century, a few early randomized clinical trials both in adjuvant and advanced settings (Mouridsen et al. 2003; Thürlimann et al. 2004) showed significant superiority of a third generation of

AIs compared to tamoxifen. This new class of drugs, given orally, prevented ER signaling by inhibiting the synthesis of estradiol. Specifically, they decreased local estrogen production by inhibiting an enzyme called aromatase in fat tissue, where this enzyme changes other hormones into estrogens. However, these drugs did not stop the ovaries from producing estrogens (Patel and Bihani 2018; Kharb et al. 2020). Third-generation AIs include steroidal (letrozole and anastrozole) and non-steroidal (exemestane) compounds. In the major International Societies guidelines, they replaced TAM as the recommended first-line HT, and TAM was the second-line option (NCCN guidelines n.d.; ESMO guidelines n.d.). At the same time, a new molecular classification divided breast cancer into four subtypes. This classification included luminal A (ER + PR+, HER2/neu negative, Ki-67 low), luminal B (ER + PR+, HER2/neu negative, Ki-67 high), HER2+ (ER+, PR+/−, Ki-67 high/low, HER2+), and triple-negative (ER−, PR−, HER2/neu−) subtypes. In 2002, fulvestrant, the first of a new class of drugs termed selective estrogen receptor downregulators (SERDs), received FDA approval for the treatment of tamoxifen-resistant breast cancer patients (Nathan and Schmid 2017). This class of drugs prevents ER signaling by degrading ER proteins and inhibiting trophic activity and does not have agonistic estrogenic action (Patel and Bihani 2018). From 1990 to over 2000, a few investigational studies and some pilot clinical studies based on the use of immunotherapy (IT) and involving cytokines (mainly interleukin-2, bone marrow CSGF, and alpha-/beta-interferons) given alone or in combination with other conventional therapies were conducted more often in ER+ breast cancer (Nicolini et al. 2006). Most of these studies did not achieve significant improvement in survival, and these cytokines did not definitely enter clinical practice for breast cancer treatment. In this chapter, the current NCCN guidelines on the use of HT, IT, and hormone immunotherapy (HIT) combinations in advanced breast cancer are briefly reported. The following section is devoted to experimental findings supporting the relationships between tumor growth and immune evasion. Then, clinical data suggesting the usefulness of HT and/or maintenance IT following conventional systemic therapy are described. Finally, the rationale for an efficacious combination of a novel IT with conventional HT (HIT) in the largest subset of ER+ breast cancer patients is introduced and discussed.

## 2 Current Guidelines for HT, IT, and Their Combination in Advanced Breast Cancer

HT, IT, and HIT recommendations/options according to the NCCN 2020 guidelines for invasive breast cancer (NCCN guidelines n.d.) are summarized here. Both in adjuvant and in locoregional recurrent or stage IV (M1) settings, HT is the mainstay treatment for ER+ breast cancer. Despite the increasing interest in using IT against cancer, only anti-HER family receptor monoclonal antibodies (Abs) (trastuzumab, pertuzumab) in HER2-positive breast cancer (Zhu and Joy 2017; Gianni et al. 2012) and monoclonal Abs PD-L1 inhibitors (atezolizumab) combined with chemotherapy (CT) in triple-negative breast cancer (TNBC) subtypes are recommended (Gonzalez-Ericsson et al. 2020; Kwa and Adams 2018). Trastuzumab and pertuzumab inhibit tumor growth by blocking HER family receptors. However, they also mediate antibody-dependent cellular cytotoxicity (ADCC).

## 2.1 Systemic Therapy in the Adjuvant Setting (Table 1)

HT
Premenopausal Women

In premenopausal women, tamoxifen for 5 years ± ovarian suppression or ablation and an AI (exemestane, anastrozole, letrozole) for 5 years plus ovarian suppression or ablation are the two main treatment options. Among patients who became postmenopausal and had received tamoxifen, an AI for 5 additional years is recommended or 5 additional years to complete 10 years with tamoxifen. In those who remained premenopausal, one can consider tamoxifen for an additional 5 years to complete 10 years or provide no further HT.

Postmenopausal Women

In postmenopausal women, AIs for 2–3 to 5 years are recommended, and tamoxifen for 2–3 to 6 years is the alternative option. Thereafter, AI for 5 years can be continued for an additional 3–5 years, whereas AI for 2–3 years is replaced with tamoxifen to complete 5 years of HT. Similarly, tamoxifen for 2–3 years can be continued with AI to complete 5 years of HT or with up to 5 years of AI. Tamoxifen for 4.5–6 years can be replaced by an AI for 5 years or continued for an additional 5 years to complete 10 years of tamoxifen. In women with a contraindication or who decline or are intolerant to AIs, tamoxifen for 5–10 years is the recommended alternative option.

Node-negative (N0), ER+, and/or PR+, HER2/neu− patients
HT or chemotherapy (CT) followed by HT

In patients who are pT1–T2 or T3 and pN0 with tumor ≤0.5 cm, HT only must be considered. In those with tumors ≥0.5 cm, the 21-gene RT PCR assay, if possible, is a helpful tool. If performed and the recurrence score (RS) is <26, HT is recommended. If the RS is 26–30, HT and CT followed by HT are two options, and if RS is >31, CT followed by HT is preferred. In the absence of the 21-gene RT PCR assay, HT and CT followed by HT are recommended.

Node-positive (N+), ER+, and/or PR+, HER2/neu− patients

In patients with pN1mi (≤ 2 mm axillary node metastasis) or N1 and less than four nodes, clinical characteristics, tumor stage, and pathology must be considered for the

**Table 1** Systemic adjuvant endocrine therapy of invasive breast cancer: hormone therapy (HT), chemotherapy (CT) plus immunotherapy (IT) (NCCN guidelines n.d., version 5.2020)

| | | |
|---|---|---|
| *Systemic adjuvant HT* | | |
| Premenopausal | TAM for 5 years ± ovarian suppression or ablation | Post: AI for 5 years to complete 10 years or consider TAM for additional 5 years to complete 10 years |
| | | Pre: Consider TAM for additional 5 years to complete 10 years or no further HT |
| | AI for 5 years + ovarian suppression or ablation | Post: Consider AI for additional 3–5 years |
| Postmenopausal | AI for 5 years | Consider AI for additional 3–5 years |
| | Or AI for 2–3 years | TAM for 2–3 years to complete 5 years AI |
| | Or TAM for 2–3 years | AI for further 2–3 years to complete 5 years HT, or up to 5 years AI |
| | Or TAM for 4.5–6 years | AI for 5 years or consider TAM for additional 5 years to complete 10 years HT |
| | If patient cannot receive AI (contraindications, intolerance, refusal) | TAM for 5 years or consider TAM for up to 10 years |
| *ER+ PR+ HER2+* | | |
| HT or CT plus anti HER2 abs (IT) and HT | | |
| pT1, pT2, or pT3; and pN0 or pN1mi (≤ 2 mm axillary lymph nodes) | Tumor ≤0.5 cm | HT ± CT with trastuzumab (pN0) |
| | | HT or CT with trastuzumab and HT (pN1mi) |
| | Tumor 0.6–1 cm | HT or CT with trastuzumab and HT |
| | Tumor >1 cm | CT with trastuzumab and HT |
| N+ (1 or more ipsilateral axillary metastases >2 mm) | | CT with trastuzumab and HT |
| | | Or CT with trastuzumab + pertuzumab and HT |

*TAM* tamoxifen, *AI* aromatase inhibitor, *anti-HER2 Abs (IT)* anti-HER2 receptor family monoclonal antibodies

initial decision-making for adjuvant CT. If the patient, based on these characteristics, is not a candidate for CT, HT alone is recommended. In this instance, in premenopausal patients, the magnitude of the benefit of ovarian ablation by surgical removal or by radiation is the same as that achieved with CMF alone. If the patient is a candidate for CT, prognosis and CT benefit should be assessed by gene expression assays. If so, HT and CT followed by HT are the two options. If the patient is a candidate for CT but a gene expression assay is not available, clinical and pathologic features must be used for decision-making. In this instance, and in patients with four or more than four ipsilateral nodes >2 mm, CT followed by HT is the only option.

ER− and PR−, HER2/neu +
CT plus anti-HER family receptor monoclonal Abs

In patients who are pT1–T2 or T3 and pN0 or pN1mi (≤ 2 mm axillary node metastasis) with tumors ≤0.5 cm, CT with trastuzumab must be considered. Similarly, in patients with tumors 0.6–1 cm and tumors >1 cm, CT with trastuzumab is recommended. In those with N+ (one or more ipsilateral metastases >2 mm), CT with trastuzumab and CT with trastuzumab plus pertuzumab are the two recommended options.

ER+ and/or PR+, HER2/neu-positive patients (Table 1)
HT or CT plus anti-HER family receptor monoclonal Abs and HT

In pT1, pT2, or pT3 and pN0 or pN1mi (≤ 2 mm axillary node metastasis) patients with tumor ≤0.5 cm, HT ± CT with trastuzumab (pN0) and HT or CT plus trastuzumab and HT (pN1mi) are recommended. In patients with tumors 0.6–1 cm, HT or CT with trastuzumab and HT and the latter in patients with tumor >1 cm should be administered. In N+ patients (one or more ipsilateral metastasis >2 mm), CT with trastuzumab and HT or CT with trastuzumab plus pertuzumab and HT should be given.

## 2.2 Systemic Therapy in Locoregional Recurrent or Stage IV (M1) Breast Cancer (Table 2)

The preferred options of first-line HT are as follows:

-AI plus cyclin D-dependent kinase (CDK) 4/6 inhibitor (abemaciclib, palbociclib, ribociclib), SERD (fulvestrant) ± nonsteroidal AI (NSAI) (anastrozole, letrozole), fulvestrant + CDK4/6 inhibitor, NSAI, SERM (tamoxifen or toremifene), or steroidal aromatase inactivator (exemestane).

Second and subsequent lines are fulvestrant + CDK4/6 inhibitor if CDK4/6 inhibitor was not previously used. For PIK3CA-mutated tumors, consider additional targeted therapy options: everolimus + HT (exemestane, fulvestrant, tamoxifen), NSAI, steroidal aromatase inactivator (exemestane), SERD, and SERM.

ER+ and/or PR+ HER2neu− patients
One or more HT lines and then CT

In patients with visceral crisis, initial CT should be considered, and therapy should be continued until progression or unacceptable toxicity. In those without visceral crisis and prior HT within 1 year, if premenopausal, ovarian ablation or suppression plus systemic therapy is recommended (in case of progression on initial HT, switch to a different HT option). If they are postmenopausal, systemic therapy is recommended (in case of progression on initial HT, switch to a different HT option).

**Table 2** Systemic therapy in recurrent or stage IV (M1) breast cancer: hormone therapy (HT) or hormone immunotherapy (HIT), or immunotherapy (IT) ± chemotherapy (CT) (NCCN guidelines n.d., version 5.2020)

| | | |
|---|---|---|
| ER+ and/or PR+ HER2− postmenopausal or premenopausal receiving ovarian ablation or suppression | *HT* | Continue therapy until progression or unacceptable toxicity |
| | AI + CDK4/6 inhibitor (palbociclib, ribociclib, abemaciclib) | |
| | SERD (fulvestrant) ± NSAI (anastrozole, letrozole) | |
| | SERD (fulvestrant) + CDK4/6 inhibitor | |
| | NSAI | |
| | SERM (tamoxifen, toremifene) | |
| | Steroidal aromatase inactivator (exemestane) | |
| ER+ and/or PR+ HER2+ postmenopausal or premenopausal receiving ovarian ablation or suppression | *CT or HT + anti-HER2 Abs (IT)* | |
| | Trastuzumab ± pertuzumab + docetaxel | |
| | Trastuzumab ± pertuzumab + paclitaxel | |
| | Ado-trastuzumab emtansine (T-DM1) | |
| | FAM trastuzumab deruxtecan-nxki | |
| | AI ± trastuzumab | |
| | AI ± lapatinib | |
| | AI ± lapatinib + trastuzumab | |
| | Fulvestrant ± trastuzumab | |
| | Tamoxifen ± trastuzumab | |

*ER* estrogen receptor, *PR* progesterone receptor, *AI* aromatase inhibitor, *CDK* cyclin-dependent kinase, *SERD* selective estrogen receptor downregulator, *SERM* selective estrogen receptor modulator, *anti HER2 Abs (IT)* anti HER2 receptor family monoclonal antibodies

All patients should continue HT therapy until progression or unacceptable toxicity. In those without visceral crisis and no prior HT within 1 year, if premenopausal, ovarian ablation and suppression plus systemic therapy or SERMs are the options. If they are postmenopausal, systemic therapy is recommended. All these patients should continue therapy until progression or unacceptable toxicity. In patients with progression or unacceptable toxicity on first-line HT, if not refractory to HT, consider an additional line of HT ± targeted therapy or systemic therapy. In patients with no clinical benefit following three sequential HT regimens or symptomatic visceral disease, systemic therapy is recommended. For those with visceral crisis, progression, or unacceptable toxicity on first-line systemic therapy, another line of systemic therapy is recommended. However, most of these patients will be candidates for many lines of systemic therapy for palliation of advanced disease. Therefore, in these patients, based on careful assessment of the value of the ongoing treatment, the risks and benefits of an additional line of systemic therapy, performance status, and preferences, no further cytotoxic therapy and continued supportive care can be another option.

ER+ and/or PR+ HER2neu + patients

The preferred HT plus anti-HER family receptor monoclonal Ab options are as follows:

-AI± trastuzumab, AI± apatinib, AI± lapatinib + trastuzumab, fulvestrant ± trastuzumab, or tamoxifen ± trastuzumab

The preferred CT regimens plus anti-HER family receptor monoclonal Abs are as follows:

Pertuzumab + trastuzumab+docetaxel or pertuzumab + trastuzumab+paclitaxel.

The CT and/or HT plus anti-HER family receptor monoclonal Abs or HT options are trastuzumab+docetaxel, trastuzumab+paclitaxel, pertuzumab + trastuzumab+ docetaxel, pertuzumab+trastuzumab+ paclitaxel or ado-trastuzumab emtansine (T-DM1), and fam-trastuzumab deruxtecan-nxki or HT (if prior HT within 1 year, a different HT must be considered, if premenopausal, ovarian ablation or suppression should also be considered) with AI± trastuzumab, AI± lapatinib, AI± lapatinib + trastuzumab, fulvestrant ± trastuzumab, and tamoxifen ± trastuzumab. The treatment should be continued until progression or unacceptable toxicity. If progression on first-line HT occurs, another line of HT, if not endocrine resistant, must be considered ± HER2/neu-targeted therapy. If there is no clinical benefit after three sequential HT regimens ± HER2/neu-targeted therapy or symptomatic visceral disease, systemic CT therapy + HER2/neu-targeted therapy until progression must be considered. In the case of progression on first-line CT plus HER2/neu-targeted therapy, another line of CT plus HER2/neu-targeted therapy must be considered, and HER2/neu-targeted therapy should be continued until progression. Most of the abovementioned patients are candidates for multiple lines of systemic therapy to palliate advanced breast cancer. Therefore, in these patients, based on careful assessment of the ongoing treatment, the risks and benefits of an additional line of systemic therapy, performance status, and preferences, no further HER2/neu-targeted therapy and continued supportive care can be another option.

HR-negative HER2/neu (TNBC) patients with PD-L1 expression ≥1% on tumor-infiltrating Immune Cells.

CT plus a PD-L1 inhibitor, in particular atezolizumab+albumin-bound paclitaxel, is the recommended option.

ER− and/or PR−, HER2/neu+

Systemic therapy plus anti-HER family receptor monoclonal Abs with pertuzumab plus trastuzumab plus docetaxel or paclitaxel or ado-trastuzumab emtansine (T-DM1) or fam-trastuzumab deruxtecan-nxki or trastuzumab plus CT regimen or other HER2/neu-targeted therapies are recommended. These options must be continued until progression or unacceptable toxicity. At progression, another line of systemic therapy plus HER2/neu-targeted therapy is the option. Most of the abovementioned patients are candidates for multiple lines of systemic therapies to palliate advanced breast cancer. Therefore, in these patients, based on careful assessment of the ongoing treatment, the risks and benefits of an additional line of systemic therapy, performance status, and preferences, no further HER2/neu targeted therapy and continued supportive care can be another option.

## 3 Experimental Studies Suggest a Close Relationship Between Tumor Growth and the G0–G1 State and Immune Evasion

### 3.1 Tumor Growth and Immune Evasion

We have reported a few experimental studies in which proliferation and tumor growth were closely linked to immune evasion in breast cancer (Nicolini et al. 2018). Accordingly, other more recent investigational studies showing this relationship are briefly described here. Aneuploidy, also termed somatic copy number alterations (SCNAs), is widespread in cancer and is thought to drive tumorigenesis. In a previous study (Davoli et al. 2017) analyzing 12 human cancer types in highly aneuploid tumors, reduced expression of markers of cytotoxic infiltrating immune cells, namely, $CD8^+$ T cells, together with increased expression of cell proliferation markers was found. Different types of SCNAs predicted the proliferation and immune signatures, implying distinct underlying mechanisms. Additionally, it was found that tumor aneuploidy inversely correlates with patient survival. The authors concluded that aneuploidy "together with other tumor characteristics such as tumor mutational load, may thus help identify patients most likely to respond to immunotherapy." In a review (Hutcheson et al. 2015) considering that the loss of retinoblastoma (RB) promotes tumor etiology and progression and that malignant tumors are characterized by aberrant cell division, the authors highlighted that much research has focused on the ability of RB to govern the cell cycle by repressing proliferation-related genes. In this research, it is progressively understood that RB is a relevant mediator of many functions and that one area that is gaining increasing interest is the emerging role for RB in regulating diverse features of immune function. Based upon these findings, the authors suggest that "RB is more than simply a regulator of cellular proliferation; it is at the crossroads of proliferation and the immune response." In another investigational study (Tauriello et al. 2018) carried out in mice with progressive liver metastatic disease, blockade of TGFβ signaling rendered tumors susceptible to anti-PD-1-PD-L1 therapy. The oncogenic role of TGFβ signaling in promoting tumor proliferation and progression is widely known (Gulei et al. 2017). However, these data showed that increased TGFβ in the tumor microenvironment (TME) was also a primary mechanism of immune evasion, promoting T-cell exclusion and inhibiting the acquisition of the $T_H1$ effector phenotype. A previous study (Su et al. 2020) focused on fat mass and obesity-associated protein (FTO), an RNA $N^6$-methyladenosine ($m^6A$) demethylase that plays oncogenic roles in various cancers. In this research, two potent small-molecule FTO inhibitors that exhibit strong antitumor effects in multiple types of cancers were reported. It was found that genetic depletion and pharmacological inhibition of FTO strongly decreased leukemia stem/initiating cell self-renewal and reprogrammed the immune response by suppressing the expression of immune checkpoint genes, especially LILRB4. FTO inhibition sensitized leukemia cells to T-cell cytotoxicity and overcame hypomethylating agent-induced immune evasion. The authors concluded that their findings show that FTO plays crucial roles in cancer stem cell self-renewal and immune evasion and underscore the potential of targeting FTO for cancer therapy.

## 3.2 G0–G1 State and Immunosuppressive TME Reversion Induced by Antiestrogens

Estrogens promote tumor growth and immune evasion in the TME and play a role in the development and progression of multiple cancers, including breast cancer (Folkerd and Dowsett 2010). In in vivo experimental models (Osborne 1994; Wolf et al. 1994; Doisneau-Sixou et al. 2003), antiestrogen therapy accounts for a prolonged G0–G1 state (resting state) of estrogen-dependent cancer cells. Therefore, a cytostatic rather than a cytotoxic-cytocidal effect occurs. In the TME, antiestrogens inhibit tumor growth and progression and downregulate ER-mediated immune suppression (immunostat). Therapies interfering with E2 signaling, such as SERMs or SERDs and AIs, have been seminal in reducing ER-positive breast cancer mortality over the past three decades (Cancer Facts & Figures 2017). In a study by Welboren et al. (2009) that confirmed the findings reported by Frasor et al. (Frasor et al. 2004, 2006), it was shown that most of the E2-upregulated genes were antagonized by tamoxifen, which mainly showed agonistic behavior on E2-downregulated genes (Nicolini et al. 2018). In addition to the antiproliferative action, targeted inhibition of the E2 pathway may act as a novel strategy to enhance the effects of immunotherapies and reverse the immune imbalance within the TME. Therefore, antiestrogen therapy increases the response in E2-sensitive tumors, as it not only inhibits tumor growth and progression, mediated by ERs via the "genomic" and "nongenomic" pathways (Hoskins et al. 2009), but also reverses an immunosuppressive TME. Based on these findings, a paradigm shift has been proposed toward utilizing antiestrogen therapy to target the immunosuppressive TME, thereby increasing the efficacy and duration of response of current immunotherapies (Welte et al. 2017).

## 4 Usefulness of Maintenance IT in Patients Showing Clinical Benefit During Conventional CT or with Minimal Residual Disease (MRD)

In a pilot study (Recchia et al. 2008), maintenance IT with low-dose IL-2 and 13-cis-retinoic acid was administered to 100 consecutive metastatic breast cancer patients with a clinical benefit (complete response (CR)+, partial response (PR)+, stable disease (SD)) following six to eight courses of CT induction. There were 68 ER+ and/or Pgr + patients, all of whom, after CT induction, received, in addition to maintenance IT, HT with luteinizing hormone-releasing hormone (LHRH) analogs or letrozole according to whether they were pre- or postmenopausal. In the 100 patients, the median PFS and OS were 37.1 and 57.5 months, respectively. PFS and OS were 44.7 and 64.5, respectively, in the 68 patients with ER+ tumors compared with 32.7 and 51.4, respectively, in the 23 patients with ER− tumors. In this study, the authors highlighted "a sustained improvement in lymphocytes, NKs, and CD4+/CD8+ ratio with respect to baseline values." In an earlier pilot study from the same research group (Recchia et al. 1998), maintenance IT with IFN-β, retinyl palmitate, and tamoxifen until progression was administered to 23 metastatic breast cancer patients who had achieved a clinical response (11 CR and 12 PR) following 6 courses of induction and 2 successive courses of consolidating CT. All 23 patients were unresponsive to HT. The PFS and OS were 31.4 and 44 months, respectively, in the 23 responders (CR + PR). The OS was 66 months (with a 9-year survival rate of 34%) in the 11 complete responders and 17 months in the nonresponders (7 with SD and 6 with progressive disease). In the former of these 2 mentioned pilot studies, all 68 ER+ patients received maintenance IT without any previous clinical evaluation of response to hormone therapy. In the latter, all 23 of the recruited patients were unresponsive to HT. In both of these pilot studies, a well-documented immune modulation and antiproliferative action by IFNs and retinoids were shown. In particular, the capacity of retinoids to increase the number of IL-2 receptors and peripheral blood lymphoid cells expressing the surface markers of Th cells (Prabhala et al. 1991) and the function of retinoids, which facilitate the differentiation of immature myeloid suppressor cells (Gr1+ CD115−), were reported (Huang et al. 2006). The biological activity of retinoids is mediated by nuclear retinoic acid receptors (RARs) and retinoid X receptors (RXRs), which are ligand-activated transcription factors and are basally present in breast cancer cells. More recently, the relevant role of retinoids for breast cancer chemoprevention and treatment, because of their ability to induce cell differentiation and growth suppression, has been highlighted (Garattini et al. 2014; Seo et al. 2015). ER+ breast tumors are also RAR sensitive and are mainly activated by all trans-retinoic acids and 13-cis-retinoic acid (Garattini et al. 2014). Overall, in these two pilot studies, a longer than expected clinical benefit and OS were observed, likely because the hormone and/or immune modulation following successful antiproliferative cytotoxic CT favored a cytotoxic immune response. Accordingly, in a recent report (Nicolini et al. 2019b), we described a breast cancer patient who underwent atypical lung resection of a 13 mm metastatic nodule of the middle lung lobe, which occurred approximately

1 year after she had been operated on for primary breast cancer during adjuvant tamoxifen treatment. The immunophenotype from postoperative histology confirmed that the primary breast cancer was the probable origin and ER, PR, and Ki67 were 30%, negative, and 50%, respectively. This patient, who had likely become hormone-resistant soon after the removal of the metastatic pulmonary nodule, was continuously given 3 months of cycles of oral CT (vinorelbine and capecitabine) followed by 9 months of IT. This IT aimed to promote a cytotoxic immune response through Treg depletion and by inhibiting macrophages and MDSC-mediated immune suppression (Nicolini et al. 2019b). To date, 54 months following the removal of the metastatic pulmonary nodule, no sign of relapse has occurred. Overall, these findings suggest that clinical benefit following systemic conventional therapy and MRD are two main conditions favoring a cytotoxic immune response to maintenance IT carried out with cytokines (IFNs or interleukins) and retinoids or by inhibiting immune suppression.

## 5 A Prolonged Antiestrogen-Induced G0–G1 State Concomitant with an Increased Cytotoxic Immune Response During a New Schedule of HIT in Metastatic, Endocrine-Dependent Breast Cancer

In the last decade, we have reported very promising results (Nicolini and Carpi 2005; Nicolini et al. 2005, 2008) from an open pilot clinical study in metastatic endocrine-dependent breast cancer where conventional antiestrogen therapy was combined with cyclic immune stimulation with interferon beta interleukin-2 sequence. In our pilot study, only patients showing a clinical benefit (CR + PR + SD) during antiestrogen therapy were selected to receive the additional IT (Nicolini and Carpi 2005). In particular, all enrolled patients received a clinical benefit in an initial 2- to 6-month interval (induction time) during which they were given antiestrogen alone. The IT, which was then administered in addition to antiestrogens, consisted of 3,000,000 IU interferon beta every other day, i.m., 3 times a week for 4 weeks, followed by interleukin-2 s.c. at 3,000,000 IU every other day 3 times a week for an additional 4 weeks. In the successive 2 weeks, all patients received antiestrogens only and then started again with the same schedule (Table 3). The results and their interpretation have been published previously (Nicolini et al. 2014, 2015, 2016). In an early report (Nicolini and Carpi 2005) in 31 endocrine-dependent breast cancer patients, PFS and median OS since the diagnosis of distant metastases were 33 and 94 months, respectively. In 24 patients with a higher rate of hormone dependency (55% ER + PR+), the median OS was 98.5 months, and the median PFS was 45.5 months compared with the remaining 7 subjects with a lower rate of hormone dependency (20% ER+ Pgr+), in whom the median OS was 37 months and median PFS was 20 months (Nicolini et al. 2014, 2016). It is noteworthy that 16% of the patients have survived for more than 10 years in complete remission. During the clinical benefit, we showed "laboratory evidence of the effect of IT as well as that hormone resistance occurs at the progression of the disease concomitantly with a

**Table 3** An immunotherapy (IT) schedule in metastatic breast cancer patients in clinical benefit during first-line hormone therapy (HT)

| Drug | Route of administration | Dose | Days |
|---|---|---|---|
| First-line tamoxifen/ toremifene | Orally | 20 mg/60 mg per day | Continuously |
| Beta-interferon | Subcutaneously | 3,000,000 IU three times a week | Weeks 1–4 |
| Interleukin-2 | Subcutaneously | 3,000,000 IU three times a week | Weeks 5–8 |

After a 2-week rest (weeks 9–10), this cycle of immunotherapy with the same cytokines was repeated until progression (Nicolini et al. 2020)

laboratory pattern compatible with immune inhibition" (Nicolini et al. 2014, 2007). We hypothesized that the prolonged tamoxifen-induced G0–G1 state and the concomitant tumor burden decrease in responders downregulated the main hallmarks sustaining tumor growth and immune evasion, thus favoring immune stimulation by the INF-β/IL-2 sequence. In a successive report (Nicolini et al. 2019a), 42 patients were recruited. In 28 of them, it was possible to define a luminal (ER+) molecular subtype, and the subtype was nonluminal (ER−) in 9 others. The median PFS from the beginning of HT in the 42 studied patients was 33 months, and the median OS was 82 months. The cumulative survival rate at 5 and 10 years was 0.69 ± 0.07 and 0.15 ± 0.06, respectively. In the luminal subtype, the median PFS was 33 months, and the median OS was 91 months. In the nonluminal subtype, the median PFS was 32 months, and the median OS was 59 months. We concluded that these data confirmed the previously reported findings and that a relevant, prolonged PFS and OS compared with that expected in similar populations occurred both in luminal and in nonluminal molecular subtypes. The pilot study clinical findings were confirmed by a larger, observational, retrospective, 2:1 ratio control-case study, where the 42 studied patients (cases) were assessed together with 89 comparable controls. The 89 controls were treated with conventional first-line HT alone, without IT, and a few of them additionally had received cyclin kinase inhibitors (CKIs) (Nicolini et al. 2020). The median PFS was significantly longer in the cases than in controls, 33.1 (95% CI 24.5–41.8) vs. 18 (95% CI 12.1–23.8) months ($p < 0.0001$). Additionally, the median OS was significantly longer in the cases, 81 vs. 62 (95% CI 48.1–75.9) months ($p < 0.0029$) (Table 4). An update (paper submitted) of the study involving the same 42 cases and 6 more controls ($n = 95$ controls) with more prolonged follow-up confirmed the initial findings. In particular, median PFS and OS remained significantly longer in the 42 studied patients who additionally had received the INF-β/IL-2 sequence than in the 95 controls (median time 33 vs. 18 months, $p = 0.002$ and 81 vs. 62 months, $p = 0.019$). Overall, the principal prognostic/predictive characteristics (hormone receptor and HER2 status, AIs, biological therapy, and visceral involvement) except DFI were in favor of controls. Thus, both endpoints maintained a significant difference in favor of the 42 cases when all these factors were included in the adjusted statistical analysis. Moreover, the HRs for PFS

**Table 4** PFS and OS unadjusted and adjusted for disease-free interval (DFI), hormone receptor and HER2 status, visceral involvement, and biological therapy in patients treated with hormone immunotherapy (HIT, 0) compared with hormone therapy alone (HT, 1)

| Endpoint | Unadjusted analysis | | Adjusted analysis | | |
|---|---|---|---|---|---|
| | HR (95% CI) | *P* value | RC | HR (95% CI) | *P* value |
| PFS | 2.347 (1.556–3.541) | <0.0001 | 1.128 | 3.090 (1.685–5.668) | 0.0002 |
| OS | 1.874 (1.067–3.291) | 0.029 | 0.764 | 2.147 (0.943–4.891) | 0.069 |

*PFS* progression-free survival, *OS* overall survival, *RC* regression coefficient, HIT, 0; HT, 1 (Nicolini et al. 2020)

and OS were 1.902 and 1.684 in the unadjusted analysis and increased to 2.533 and 2.158, respectively, in adjusted statistical analysis in favor of cases. Interestingly, in the 42 cases, good HIT tolerability was confirmed, and no grade 3–4 adverse events were observed. Grade 0–1 or grade 1–2 flu-like syndrome and injection site reaction were the most common side effects, occurring in 60% and 93% of the cases, respectively. In one recent paper (Nicolini et al. 2019b), we also updated one of the six metastatic long-term breast cancer survivors in CR during HIT (cumulative survival at 10 years 0.16, SE 0.15 ± 0.06). This patient, who underwent whole-body computed tomography (CT), was disease-free 138 months after the initial occurrence of CR. In the last 6 years, the immune-stimulating cycles of the INF-β/IL-2 sequence were interrupted for 6 months per year, and in the last 5 years, HT was also interrupted together with the INF-β/IL-2 cycles. Instead, all other long-term survivors had previously received the experimental HIT until progression. In this patient in prolonged CR, the recovery of the innate and adaptive cytotoxic immune response induced by HIT may be enough to maintain disseminated cancer cell (DCC) dormancy even when HIT was withdrawn.

## 6 The Potential Rationale of a New HIT Schedule

The addition of immune therapies, including interleukin-2, to antiestrogens in breast cancer patients has been investigated in some experimental studies conducted in human breast cancer cell lines (Albertini et al. 1992; Dabrosin et al. 2003). The results from our pilot and other studies are consistent with the higher dormancy signature scores found in ER-positive breast cancer cell lines and primary breast cancers compared to ER-negative cell lines and tumors (Kim et al. 2012). Moreover, TGFβ2 downregulation by estradiol has been shown in human MCF-7 breast cancer cell lines (Frasor et al. 2004). Recruitment of HDACs that induce H3 and H4 deacetylation in the pS2 and c-myc promoters is a reported mechanism of action of tamoxifen (Liu and Bagchi 2004). Tamoxifen also induces ERα conformational change favoring ERα interaction with corepressors. The repression of ERα-mediated transcription induces a compacted heterochromatin structure, cell cycle arrest, and cell death (Légaré and Basik 2016). Interestingly, in Drosophila wings, together with a repressive chromatin structure and silencing of the cell cycle in postmitotic cells, a concomitant robust G0 state has been reported, where cells became strongly refractory to cell cycle reentry in contrast to a more flexible G0 state (Liu and Bagchi 2004; Ma and Buttitta 2017). Consistent with our laboratory data (Nicolini et al. 2014, 2007), active immune stimulation with the INF-β/IL-2 sequence boosted the cytotoxic immune response, which likely synergized to further stabilize the tamoxifen-induced G0–G1 state. In earlier reports (Nicolini et al. 2015; Carpi et al. 2009), we proposed the following mechanistic interpretation: "within the TME, stromal cells, infiltrating lymphocytes and tumor cells foster tumor growth and immune evasion through a complex network of autocrine and paracrine loops mediated by cytokines and growth factors. In metastatic disease in clinical benefit during anti-estrogen, a stable or decreased tumor burden and a lower genetic instability due to quiescent state (G0–G1 state) of tumor cells are also likely to reduce immune evasion by the downregulation of immune escape and immune inhibition. This favors the immune attack stimulated by the sequential administration of IFN-β and IL-2, which, by synergizing with antiestrogen therapy, can delay hormone resistance and clinically result in a prolonged response or prolonged stable disease. The larger the tumor working portion is, the higher its biological aggressiveness. In those with shorter median survival and a likely lower hormone dependence, an expected greater and more aggressive portion of the tumor burden worked with a higher production of cytokines and growth factors. This can explain the earlier occurrence and perhaps more effective immune inhibition during the progression of metastatic disease in those surviving for <5 years. On the other hand, in those surviving for >5 years and with higher levels of hormone dependence, an expected smaller and probably less aggressive portion of the tumor burden worked with lower production of cytokines and growth factors. This may have permitted the immune system to work more effectively for a longer time" (Fig. 1).

## 7 Conclusion

Currently, first-line HT with AIs or fulvestrant is recommended in ER+ HER2− metastatic breast cancer patients (Najim et al. 2019). Acquired resistance occurs in approximately 30–50% of ER+ breast cancer patients during first-line HT, and additional or substitutive treatment is required to maintain the clinical benefit (Nayar et al. 2019; AlFakeeh and Brezden-Masley 2018). Recently, the CCND1-CDK4/6-RB pathway, which is innately fundamental for cell cycle control and governs whether a cell moves on or arrests at the G1-S phase, has been recognized as a helpful molecular target to prolong the clinical benefit of ER + HER2– luminal metastatic breast cancer during first-line HT (Piezzo et al. 2020). Therefore, following successful investigational clinical trials, some CDK 4/6 inhibitors (CKIs), particularly ribociclib (Hortobagyi et al. 2018), palbociclib (Finn et al. 2020), and abemaciclib (Johnston et al. 2019), have received FDA approval. Therefore, CKIs in combination with AIs or fulvestrant are the mainstay as first-line treatment of

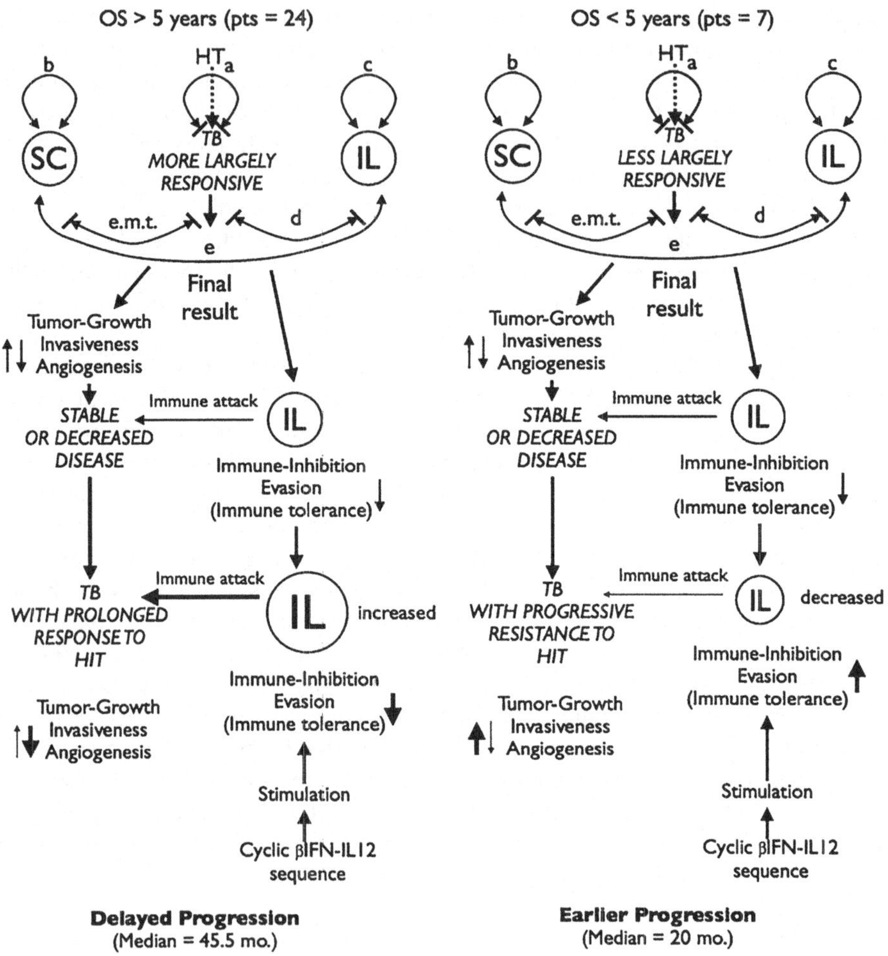


**Fig. 1** In the tumor microenvironment (TME), infiltrating lymphocytes and stromal and tumoral cells play a central role in a complex network interconnected by autocrine (*a, b, c*) and paracrine (*d, e, e.m.t.*) loops formed by pathological molecular pathways that foster tumor growth and immune evasion. *OS* overall survival, *HT* hormone therapy, *TB* tumor burden, *SC* stromal cell, *IL* interleukin, *e.m.t.* epithelial to mesenchymal transition, *HIT* hormone immunotherapy. (Modified from Nicolini A, et al. Curr Drug Targets. 2016;17(10):1127–39. Figure 1)

ER + HER2– metastatic breast cancer patients. In ER+ HER2– luminal breast cancer compared with TNBC and HER2+ breast cancer subtypes, the lowest level of PDL1 expression (Cimino-Mathews et al. 2016; Rugo et al. 2018; Nanda et al. 2016), a lower TML (Vogelstein et al. 2013; Cancer Genome Atlas Network 2012; Lee et al. 2016), and a relatively low level of TILs have been reported. This finding allowed to term ER-positive breast cancer, including ER+ HER2– breast cancer, as being immunologically "cold" and not amenable for the use of IT. Despite this status, estrogen has been linked to inflammation in the TME (Nagari and Franco 2015; Quigley et al. 2017; Rothenberger et al. 2018), and some evidence indicates that immunity and inflammation may also be implicated in the biology of this subtype (Dieci et al. 2016). Furthermore overall, data accumulated during our two clinical studies (pilot and observational) robustly indicated that, unlike the ongoing common thought, ER-positive breast cancer is a molecular subtype where successful active immune manipulation can be attained, and the proposed strategy is likely the best approach. In the last decade, many experimental studies have provided further support for our initial hypothesis. In fact, it has been reported that in E2-sensitive tumors, not only does antiestrogen therapy inhibit tumor growth and progression mediated by ERs via the "genomic" and "nongenomic" pathways (Hoskins et al. 2009), but it also reverses an immunosuppressive TME. In clinical trials carried out with the addition of CKIs to AIs or fulvestrant, the median PFS ranged from 25.3 months with ribociclib (Hortobagyi et al. 2018) to 28.2 months with abemaciclib (Johnston et al. 2019). In ribociclib and abemaciclib trials, the median OS has not yet been reported. However, mature data in the palbociclib trial (Finn et al. 2020) did not show any significant difference in treated patients vs. controls (34.5 vs. 37.5 months). In the arm additionally treated with ribociclib, palbociclib, and abemaciclib, grade 3–4 AEs from any cause have been reported in >10%, >15%, and 58% of patients, respectively. In particular, grade 3 neutropenia occurred in >50% of patients additionally receiving ribociclib or palbociclib and in 22% receiving abemaciclib. In our observational 2:1 control-case study, in the 42 cases treated with SERMs/SERDs plus IT, no grade 3–4 AEs occurred, and grade 1–2 flu-like syndrome and injection site reaction were confirmed as the only two more commonly observed serious AEs (Nicolini and Carpi 2005). Furthermore, our proposed IT is 8–18 times cheaper than the CKIs, and the low cost of both immunological drugs involved in the proposed HIT schedule likely accounts for the disinterest of pharmaceutical companies. In the last two decades, pharmaceutical companies have focused on building up so-called "targeted" therapies/drugs. This process entails prolonged investigation, and usually the obtained drugs are expensive with limited clinical benefit and/or heavy side effects when they enter into clinical practice (Nicolini 2017). This issue was the main reason for the development of the Repurposing Drugs in Oncology (ReDO) project, which is an ongoing collaborative project that was launched in 2014 exclusively aimed at the potential use of licensed noncancer medications as sources of new cancer therapeutics (Pantziarka et al. 2018). However, in this novel research field, less than 4% of clinical trials have received financial support from a commercial sponsor or government institution (Pantziarka et al. 2018). Overall, this pattern raises a major concern for advances in oncology. Currently, most researchers take part in collaborative groups formed within the main oncologic centers usually involved in investigational clinical trials sponsored by the major pharmaceutical companies and addressing the more profitable targeted drugs. However, a few others worldwide work outside of these collaborative groups, and their findings, even though brilliant, often do not turn into validation projects due to a lack of sufficient financial appeal and specious obstacles. This issue, in addition to being unfair, accounts for a significant delay in the progression of oncologic research. Therefore, it is the time that the free scientific community removes this hurdle.

**Conflict of Interest** The authors declare they do not have potential conflicts of interest.
All patients involved in research conducted and mentioned by the authors gave written, witnessed, informed consent.

## References


Albertini MR, Gibson DF, Robinson SP, Howard SP, Tans KJ, Lindstrom MJ, Robinson RR, Tormey DC, Jordan VC, Sondel PM (1992) Influence of estradiol and tamoxifen on susceptibility of human breast cancer cell lines to lysis by lymphokine-activated killer cells. J Immunother 11(1):30–39

AlFakeeh A, Brezden-Masley C (2018) Overcoming endocrine resistance in hormone receptor-positive breast cancer. Curr Oncol 25(Suppl 1):S18–S27

Cancer Facts & Figures (2017) American Cancer Society, Atlanta

Cancer Genome Atlas Network (2012) Comprehensive molecular portraits of human breast tumours. Nature 490(7418):61–70

Carey LA, Perou CM, Livasy CA, Dressler LG, Cowan D, Conway K, Karaca G, Troester MA, Tse CK, Edmiston S, Deming SL, Geradts J, Cheang MC, Nielsen TO, Moorman PG, Earp HS, Millikan RC (2006) Race, breast cancer subtypes, and survival in the Carolina Breast Cancer Study. JAMA 295(21):2492–2502

Carpi A, Nicolini A, Antonelli A, Ferrari P, Rossi G (2009 Dec) Cytokines in the management of high risk or advanced breast cancer: an update and expectation. Curr Cancer Drug Targets 9(8): 888–903

Cimino-Mathews A, Thompson E, Taube JM, Ye X, Lu Y, Meeker A, Xu H, Sharma R, Lecksell K, Cornish TC, Cuka N, Argani P, Emens LA (2016) PD-L1 (B7-H1) expression and the immune tumor microenvironment in primary and metastatic breast carcinomas. Hum Pathol 47(1):52–63

Dabrosin C, Palmer K, Gauldie J (2003) Oestradiol enhances tumour regression induced by B7-1/IL-2 adenoviral gene transfer in a murine model of breast cancer. Br J Cancer 89(2):385–390

Davoli T, Uno H, Wooten EC, Elledge SJ (2017) Tumor aneuploidy correlates with markers of immune evasion and with reduced response to immunotherapy. Science 355(6322):eaaf8399

Dieci MV, Griguolo G, Miglietta F, Guarneri V (2016) The immune system and hormone-receptor positive breast cancer: is it really a dead end? Cancer Treat Rev 46:9–19

Doisneau-Sixou SF, Sergio CM, Carroll JS, Hui R, Musgrove EA, Sutherland RL (2003) Estrogen and antiestrogen regulation of cell cycle progression in breast cancer cells. Endocr Relat Cancer 10(2):179–186

ESMO guidelines (n.d.) https://www.esmo.org/Guidelines/Breast-Cancer. Accessed 15 May 2022

Finn RS, Boer K, Bondarenko I, Patel R, Pinter T, Schmidt M, Shparyk YV, Thummala A, Voitko N, Bananis E, McRoy L, Wilner K, Huang X, Kim S, Slamon DJ, Ettl J (2020 Sept) Overall survival results from the randomized phase 2 study of palbociclib in combination with letrozole versus letrozole alone for first-line treatment of ER+/HER2− advanced breast cancer (PALOMA-1, TRIO-18). Breast Cancer Res Treat 183(2):419–428

Folkerd EJ, Dowsett M (2010) Influence of sex hormones on cancer progression. J Clin Oncol 28(26):4038–4044

Frasor J, Stossi F, Danes JM, Komm B, Lyttle CR, Katzenellenbogen BS (2004) Selective estrogen receptor modulators: discrimination of agonistic versus antagonistic activities by gene expression profiling in breast cancer cells. Cancer Res 64(4):1522–1533

Frasor J, Chang EC, Komm B, Lin CY, Vega VB, Liu ET, Miller LD, Smeds J, Bergh J, Katzenellenbogen BS (2006 Jul 15) Gene expression preferentially regulated by tamoxifen in breast cancer cells and correlations with clinical outcome. Cancer Res 66(14):7334–7340

Garattini E, Bolis M, Garattini SK, Fratelli M, Centritto F, Paroni G, Gianni' M, Zanetti A, Pagani A, Fisher JN, Zambelli A, Terao M (2014) Retinoids and breast cancer: from basic studies to the clinic and back again. Cancer Treat Rev 40(6):739–749

Gianni L, Pienkowski T, Im YH, Roman L, Tseng LM, Liu MC, Lluch A, Staroslawska E, de la Haba-Rodriguez J, Im SA, Pedrini JL, Poirier B, Morandi P, Semiglazov V, Srimuninnimit V, Bianchi G, Szado T, Ratnayake J, Ross G, Valagussa P (2012 Jan) Efficacy and safety of neoadjuvant pertuzumab and trastuzumab in women with locally advanced, inflammatory, or early HER2-positive breast cancer (NeoSphere): a randomised multicentre, open-label, phase 2 trial. Lancet Oncol 13(1):25–32

Globocan (n.d.) https://gco.iarc.fr

Gonzalez-Ericsson PI, Stovgaard ES, Sua LF, Reisenbichler E, Kos Z, Carter JM, Michiels S, Le Quesne J, Nielsen TO, Laenkholm AV et al (2020) The path to a better biomarker: application of a risk management framework for the implementation of PD-L1 and TILs as immuno-oncology biomarkers in breast cancer clinical trials and daily practice. J Pathol 250(5): 667–684

Gulei D, Mehterov N, Ling H, Stanta G, Braicu C, Berindan-Neagoe I (2017) The "good-cop bad-cop" TGF-beta role in breast cancer modulated by non-coding RNAs. Biochim Biophys Acta Gen Subj 1861(7):1661–1675

Hayes DF, Van Zyl JA, Hacking A, Goedhals L, Bezwoda WR, Mailliard JA, Jones SE, Vogel CL, Berris RF, Shemano I et al (1995) Randomized comparison of tamoxifen and two separate doses of toremifene in postmenopausal patients with metastatic breast cancer. J Clin Oncol 13(10): 2556–2566

Hortobagyi GN, Stemmer SM, Burris HA, Yap YS, Sonke GS, Paluch-Shimon S, Campone M, Petrakova K, Blackwell KL, Winer EP, Janni W, Verma S, Conte P, Arteaga CL, Cameron DA, Mondal S, Su F, Miller M, Elmeliegy M, Germa C, O'Shaughnessy J (2018) Updated results from MONALEESA-2, a phase III trial of first-line ribociclib plus letrozole versus placebo plus letrozole in hormone receptor-positive, HER2-negative advanced breast cancer. Ann Oncol 29(7):1541–1547

Hoskins JM, Carey LA, McLeod HL (2009) CYP2D6 and tamoxifen: DNA matters in breast cancer. Nat Rev Cancer 9(8):576–586

Huang B, Pan PY, Li Q, Sato AI, Levy DE, Bromberg J, Divino CM, Chen SH (2006) Gr-1+CD115+ immature myeloid suppressor cells mediate the development of tumor-induced T regulatory cells and T-cell anergy in tumor-bearing host. Cancer Res 66(2):1123–1131

Hutcheson J, Witkiewicz AK, Knudsen ES (2015) The RB tumor suppressor at the intersection of proliferation and immunity: relevance to disease immune evasion and immunotherapy. Cell Cycle 14(24):3812–3819

Johnston S, Martin M, Di Leo A, Im SA, Awada A, Forrester T, Frenzel M, Hardebeck MC, Cox J, Barriga S, Toi M, Iwata H, Goetz MP (2019) MONARCH 3 final PFS: a randomized study of abemaciclib as initial therapy for advanced breast cancer. NPJ Breast Cancer 5:5

Kharb R, Haider K, Neha K, Yar MS (2020) Aromatase inhibitors: role in postmenopausal breast cancer. Arch Pharm (Weinheim) 353(8):e2000081

Kim RS, Avivar-Valderas A, Estrada Y, Bragado P, Sosa MS, Aguirre-Ghiso JA, Segall JE (2012) Dormancy signatures and metastasis in estrogen receptor positive and negative breast cancer. PLoS One 7(4):e35569

Kwa MJ, Adams S (2018) Checkpoint inhibitors in triple-negative breast cancer (TNBC): where to go from here. Cancer 124(10):2086–2103

Lee HJ, Song IH, Park IA, Heo SH, Kim YA, Ahn JH, Gong G (2016) Differential expression of major histocompatibility complex class I in subtypes of breast cancer is associated with estrogen receptor and interferon signaling. Oncotarget 7(21):30119–30132

Légaré S, Basik M (2016) Minireview: the link between ERα corepressors and histone deacetylases in tamoxifen resistance in breast cancer. Mol Endocrinol 30(9):965–976. https://doi.org/10.1210/me.2016-1072

Liu XF, Bagchi MK (2004) Recruitment of distinct chromatin-modifying complexes by tamoxifen-complexed estrogen receptor at natural target gene promoters in vivo. J Biol Chem 279(15): 15050–15058

Ma Y, Buttitta L (2017) Chromatin organization changes during the establishment and maintenance of the postmitotic state. Epigenetics Chromatin 10(1):53

Mouridsen H, Gershanovich M, Sun Y, Perez-Carrion R, Boni C, Monnier A, Apffelstaedt J, Smith R, Sleeboom HP, Jaenicke F, Pluzanska A, Dank M, Becquart D, Bapsy PP, Salminen E, Snyder R, Chaudri-Ross H, Lang R, Wyld P, Bhatnagar A (2003) Phase III study of letrozole versus tamoxifen as first-line therapy of advanced breast cancer in postmenopausal women: analysis of survival and update of efficacy from the International Letrozole Breast Cancer Group. J Clin Oncol 21(11):2101–2109

Nagari A, Franco HL (2015) Interplay between inflammatory and estrogen signaling in breast cancer. Cytokine 76(2):588–590

Najim O, Seghers S, Sergoynne L, Van Gaver H, Papadimitriou K, Wouters K, Trinh XB, Huizing MT, Tjalma W (2019) The association between type of endocrine therapy and development of estrogen receptor-1 mutation(s) in patients with hormone-sensitive advanced breast cancer: a systematic review and meta-analysis of randomized and non-randomized trials. Biochim Biophys Acta Rev Cancer 1872(2):188315

Nanda R, Chow LQ, Dees EC, Berger R, Gupta S, Geva R, Pusztai L, Pathiraja K, Aktan G, Cheng JD, Karantza V, Buisseret L (2016) Pembrolizumab in patients with advanced triple-negative breast cancer: phase Ib KEYNOTE-012 study. J Clin Oncol 34(21):2460–2467

Nathan MR, Schmid P (2017) A review of fulvestrant in breast cancer. Oncol Ther 5(1):17–29

Nayar U, Cohen O, Kapstad C, Cuoco MS, Waks AG, Wander SA, Painter C, Freeman S, Persky NS, Marini L, Helvie K, Oliver N, Rozenblatt-Rosen O, Ma CX, Regev A, Winer EP, Lin NU, Wagle N (2019) Acquired HER2 mutations in ER+ metastatic breast cancer confer resistance to estrogen receptor-directed therapies. Nat Genet 51(2):207–216

NCCN guidelines (n.d.) https://www.nccn.org/professionals/physician_gls/pdf/breast.pdf. Accessed 15 May 2022

Nicolini A (2017) Biomarkers and the 'target therapies' in oncology, research and business. Biomark Med 11(3):217–219

Nicolini A, Carpi A (2005) Beta-interferon and interleukin-2 prolong more than three times the survival of 26 consecutive endocrine dependent breast cancer patients with distant metastases: an exploratory trial. Biomed Pharmacother 59(5):253–263

Nicolini A, Carpi A, Rossi G (2005) An immunotherapy schedule in endocrine-dependent metastatic breast cancer: correlation between clinical course and immunologic parameters. J Immunother 28(3):276–279

Nicolini A, Carpi A, Rossi G (2006) Cytokines in breast cancer. Cytokine Growth Factor Rev 17(5): 325–337

Nicolini A, Carpi A, Rossi G (2007) Relationship of cellular immunity, cytokines and CRP with clinical course in breast cancer patients with endocrine-dependent distant metastases treated with immunotherapy. Cancer Lett 251(2):330–338

Nicolini A, Carpi A, Ferrari P, Rossi G (2008) Immunotherapy prolongs the serum CEA-TPA-CA15.3 lead time at the metastatic progression in endocrine-dependent breast cancer patients: a retrospective longitudinal study. Cancer Lett 263(1):122–129

Nicolini A, Rossi G, Ferrari P, Carpi A (2014 Mar) Clinical and laboratory patterns during immune stimulation in hormone responsive metastatic breast cancer. Biomed Pharmacother 68(2): 171–178

Nicolini A, Ferrari P, Kotlarova L, Rossi G, Biava PM (2015) The PI3K-AKt-mTOR pathway and new tools to prevent acquired hormone resistance in breast cancer. Curr Pharm Biotechnol 16(9):804–815

Nicolini A, Carpi A, Ferrari P, Biava PM, Rossi G (2016) Immunotherapy and hormone-therapy in metastatic breast cancer: a review and an update. Curr Drug Targets 17(10):1127–1139

Nicolini A, Ferrari P, Rossi G, Carpi A (2018) Tumour growth and immune evasion as targets for a new strategy in advanced cancer. Endocr Relat Cancer 25(11):R577–R604

Nicolini A, Barak V, Biava P, Ferrari P, Rossi G, Carpi A (2019a) The use of immunotherapy to treat metastatic breast cancer. Curr Med Chem 26(6):941–962

Nicolini A, Ferrari P, Morganti R, Carpi A (2019b) Treatment of metastatic or high-risk solid cancer patients by targeting the immune system and/or tumor burden: six cases reports. Int J Mol Sci 20(23):5986

Nicolini A, Rossi G, Ferrari P, Morganti R, Carpi A (2020) A new immunotherapy schedule in addition to first-line hormone therapy for metastatic breast cancer patients in a state of clinical benefit during hormone therapy. J Mol Med (Berl) 98(3):375–382

Osborne CK (1994) Interactions of tamoxifen with cytotoxic chemotherapy for breast cancer. In: Jordan VC (ed) Long-term tamoxifen treatment for breast cancer. The University of Wisconsin Press, Madison, pp 181–196

Pantziarka P, Verbaanderd C, Sukhatme V, Rica Capistrano I, Crispino S, Gyawali B, Rooman I, Van Nuffel AM, Meheus L, Sukhatme VP, Bouche G (2018) ReDO_DB: the repurposing drugs in oncology database. Ecancermedicalscience 12:886

Patel HK, Bihani T (2018) Selective estrogen receptor modulators (SERMs) and selective estrogen receptor degraders (SERDs) in cancer treatment. Pharmacol Ther 186:1–24

Piezzo M, Cocco S, Caputo R, Cianniello D, Gioia GD, Lauro VD, Fusco G, Martinelli C, Nuzzo F, Pensabene M, De Laurentiis M (2020) Targeting cell cycle in breast cancer: CDK4/6 inhibitors. Int J Mol Sci 21(18):6479

Prabhala RH, Garewal HS, Hicks MJ, Sampliner RE, Watson RR (1991) The effects of 13-cis-retinoic acid and beta-carotene on cellular immunity in humans. Cancer 67(6): 1556–1560

Quigley DA, Tahiri A, Lüders T, Riis MH, Balmain A, Børresen-Dale AL, Bukholm I, Kristensen V (2017) Age, estrogen, and immune response in breast adenocarcinoma and adjacent normal tissue. Onco Targets Ther 6(11):e1356142

Recchia F, Frati L, Rea S, Torchio P, Sica G (1998) Minimal residual disease in metastatic breast cancer: treatment with IFN-beta, retinoids, and tamoxifen. J Interf Cytokine Res 18(1):41–47

Recchia F, Sica G, Candeloro G, Necozione S, Bisegna R, Bratta M, Rea S (2008) Maintenance immunotherapy in metastatic breast cancer. Oncol Rep 20(5):1173–1179

Rothenberger NJ, Somasundaram A, Stabile LP (2018) The role of the estrogen pathway in the tumor microenvironment. Int J Mol Sci 19(2):611

Rugo HS, Delord JP, Im SA, Ott PA, Piha-Paul SA, Bedard PL, Sachdev J, Le Tourneau C, van Brummelen EMJ, Varga A, Salgado R, Loi S, Saraf S, Pietrangelo D, Karantza V, Tan AR (2018) Safety and antitumor activity of pembrolizumab in patients with estrogen receptor-positive/human epidermal growth factor receptor 2-negative advanced breast cancer. Clin Cancer Res 24(12):2804–2811

Senkus E, Cardoso F, Pagani O (2014) Time for more optimism in metastatic breast cancer? Cancer Treat Rev 40(2):220–228

Seo HS, Woo JK, Shin YC, Ko SG (2015) Identification of biomarkers regulated by rexinoids (LGD1069, LG100268 and Ro25-7386) in human breast cells using Affymetrix microarray. Mol Med Rep 12(1):800–818

Shagufta AI (2018) Tamoxifen a pioneering drug: an update on the therapeutic potential of tamoxifen derivatives. Eur J Med Chem 143:515–531

Su R, Dong L, Li Y, Gao M, Han L, Wunderlich M, Deng X, Li H, Huang Y, Gao L, Li C, Zhao Z, Robinson S, Tan B, Qing Y, Qin X, Prince E, Xie J, Qin H, Li W, Shen C, Sun J, Kulkarni P, Weng H, Huang H, Chen Z, Zhang B, Wu X, Olsen MJ, Müschen M, Marcucci G, Salgia R, Li L, Fathi AT, Li Z, Mulloy JC, Wei M, Horne D, Chen J (2020) Targeting FTO suppresses cancer stem cell maintenance and immune evasion. Cancer Cell 38(1):79–96.e11

Tauriello DVF, Palomo-Ponce S, Stork D, Berenguer-Llergo A, Badia-Ramentol J, Iglesias M, Sevillano M, Ibiza S, Cañellas A, Hernando-Momblona X, Byrom D, Matarin JA, Calon A, Rivas EI, Nebreda AR, Riera A, Attolini CS, Batlle E (2018) TGFβ drives immune evasion in genetically reconstituted colon cancer metastasis. Nature 554(7693):538–543
Thürlimann B, Hess D, Köberle D, Senn I, Ballabeni P, Pagani O, Perey L, Aebi S, Rochlitz C, Goldhirsch A (2004) Anastrozole ('arimidex') versus tamoxifen as first-line therapy in postmenopausal women with advanced breast cancer: results of the double-blind cross-over SAKK trial 21/95--a sub-study of the TARGET (tamoxifen or 'arimidex' randomized group efficacy and tolerability) trial. Breast Cancer Res Treat 85(3):247–254
Vogel CL, Johnston MA, Capers C, Braccia D (2014) Toremifene for breast cancer: a review of 20 years of data. Clin Breast Cancer 14(1):1–9
Vogelstein B, Papadopoulos N, Velculescu VE, Zhou S, Diaz LA Jr, Kinzler KW (2013) Cancer genome landscapes. Science 339(6127):1546–1558
Welboren WJ, van Driel MA, Janssen-Megens EM, van Heeringen SJ, Sweep FC, Span PN, Stunnenberg HG (2009) ChIP-Seq of ERalpha and RNA polymerase II defines genes differentially responding to ligands. EMBO J 28(10):1418–1428
Welte T, Zhang XH, Rosen JM (2017) Repurposing antiestrogens for tumor immunotherapy. Cancer Discov 7(1):17–19
Wolf DM, Gottardis MM, Jordan VC (1994) Tamoxifen-resistant growth. In: Jordan VC (ed) Long-term tamoxifen treatment for breast cancer. The University of Wisconsin Press, Madison, pp 199–217
Zhu X, Joy AA (2017) Targeting HER2 in advanced breast cancer. Methods Mol Biol 1652:63–77

# Discovering New Targets in Triple-Negative Breast Cancer (TNBC): The Androgen Receptor and the Estrogen Receptor β

Pia Giovannelli, Gabriella Castoria, and Antimo Migliaccio

**Abstract**

Triple-negative breast cancer (TNBC) is a breast cancer (BC) subtype characterized by the lack of estrogen receptor (ER) and progesterone receptor (PR) and the absence of type-2 epidermal growth factor receptor (HER-2) overexpression. This status makes TNBC a challenge for scientists and physicians. Many studies tried to characterize the TNBC to find new targetable proteins to cure this BC subgroup, and the results have certainly underlined its high heterogeneity as principal and undisputed trait.

TNBC is usually characterized by poor prognosis, and, to date, chemotherapy is the first-line treatment. Unfortunately, many TNBCs only partially respond to chemotherapy. Many molecules, such as EGFR, PARP, MEK, histone deacetylase (HDAC), heat shock protein 90 (HSP90), and PI3K, are mutated or hyperactivated in TNBC, thereby representing an attractive target to develop effective therapies. Another weakness of TNBC is the presence of a lot of immune cells such as tumor-infiltrating lymphocytes. This makes immunotherapy another fascinating weapon to treat this incurable cancer.

To date, many studies have been focused on other two steroid receptors belonging to the nuclear receptor superfamily: the androgen receptor (AR) and the β isoform of estrogen receptor (ERβ). The two receptors represent promising target for controlling growth and spreading of TNBC. The AR and ERβ signaling are complexes and interfere with other cellular pathways hyperactivated in TNBC (EGFR, PI3K, CDKs) thereby strengthening their position and making it the perfect candidate to develop new and effective therapies for TNBC full remission.

P. Giovannelli (✉) · G. Castoria · A. Migliaccio
Department of Precision Medicine, University of Campania "L. Vanvitelli", Naples, Italy
e-mail: pia.giovannelli@unicampania.it

 
Interdisciplinary Cancer Research, https://doi.org/10.1007/16833_2023_141
Published online: 25 February 2023

**Keywords**

Androgen receptor · Estrogen receptor β · Steroid hormones · TNBC targets · TNBC therapy · Triple-negative breast cancer (TNBC)

## 1 Introduction

Breast cancer (BC) is the most frequent cancer in women worldwide and the fifth cause of cancer death overall. Every year it has registered an increasing number of cases, and in 2020 BC overtook lung cancer as the most diagnosed cancer in the world (Globocan 2021, https://www.iarc.who.int/featured-news/breast-cancer-awareness-month-2021/). BC is considered a hormone-related tumor commonly arising in post-menopausal women; its onset is linked to genetic components, but is also influenced by factors as nulliparity, early age menarche, late age menopause, and obesity (Giovannelli et al. 2018). Due to its heterogeneity, BC is now considered not just a single disease but a collection of breast tumors each with a typical molecular feature (Rivenbark et al. 2013). Traditionally, BC is classified in four subtypes depending on the expression status of progesterone and estrogen receptors (PR and ER) and epidermal growth factor receptor 2 (HER2; Giovannelli et al. 2018). This clinical classification is useful to define the therapeutical strategies for BC patients, but fails to predict short-term patient response and long-term patient survival (Rivenbark et al. 2013). Among the four subtypes of BC, triple-negative breast cancer (TNBC) is characterized by the total absence of the three classical therapeutical targets (ER, PR, and Her2); for this reason, the only useful therapy is conventional chemotherapy.

## 2 Triple-Negative Breast Cancer

TNBC accounts for the 15–20% of diagnosed breast cancers. It represents the most aggressive BC subtype, shows the higher incidence in young and in African American women, and is generally associated with an adverse prognosis (Giovannelli et al. 2018).

This BC subtype is defined by the lack of ER, PR, and HER2 amplification and is often associated with the germline mutations in the *brca*1 gene. These parameters allowed to identify TNBC as a single disease. In the last decade, scientists analyzed this disease from different points of view such as molecular, histological, genetic, and clinical profiles, thereby highlighting its amazing heterogeneity (Ensenyat-Mendez et al. 2021).

To identify, characterize, and assess their clinical properties, TNBCs were grouped in distinct subtypes. To date, there is more than one classification, in relation to a selected parameter; among them, the most important considers immunohistochemical or gene expression profiles.

The immunohistochemical classification analyzes the expression of a panel of 13 markers like basal (CK5/6, CK14) and luminal (CK7, 18, and 19) cytokeratins, vimentin, the epidermal growth factor receptor (EGFR) and BCL2 oncogenes, the tumor suppressor p53 and p16, and the proliferation marker Ki67. Using this approach, Elsawaf and colleagues (2013) divided TNBC in four subgroups: luminal, baso-luminal, basal A, and basal B with respect to the predominant pattern of cytokeratin expression.

The basal A subtype expresses high level of cytoplasmic p16 and high Ki-67 (high proliferation index), while the basal B subtype expresses high levels of p53. Both the subtypes include high-grade tumors are characterized by a high basal cytokeratin expression and a favorable overall survival, as compared with the baso-luminal and luminal TNBC. The baso-luminal subtype is characterized by the mixed expression of basal and luminal cytokeratins and an EGFR high expression. The luminal subtype is characterized by a high expression of luminal cytokeratins and low proliferation rate (low Ki-67; Elsawaf et al. 2013).

Considering the gene expression profile of TNBC, Lehmann and colleagues have drawn up a classification including six TNBC subtypes: mesenchymal stem-like (MSL), immunomodulatory (IM), basal-like1 (BL1), basal-like2 (BL2), mesenchymal (M), and luminal androgen receptor (LAR). Five years later, the same authors modified it because of the consistent presence of tumor-infiltrating lymphocytes (TILs) and stromal cells had influenced the examined genetic profiles in two of the six identified subtypes. The new classification divided TNBC in four subtypes: basal-like 1 (BL1), basal-like 2 (BL2), mesenchymal (M), and luminal androgen receptor (LAR). The BL1 subtype expresses a high number of genes related to cell cycle and DNA damage response, while the BL2 is characterized by the elevated expression of myo-epithelial markers and genes related to the growth factor signaling. The M subtype displays genetic patterns responsible for mesenchymal differentiation and cell motility, while the LAR subtype is prevalently characterized by the expression of the androgen receptor (AR) and the genes involved in hormone-related and inflammatory pathways (Lehmann et al. 2016; Borri and Granaglia 2021).

These four TNBC subtypes show different clinical characteristics. BL1 includes cancers with a higher grade but lower stage and an increased patient overall and relapse-free survival and displays a higher response to standard chemotherapy. LAR cancers are characterized by an increased regional spread and formation of distant metastases to the bone, while M subtypes metastasize to the lung.

The histopathological classification of TNBC includes lesions characterized by different pathologic grades. The lesions with a high pathologic grade are the most numerous and are represented by metaplastic and medullary carcinomas and carcinomas with an apocrine differentiation. The low-grade lesions are represented by secretory and salivary-gland type adenoid-cystic carcinomas and are a small group. Carcinomas with apocrine differentiation are prevalently characterized by the expression of AR and PI3K pathway mutations. In these carcinomas, the role of AR as prognostic factor is still debated (Geyer et al. 2017). Metaplastic, medullary, and adenoid cystic carcinoma have basal-like features, but different metaplastic cancers show characteristic of M subtype.

## 2.1 TNBC Therapies

TNBC is usually characterized by poor prognosis and lacks a wide choice of therapeutic agents for the absence of targetable hormone receptors and HER2 expression. Therefore, its better knowledge is a very intriguing challenge for scientists.

***Chemotherapy*** The election therapy for TNBC is chemotherapy used both in early and in advanced stages of the disease. After chemotherapy treatments, however, TNBC relapses within 1–3 years, and often metastasizes. Nevertheless, patients with TNBC respond to chemotherapy better than the other BC patients, but there are no data about the major effectiveness of adjuvant or neoadjuvant chemotherapy in these patients (Bianchini et al. 2016). Currently, there are no preferred standard forms of chemotherapy for TNBC patients. The most effective treatments are based on the use of taxanes and anthracyclines, in mono- or combo-therapy. The latter approach improves the response rate as compared with the monotherapy, although it does not improve patient survival because of the toxic effects (Foulkes et al. 2010). The platinum compounds have been tested to improve the classical chemotherapy. Patients with a BRCA1 and 2 respond better to carboplatin than to docetaxel. Neither in adjuvant nor in neoadjuvant chemotherapy, there are significant data concerning the efficiency of platinum, probably because of the added toxicity and an unclear correlation between an improved pathological complete response (pCR) rate and improved event-free survival in neoadjuvant settings (Bianchini et al. 2016).

## 2.2 Targeted Therapies

Given the amazing heterogeneity of TNBC, it is necessary to discover new and different molecular targets useful to develop personalized, tumor-specific, and efficient treatments. Many TNBCs only partially respond to chemotherapy and present alterations of different molecules under clinical investigations, such as EGFR, PARP, MEK, histone deacetylase (HDAC) and BET, heat shock protein 90 (HSP90), and PI3K.

***PI3K Inhibitors*** TNBC frequently exhibit mutations in a "hot point" region of the *pik3ca* gene, encoding the catalytic subunit α of the PI3K. These activating mutations are often associated with the loss of PTEN (Shah et al. 2012). Considering that the PI3K/AKT/mTOR signaling pathway is an oncogenic driver in TNBC, it became evident that inhibition of this cascade is a very promising strategy. The PI3K inhibitor, BKM120, inhibits the tumor growth in TNBC patient derived xenograft models, and its action is consistent in mice PTEN positive and lower in PTEN-null mice (Targeted Therapeutic Strategies for Triple-Negative Breast Cancer – PubMed). This pathway can be also blocked by using inhibitors acting on the other members such as Akt or mTOR. The rapamycin, a mTORCH1 inhibitor, as well as other inhibitors reduces metastases in TNBC.

***EGFR Inhibitors*** The epidermal growth factor receptor (EGFR) is a transmembrane glycoprotein endowed with a tyrosine kinase activity, which plays a pivotal role in cell proliferation and survival, cell migration and invasion, adhesion, apoptosis, epithelial to mesenchymal transition, differentiation, and angiogenesis. EGFR is closely related to carcinogenesis and tumor progression, and inversely correlated to a good prognosis in TNBC patients (Viale et al. 2009). The receptor tyrosine kinase activity can be inhibited by two different groups of drugs: inhibitors of tyrosine kinase (TKI, such as lapatinib, gefitinib, erlotinib) and monoclonal antibodies (such as cetuximab, panitumumab), long standing available and clinically approved. For these reasons, different clinical trials were yet started in TNBC patients with unsatisfactory results. Phase-2 clinical trials using cetuximab alone or plus carboplatin, or cisplatin alone or plus cetuximab, respectively, demonstrated that, while the combination between carboplatin and cetuximab has no effects on TNBC progression, the cisplatin and cetuximab combinatorial therapy contributes to slightly increase the progression-free survival (Bianchini et al. 2016). By contrast, preclinical observations show that inhibition of EGFR impairs TNBC cell growth, with a concomitant inhibition of EMT. These effects, however, might be non-specific, since erlotinib inhibits other kinases (Targeted Therapeutic Strategies for Triple-Negative Breast Cancer – PubMed). Collectively, these findings indicate that additional studies are needed to establish whether blockade of EGFR can be used in TNBC treatment.

***MEK Inhibitors*** Ras activity plays an undeniable role in TNBC progression (Galiè 2019). Several TNBC derived cell lines exhibit hyperactivation of the Ras/MAPKs pathway and are, hence, sensitive to MEK inhibitors. Activation of this pathway is less frequently caused by oncogenic mutation in Ras and Raf (Barretina et al. 2012; Giltnane and Balko 2014), while overexpression or activation of growth factor receptors seem to be one of the main responsible in Ras/MAPKs pathway aberrant activation (Bianchini et al. 2016). Again, the genetic or epigenetic loss of DUSP4, a negative regulator of this pathway, has been proposed to explain the Ras/ERK axis hyperactivation (Balko et al. 2013). MEK inhibitors, eventually combined with other drugs, might represent a good therapeutical strategy in TNBC. Duncan and colleagues demonstrated that, despite the good inhibitory effect, the use of MEK inhibitor induces the synthesis and activation of additional growth factor receptors, thereby causing resistance to therapy (Duncan et al. 2012). As such, a combinatorial treatment with MEK inhibitors and monoclonal antibodies or small molecules targeting the tyrosine-receptor kinases might be a good solution to overcome the drug resistance (Bianchini et al. 2016). In addition, combination of chemotherapy and MEK inhibitors or other targeted drugs is currently under clinical investigation. Infante and colleagues have demonstrated that patients with metastatic TNBC completely respond to a therapy which combines the gemcitabine with trametinib, a potent MEK inhibitor (Infante et al. 2013). These results, however, derive from a single study and must be confirmed by other data.

***PARP Inhibitors*** Because of the deficiency of BRCA1 and 2 as well as other components of homologous recombination in many human cancers, PARP inhibitors are widely used as they lead to a severe and selective toxicity called "synthetic lethality" (Turner et al. 2008). In the majority of studies, mostly phases II and III trials, PARP inhibitors, particularly Olaparib, reduce the growth of TNBC carrying BRCA 1 and 2 mutations when used as single molecule. By contrast, their effect seems less impressive and more unclear when used in combination with chemotherapy (Bianchini et al. 2016).

***Anti-androgen Therapy*** In the last years, the AR is gaining a prominent position in TNBC. One of these, LAR (Luminal -AR), is characterized by the AR expression. AR is also expressed in the other TNBC subtypes, even if at lower frequency. The pivotal role of AR is underlined by another simplistic classification made by scientists, distinguishing TNBC in AR-positive and quadruple negative BC (QNBC (Bhattarai et al. 2020)). Even if the action of AR in TNBC is not completely understood, different clinical trials are yet started. The next paragraphs aim, however, to profoundly discuss this matter.

***Other Molecular Targets*** Other molecular targets in TNBC are under investigation. Among them, there are proteins involved in genetic and epigenetic processes such as DNA methylases (DNMT) and histone deacetylases (HDAC), and protein involved in cell cycle progression such as some cyclin-dependent kinases (CDKs) or polo-like kinases (PLKs) or the monopolar spindle 1 (MPS1), also known as TTK1.

DNMTs work by adding a methyl group to the CpG dinucleotides. Among the different DNMT, DNMT1 is highly expressed in TNBC subtype where it is associated with poor survival. In TNBC, DNMT1 represses the ER expression, promotes the epithelial-mesenchymal transition (EMT) required for metastasis, induces cellular autophagy, and promotes the stem cell growth. In these cancers, DNMT1 represents an epigenetic target, and its inhibitors exert anti-tumorigenic effects (Wong 2021).

The HDACs deacetylase the histones thereby inducing a chromatin condensation and inhibiting the gene expression. In particular, the negative regulation of transcription of the tumor suppressor genes by HDAC activation stimulates tumor cell migration and invasion, angiogenesis, and proliferation (Targeted Therapeutic Strategies for Triple-Negative Breast Cancer – PubMed). The HDAC inhibitor, valproic acid, in combination with rapamycin, promotes the estrogen receptor gene (ESR1) expression in TNBC (Sulaiman et al. 2018). Again, HDAC inhibitors increase the expression of PDL1 and HLA-DR, while reduce the proportion on T-cells CD4 Foxp3$^{+}$, thereby enhancing the in vivo response to PD-1/CTLA-4 in TNBC mouse models. This allows the reduction of tumor growth and the increase of survival, associated with an increased T-cell tumor infiltration (Terranova-Barberio et al. 2017). These results evidence how HDAC inhibitors sustain and reinforce the immunotherapy in TNBC.

CDKs are involved in cell cycle regulation and progression, and their inhibitors prevent the cell proliferation and tumor growth. Unfortunately, few papers have investigated the effects of the use of CDK inhibitors in TNBC. Additionally, the frequent loss of Rb in TNBC reduces the effect of CDK inhibitors and makes more complex this plot. The efficacy of CDK inhibitors in TNBC might be reinforced by the combining use of AR or PI3K inhibitors (Targeted Therapeutic Strategies for Triple-Negative Breast Cancer – PubMed). Anyway, phase I and II trials testing the efficiency of CDK4/6 blockade, with or without other agents, are ongoing (Targeted Therapeutic Strategies for Triple-Negative Breast Cancer – PubMed).

The cell cycle regulators TTKs are involved in control of the spindle assembly checkpoint, thereby ensuring integrity and stability of genome in mitosis (Daniel et al. 2011). A gene expression meta-analysis of TNBCs identified an "aggressiveness"-associated gene expression signature, enriched for TTK genes. This study also confirmed the dependency of TNBC cells on TTK (Thu et al. 2018). As such, they represent druggable targets in TNBC, and their chemical blockade is currently available. Several TTK inhibitors are currently being evaluated in early phase clinical trials as single agents or in combination with taxane therapy, thereby demonstrating the therapeutic potential of these compounds (Thu et al. 2018). PLK-4 is the director of centriole duplication and has a pivotal role in maintaining the numerical integrity of centriole and centrosomes. PLK4 overexpression is common in cancer, including BCs, where it correlates with the disease aggressiveness. PLK4 was identified as a promising target with efficacy in TNBC. The TLK4 inhibition potentiates aneuploid and genomic instability and leads to cell death (Dominguez-Brauer et al. 2015). A novel inhibitor of PLK4, CFI-400945, in combination with radiation, exhibits anticancer effect in TNBC cell organoids. TTK inhibition, however, also kills non-transformed cells, and patient stratification strategies would greatly improve the applicability of TTK inhibitors in the clinic.

## 2.3 Immunotherapy

Immunotherapy represents a revolutionary technique to cure different types of cancers. The presence of numerous immune cells, such as tumor-infiltrating lymphocytes, is already a good prognostic factor (Borri and Granaglia 2021) and can be also used as weakness of TNBC.

Recruitment and re-activation of immune cells of the host is the aim of immunotherapy against cancer cells. The presence of lymphocytes in cancerous tissues causes the apoptotic death of tumor cells and the release of cancer cell antigens recognized by dendritic cells. Unfortunately, cancer cells use the immune checkpoints for immune escape. One of the molecules frequently targeted by immunotherapy is PD-L1. PD-1 (Programmed Cell Death-1)/PD-L1 (Programmed Cell Death Ligand-1) is a common immune checkpoint targeted by therapeutic agents that prevent the T- lymphocytes CD8+ de-activation and promote tumor regression by the increase of immunological response against cancer cells (Edechi et al. 2019). The role of PD-L1 in TNBC is still debated, but some studies demonstrate that it is largely expressed and associated with a better free-disease survival (Botti et al. 2017; Borri and Granaglia 2021). Treatment of TNBC with PD-L1 inhibitors, such as atezolizumab, is associated with a strong pathological response of cancer cells treated with chemotherapeutic drugs (Molinero et al. 2019; Schmid et al. 2020b). Furthermore, the high tumor mutational burden (TMB), characterizing the TNBC subtype, has been linked with an increased responsiveness to immune-checkpoints inhibitors (Li et al. 2021).

The efficacy of these inhibitors is higher in TNBC when administered early in the disease course, possibly due to the immune escape mechanisms during the advancement of the disease (Li et al. 2021). Based on these data, different randomized trials are now available. Of five randomized trials (IMpassion 031 (NCT03197935), KEYNOTE-522 (NCT03036488), GeparNuevo (NCT02685059), NeoTRIPaPDL1 (NCT02620280), and I-SPY2 (NCT01042379)) testing the administration of an anti PD-1/PD-L1 agent in combination with neoadjuvant chemotherapy, two (I-SPY2 and KEYNOTE-522) significantly improved the pathological complete response rate compared with immunotherapy alone (Schmid et al. 2020a, b; Li et al. 2021). In these two trials, the anti PD-1 administered was pembrolizumab. Based on these compelling results, on July 26, 2021, the Food and Drug Administration (FDA) approved pembrolizumab, in combination with chemotherapy, for early stage, high-risk TNBC (Shah et al. 2022).

In sum, considering the frequent and elevated expression on PD-L1 in TNBC stromal and epithelial cells, the immunotherapy represents a powerful weapon to treat this incurable cancer.

## 2.4 AR in Triple-Negative Breast Cancer

### 2.4.1 AR: Structure and Function

AR belongs to the sex steroid receptors family. As such, it controls the transcription of genes containing, in their promoter, specific sequences responsive to the androgen receptor, also called androgen responsive elements (ARE). The receptor is encoded by a gene localized on the chromosome Xq11-12 (Giovannelli et al. 2018), consisting of 8 exons (Saranyutanon et al. 2019), and it is made of 919 amino acids and three functional domains: an amino-terminal domain (NTD) ranging from the residue 1 to 555, a DNA binding domain (DBD) ranging from the 555 to the 623 residues, and the carboxyl-terminal ligand binding domain (LBD), ranging from the 665 to the 919 residues. Exon 1 encodes for the NTD, the DBD is encoded by the exons 2 and 3, while the LBD is encoded by exons 4–8 (Saranyutanon et al. 2019). The different transcription of the 8 exons produces the AR splice variants, fully or partially lacking the DBD or the LDB. AR-V1, V7, and V9 completely lack the LBD, while AR-V567 possesses a shorter LBD and retains the ability to bind DNA (Liu et al. 2013; Saranyutanon et al. 2019). The amino-terminal domain contains a variable number of CAG (glutamine) repeats that influence the folding as well as the structure of this domain and alter the AR-transcription activity (Werner et al. 2006).

When unbounded to its ligand, AR is localized in cytoplasm, complexed with chaperones and heat shock proteins (HSP90 and HSP70); the activation by androgens induces a conformational change which allows the AR dimerization and nuclear translocation. Once in nucleus, the AR dimers regulate the transcription of target genes. This modality of action, typical of all the steroid receptors (SR), is called "transcriptional" or "genomic" and occurs in a timeframe ranging from 30 minutes to several hours.

AR and the other SRs can also act through the so-called "non-genomic" or "rapid" actions. These actions occur in seconds or minutes, are insensitive to modulators of transcription and regulate different processes such as cell proliferation and growth, survival, cell migration and invasion, and apoptosis.

In this case, circulating androgens activate their receptor, thereby inducing the rapid activation of cytoplasmic transduction pathways controlled by extranuclear proteins such as Src, p85-PI3K, PKC, filamin A, and other effectors (Castoria et al. 2017).

In many cell types, genomic and non-genomic actions of steroid receptors collaborate to control, at different levels, a wide range of cellular processes. In other cell types, SRs prevalently act through non-genomic actions, probably for the receptor's failure to dimerize because of its very low expression level (Castoria et al. 2017).

### 2.4.2 AR in Breast Cancer: An Overview

Since AR is expressed in a high percentage (50–70%) of BC, the knowledge of its role in these tumors and the possibility to use it as target for BC therapy is an area of active investigation (Kono et al. 2017).

Besides the wild type (wt) isoform, different BC-derived cell lines exhibit other AR isoforms. In MCF and T47D cells, the presence of a membrane AR (mAR) was revealed, while TNBC derived MDA MB 231 and MDA MB453 cell lines express the AR V7 and AR45, a splice variant lacking the exon 1 and expressing a N-terminal extension of 7 amino acids that inhibits the AR function (Moore et al. 2012; Antonarakis et al. 2014). MDA MB453 cells also express a mutated form of AR, AR865H (glutamine to histidine substitution), with a reduced sensitivity to 5a-dihydrotestosterone (DHT; Moore et al. 2012). The active splice variant AR V7 is also expressed in BC circulating tumor cells (CTCs), and its expression associates with the number of bone metastases (Aceto et al. 2018).

The prognostic value of AR in breast cancer is strictly linked to the BC subtype and molecular context. Different studies demonstrated that there is no relationship between the length of the poly-Q sequence and the BC risk. By contrast, Rebbeck and colleagues (1999) observed that women have a major risk to develop BC when CAG are repeated more than 28 times. AR is expressed in normal breast tissues, and its expression seems to decrease as the ductal carcinoma in situ progresses towards invasive cancer. In this last BC group, AR is a positive prognostic factor when co-expressed with ER and PR, while there are no studies indicating the AR-prognostic value in in situ BC (Salvi et al. 2020). In ER-positive BC, AR-expression is a more favorable prognostic factor than in ER-negative BC, and, in general, patients with a discordant AR and ER expression have a worse prognosis than patients expressing both the SRs (Kono et al. 2017). This relationship can be explained by investigating the different activity of AR in ER-positive and ER-negative BCs. In ER-positive BC, AR binds the estrogen responsive elements (ERE) in nucleus, thus inducing apoptosis. In ER-negative BC, AR recognizes the ARE and induces cell proliferation. Studies have demonstrated that, more than the simple co-expression of AR and ER, it's their relative nuclear expression to determine the final phenotype. A ratio of 2, by immunohistochemical staining, indicates a shorter disease-free survival of patients than a ratio less than 2. This suggests that in ER positive cancers, AR overexpression is associated with hormone-resistance. Lamb and colleagues have reported that not only the AR/ER ratio, but also the AR/PR ratio has a reliable predictive value in luminal BC subtypes (Lamb et al. 2019).

In ERBB2-positive BC, the AR expression seems to be a favorable marker. Its positivity has been associated to ER and PR expression, low Ki-67, smaller and less aggressive tumor subtypes (Kono et al. 2017). On the other, in all ERBB2-positive BC not-expressing ER, AR positively regulates cell proliferation and cancer growth (Ni et al. 2011; Kono et al. 2017). The expression of AR in luminal A and B BC is higher than in HER2 enriched and TNBC, and this trend is similar in primary tumors and metastases (Bronte et al. 2018; Salvi et al. 2020).

In TNBC, AR seems to act in a more homogeneous way and control growth and spreading (Salvi et al. 2020). Even if its role as predictive marker is controversial, its expression makes it a very exploitable therapeutic target and provides new opportunities to cure the subgroup of AR-positive TNBC.

In sum, the role of AR in BC is extremely variable and depends on the BC subset in which it is evaluated (Salvi et al. 2020). The receptor, indeed, might act in an opposite way when expressed in ER positive or negative BC, and exploring its role becomes indispensable to improve cancer drug development.

### 2.4.3 The Role of AR in TNBC

Depending on the parameters chosen to define its positivity, AR is expressed in 10–35% of TNBCs (Kono et al. 2017). The receptor represents a target and a promising biomarker for TNBC, mostly in LAR subtype, but also in a little group of non-LAR TNBCs (Borri and Granaglia 2021). The AR expression in LAR cancer is related to the age of patients and is higher in older women (Jongen et al. 2019), but there is no association between AR and menopausal status. Again, there is no correlation between AR and body mass index (BMI), while a strong correlation between AR and diabetes exists (Collina et al. 2016).

The prognostic role of AR in TNBC is still discussed and data in literature reach opposite conclusions. Some studies evidence that AR is related to poor diagnosis (Jiang et al. 2016) and its expression level is positively associated with high-grade tumors, tumor size, and lymph node metastasis (Shen et al. 2017). Other studies conclude that the receptor expression is associated with a lower risk of recurrence (Wang et al. 2020) and a better overall survival (Thike et al. 2014; Kim et al. 2015). On the other hand, the LAR subtype has the liability to be less responsive to chemotherapy than the other TNBC, commonly considered more sensitive to chemotherapy for the high expression of proliferation-related genes. This feature makes the AR an even more desirable marker to develop new therapies and makes necessary to better know its actions in TNBCs.

AR can act both through genomic and non-genomic action in TNBC, thus regulating cell proliferation and apoptosis, migration, and invasion in these cancers (Giovannelli et al. 2019).

In LAR subtype-derived MDA MB453 cells, AR controls proliferation by transcriptional mechanisms. On DHT stimulation, AR positively regulates the Wnt7 b transcription, thus activating the Wnt/β-catenin pathway (Ni et al. 2011). The activated β-catenin translocates in nucleus where it binds AR. Here, the AR/β-catenin complex, in concert with the transcriptional factor FOXA1, recognizes the HER3 promoter and stimulates its transcription thus triggering the cell proliferation. Addition of bicalutamide, a nonsteroidal anti-AR drug, inhibits the DHT-induced cell growth (Ni et al. 2011).

In both MDA MB453 and basal claudin low BC derived MDA MB231cells, DHT treatment induces the association between AR and β-catenin that increases the transcription of the oncogene c-Myc in nucleus. The antiandrogen bicalutamide reverts this effect (Huang et al. 2017).

In TNBC-derived MDA MB231 and Hs578T cells, the DHT-triggered proliferation is controlled by the inhibition of the GPER transcription. On androgen stimulation, the AR recognizes AREs localized in the promoter of the GPER gene thus downregulating its transcription. The GPER agonist, G1, inhibits the DHT-induced cell proliferation, indicating a mutual antagonizing action of the two receptors (Shen et al. 2017).

AR also influences the behavior of TNBC through non-transcriptional actions. Treatment of TNBC cells with the synthetic and stable androgen, R1881, rapidly activates various signaling effectors, such as the non-receptor kinase Src, PI3K, or filamin A (FilA), and induces cell migration and invasion (Giovannelli et al. 2019), as well as the increase in TNBC-derived spheroids size (unpublished data). The AR-dependent activation of PI3K pathway might also regulate FOXO1 protein phosphorylation status, thus controlling proliferation of TNBC cells (Ni et al. 2011). In TNBC, there is a correlation between AR and FOXO1 expression. Both are co-expressed in 15% of TNBCs (Guiu et al. 2018), and this co-expression is associated with a shorter disease-free survival (Mangia et al. 2019). Furthermore, the gene encoding the catalytic subunit of PI3K is mutated in 23.7% of TNBC (Arafeh and Samuels 2019). This activating mutation, PIK3CA, is mostly associated to the LAR subtype and, in general, with the AR expression in TNBC (Lehmann et al. 2014). AR and FOXO1 positive TNBC-subgroups are associated with a higher rate of PIK3CA mutations. Thus, it might be important to evaluate the status of the three pathways' components to use a combined therapy with antiandrogens and PIK3CA inhibitors in these BC.

Figure 1 depicts the "genomic" and "non-genomic" actions controlled by AR.

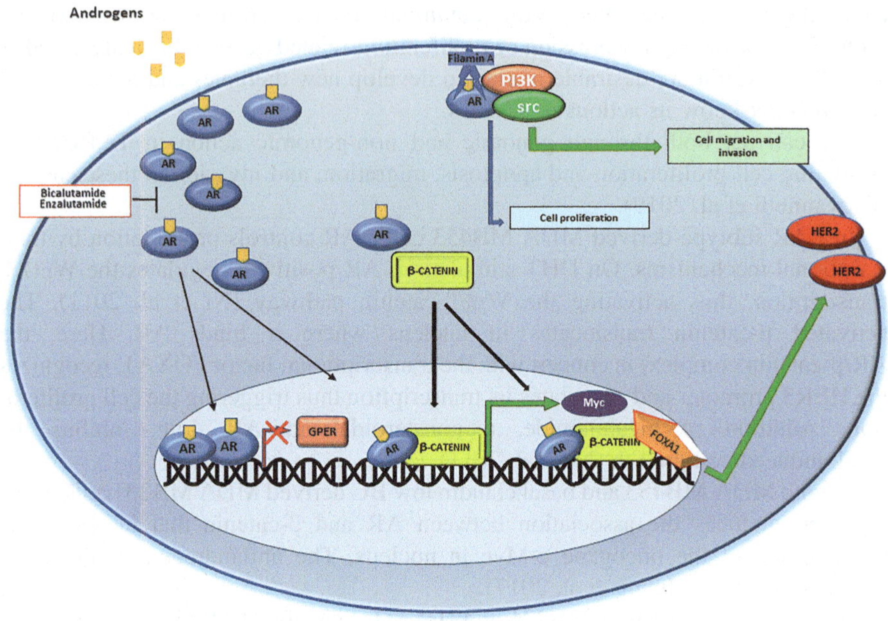


**Fig. 1** A graphical representation of "genomic" and "non-genomic" actions controlled by AR in TNBC

### 2.4.4 Androgen Targeted Therapies

***Monotherapies*** The standard therapies in TNBC treatment offer limited options aside from traditional cytotoxic chemotherapy agents. After the discovery that AR guides the growth and spreading of TNBC, treatments targeting the androgen-signaling pathway have made possible to care a greater number of patients with advanced or metastatic TNBC by using effective and less toxic endocrine agents. The therapies based on the AR antagonism use specific drugs inhibiting the AR activation or the androgen synthesis. The use of specific anti-androgens, such as the classical Bicalutamide (sold under the brand name of Casodex) or the newest Enzalutamide, a second-generation anti-androgen, is effective not only on TNBC cells or artificial models but also in clinical trials. By competing with androgens for the AR binding site, Bicalutamide inhibits the AR activation. The ability of this compound to reduce the androgen triggered action in TNBC models is widely described (Zhu et al. 2016; Huang et al. 2017; Giovannelli et al. 2019; Kong et al. 2020), and there are different phase II clinical trials supporting the therapeutical role of this drug in TNBC patient's management. The treatment with 150mg anti-androgen, orally daily administered to AR-positive TNBC patients, results in 6-month clinical benefit rate of 19% (Gucalp et al. 2013). Anyway, the daily administration of Bicalutamide was well tolerated, and the most common adverse effects were elevation of liver function tests, fatigue, hot flashes, and limb edema.

Enzalutamide, a new generation AR inhibitor, represents another valuable AR. The mechanism of action of this drug is threefold. It is a potent competitive binder of androgen, impedes the AR nuclear translocation, and, if the binding with AR occurs in nucleus, inhibits the AR recognizing of ARE on DNA, thereby preventing gene transcription (Saad 2013). This treatment has shown a clinical benefit rate of 35% at 17 weeks and 29% at 24 weeks in a phase II clinical trial enrolling AR-positive TNBC patients (Traina et al. 2015; Kono et al. 2017).

Another efficient monotherapy involves the use of CYP17 inhibitors. CYP17 is a key enzyme for androgen synthesis, and its inhibition causes the decrease of circulating androgens. Abiraterone is a selective and irreversible CYP17 inhibitor. This agent is safe and well tolerated (Reid et al. 2008, 17), and it has been used in association with prednisone in a phase II clinical trial for AR-positive TNBC patients, producing a 6-month clinical benefit rate of 20%, a value lower than its predefined (25%) end-point (Bonnefoi et al. 2016). Nevertheless, abiraterone is effective for a selected subset of AR-positive TNBC, and several ongoing phase I and II clinical trials are, indeed, investigating the effectiveness of CYP17 inhibitors, alone or combined with other drugs (Mina et al. 2017).

Seviteronel (INO-464) is another inhibitor used in AR-targeted therapy. It is more effective than abiraterone acetate because it not only suppresses the CYP17 but also possesses antagonistic effects against AR. In phase I studies, this compound has been well tolerated both in men with castration-resistant prostate cancer (CRPC) and in women with ER+ BC or TNBC. There is hope that this novel agent will be effective in patients with AR+ TNBCs (Michmerhuizen et al. 2020).

In conclusion, preclinical and clinical studies highlight the efficiency of anti-androgens in AR-positive TNBC. Obviously, the quoted results are exemplificative of a wide number of other ongoing and completed clinical trials, such as NCT01889238, NCT01842321, NCT00755885, NCT01808040, NCT01990209, NCT02580448, NCT03383679, NCT02348281, NCT02130700, and NCT02067741 (Michmerhuizen et al. 2020).

***Combined Therapies*** The use of antiandrogen therapies, in combination with other drugs already used in TNBC such as PI3K and CDK inhibitors as well as chemotherapeutic agents, seems to be a winning move to cure this BC subtype. Many studies have demonstrated that CDK4 and 6 inhibitors such as palbociclib inhibit TNBC cell proliferation and encourage to start a Phase I/II clinical trials in which combinatorial therapy with CDK4/6 inhibitors and androgen deprivation treatment will be evaluated (Mina et al. 2017). The PI3K/AKT/mTOR axis is also often altered in TNBC. Preclinical evidence shows that combinatorial therapies using AR and PI3K or AR and rapamycin, a mTOR inhibitor, exert an additive inhibitory effect in TNBC cell lines and xenograft models. As consequence, an ongoing Phase I trial is evaluating the efficiency of combinatorial treatments with abiraterone and PI3K or mTOR inhibitors in AR-positive TNBC (Mina et al. 2017).

## 2.5 Estrogen Receptor β in Triple-Negative Breast Cancer

### 2.5.1 Estrogen Receptor β: Generalities

Estrogen receptor β (ERβ) is another member of SR family. It consists of 530 amino acids and is encoded by the ESR2 gene, located on chromosome 14. ERβ is structured in six domains: the N-terminal domain, containing the activation function domain 1, the DNA binding domain (DBD), the ligand binding domain (LBD) endowed with the second activation function domain, AF-2. The AF domains are responsible for transcriptional activity regulation of ERβ but, while the AF1 works in a hormone-independent way, the AF2 requires the presence of the ligand. The synergism of both the AF domains allows the full transcriptional activity of the receptor. The DBD and LBD domains are connected by a hinge region, a flexible bridge containing nuclear localization signaling. The DBD and the LBD show a high sequence homology with the most famous ERα, while the NTD of ERβ is shorter than that of ERα and shows a poor sequence homology with the corresponding domain of the alpha isoform (Kumar et al. 2011). The distribution of ERβ in tissues and the signaling response is different between the two ER receptors, but the physiological actions of estrogens are executed by both the receptors (Kumar et al. 2011) that have similar but not identical functions. ERβ is expressed in five different variants named from ERβ1 to ERβ5. ERβ1 represents the firstly discovered full-length ERβ. The other four variants derive from an alternative splicing of the exon 8 and can be detected both in normal and in cancer tissues where they play different functions (Monaco et al. 2021). The ERβ tissue localization is wider than that of ERα and includes brain, testis, prostate, ovary, uterus, lung, gastrointestinal tract, subcutaneous adipose tissue, luminal, and myoepithelial cells of normal breast. ERβ1, ERβ2, and ERβ5 are mainly expressed in breast tissues while ERβ3 is typical of prostate tissues. Regarding the cellular localization, ERβ resides in nucleus and cytoplasm, but can be also found at peri-membrane and inside mitochondria (Yaşar et al. 2017). ERβ participates in both physiological and pathological processes and acts by genomic and non-genomic pathways, such as the other members of its family. In its unbounded inactive state, ERβ is complexed with cytoplasmic chaperones, such as the heat shock protein 90 (Hsp90), and it becomes activated, leading to different outcomes upon estrogen binding (Monaco et al. 2021).

### 2.5.2 The Role of ERβ in BC: An Overview

ERβ is expressed at high level in most normal breast epithelial cells and is present in 20–30% of breast cancers (Hawse et al. 2008). Its expression is not limited to cancer cells, but it is also present in different cells of cancer microenvironment such as endothelial cells, BC infiltrating lymphocytes, and fibroblasts. Many published studies analyze the role of ERβ (ERβ1) in BC. By influencing both "genomic" and "non-genomic" pathways, this receptor controls proliferation, migration and invasion, responsiveness to therapy, and mitochondrial functions in BC. The PI3K/AKT pathway controls the expression and activity of ERβ, which in turn controls the pathway activity in a feedback loop. In general, this transduction pathway regulates migration, proliferation, survival, apoptosis, invasion, as well as angiogenesis in BC (Zhou and Liu 2020), and is negatively regulated by phosphatase and tensin homolog (PTEN). The increase in ERβ level results in an increase of PTEN levels, leading to tamoxifen-resistance (Treeck et al. 2020) and a reduction of the HER2/HER3 transduction pathway (Zhou and Liu 2020).

ERβ inhibits cell cycle progression by altering both cyclins' levels and cyclin/CDK complexes inhibitors. ERβ knockdown is responsible for the induction of the cell cycle progression causing the cyclin A expression (Treeck et al. 2010). This inhibitory role of ERβ in BCs (Piperigkou et al. 2016) is confirmed by the observation that a combinatorial therapy with ERβ agonist and letrozole, a non-steroidal aromatase inhibitor, reduces the growth of tumors in mice. After treatment, the cells of the cancerous mass show increased levels of ERβ and cyclin/cdk inhibitors, p21 and p27, and a decreased level of cyclin $D_1$ with the consequent cell cycle arrest in G1/S phase (Nair et al. 2011). ERβ can also inhibit the cell cycle into the G2 phase by repressing the transcription of different genes related to the cell cycle progression such as c-Myc, cyclin A, $D_1$ (Paruthiyil et al. 2004), and $B_1$ (Paruthiyil et al. 2011) and increasing the expression of genes related to the cell cycle inhibition, such as p21 and p27 (Tong et al. 2002) or GADD45A and BTG2 (Paruthiyil et al. 2011).

ERβ regulates migration and invasion of BC by acting through different pathways. The receptor might suppress the metastatic phenotype by decreasing the expression of a set of genes related to cell invasion and extracellular matrix (Schüler-Toprak et al. 2016). Furthermore, ERβ increases the integrin α1 and vinculin expression level, thereby positively regulating the adhesion and negatively controlling BC cell migration (Lindberg et al. 2010). Again, ERβ inhibits TNBC cell motility through the EGFR/insulin-like growth factor I receptor (IGF-I)/JAK/STAT signaling pathway. Again, ERβ inhibits or reduces the growth of cancer by acting on mitochondria, as the expression of ERβ reduces both the mitochondrial biogenesis and the fission event (Zhou and Liu 2020).

ERβ2 is the most expressed variant in BC, while ERβ3 has never been revealed both in commercially available BC cell lines and human specimens, and its expression seems to be restricted to the testes (Moore et al. 1998). ERβ2 shows an undetectable affinity for 17-β-estradiol; thus it has no ligand-dependent transcriptional activity (Kim et al. 2015), and its role depends on the cellular context. It is often expressed in ERα-negative BC, while it is responsible for a strong response to tamoxifen therapy when expressed in ERα-positive BC (Monaco et al. 2021). ERβ4 and ERβ5 are two truncated variants, expressed in BC with a more open conformation in the ligand binding pocket (Moore et al. 1998). As consequence, they bind ligands in an unstable way thus lowering the binding affinity. Both these variants enhance the ligand bounded ERβ1 transactivation, by heterodimerizing with it (Leung et al. 2006). Their role in BC is quite different, since ERβ4 might transform normal mammary epithelial cells, while ERβ2 and ERβ5 support the tumor growth in a hypoxic environment and trigger the c-Myc transcription, two events that correlate with a bad BC prognosis (Faria et al. 2018).

### 2.5.3 The Role of ERβ in Triple-Negative Breast Cancer

TNBC is difficult to treat because of the paucity of molecular targets and its extreme heterogeneity. For this reason, the discovery of new signatures of TNBC is an exciting event. The limited number of studies in literature describes ERβ as a tumor suppressor in TNBC. Most of the target genes controlled by the receptor are ligand dependent, while only a limited quantity (about 20%) is ligand-independent (Monaco et al. 2021). Treatment with estrogen or LY500307, an ERβ agonist, induces changes in gene expression profile. A wide group of genes controlled by the ligand activated ERβ is linked to cell cycle. The receptor negatively controls the transcription of cyclin B and cyclin H but also cyclin-dependent kinase (CDK1), while upregulating the transcription of genes involved in cell cycle inhibition, such as p21 (Monaco et al. 2021). The ligand-mediated growth inhibitory effect is also due to the transcriptional regulation of genes involved in the Wnt/β-catenin pathway, such as Wnt4, the non-canonical Wnt ligand, or the β-catenin interacting protein, CDH1 (Shanle et al. 2013; Monaco et al. 2021). Low levels of mitochondrial ERβ (mitoERβ) in TNBC specimens have been associated with increased tumor recurrence after surgery. This relationship might be due to the direct interaction of ERβ with Gpr75 and the subsequent receptor translocation in mitochondria. In this location, ERβ would activate mitochondrial function and inhibit TNBC proliferation (Song et al. 2019).

ERβ also looks as an antimetastatic factor in TNBC. After binding with agonists, ERβ interferes with different pathways, including TGFβ and EGF, inhibits migration as well as invasiveness, and elicits potent anticancer effects in TNBC cells. The ligand-mediated activation of ERβ increases the cystatin levels by inducing their gene expression. This results in the inhibition of canonical TGFβ signaling, thereby blocking the metastatic phenotype in TNBC patients (Reese et al. 2018). In TNBC models, ERβ knockdown leads to an incorrect TGFβ activation that increases cell migration and invasion (Schüler-Toprak et al. 2016). The ligand-activated ERβ negatively controls the EGFR transcription as well as the insulin-like growth factor II (IGF-II) mRNA-binding protein3 (IMP3) and inhibits cell migration in TNBC cells (Samanta et al. 2012). Again, ERβ reduces the metastatic potential of TNBC by preventing the oncogenic functions mediated by the mutated p53. In particular, ERβ binds to and counteracts the transcriptional activity of mutant p53 at the promoters of genes that regulate metastasis and EMT (Bado et al. 2016).

Aside to the previously listed genomic actions, ERβ can also work in a "non-genomic" manner to control TNBC progression. In many cases, the worse prognosis and outcome of these tumors are linked to PI3K/AKT pathway upregulation, often caused by the loss of function of PTEN. ERβ might interfere with the PI3K/AKT, downregulating the pAKT level, thereby slowing or inhibiting the cancer progression (Wang et al. 2015). In this way, in aggressive TNBCs, the ERβ1 overexpression is associated with a favorable prognosis and a high overall survival. In different TNBC cells, the downregulation of PI3K/AKT pathway, due to activated ERβ, increases the tumor sensitivity to doxorubicin, a chemotherapeutic agent (Lei et al. 2020). By acting through non-genomic actions, ERβ also interferes with EGFR/MAPK signal pathway and increases the sensitivity to raloxifene and tamoxifen, two

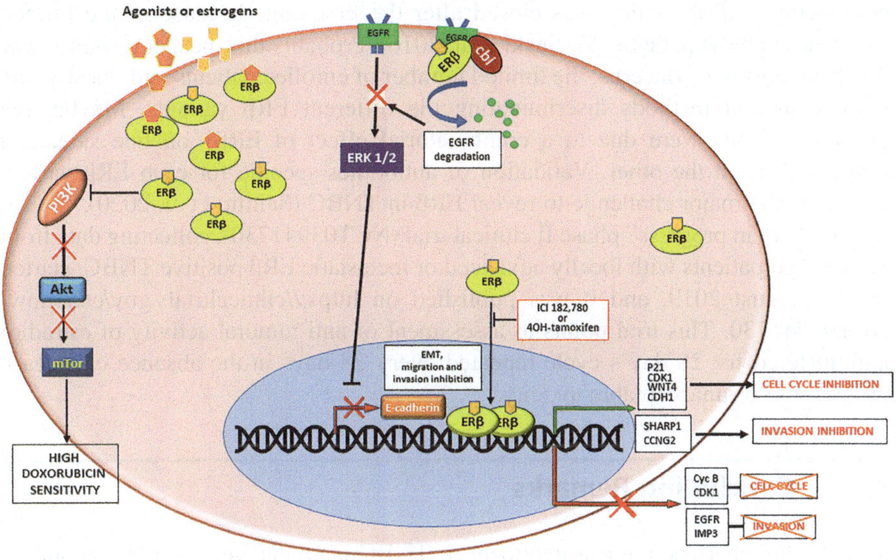


**Fig. 2** A graphical representation of "genomic" and "non-genomic" action controlled by ERβ in TNBC

commonly used drugs in TNBC cells (Greish et al. 2019). Again, ERβ interferes with EMT and cell survival by activating rapid signals in TNBC cells. The ERβ1 expression is positively related to the E-cadherin expression both in cellular models and TNBC specimens. Furthermore, SR also increase the Cbl-dependent EGFR degradation. In this way, the EGFR pathway is inhibited and the EMT is impaired (Thomas et al. 2012).

Taken together, the data so far presented indicate that by both genomic and non-genomic action, the overall outcome of ERβ activation depends on the relative expression of its variants, with ERβ1 showing antioncogenic properties, while ERβ2 and 5 promoting the tumor growth and spreading. Figure 2 briefly illustrates "genomic" and "non-genomic" actions controlled by ERβ.

### 2.5.4 Estrogen Receptor β Targeted Therapy in TNBC

Although ERβ is expressed in a high TNBC percentage (up than 50%), in vivo studies and clinical trials focused on the use of this receptor as a potential marker in TNBC therapy are limited. This is probably due to the very recent interest of the scientific world on this topic. Anyway, all the data published in literature unanimously show that the use of ERβ1 agonists and the use of for ERβ2 and 5 antagonists reduce growth, invasiveness, and spreading of TNBC. Furthermore, the ERβ agonist activation increases the sensitivity of TNBC models to chemotherapeutic agents.

More difficult is to find studies using agonists of ERβ in TNBC patients. A phase II study evaluates the effect of high estradiol concentration (10 mg oral 3 times daily) in patients with advanced TNBC. However, the obtained results were not convincing, and the study was closed after the first stage because of the limited responses in these patients (Wisinski et al. 2016). Probably this meager response was due to at least two concerns, the limited number of enrolled patients and, mostly, the scant detection methods discriminating the different ERβ variants. Maybe, the observed results were due to a combinatorial effect of ERβ1 on one side, and ERβ2 and 5, on the other. Validation of antibodies specific for each ERβ variant is, hence, the major challenge to reveal ERβ in TNBC (Sellitto et al. 2020).

Another "in progress" phase II clinical trial (NCT03941730), collecting data from 38 enrolled patients with locally advanced or metastatic ERβ positive TNBC, started on 28 August 2019, and it was published on https://clinicaltrials.gov/ct2/show/NCT03941730. This trial points to assessment of anti-tumoral activity of estradiol administered for 28 day's cycle repeated every 28 days in the absence of disease progression or unacceptable toxicity.

## 3 Concluding Remarks

To date, the pharmacological treatment of TNBC is an undeniable problem and an open challenge for oncologists. A *plethora* of molecular targets are being identified, but there is not a conclusive and efficient therapeutical approach to cure this highly malignant and heterogeneous pathology.

The analysis of the role of AR and the most recent discovery of ERβ expression in TNBC have opened new horizons. Both the receptors work in transcriptional and non-transcriptional manners, thereby influencing the behavior of TNBC. AR controls growth and invasiveness of TNBC and seems to be a promising target for developing new cures. The relevant role of AR in TNBC is also underlined by the worst prognosis of quadruple negative BC (QNBC), a particular subgroup of TNBC lacking AR (Huang et al. 2020). Besides to AR, ERβ seems to be an equally promising target in TNBC therapy. Three variants of this receptor are prevalently expressed in TNBC, and, while the most expressed ERβ1 variant has a tumor suppressor function, the other two variants, ERβ2 and 5, show pro-oncogenic activities. The analyzed data not only underline the need to deepen the knowledge of ERβ function in TNBC, but also highlight the need of more appropriate and specific detection methods for this receptor, to achieve a clearer patient's stratification and select the best therapeutical approach.

The two steroid receptors belong to the same family, and it is conceivable that they work in cooperation to modulate TNBC progression. After androgen stimulation, AR positively regulates the ERβ expression, by binding specific ARE localized in its promoter. The increased ERβ expression enhances the anti-metastatic power of this receptor (Song et al. 2017) and rises the TNBC responsiveness to both chemotherapeutic agents and anti-androgens (Anestis et al. 2019). In TNBC cells, the transient expression of ERβ positively regulates PTEN thereby reducing the pAKT and the AR activation. The indirect inhibition of the AR signaling by ERβ causes the reduction of proliferation and metastatic potential of TNBC cells. Besides to this indirect control of the AR actions, the high ERβ level directly increases the antiandrogen responsiveness of TNBC cells. In AR-positive TNBC cells overexpressing ERβ, the antiandrogen treatment inhibits cell migration and fosters apoptosis. Just like an antiandrogenic drug, the ERβ impairs AR nuclear translocation probably by the direct interaction with it and the formation of ERβ/AR heterodimers. The androgen stimulation disrupts this association to the advantage of the AR/AR homodimers formation, while the enzalutamide treatment raises the ERβ/AR association and reinforces the role of ERβ as tumor suppressor in TNBC.

The tight relationship between AR and ERβ in TNBC deserves further investigations because it might represent the key to discover new personalized therapies for TNBC.

**Funding Statement** This work is supported by the Italian Ministry of University and Scientific Research (P.R.I.N. 2017EKMFTN_002 to G.C.) and VALERE (Vanvitelli per la Ricerca Program to P. G. and GoMAGIC to A.M.).

**Conflict of Interest Statement** The authors declare that they have no conflict of interest in relation to the work described.

## References

Aceto N, Bardia A, Wittner BS, Donaldson MC, O'Keefe R, Engstrom A et al (2018) AR expression in breast cancer CTCs associates with bone metastases. Mol Cancer Res 16:720–727. https://doi.org/10.1158/1541-7786.MCR-17-0480

Anestis A, Sarantis P, Theocharis S, Zoi I, Tryfonopoulos D, Korogiannos A et al (2019) Estrogen receptor beta increases sensitivity to enzalutamide in androgen receptor-positive triple-negative breast cancer. J Cancer Res Clin Oncol 145:1221–1233. https://doi.org/10.1007/s00432-019-02872-9

Antonarakis ES, Lu C, Wang H, Luber B, Nakazawa M, Roeser JC et al (2014) AR-V7 and resistance to enzalutamide and abiraterone in prostate cancer. N Engl J Med 371:1028–1038. https://doi.org/10.1056/NEJMoa1315815

Arafeh R, Samuels Y (2019) PIK3CA in cancer: the past 30 years. Semin Cancer Biol 59:36–49. https://doi.org/10.1016/j.semcancer.2019.02.002

Bado I, Nikolos F, Rajapaksa G, Gustafsson J-Å, Thomas C (2016) ERβ decreases the invasiveness of triple-negative breast cancer cells by regulating mutant p53 oncogenic function. Oncotarget 7:13599–13611. https://doi.org/10.18632/oncotarget.7300

Balko JM, Schwarz LJ, Bhola NE, Kurupi R, Owens P, Miller TW et al (2013) Activation of MAPK pathways due to DUSP4 loss promotes cancer stem cell-like phenotypes in basal-like breast cancer. Cancer Res 73:6346–6358. https://doi.org/10.1158/0008-5472.CAN-13-1385

Barretina J, Caponigro G, Stransky N, Venkatesan K, Margolin AA, Kim S et al (2012) The cancer cell line encyclopedia enables predictive modelling of anticancer drug sensitivity. Nature 483: 603–607. https://doi.org/10.1038/nature11003

Bhattarai S, Saini G, Gogineni K, Aneja R (2020) Quadruple-negative breast cancer: novel implications for a new disease. Breast Cancer Res 22:127. https://doi.org/10.1186/s13058-020-01369-5

Bianchini G, Balko JM, Mayer IA, Sanders ME, Gianni L (2016) Triple-negative breast cancer: challenges and opportunities of a heterogeneous disease. Nat Rev Clin Oncol 13:674–690. https://doi.org/10.1038/nrclinonc.2016.66

Bonnefoi H, Grellety T, Tredan O, Saghatchian M, Dalenc F, Mailliez A et al (2016) A phase II trial of abiraterone acetate plus prednisone in patients with triple-negative androgen receptor positive

locally advanced or metastatic breast cancer (UCBG 12-1). Ann Oncol 27:812–818. https://doi.org/10.1093/annonc/mdw067
Borri F, Granaglia A (2021) Pathology of triple negative breast cancer. Semin Cancer Biol 72:136–145. https://doi.org/10.1016/j.semcancer.2020.06.005
Botti G, Collina F, Scognamiglio G, Rao F, Peluso V, De Cecio R et al (2017) Programmed Death Ligand 1 (PD-L1) tumor expression is associated with a better prognosis and diabetic disease in triple negative breast cancer patients. Int J Mol Sci 18:E459. https://doi.org/10.3390/ijms18020459
Bronte G, Bravaccini S, Ravaioli S, Puccetti M, Scarpi E, Andreis D et al (2018) Androgen receptor expression in breast cancer: what differences between primary tumor and metastases? Transl Oncol 11:950–956. https://doi.org/10.1016/j.tranon.2018.05.006
Castoria G, Auricchio F, Migliaccio A (2017) Extranuclear partners of androgen receptor: at the crossroads of proliferation, migration, and neuritogenesis. FASEB J 31:1289–1300. https://doi.org/10.1096/fj.201601047R
Collina F, Cerrone M, Peluso V, Laurentiis MD, Caputo R, Cecio RD et al (2016) Downregulation of androgen receptor is strongly associated with diabetes in triple negative breast cancer patients. Am J Transl Res 8:3530–3539
Daniel J, Coulter J, Woo J-H, Wilsbach K, Gabrielson E (2011) High levels of the Mps1 checkpoint protein are protective of aneuploidy in breast cancer cells. Proc Natl Acad Sci U S A 108:5384–5389. https://doi.org/10.1073/pnas.1007645108
Dominguez-Brauer C, Thu KL, Mason JM, Blaser H, Bray MR, Mak TW (2015) Targeting mitosis in cancer: emerging strategies. Mol Cell 60:524–536. https://doi.org/10.1016/j.molcel.2015.11.006
Duncan JS, Whittle MC, Nakamura K, Abell AN, Midland AA, Zawistowski JS, Johnson NL, Granger DA, Jordan NV, Darr DB, Usary J, Kuan PF, Smalley DM, Major B, He X, Hoadley KA, Zhou B, Sharpless NE, Perou CM, Kim WY, Gomez SM, Chen X, Jin J, Frye SV, Earp HS, Graves LM, Johnson GL (2012). Dynamic reprogramming of the kinome in response to targeted MEK inhibition in triple-negative breast cancer. Cell 149(2):307–321. https://doi.org/10.1016/j.cell.2012.02.053. PMID: 22500798; PMCID: PMC3328787
Edechi CA, Ikeogu N, Uzonna JE, Myal Y (2019) Regulation of immunity in breast cancer. Cancer 11:E1080. https://doi.org/10.3390/cancers11081080
Elsawaf Z, Sinn H-P, Rom J, Bermejo JL, Schneeweiss A, Aulmann S (2013) Biological subtypes of triple-negative breast cancer are associated with distinct morphological changes and clinical behaviour. Breast Edinb Scotl 22:986–992. https://doi.org/10.1016/j.breast.2013.05.012
Ensenyat-Mendez M, Llinàs-Arias P, Orozco JIJ, Íñiguez-Muñoz S, Salomon MP, Sesé B et al (2021) Current triple-negative breast cancer subtypes: dissecting the most aggressive form of breast cancer. Front Oncol 11:681476. https://doi.org/10.3389/fonc.2021.681476
Faria M, Karami S, Granados-Principal S, Dey P, Verma A, Choi DS et al (2018) The ERβ4 variant induces transformation of the normal breast mammary epithelial cell line MCF-10A; the ERβ variants ERβ2 and ERβ5 increase aggressiveness of TNBC by regulation of hypoxic signaling. Oncotarget 9:12201–12211. https://doi.org/10.18632/oncotarget.24134
Foulkes WD, Smith IE, Reis-Filho JS (2010) Triple-negative breast cancer. N Engl J Med 363:1938–1948. https://doi.org/10.1056/NEJMra1001389
Galiè M (2019) RAS as supporting actor in breast cancer. Front Oncol 9:1199. https://doi.org/10.3389/fonc.2019.01199
Geyer FC, Pareja F, Weigelt B, Rakha E, Ellis IO, Schnitt SJ et al (2017) The spectrum of triple-negative breast disease: high- and low-grade lesions. Am J Pathol 187:2139–2151. https://doi.org/10.1016/j.ajpath.2017.03.016
Giltnane JM, Balko JM (2014) Rationale for targeting the Ras/MAPK pathway in triple-negative breast cancer. Discov Med 17:275–283
Giovannelli P, Di Donato M, Galasso G, Di Zazzo E, Bilancio A, Migliaccio A (2018) The androgen receptor in breast cancer. Front Endocrinol 9:492. https://doi.org/10.3389/fendo.2018.00492

Giovannelli P, Di Donato M, Auricchio F, Castoria G, Migliaccio A (2019) Androgens induce invasiveness of triple negative breast cancer cells through AR/Src/PI3-K complex assembly. Sci Rep 9:4490. https://doi.org/10.1038/s41598-019-41016-4

Greish K, Nehoff H, Bahman F, Pritchard T, Taurin S (2019) Raloxifene nano-micelles effect on triple-negative breast cancer is mediated through estrogen receptor-β and epidermal growth factor receptor. J Drug Target 27(8). Available at: https://www.tandfonline.com/doi/abs/10.1080/1061186X.2019.1566341?journalCode=idrt20. Accessed 23 Feb 2022

Gucalp A, Tolaney S, Isakoff SJ, Ingle JN, Liu MC, Carey LA et al (2013) Phase II trial of bicalutamide in patients with androgen receptor-positive, estrogen receptor-negative metastatic breast cancer. Clin Cancer Res 19:5505–5512. https://doi.org/10.1158/1078-0432.CCR-12-3327

Guiu S, Mollevi C, Charon-Barra C, Boissière F, Crapez E, Chartron E et al (2018) Prognostic value of androgen receptor and FOXA1 co-expression in non-metastatic triple negative breast cancer and correlation with other biomarkers. Br J Cancer 119:76–79. https://doi.org/10.1038/s41416-018-0142-6

Hawse JR, Subramaniam M, Monroe DG, Hemmingsen AH, Ingle JN, Khosla S et al (2008) Estrogen receptor beta isoform-specific induction of transforming growth factor beta-inducible early gene-1 in human osteoblast cells: an essential role for the activation function 1 domain. Mol Endocrinol Baltim Md 22:1579–1595. https://doi.org/10.1210/me.2007-0253

Huang R, Han J, Liang X, Sun S, Jiang Y, Xia B et al (2017) Androgen receptor expression and bicalutamide antagonize androgen receptor inhibit β-catenin transcription complex in estrogen receptor-negative breast cancer. Cell Physiol Biochem 43:2212–2225. https://doi.org/10.1159/000484300

Huang M, Wu J, Ling R, Li N (2020) Quadruple negative breast cancer. Breast Cancer 27:527–533. https://doi.org/10.1007/s12282-020-01047-6

Infante JR, Papadopoulos KP, Bendell JC, Patnaik A, Burris HA, Rasco D et al (2013) A phase 1b study of trametinib, an oral Mitogen-activated protein kinase kinase (MEK) inhibitor, in combination with gemcitabine in advanced solid tumours. Eur J Cancer Oxf Engl 1990(49):2077–2085. https://doi.org/10.1016/j.ejca.2013.03.020

Jiang H-S, Kuang X-Y, Sun W-L, Xu Y, Zheng Y-Z, Liu Y-R et al (2016) Androgen receptor expression predicts different clinical outcomes for breast cancer patients stratified by hormone receptor status. Oncotarget 7:41285–41293. https://doi.org/10.18632/oncotarget.9778

Jongen L, Floris G, Wildiers H, Claessens F, Richard F, Laenen A et al (2019) Tumor characteristics and outcome by androgen receptor expression in triple-negative breast cancer patients treated with neo-adjuvant chemotherapy. Breast Cancer Res Treat 176:699–708. https://doi.org/10.1007/s10549-019-05252-6

Kim Y, Jae E, Yoon M (2015) Influence of androgen receptor expression on the survival outcomes in breast cancer: a meta-analysis. J Breast Cancer 18:134–142. https://doi.org/10.4048/jbc.2015.18.2.134

Kong Y, Qu F, Yuan X, Yan X, Yu W (2020) Effect of Bicalutamide on the proliferation and invasion of human triple negative breast cancer MDA-MB-231 cells. Medicine (Baltimore) 99:e19822. https://doi.org/10.1097/MD.0000000000019822

Kono M, Fujii T, Lim B, Karuturi MS, Tripathy D, Ueno NT (2017) Androgen receptor function and androgen receptor-targeted therapies in breast cancer: a review. JAMA Oncol 3:1266–1273. https://doi.org/10.1001/jamaoncol.2016.4975

Kumar R, Zakharov MN, Khan SH, Miki R, Jang H, Toraldo G et al (2011) The dynamic structure of the estrogen receptor. J Amino Acids 2011:812540. https://doi.org/10.4061/2011/812540

Lamb CA, Vanzulli SI, Lanari C (2019) Hormone receptors in breast cancer: more than estrogen receptors. Medicina (Mex) 79:540–545

Lehmann BD, Bauer JA, Schafer JM, Pendleton CS, Tang L, Johnson KC et al (2014) PIK3CA mutations in androgen receptor-positive triple negative breast cancer confer sensitivity to the combination of PI3K and androgen receptor inhibitors. Breast Cancer Res 16:406. https://doi.org/10.1186/s13058-014-0406-x

Lehmann BD, Jovanović B, Chen X, Estrada MV, Johnson KN, Shyr Y et al (2016) Refinement of triple-negative breast cancer molecular subtypes: implications for neoadjuvant chemotherapy selection. PLoS One 11:e0157368. https://doi.org/10.1371/journal.pone.0157368

Lei S, Fan P, Wang M, Zhang C, Jiang Y, Huang S et al (2020) Elevated estrogen receptor β expression in triple negative breast cancer cells is associated with sensitivity to doxorubicin by inhibiting the PI3K/AKT/mTOR signaling pathway. Exp Ther Med 20:1630–1636. https://doi.org/10.3892/etm.2020.8809

Leung Y-K, Mak P, Hassan S, Ho S-M (2006) Estrogen receptor (ER)-beta isoforms: a key to understanding ER-beta signaling. Proc Natl Acad Sci U S A 103:13162–13167. https://doi.org/10.1073/pnas.0605676103

Li Y, Zhan Z, Yin X, Fu S, Deng X (2021) Targeted therapeutic strategies for triple-negative breast cancer. Front Oncol 11:731535. https://doi.org/10.3389/fonc.2021.731535

Lindberg K, Ström A, Lock JG, Gustafsson J-A, Haldosén L-A, Helguero LA (2010) Expression of estrogen receptor beta increases integrin alpha1 and integrin beta1 levels and enhances adhesion of breast cancer cells. J Cell Physiol 222:156–167. https://doi.org/10.1002/jcp.21932

Liu G, Sprenger C, Sun S, Epilepsia KS, Haugk K, Zhang X et al (2013) AR variant ARv567es induces carcinogenesis in a novel transgenic mouse model of prostate cancer. Neoplasia 15: 1009–1017. https://doi.org/10.1593/neo.13784

Mangia A, Saponaro C, Vagheggini A, Opinto G, Centonze M, Vicenti C et al (2019) Should tumor infiltrating lymphocytes, androgen receptor, and FOXA1 expression predict the clinical outcome in triple negative breast cancer patients? Cancer 11:E1393. https://doi.org/10.3390/cancers11091393

Michmerhuizen AR, Chandler B, Olsen E, Wilder-Romans K, Moubadder L, Liu M et al (2020) Seviteronel, a novel CYP17 lyase inhibitor and androgen receptor antagonist, radiosensitizes AR-positive triple negative breast cancer cells. Front Endocrinol 11:35. https://doi.org/10.3389/fendo.2020.00035

Mina A, Yoder R, Sharma P (2017) Targeting the androgen receptor in triple-negative breast cancer: current perspectives. OncoTargets Ther 10:4675–4685. https://doi.org/10.2147/OTT.S126051

Molinero L, Li Y, Chang C-W, Maund S, Berg M, Harrison J et al (2019) Tumor immune microenvironment and genomic evolution in a patient with metastatic triple negative breast cancer and a complete response to atezolizumab. J Immunother Cancer 7:274. https://doi.org/10.1186/s40425-019-0740-8

Monaco A, Licitra F, Di Gisi M, Galasso G, Di Donato M, Giovannelli P et al (2021) ERβ in triple-negative breast cancer: emerging concepts and therapeutic possibilities. Endocrine 2:356–365. https://doi.org/10.3390/endocrines2030033

Moore JT, McKee DD, Slentz-Kesler K, Moore LB, Jones SA, Horne EL et al (1998) Cloning and characterization of human estrogen receptor β isoforms. Biochem Biophys Res Commun 247: 75–78. https://doi.org/10.1006/bbrc.1998.8738

Moore NL, Buchanan G, Harris JM, Selth LA, Bianco-Miotto T, Hanson AR et al (2012) An androgen receptor mutation in the MDA-MB-453 cell line model of molecular apocrine breast cancer compromises receptor activity. Endocr Relat Cancer 19:599–613. https://doi.org/10.1530/ERC-12-0065

Nair HB, Kirma NB, Ganapathy M, Vadlamudi RK, Tekmal RR (2011) Estrogen receptor-β activation in combination with letrozole blocks the growth of breast cancer tumors resistant to letrozole therapy. Steroids 76:792–796. https://doi.org/10.1016/j.steroids.2011.02.038

Ni M, Chen Y, Lim E, Wimberly H, Bailey ST, Imai Y et al (2011) Targeting androgen receptor in estrogen receptor-negative breast cancer. Cancer Cell 20:119–131. https://doi.org/10.1016/j.ccr.2011.05.026

Paruthiyil S, Parmar H, Kerekatte V, Cunha GR, Firestone GL, Leitman DC (2004) Estrogen receptor beta inhibits human breast cancer cell proliferation and tumor formation by causing a G2 cell cycle arrest. Cancer Res 64:423–428. https://doi.org/10.1158/0008-5472.can-03-2446

Paruthiyil S, Cvoro A, Tagliaferri M, Cohen I, Shtivelman E, Leitman DC (2011) Estrogen receptor β causes a G2 cell cycle arrest by inhibiting CDK1 activity through the regulation of cyclin B1, GADD45A, and BTG2. Breast Cancer Res Treat 129:777–784. https://doi.org/10.1007/s10549-010-1273-5

Piperigkou Z, Bouris P, Onisto M, Franchi M, Kletsas D, Theocharis AD et al (2016) Estrogen receptor beta modulates breast cancer cells functional properties, signaling and expression of matrix molecules. Matrix Biol 56:4–23. https://doi.org/10.1016/j.matbio.2016.05.003

Rebbeck TR, Kantoff PW, Krithivas K, Neuhausen S, Blackwood MA, Godwin AK, Daly MB, Narod SA, Garber JE, Lynch HT, Weber BL, Brown M (1999) Modification of BRCA1-associated breast cancer risk by the polymorphic androgen-receptor CAG repeat. Am J Hum Genet 64(5):1371–1377. https://doi.org/10.1086/302366. PMID: 10205268; PMCID: PMC1377873

Reese JM, Bruinsma ES, Nelson AW, Chernukhin I, Carroll JS, Li Y et al (2018) ERβ-mediated induction of cystatins results in suppression of TGFβ signaling and inhibition of triple-negative breast cancer metastasis. Proc Natl Acad Sci 115:E9580–E9589. https://doi.org/10.1073/pnas.1807751115

Reid AH, Attard G, Barrie E, de Bono JS (2008) CYP17 inhibition as a hormonal strategy for prostate cancer. Nat Clin Pract Urol 5:610–620. https://doi.org/10.1038/ncpuro1237

Rivenbark AG, O'Connor SM, Coleman WB (2013) Molecular and cellular heterogeneity in breast cancer: challenges for personalized medicine. Am J Pathol 183:1113–1124. https://doi.org/10.1016/j.ajpath.2013.08.002

Saad F (2013) Evidence for the efficacy of enzalutamide in postchemotherapy metastatic castrate-resistant prostate cancer. Ther Adv Urol 5:201–210. https://doi.org/10.1177/1756287213490054

Salvi S, Bonafè M, Bravaccini S (2020) Androgen receptor in breast cancer: a wolf in sheep's clothing? A lesson from prostate cancer. Semin Cancer Biol 60:132–137. https://doi.org/10.1016/j.semcancer.2019.04.002

Samanta S, Sharma VM, Khan A, Mercurio AM (2012) Regulation of IMP3 by EGFR signaling and repression by ERβ: implications for triple negative breast cancer. Oncogene 31:4689–4697. https://doi.org/10.1038/onc.2011.620

Saranyutanon S, Srivastava SK, Pai S, Singh S, Singh AP (2019) Therapies targeted to androgen receptor signaling axis in prostate cancer: progress, challenges, and hope. Cancer 12:E51. https://doi.org/10.3390/cancers12010051

Schmid P, Cortes J, Pusztai L, McArthur H, Kümmel S, Bergh J et al (2020a) Pembrolizumab for early triple-negative breast cancer. N Engl J Med 382:810–821. https://doi.org/10.1056/NEJMoa1910549

Schmid P, Rugo HS, Adams S, Schneeweiss A, Barrios CH, Iwata H et al (2020b) Atezolizumab plus nab-paclitaxel as first-line treatment for unresectable, locally advanced or metastatic triple-negative breast cancer (IMpassion130): updated efficacy results from a randomised, double-blind, placebo-controlled, phase 3 trial. Lancet Oncol 21:44–59. https://doi.org/10.1016/S1470-2045(19)30689-8

Schüler-Toprak S, Häring J, Inwald EC, Moehle C, Ortmann O, Treeck O (2016) Agonists and knockdown of estrogen receptor β differentially affect invasion of triple-negative breast cancer cells in vitro. BMC Cancer 16:951. https://doi.org/10.1186/s12885-016-2973-y

Sellitto A, D'Agostino Y, Alexandrova E, Lamberti J, Pecoraro G, Memoli D et al (2020) Insights into the role of estrogen receptor β in triple-negative breast cancer. Cancer 12:1477. https://doi.org/10.3390/cancers12061477

Shah SP, Roth A, Goya R, Oloumi A, Ha G, Zhao Y et al (2012) The clonal and mutational evolution spectrum of primary triple-negative breast cancers. Nature 486:395–399. https://doi.org/10.1038/nature10933

Shah M, Osgood CL, Amatya AK, Fiero MH, Pierce WF, Nair A et al (2022) FDA approval summary: pembrolizumab for neoadjuvant and adjuvant treatment of patients with high-risk early-stage triple-negative breast cancer. Clin Cancer Res:OF1–OF5. https://doi.org/10.1158/1078-0432.CCR-22-1110

Shanle EK, Zhao Z, Hawse J, Wisinski K, Keles S, Yuan M et al (2013) Research resource: global identification of estrogen receptor β target genes in triple negative breast cancer cells. Mol Endocrinol Baltim Md 27:1762–1775. https://doi.org/10.1210/me.2013-1164

Shen Y, Yang F, Zhang W, Song W, Liu Y, Guan X (2017) The androgen receptor promotes cellular proliferation by suppression of G-protein coupled estrogen receptor signaling in triple-negative breast cancer. Cell Physiol Biochem 43:2047–2061. https://doi.org/10.1159/000484187

Song W, Tang L, Xu Y, Sun Q, Yang F, Guan X (2017) ERβ1 inhibits metastasis of androgen receptor-positive triple-negative breast cancer by suppressing ZEB1. J Exp Clin Cancer Res 36:75. https://doi.org/10.1186/s13046-017-0545-x

Song I-S, Jeong YJ, Jeong SH, Kim JE, Han J, Kim T-H et al (2019) Modulation of mitochondrial ERβ expression inhibits triple-negative breast cancer tumor progression by activating mitochondrial function. Cell Physiol Biochem 52:468–485. Available at: https://www.cellphysiolbiochem.com/Articles/000034/. Accessed 26 Feb 2022

Sulaiman A, McGarry S, Lam KM, El-Sahli S, Chambers J, Kaczmarek S et al (2018) Co-inhibition of mTORC1, HDAC and ESR1α retards the growth of triple-negative breast cancer and suppresses cancer stem cells. Cell Death Dis 9:815. https://doi.org/10.1038/s41419-018-0811-7

Targeted Therapeutic Strategies for Triple-Negative Breast Cancer – PubMed. Available at: https://pubmed.ncbi.nlm.nih.gov/34778045/. Accessed 28 Feb 2022

Terranova-Barberio M, Thomas S, Ali N, Pawlowska N, Park J, Krings G et al (2017) HDAC inhibition potentiates immunotherapy in triple negative breast cancer. Oncotarget 8:114156–114172. https://doi.org/10.18632/oncotarget.23169

Thike AA, Yong-Zheng Chong L, Cheok PY, Li HH, Wai-Cheong Yip G, Huat Bay B et al (2014) Loss of androgen receptor expression predicts early recurrence in triple-negative and basal-like breast cancer. Mod Pathol 27:352–360. https://doi.org/10.1038/modpathol.2013.145

Thomas C, Rajapaksa G, Nikolos F, Hao R, Katchy A, McCollum CW et al (2012) ERβ1 represses basal-like breast cancer epithelial to mesenchymal transition by destabilizing EGFR. Breast Cancer Res 14:R148. https://doi.org/10.1186/bcr3358

Thu K, Soria-Bretones I, Mak T, Cescon D (2018) Targeting the cell cycle in breast cancer: towards the next phase. Cell Cycle 17:1871–1885. https://doi.org/10.1080/15384101.2018.1502567

Tong D, Schuster E, Seifert M, Czerwenka K, Leodolte S, Zeillinger R (2002) Expression of estrogen receptor beta isoforms in human breast cancer tissues and cell lines. Breast Cancer Res Treat 71:249–255. https://doi.org/10.1023/a:1014465916473

Traina TA, Miller K, Yardley DA, O'Shaughnessy J, Cortes J, Awada A et al (2015) Results from a phase 2 study of enzalutamide (ENZA), an androgen receptor (AR) inhibitor, in advanced AR+ triple-negative breast cancer (TNBC). J Clin Oncol 33:1003–1003. https://doi.org/10.1200/jco.2015.33.15_suppl.1003

Treeck O, Lattrich C, Springwald A, Ortmann O (2010) Estrogen receptor beta exerts growth-inhibitory effects on human mammary epithelial cells. Breast Cancer Res Treat 120:557–565. https://doi.org/10.1007/s10549-009-0413-2

Treeck O, Schüler-Toprak S, Skrzypczak M, Weber F, Ortmann O (2020) Knockdown of PTEN decreases expression of estrogen receptor β and tamoxifen sensitivity of human breast cancer cells. Steroids 153:108521. https://doi.org/10.1016/j.steroids.2019.108521

Turner NC, Lord CJ, Iorns E, Brough R, Swift S, Elliott R et al (2008) A synthetic lethal siRNA screen identifying genes mediating sensitivity to a PARP inhibitor. EMBO J 27:1368–1377. https://doi.org/10.1038/emboj.2008.61

Viale G, Rotmensz N, Maisonneuve P, Bottiglieri L, Montagna E, Luini A et al (2009) Invasive ductal carcinoma of the breast with the "triple-negative" phenotype: prognostic implications of EGFR immunoreactivity. Breast Cancer Res Treat 116:317–328. https://doi.org/10.1007/s10549-008-0206-z

Wang J, Zhang C, Chen K, Tang H, Tang J, Song C et al (2015) ERβ1 inversely correlates with PTEN/PI3K/AKT pathway and predicts a favorable prognosis in triple-negative breast cancer. Breast Cancer Res Treat 152:255–269. https://doi.org/10.1007/s10549-015-3467-3

Wang J, Zhang Y, Wei C, Gao X, Yuan P, Gan J et al (2020) Prognostic value of androgen receptor splice variant 7 in the treatment of metastatic castration-resistant prostate cancer: a systematic review and meta-analysis. Front Oncol 10:562504. https://doi.org/10.3389/fonc.2020.562504
Werner R, Holterhus P-M, Binder G, Schwarz H-P, Morlot M, Struve D et al (2006) The A645D mutation in the hinge region of the human androgen receptor (AR) gene modulates AR activity, depending on the context of the polymorphic glutamine and glycine repeats. J Clin Endocrinol Metab 91:3515–3520. https://doi.org/10.1210/jc.2006-0372
Wisinski KB, Xu W, Tevaarwerk AJ, Saha S, Kim K, Traynor A et al (2016) Targeting estrogen receptor beta in a phase 2 study of high-dose estradiol in metastatic triple-negative breast cancer: a Wisconsin Oncology Network Study. Clin Breast Cancer 16:256–261. https://doi.org/10.1016/j.clbc.2016.03.005
Wong KK (2021) DNMT1: a key drug target in triple-negative breast cancer. Semin Cancer Biol 72:198–213. https://doi.org/10.1016/j.semcancer.2020.05.010
Yaşar P, Ayaz G, User SD, Güpür G, Muyan M (2017) Molecular mechanism of estrogen-estrogen receptor signaling. Reprod Med Biol 16:4–20. https://doi.org/10.1002/rmb2.12006
Zhou Y, Liu X (2020) The role of estrogen receptor beta in breast cancer. Biomark Res 8:39. https://doi.org/10.1186/s40364-020-00223-2
Zhu A, Li Y, Song W, Xu Y, Yang F, Zhang W et al (2016) Antiproliferative effect of androgen receptor inhibition in mesenchymal stem-like triple-negative breast cancer. Cell Physiol Biochem 38:1003–1014. https://doi.org/10.1159/000443052

# Growth Factor Receptor Implications in Breast Cancer: Prospects for Their Molecular Transactivation in the Future and Obstacles for Target Therapy

Esteban Acosta-Ramos, Mariana Segovia-Mendoza, and Jesus A. Olivares-Reyes

**Abstract**

Cancer is one of the diseases with the highest incidence worldwide. In particular, breast cancer (BC) has been characterized by increasing numbers around the world. Different studies have addressed the involvement of different types of receptors with tyrosine kinase activity (RTKs) in the pathophysiology of this neoplasia. However, the joint participation of RTKs with seven transmembrane G protein-coupled receptors (GPCRs) has attracted attention in recent years since it can generate receptor activation independently of a specific ligand binding, a process known as receptor transactivation. Thus, this chapter describes the biological function, signaling pathways, and molecular interactions of RTKs and GPCRs to understand their pro-tumoral processes in BC. In addition, this chapter aims to expand the knowledge about the molecular mechanisms of this receptor interaction to propose therapeutic strategies for different BC subtypes and to evoke the search for personalized therapies for this disease based on RTK and GPCR transactivation.

E. Acosta-Ramos · J. A. Olivares-Reyes (✉)
Departamento de Bioquimica, Centro de Investigacion y de Estudios Avanzados del IPN, Mexico City, Mexico
e-mail: esteban.acosta@cinvestav.mx; jolivare@cinvestav.mx

M. Segovia-Mendoza (✉)
Departamento de Farmacologia, Facultad de Medicina, Universidad Nacional Autonoma de Mexico, Mexico City, Mexico
e-mail: mariana.segovia@facmed.unam.mx

 
Interdisciplinary Cancer Research, https://doi.org/10.1007/16833_2024_281
Published online: 21 June 2024

**Keywords**

Breast cancer · Growth factor receptor · Target therapy · Transactivation

# 1 Introduction

Breast cancer (BC) is a heterogeneous disease with diverse molecular subtypes, which are characterized by distinct biological features, clinical outcomes, and treatment responses. The molecular classification of BC enables the employment of more personalized and targeted therapeutic approaches. This chapter explores the key biomarkers associated with each BC subtype, highlighting their differences and emphasizing the critical role of the tyrosine kinase receptor (RTK) activity and their transactivation with G-protein-coupled receptors (GPCRs) in BC development and progression (Torre et al. 2016).

## 1.1 Incidence

BC is currently the second most common malignant neoplasia around the world. Since 2022, the World Health Organization (WHO) has reported the highest incidence rate of BC, with an estimated 2.3 million new cases, and one of the main causes of death in women with 666,000 deaths (Ferlay et al. 2021, 2024; Sung et al. 2021). Since 1990 to date, the prevalence of BC diagnosis in women has been growing by an average of two to threefold per year. In addition, the prognostic of this disease for 2040 estimates an incidence of over 40%, representing three million new cases yearly. Developed countries, such as Australia/New Zealand, Western Europe, Northern America, and Northern Europe, have the highest rates of incidence of ~>80 per 100,000 females, whereas non-developmental countries like Central America, Eastern and Middle Africa, and South Central Asia have the lowest rates of incidence with ~<40 per 100,000 females, where BC is considered as a global health concern.

The overgrowing incidence of BC is due to its multiple risk factors, which are classified as modifiable and non-modifiable (Table 1). Modifiable risk factors depend on the woman's lifestyle, while non-modifiable factors are related to the person's genetics.

## 1.2 Risk Factors

The principal risk factor for BC development is gender, which is related to the highest prevalence in the diagnostic female population (95% of all cases reported). The second most common risk factor is age, associated with hormonal changes in aging. In addition, in the highest-income regions like the United Kingdom, it is reported that the average age of BC diagnosis is around 80–84 years old, while the

**Table 1** Risk factors associated with BC development

| Modifiable | Non-modifiable |
|---|---|
| Hormonal replacement therapy | Female sex |
| Physical activity | Older age |
| Overweight/obesity | Family history (breast/ovarian cancer) |
| Alcohol consumption | Genetic mutations |
| Smoking | Race/ethnicity |
| Drug intake | Menarche/menstrual period and menopause |
| | Density of breast tissue |

United States reports a median age of diagnosis of 62 years old (Giaquinto et al. 2022; Baker and Mansfield 2023). However, there is a tendency to increase BC incidence in young adult women. In contrast, in undeveloped countries like Mexico, BC diagnosis and prevalence are more frequent in women under 40 years old, and it is related to modifiable risk factors like overweight/obesity, alcohol intake, and smoking (Harbeck et al. 2019; Villarreal-Garza et al. 2020; Cathcart-Rake et al. 2021).

On the other hand, at least 10% of BC cases are associated with family history, in which the genetic predisposition has become relevant in recent years. Related to the above, it has been described that several gene mutations and deletions, for instance, BC gene 1 (*BCRA1*) and 2 (*BCRA2*), cause alteration in the repair DNA damage protein syntheses such as ATM, CEK2, PALB2, PTEN, STTKII, and TP53 (Balmaña et al. 2011; Taylor et al. 2018; Forbes et al. 2019).

Given the income-growing incidence, prevalence, and mortality of this disease, the principal aim of the WHO and healthcare institutions is to focus on reducing BC with timely diagnosis and proper treatment (Torre et al. 2016).

## 2 Definition and Molecular Classification of BC

### 2.1 BC Classification

BC is an adenocarcinoma that is developed in the breast tissue. It is characterized by a malignant neoplasm that arises in the terminal ductal-lobular units of the epithelial tissue on the breast. In general, BC is considered a heterogeneous disease, composed of glandular epithelial cells, called galactophores, responsible for milk production during lactation periods, and epithelial cells that make up the lactiferous duct, which is responsible for interconnecting the galactophores with the nipple. Due to the BC cellular heterogeneity, histological and molecular analyses are required to establish tumor typing and patient treatment. Histological analyses are based on the identification of morphological aspects, establishing four main cellular subtypes of BC according to their cellular origin place (ductal or lobular cells) and the tumor location (in a certain site [in situ] or its spread along the tissue [invasive]). However, these are still subdivided by the architecture that some tumors may present (do Nascimento and Otoni 2020; Zewdie et al. 2021).

## 2.2 Histological Subtypes

Histological analysis identifies specific cellular subtypes such as ductal carcinoma in situ (DCIS), invasive ductal cancer (IDC), lobular carcinoma in situ (LCIS), and invasive lobular carcinoma (ILC). These subtypes differ based on the tumor location and cancer cell behavior within breast tissue.

The DCIS is considered a benign tumor characterized by uncontrolled cell proliferation in the milk ducts without spreading to the surrounding tissues. However, DCIS tumors can progress to an IDC, which has a high cell migration rate to the adjacent tissue parenchyma. On the other hand, the LCIS is considered a benign lesion and originates in the milk-producing lobular cells. However, the LCIS development can progress to an ILC, which is defined as a neoplasm with a great cellular migratory capacity (Makki 2015).

## 2.3 Molecular Subtypes

Besides the histological characteristics of BC, molecular protein analysis is essential to establish biomarkers for therapy. Hormone receptors, including estrogen receptor (ER), progesterone receptor (PR), and human epidermal growth factor receptor 2 (HER2), play a central role in the molecular subtyping of BC classification. Importantly, the proliferative protein Ki67 is another biomarker involved in the proliferative rate tumor identification (Tsang and Tse 2020; Gamble et al. 2021). The presence or absence of these receptors and their expression levels contribute to the treatment strategies and patient outcome BC.

Nevertheless, ongoing research offers more precise diagnostics and therapeutic interventions (Johnson et al. 2020). These markers' expression profiles categorize BC into four subtypes: luminal A, luminal B, HER2-enriched, and triple negative (TN). Next, the main characteristics of each subtype are described in Fig. 1.

The luminal A subtype is characterized by ER and PR overexpression with the absence of HER2. In addition, the Ki67 expression in this subtype is associated with a better patient prognosis. Meanwhile, the luminal B type also expresses both ER and PR; however, it could have the presence or absence of HER2 protein levels and low levels of Ki67. Luminal B is more aggressive than luminal A tumors and has a poorer patient prognosis.

The HER2 subtype does not express any other hormone receptor (HR), and it is more aggressive with a low survival rate. On the other hand, the TN or basal-like subtype is characterized by the absence of HR and HER2. It is the most aggressive tumor, with the highest rate of death and the lowest patient survival (Table 2) (Pinzone et al. 2004; do Nascimento and Otoni 2020; Tsang and Tse 2020).

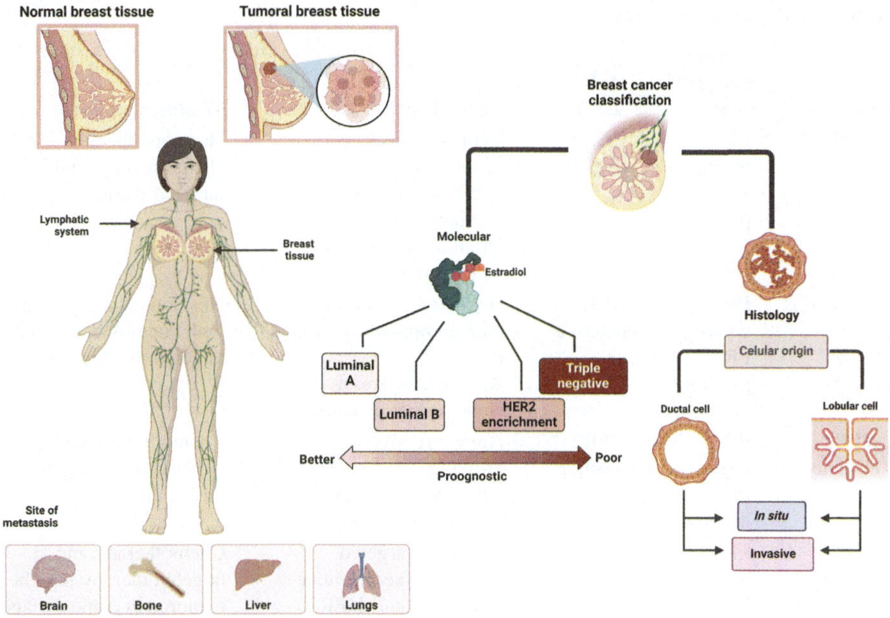


**Fig. 1** Breast cancer (BC) classification. BC is a heterogeneous disease with diverse characteristics. The figure shows normal and BC tissue and the lymphatic system, which plays a central role in the tumor cell invasion and dissemination to surrender organs like the brain, lungs, bone, and liver. Given the diversity of BC, a classification following histological and molecular characteristics has been established. The histological classification is based on the type of cell that originated the carcinoma; in this sense, BC is made up of two types of cells: epithelia-glandular cells and denominated lobular cells (LC), which form lobules involved in milk production during the lactation periods. The second one is the epithelial cells of the lining of the milk ducts, named ductal cells, which are thin tubes that interconnect the lobules with the nipple and carry milk during lactation. Both types of cells can develop carcinoma tumors and can be divided according to a defined tissue localization giving in situ or metastatic BC. On the other hand, the molecular BC classification is based on the presence or absence of different biomarkers such as hormonal receptors (HR) including estrogen receptor (ER), progesterone receptor (PR), and human epidermal growth factor receptor 2 (HER2). All these receptors establish four BC subtypes: luminal A, luminal B, HER2 enrichment, and triple negative. Patients who are diagnostic in the subtype luminal A and B have better prognostic of survival, meanwhile, HER2-enriched and triple-negative types have poor prognostic of survival. Importantly, based on the expression of these markers, BC therapy is defined

The American Cancer Society has established that the primary diagnosed BC subtype between 2015 and 2019 in the USA was the luminal A subtype, with 68% of new cases reported, followed by the TN and luminal B subtypes, both with 10%. Meanwhile, the third place was the $HER2^+$ subtype with 8% (American Cancer Society 2022). In comparison, in Mexico, luminal B was the subtype with the highest incidence, with 48% of the 29,929 new cases diagnosed; the second most common was luminal A, with 23% of the new cases, and in the same proportion of 15%, the $HER2^+$ subtype and TN. Due to the high incidence of ER-positive ($ER^+$) BC,

**Table 2** Molecular BC subtypes

| Subtype | Hormone receptor status | HER2 status | Characteristics | Treatment options |
|---|---|---|---|---|
| *Luminal A* | ER positive<br>PR positive | HER2 negative | Typically has lower proliferation rate<br>Are less aggressive compared to other subtypes<br>Often associated with better prognosis | Hormone therapy, chemotherapy, and targeted therapy |
| *Luminal B* | ER positive<br>PR positive/ negative | HER2 positive or -negative | Have higher proliferation rate, compared to luminal A<br>Are more aggressive<br>Have poorer prognosis | Hormone therapy, often with chemotherapy and targeted therapy |
| *HER2 enriched* | ER negative<br>PR negative | HER2 positive | Are aggressive and fast-growing<br>Often associated with a higher risk of recurrence | Chemotherapy and targeted therapy (anti-HER2 drugs) |
| *Triple negative* | ER negative<br>PR negative | HER2 negative | Lacks for targeted therapies like hormone therapy or anti-HER2 drugs<br>Are more aggressive<br>Has a higher likelihood of recurrence | Chemotherapy and targeted therapy may be explored in clinical trials |

different research groups have studied and characterized the role of estrogen in the induction and progression of these tumors (Cejudo-Arteaga et al. 2022).

## 2.4 Estrogen Receptor (ER)

ER is a crucial biomarker in BC; this protein plays a central role in developing and progressing HR-positive subtypes. In breast tissue, two types of ERs, nuclear and membrane receptors, have been identified. The first type belongs to the nuclear receptor superfamily, acting as a transcription factor and promoting the expression of some genes involved in the progression of the disease. It is estimated that around 70–80% of BC cases express this receptor. The second one corresponds to the membrane receptor. In the text below, we describe the main biological characteristics of both types of receptors (Chen et al. 2022; Clusan et al. 2023).

## 2.5 Nuclear ER

Nuclear ERs are expressed in two isoforms: alpha (ERα) and beta (ERβ), with molecular weights of 66 kDa and 60 kDa, respectively. Different genes located on chromosome 6 (*ESR1*/ERα) and chromosome 14 (*ESR2*/ERβ) encode these isoforms. Both receptors consist of the same six domains (designated by letters A to F); the A/B domain contains the N-terminal and the activation function 1 (AF-1) region, and the C domain is involved in DNA binding through zinc finger structures, necessary for receptor dimerization and recognition of estrogen response elements (EREs). The E domain contains the ligand-binding region, which is conformed by twelve α helices and a β sheet structures; both α helices and β sheet form a ligand-binding cavity containing a dynamic region (H12) composed of leucine (Leu) at position three, two of them separated by two aleatory aminoacid (X), Leu-Leu-XX-Leu motif for estrogen interaction, and domain F with the C-terminal region, and share a 59% homology in their structure (Kumar et al. 2011).

The ERα plays a crucial role in mediating the effects of estradiol (E2) in various tissues, including the breast. This receptor is located at cytoplasmatic and nuclear levels but can also stay anchored in the cell membrane (Anstead et al. 1997) (Lin et al. 2013). Ligand-receptor interaction leads to ERα conformational changes, including phosphorylation at serine (Ser), threonine (Thr), and tyrosine (Tyr) residues, uncoupling from the regulatory protein, dimerization, and exposure of the translocation sequence to the nucleus, allowing to carry out the genomic and non-genomic effects (Yasar et al. 2017; Katzenellenbogen et al. 2018).

In this regard, the genomic effects of active ERα dimers impact the transcription factor receptor activity, which is triggered by two mechanisms: the first is by its binding to specific DNA sequences known as estrogen response elements (EREs). The EREs are well-established consensus DNA sequences composed of two inverted sequences separated by three random nucleotides, GGTCAnnnTGACC, which facilitates the transcription of certain genes. The second mechanism is the association of active ERα with coactivator proteins, such as activator protein 1 (AP-1), coenzyme A, or Sp1 transcription factor protein, allowing gene expression without interacting with EREs. These lead to the modulation of gene expression involved in cell growth, differentiation, and tumoral processes (Fuentes and Silveyra 2019).

The non-genomic effects of nuclear ERs are also known as rapid action mechanisms, which are defined by the activation of various signaling pathways, such as the release of growth factors like the connective tissue growth factor (CTGF), insulin-like growth factor 1 (IGF-1), and epidermal growth factor (EGF), as well as the NF-κB, Notch, mitogen-activated protein kinases (MAPKs), phosphoinositide 3-kinase (PI3K)/Akt, AMP-activated protein kinase (AMPK), pathways, and also activated the hypoxia-inducible factor-1 (HIF-1), protein kinase C (PKC), and protein kinase A (PKA). All are involved in cellular metabolism, cell migration, invasion, and metastasis (Stefkovich et al. 2018).

## 2.6 Membranal ER

Membrane ER belongs to the superfamily of G-protein-coupled receptors (GPCR) referred to as G protein estrogen receptor (GPER), which is characterized by having seven transmembrane domains that are associated with heterotrimeric G proteins at their intracellular carboxyl-terminal end. GPER is encoded by the chromosome 7 p22.3, with a molecular weight of 41 kDa. It is expressed in breast and ovarian tissue epithelial-glandular cells, in bone, nervous, muscular, and immune cells. However, breast and ovarian tumors have a higher GPER expression compared to non-neoplastic cells. GPER is associated with cell proliferation, survival, invasion, and metastatic processes (Luo and Liu 2020). GPER is located on the cellular membrane and the endoplasmic reticulum and Golgi apparatus membranes. It is involved in the rapid cell-signaling pathways modulation. In addition, GPER through its interaction with heterotrimeric G proteins, particularly Gs, Gi, and Gq, is responsible for the adenylyl cyclase and phospholipase C-β (PLC-β) activation and regulation; it can promote the Gβγ signaling pathway which results in the PI3K-Akt cascade signaling activation, nitric oxide (NO) formation, and calcium, sodium, and potassium ion channels regulation (Chan et al. 2020; Hirtz et al. 2022).

Both ERα and GPER have been identified in $ER^+$ tumors and are involved in tumor growth and survival. For the above, many hormonal therapies have been developed, such as tamoxifen or aromatase inhibitors, that target the ERα pathway. Interestingly, GPER can be activated by the drugs used in hormonal therapies, evoking its biological responses (Liu et al. 2020a, c; Arterburn and Prossnitz 2023).

## 2.7 Relationship Between ER and HER2

Besides that, molecular classification is used to establish the BC subtype and treatment for each patient; however, distinct trials have reported that approximately 10% of the patients in therapy against ER present an increase in the expression of HER2 and acquire a resistant state to the hormonal treatment. Furthermore, this has also been reported in patients in treatment against HER2, which causes an increase in the expression of hormone receptors (Liu et al. 2020a, b, c; Arterburn and Prossnitz 2023). Some clinical trials have reported different mechanisms for antihormone therapies or HER2 target therapies which are based on the increased ER, GPER, PR, and RTK expression including IGF-1 receptor (IGF-1R), insulin receptor (IR), EGF receptor (EGFR), and vascular epidermic growth factor receptor (VEGF), which are related to multifactorial molecular mechanism involved in the develop and progression of the tumors (Liu et al. 2014; Butti et al. 2018).

As mentioned above, it has been exposed that ER can influence the HER2 and other RTKs expression in $ER^+$ BC tumors; for example, there is an increase in HER2 expression and activity after the exposition to estrogen; this phenomenon is referred to as "crosstalk" between ER and HER2 signaling pathways (Pegram et al. 2023).

The relationship between ER and HER2 is complex and not fully understood. It can also vary among different BC cases and be influenced by various factors, including the genetic and molecular characteristics of the tumor and the tumor microenvironment. However, it has also become relevant in the crosstalk mechanism with other RTKS (Schettini et al. 2016).

Clinically, understanding the expression of both ER and RTKs is crucial for determining the appropriate treatment approach. However, it is worth noting that advancements in BC research continue to provide a deeper understanding of the molecular mechanisms underlying, leading to more personalized and effective treatment strategies (Giuliano et al. 2013).

## 3 Receptor Tyrosine Kinases

### 3.1 Structure and Classification

Receptor tyrosine kinases (RTKs) constitute the second largest family of integral membrane proteins, with intrinsic Tyr kinase activity. From a structural point of view, RTKs have been classified into two main groups: monomeric RTKs, which are single-chain receptors that are activated by ligand-induced dimerization, and dimeric RTKs, which are $\alpha_2\beta_2$ heterotetramers, linked by disulfide bridges between the two $\alpha$-chains and the two $\beta$-chains, whose representatives are the insulin receptor (IR), the IGF receptor (IGFR), and the insulin receptor–related receptor (IRR) (IR subfamily).

All the monomeric RTK proteins have a highly conserved structure, with an extracellular N-terminal ligand-binding domain (ECD), a single transmembrane $\alpha$-helix domain (TMD), an intracellular juxtamembrane domain (JMD), and a cytoplasmic region, which includes a Tyr kinase domain (TKD) and a Tyr-rich C-terminal tail (Trenker and Jura 2020). The TKD shares a conserved architecture with most RTKs, contrary to the ECDs, which are highly different between the receptor subfamilies, varying in size and composition of individual subdomains. ECD diversity allows receptors to recognize structurally distinct protein ligands (Trenker and Jura 2020). In this context, distinct structural components located in the ECD, such as immunoglobulin (Ig)-like domains, fibronectin III–like repeats, cysteine-rich motifs, Leu-rich, glycine (Gly)-rich repeats/motifs, and EGF-like domains, are critical to the ligand-binding properties of RTKs (Critchley et al. 2018; Batool et al. 2023).

In general, ligand binding triggers RTK activation by inducing receptor dimerization, leading to the activation of the intracellular TKD that hydrolyzes bound ATP, leading to the transfer of the $\gamma$-phosphate group onto the hydroxyl group within the same monomer (cis), or in the adjacent RTK monomer (trans) (Ullrich and Schlessinger 1990; Lemmon and Schlessinger 2010). The phosphorylated sites recruit downstream signaling molecules containing Src homology-2 (SH2) and phosphotyrosine-binding (PTB) domains (Lemmon and Schlessinger 2010). Thus, autophosphorylation of Tyr residues on RTKs is an initial and critical step in the transmission of information to different signaling pathways to subsequently regulate different physiological responses (Critchley et al. 2018).

Dimerization can involve two identical receptors (homodimerization) or between different members of the same receptor subfamily (heterodimerization). Heterodimerization of RTKs increases the set of ligands that can be recognized by each receptor alone, expanding the range of signaling pathways that can be activated by a homodimeric receptor (Lemmon and Schlessinger 2010).

Deregulation of the active intermediate state can have harmful effects on ligand-independent signaling, leading to multiple pathologies, including cancer, developmental anomalies, and metabolic disorders. Therefore, strict control is required to prevent spontaneous phosphorylation of RTKs within the ATP-rich and kinetically favorable intracellular environment (Lin et al. 2020). Additionally, various types of RTKs, such as epidermal growth factor receptors (EGFRs), vascular endothelial growth factor receptors (VEGFRs), platelet-derived growth factor receptors (PDGFRs), IGFRs, and fibroblast growth factor receptors (FGFRs), are overexpressed in different types of malignancies, including BC. Elevated levels of RTKs are also associated with increased BC aggressiveness and decreased disease-free survival (Butti et al. 2018). In addition, it has been identified that specific RTK mutations and anomalous activation of their intracellular signaling pathways have been directly associated with inflammation, angiogenesis, and cancer.

It has been identified that the human genome encodes 58 RTKs, divided into 20 subfamilies according to their ligand bind, sequence homology, and structure (Lemmon and Schlessinger 2010; Wheeler and Yarden 2015; Critchley et al. 2018). The first six subfamilies, which include epidermal growth factor receptors (EGFRs/Erb, class I), insulin and the insulin-like growth factor receptors (IRs and IGFRs, class II), platelet-derived growth factor receptors (PDGFRs, class III), fibroblast growth factor receptors (FGFRs, class IV), vascular endothelial growth factor receptors (VEGFRs, class V), and hepatocyte growth factor receptors (HGFRs/MET, class VI), are closely associated to oncological diseases (Ségaliny et al. 2015).

RTKs, like other transmembrane receptors, are subject to precise regulation through mechanisms that involve their endocytosis through clathrin-coated vesicles and posttranslational modifications, such as phosphorylation on Ser and Thr residues that uncouple the RTK from signaling proteins, and ubiquitination to direct them to degradation. The participation of Tyr phosphatases, such as PTP1B, has also been identified as crucial regulators, which dephosphorylate the receptor, promoting its inactivation.

## 3.2 Role of Receptor Tyrosine Kinase (RTK) Signaling in BC Progression: MAPKs and PI3K/Akt Signaling Pathways

### 3.2.1 Signaling Pathways

RTKs initiate intracellular signaling cascades by the interaction with their ligands. These signaling pathways play crucial roles in various cellular processes, including cell growth, differentiation, survival, and metabolism (Shaul and Seger 2007;

Morrison 2012). Some of the most common signaling pathways induced by RTKs include:

### 3.2.2 Mitogen-Activated Protein Kinase Pathway (MAPK)

MAPK is a central cellular signaling essential for the regulation of cell responses like cell proliferation, differentiation, survival, angiogenesis, apoptosis, tumor development, and metastasis (Shaul and Seger 2007). MAPK pathway consists of a series of protein kinases that relay signals from the cell surface to the nucleus, impacting gene expression and inducing cellular responses. In mammals, three main MAPK signaling cascades have been identified: extracellular response kinase (ERK), c-Jun N-terminal kinase (JNK/stress-activated protein kinase), and p38 MAPK. Besides, all three cascades are activated by different extracellular signals (Ma and Nicolet 2023).

### 3.2.3 ERK Cascade

ERK cascade is often initiated by the extracellular ligand binding to cell surface receptors, mostly by RTKs, sequentially leading to kinase activation. RTK activation induces their interaction with adaptor proteins such as Shc, Grb2, and the guanine nucleotide exchange protein son of sevenless (SOS), which has as target the small GTP-binding protein Ras. Once Ras is activated, it interacts with the first tier kinase which is the mitogen-activated protein kinase kinase kinase (MAPKKK), which in the case of the ERK cascade is represented by c-Raf-1, that is, a Ser/Thr kinase (Lu and Xu 2006).

The Ras/c-Raf-1 interaction leads to c-Raf-1 phosphorylation in Ser residues $Ser^{259}$ and $Ser^{338}$, causing the c-Raf-1 activation and the induction of the second tiers, promoting the activity of the mitogen-activated protein kinase (MAPKK) such as MEK1 and MEK2, which suffer again phosphorylation at $Ser^{218/222}$ residues in MEK1 and $Ser^{222/226}$ residue in MEK2. Once MEK 1 and 2 are active, lead the induction of the third tiers by the MAPK ERK phosphorylation at the $Thr^{202}$ and $Tyr^{204}$ residues. After ERK becomes fully active, it is translocated to the nucleus, where it can activate various transcription factors, including Elk-1, Sap-1a, and transcriptional intermediary factor 1 alfa (TIF-Iα), resulting in the gene expression modulation and the control of cellular responses such as cell cycle progression and cell survival, among others (Martin-Vega and Cobb 2023).

### 3.2.4 JNK Cascade

Besides the MAPK-ERK signaling, the JNK cascade is typically activated in extracellular stress signals, such as UV radiation, heat shock, inflammatory cytokines (e.g., TNF-α), and growth factors.

The cellular processes associated with JNK are metabolism, growth, and proliferation. In addition, the nature of the extracellular signal leads to the activation of RTKs or GPCRs. RTKs can activate the JNK pathway by recruiting and activating the MAPKKK, also known as MAP3K, which serves as the initial kinase in the JNK cascade pathway. Some of the well-known MAP3K in the JNK pathways are the apoptosis signal-regulating kinase 1 (ASK1) and the isoform TGF-β-activated kinase-1 (TAK1) (Girnius et al. 2018).

Once the MAP3K ASK1 or TAK1 are activated, they phosphorylate and activate different MAPKK, just like MAP2K, MKK4, and MKK7. All of these kinases phosphorylate JNK in specific $Tyr^{185}$ by MKK7 and $Thr^{183}$ by MKK4, activating one of the 10 JNK isoforms. JNK isoforms are produced by the alternative splicing of the three specific genes *JNK1*, *JNK2*, and *JNK3*, where the *JNK1* and *JNK2* are ubiquitously expressed, but *JNK3* expression is restricted to the cardiac, smooth cells, and nervous system (Bogoyevitch 2006).

The activated JNK translocates to the nucleus, leading the transcription factor phosphorylation with c-Jun domain at the N-terminal; the activator protein 1 (AP-1) is a primary target, and it is phosphorylated at $Ser^{63}$ and $Ser^{73}$ residues but also can activate JunB, JunD, c-Fos, c-Myc, p53, the nuclear factor of activated T cells-4 (NFAT4), and Elk-1 transcription factors. The formation of the JNK/c-Jun transcriptional factor complex leads to the gene transcription of key regulators involved in cell proliferation, survival, apoptosis, inflammation, and other cellular processes (Stronach and Perrimon 2002; Yarza et al. 2016; Ha et al. 2019).

### 3.2.5 p38 Cascade

The p38 MAPK cascade is another important branch of the MAPK family. This pathway is activated in response to various stress stimuli, including inflammatory cytokines, environmental stress, and cellular damage, and regulates a wide range of cellular processes, including inflammation, apoptosis, cell differentiation, and cellular response to stress. Likewise, to other MAPK pathways, the p38 pathway is activated by various RTKs. The specific receptors depend on the nature of the extracellular stimulus (Pua et al. 2022). Importantly, the p38 MAPK pathway dysregulation has been implicated in various diseases, including inflammatory disorders and cancer (Kudaravalli et al. 2022).

This signaling pathway is composed of four members belonging to the kinase family and encoded by different genes: p38α (also known as MAPK14), p38β (MAPK11), p38γ (MAPK12), and p38δ (MAPK13). These four isoforms are ubiquitously expressed in the organism; however, their expression levels vary between tissues; p38β is overexpressed in neuronal tissue, whereas p38γ is highly expressed in muscle cells, and p38δ is expressed in testis, pancreas, and kidney (Sanz-Ezquerro and Cuenda 2021; Pua et al. 2022).

The MAPK p38 signaling pathway is sequentially activated in response to different stimuli, like stress stimulus initiated the cascade pathway through the activation of MAP 3 K, such as ASK1 or the transforming growth factor TAK1 binding protein 1 (TAB1). ASK1 is a well-known MAP3K for the activation of p38α in response to oxidative stress. Activated ASK1 phosphorylates and activates MAP 2 K, such as MKK3 and MKK6. These kinases are specific to the p38 pathway and play a crucial role in the activation of p38. The phosphorylation of p38 MAPK becomes fully active and leads to its translocation to the nucleus and/or cytoplasm to phosphorylate its target substrates (Martínez-Limón et al. 2020).

In the nucleus, its targets include transcription factors such as activating transcription factor 2 (ATF-2), cAMP response element-binding protein (CREB), and p53, kinases, and other regulatory proteins. These phosphorylation events mediate the cellular response to stress, inflammation, and other stimuli (Kudaravalli et al. 2022).

### 3.2.6 Phosphoinositide 3-Kinase/Akt Pathway (PI3K/Akt)

The PI3K/Akt is a pathway formed by two parts: the first one is by phosphatidylinositol/3-kinase (PI3K) and the second one is by Ser/Thr protein kinase B (PKB). In general, this pathway is typically initiated by the cell surface receptors activation, primarily by RTK for growth factors, such as IR, IGF-1R, and EGFR, as well it could be activated by GPCRs. The PI3K/Akt pathway is critically involved in cell survival, growth, and metabolism regulation. Furthermore, it has a central role in some diseases since it is often dysregulated in cancer and metabolic syndrome, like diabetes (Hemmings and Restuccia 2012).

The PI3K/Akt pathway is activated through the RTK's interaction with his ligand and leads to the recruitment and activation of the PI3K at the plasma membrane by the phosphorylation of Tyr residues. Regarding the above, there are four classes of PI3K, classified as class I, class II, class III, and class IV. Among them, the first class of PI3K is presented in all mammalian cell types, and it is composed of a heterodimeric enzyme consisting of two subunits: a regulatory subunit named p85 and a catalytic subunit named p110 (Gross and Bassell 2014; Yang et al. 2019; Skolariki et al. 2022). p85 subunit is in charge of recognizing and binding to the phosphorylated Tyr in a specific RTK by the SH2 domain. Once it is coupled to the receptor, it provides two functions to the catalytic subunit: stabilization in the membrane and induction of the kinase activity (Cortés et al. 2012). p110 subunit is the most isoform involved and studied in the BC scenario because it is encoded by the phosphatidylinositol-4,5-bisphosphate 3-kinase catalytic subunit alpha (PIK3CA) oncogene, which is overexpressed in these tumors (Zardavas et al. 2014). The p110 subunit is associated with p85 and responsible for the phosphatidylinositol 4,5-bisphosphate (PIP2) phosphorylation to generate phosphatidylinositol 3,4,5-trisphosphate (PIP3) at the plasma membrane. PIP3 serves as a second messenger that recruits Akt protein to the plasma membrane. Akt is then activated through a dual phosphorylation process through the phosphoinositide-dependent kinase 1 (PDK1), which phosphorylates Akt at $Thr^{308}$, and the mammalian target of rapamycin complex 2 (mTORC2), which in turn phosphorylates Akt at $Ser^{473}$, both phosphorylations led to a full Akt active state (Hoxhaj and Manning 2020; Jiang et al. 2020).

Once Akt is activated, it can promote the mammalian target of rapamycin complex 1 (mTORC1) activation, which is responsible for protein synthesis, cell growth, and metabolism. In addition, Akt can activate various downstream targets, including glycogen synthase kinase 3 (GSK-3), which regulates glycogen synthesis, and the forkhead box O (FoxO) transcription factor, which controls gene expression involved in apoptosis and cell cycle arrest. Furthermore, Akt can phosphorylate and inactivate pro-apoptotic proteins such as protein Bcl-2-associated death promoter (BAD), promoting cell survival (Hemmings and Restuccia 2012; Cerma et al. 2023).

Akt can also regulate the cell cycle progression by modulating cyclins and cyclin-dependent kinase inhibitors (CDKIs), such as p21 and p27. This inhibition promotes cell cycle entry, which plays an essential role in the development and progression of different types of cancer, such as BC. Besides, Akt influences DNA repair mechanisms and promotes angiogenesis by activating factors involved in these processes, such as the DNA repair enzyme Rad51 and eNOS (He et al. 2021).

Given the PI3K/Akt pathway's great role in physiological and pathological processes, it has become a primary target for new therapeutics and drugs developed against aberrant PI3K/Akt signaling pathway components, favoring resistance to different therapies, including chemotherapy (Jiang et al. 2020).

### 3.2.7 Janus Kinase/Signal Transducer and Activator of Transcription Pathway (JAK/STAT)

Another important signaling pathway involved in BC development, progression, and therapy resistance is the JAK/STAT pathway, which is primarily involved in immune responses, cell proliferation, and differentiation.

The classical activation of the JAK/STAT signaling pathway is through the cytokine or interleukin ligand binding with a specific receptor. JAK family proteins are non-transmembrane Tyr kinase mediators with a size between 120 and 140 kDa; there exist four members of this family: JAK1, JAK2, JAK3, and TYK2, which have differential tissue expression and function. Regarding the above, JAK1, JAK2, and TYK2 are present in most tissues. However, JAK3 is present only in hematopoietic cells. All JAK proteins contain a conserved terminal carboxyl region with kinase catalytic activity, an SH2 domain that regulates the kinase activity domain, and a FERM domain that modulates the membrane interactions (Hu et al. 2021).

Inactive JAK proteins are coupled to a monomeric receptor since, after the ligand binding, they suffer conformational changes that induce dimer formation with JAK and the respective receptor; this interaction allows the transphosphorylation on Tyr in the SH2 domain. Of note, phosphorylated Tyr residues lead the recruitment of two cascade signal transducers and activators of transcription (STAT) proteins, which have transcriptional activity. STAT protein family is composed of seven members (STAT1, STAT2, STAT3, STAT4, STA5a, STAT5b, and STAT6), all of them contain conserved domains such as the oligomerization domain, coiled-coil, DNA-binding domain, linker domain, SH2 domain, and terminal transactivation domain (TAD) (Vainchenker et al. 2008).

Of note, in the nucleus, the STAT realized this transcriptional function through two mechanisms (Clevenger 2004; Ihle et al. 1996). The first is through the recognition and interaction with a palindromic motif sequence TTCN2–4GAA, while the second mechanism is by the interaction with other transcription factors (Bromberg and Darnell 2000; Horvath et al. 1995); however, both mechanisms led to the target gene expression. It is important to remark that several studies have demonstrated the involvement of STAT 3 and 5 in the expression of some oncogenes that participate in tumorigenesis, apoptosis, and cell proliferation (Horvath et al. 1995; Bromberg and Darnell 2000; Clevenger 2004).

Interestingly, it has been reported that an alternative JAK/STAT pathway activation through RTK activation favors the JAK/STAT protein transphosphorylation. In this regard, the most reported JAK/RTK transactivation is with EGFR and PDGFR, leading to MAPK, STAT3, and STAT5a phosphorylation (Pircher et al. 1999; Vignais and Gilman 1999; Quesnelle et al. 2007; Gkouveris et al. 2014).

### 3.2.8 Notch Signaling Pathway

Notch signaling is usually involved in organ formation, tissue function, and tissue repair; nevertheless, mutations or alterations in this signaling pathway are correlated with cancer. The Notch pathway is mediated by the activation of Notch receptors, which are activated by five different ligands: the Delta-like ligands family (DLL1, DLL3, DLL4) and the Jagged ligands family (JAG1, JAG2). After the ligand binding, the Notch receptors suffer two proteolytic cleavages. The first one, known as "S2," exposes a recognition domain for the metalloproteases ADAM10, ADAM17, or ADAMTS1 (Lobry et al. 2014; Allen and Maillard 2021). The second is the S3 cleavage, which occurs by the γ-secretase complex. It forms a substrate for the amino-terminal Notch recognition, leading to nuclear translocation with the final gene expression regulation. Of note, the Notch pathway has been associated with mammary gland development in breast tissue, and its aberrant function is strongly associated with breast tumorigenesis (De Francesco et al. 2018; Bai et al. 2020).

Interestingly, a non-canonical Notch signaling pathway activation has been reported, which is mediated without ligand binding. This form of Notch activation has been mainly observed on immune cells through the interaction with different proteins, including NF-κB, mTORC, PTEN, Akt, Wnt, Hippo, or TGF-β, at cytoplasmic and/or nuclear levels (Lobry et al. 2014; Zhou et al. 2022). On the other hand, the Notch pathway can promote the properties of malignant BC cells by interacting with the PI3K-Akt pathway, favoring the epithelial-mesenchymal transition (EMT) (Pupo et al. 2014; Bai et al. 2020).

### 3.2.9 Nuclear Factor-Kappa B Pathway (NF-κB)

The nuclear factor kappa-light-chain-enhancer of activated B cells (NF-κB) pathway is another complex and crucial signaling cascade involved in inflammation, immune response, cell survival, and development. The pathway is known for its extensibility, allowing it to integrate signals from a wide range of stimuli and respond to diverse cellular stresses (Devanaboyina et al. 2022).

The NF-κB pathway is composed of five structural proteins: p50 (NF-κB1), p52 (NF-κB1), p65 (RelA), RelB, and c-Rel, which form homo- or heterodimers, which function as transcriptional factors. In an inactive state, NF-κB proteins form dimers that are typically sequestered in the cytoplasm by the inhibitor of NF-κB (IκB) complex protein, which is composed of IκBα, IκBβ, IκBε, and IκBγ. The canonical activation of the NF-κB pathway is triggered by various extracellular signals, such as proinflammatory cytokines (e.g., TNF-α, IL-1), microbial products, and stress stimuli (Wang et al. 2015).

The external signals lead to the activation of IκB kinase (IKK) complex called NEMO, which includes IKKα, IKKβ, and IKKγ, and the activation is by the phosphorylation of the subunit IKKβ at $Ser^{177}$ and $Ser^{181}$. The activation of the IKK complex led to the phosphorylation of the IκB proteins, marking them for ubiquitination and subsequent degradation by the proteasome. Once the degradation of IκB occurs, the NF-κB dimers are released in the cytoplasm and translocate to the nucleus. In the nucleus, NF-κB binds to specific DNA sequences, known as κB sites, in the promoter regions of target genes, leading to the NF-κB regulates the transcription of a wide array of target genes, including those involved in inflammation, immune response, cell survival, and other cellular processes like TNFα, IL-6, Bcl-XL, Bcl-2, Bcl-XS, XIAP, and VEGF (Wang et al. 2015).

Besides, an alternative non-canonical pathway activates specific Rel proteins in response to stimuli like cellular stress, growth factors, and lipopolysaccharides (LPS), and involved receptors like RTKs and GPCR. Unlike the canonical pathway, this route required the RelB/NF-κB2 dimer through the inducible proteolytic processing of the NF-κB2 gene product. In addition, it has been reported that TNF-receptor superfamily members, such as CD40 and B cell activating factors, selectively activate NF-κB-inducing kinase (NIK) and IκB kinase 1 (IKK1). This activation leads to p100 phosphorylation and ubiquitination, causing its partial proteolytic processing and the generation of p52. Both canonical and non-canonical pathways contribute to the development and progression of cancer (Devanaboyina et al. 2022; Oh et al. 2023).

Nevertheless, it is highlighted that the NF-κB pathway is adaptable and can integrate signals from various sources. Given that NF-κB is a transcriptional factor, it is necessary to induce negative feedback mechanisms that control the intensity and duration of its signaling. One of them is the IκB protein synthesis and its association with NF-κB, preventing its nuclear translocation. Another mechanism is the regulatory proteins, such as A20, acting as negative regulators by inhibiting the IKK. However, the NF-κB pathway dysregulation is associated with various diseases, including inflammatory disorders, autoimmune diseases, and cancer (Shembade and Harhaj 2012).

Different studies in the BC context have shown that aberrant NF-κB pathway activity enhances tumor cell proliferation and survival. Nevertheless, recent findings suggest that canonical NF-κB signaling can also act as a tumor suppressor in specific cancers by transcriptionally activating the Fas ligand, hence the controversial role of these pathways in cancer (Liu et al. 2012b; Pavitra et al. 2023).

On the other hand, elevated proinflammatory cytokine levels within the tumor microenvironment can be heightened by the NF-κB canonical and noncanonical activation. In this regard, it has been reported that IL-1 can induce the MKK4 phosphorylation, a crucial step in the conversion of NF-κB p100 to its active p52 form, leading to the nuclear p52 translocation and promoting the transcription of genes involved in cell cycle and proliferation (Acuner Ozbabacan et al. 2014).

Additionally, other studies report that p65 mutations allow a continuous transcriptional activity on target genes involved in cell cycle, invasiveness, inflammation, tumorigenesis, metastasis, and ultimately fostering resistance to radio- and chemotherapy; at least, the expression of 60 related genes has been found, including *REL, RELA, GADD45B, TRAIL R2, CD40, CCL2, IL-15, CCND, CXCL12, VCAM1, COX2,* and *CXCL1*. In addition, it has found mutations in the NF-κB regulation through canonical signaling pathways, induced by IL-1, TNF-α, ROS, and LPS, and noncanonical signaling induced by inflammatory stimuli of interleukins and cytokines through IKKα in BC (Wang et al. 2015).

### 3.2.10 Src Family Kinases

Even though RTKs recruit and activate adaptor proteins for the signaling cascades, it has been reported that they can also activate intracellular protein Tyr kinases (PTKs), like Src family kinases, which are involved in several pro-tumoral processes, including cell adhesion, migration, and invasion (Kohale et al. 2022). Src family proteins have a molecular weight of 60 kDa and possess a Tyr kinase activity. In general, Src proteins are made up of seven domains: at the N-terminal is the Src homology 4 (SH4) domain, followed by a unique domain, which is linked with the SH3, and next to it is the SH2 domain; both SH3 and SH2 are involved in protein-protein substrate interactions. In addition, they have a linker domain, which connects the SH2-kinase domain with the ATP and the corresponding substrate; finally, at the end is the C-terminal domain, which is a negative regulatory region (Tatosyan and Mizenina 2000).

Of note, the Src activity is tightly regulated by several mechanisms; the principal is the Src phosphorylation at specific residues; for instance, the phosphorylation at $Tyr^{416}$ residue in the kinase domain is necessary for the Src activation. Besides, the $Tyr^{527}$ residue phosphorylation at the C-terminal promotes the association of the tail of the C-terminal with the SH2 domain, which has been related with a reduction of the Src enzymatic activity. Also, the $Tyr^{416}$ dephosphorylation is an Src downregulation mechanism. Meanwhile, the $Tyr^{527}$ dephosphorylation by PTP1B is reported as an Src upregulation mechanism (Roskoski 2004; Irtegun et al. 2013).

Src proteins are involved in various cellular processes, like cell proliferation, cell cycle progression, adhesion, and migration through cyclins and CDK regulation; in addition, Src proteins regulate cytoskeletal protein activity, leading to cell motility. Importantly, Src deregulation has been implicated in different human diseases, including cancer and inflammatory disorders. Src overactivation can promote uncontrolled cell proliferation, invasion, and metastasis, making it an attractive target for cancer therapy (Xu et al. 2012). In BC, Src activation leads to the induction of several signaling pathways, such as Ras/c-Raf-1/MEK/ERK1/2 and STAT proteins, resulting in the transcriptional activation of different genes (Martellucci et al. 2020; Luo et al. 2022).

As in other signaling pathways, RTK activation leads to the autophosphorylation of Tyr residues that serve as docking sites for proteins containing SH2 domains. In this sense, Src proteins can be associated with RTK signaling through SH2 and SH3 domains, activating the kinase activity. Thus, Src activation can be involved in various mechanisms, including phosphorylation status changes and structural rearrangements (Tatosyan and Mizenina 2000).

# 4 Participation of Specific RTKs in BC

## 4.1 RTKs and BC

An increasing number of growth factors and their receptors are involved in cancer progression and metastasis. In this context, RTKs constitute a class of receptors that play a critical role in cancer development and respond to environmental cues by starting appropriate signaling cascades in cancer cells (Butti et al. 2018).

Overexpression of various kinds of RTKs, particularly epidermal growth factor receptors (EGFRs), insulin-like growth factor receptors (IGFRs), vascular endothelial growth factor receptors (VEGFRs), platelet-derived growth factor receptors (PDGFRs), and fibroblast growth factor receptors (FGFRs), has been detected in BC (Lemmon and Schlessinger 2010; Butti et al. 2018). Furthermore, increasing RTK levels are related to the aggressiveness of BC and decreased overall survival (Lemmon and Schlessinger 2010; Butti et al. 2018).

Different studies have shown that the choice of molecular targets that determine proliferative and metastatic pathways that are usually overexpressed in breast tumors defines several clinical aspects, including the patient's diagnosis, therapeutic strategy development, as well as patient's survival (Ke and Shen 2017).

As mentioned above, an increased understanding of tumor biology leads to effective targeted therapy development. Importantly, it is highlighted that various receptors with catalytic activity, specifically with Tyr kinase activity, constitute a class of biomolecules that play an important role in cancer progression (Butti et al. 2018). RTKs are cell surface receptors with specialized structural and biological features that respond to the tumor microenvironment and evoke different signaling cascades in tumor cells (Du and Lovly 2018).

The next section describes the pathophysiological mechanisms of the main RTKs involved in BC progression. It should be noted that the choice of these molecular targets was also based on the therapeutic strategies developed for this pathology.

## 4.2 Role of Epidermal Growth Factor Receptor (EGFR) in BC

Epidermal growth factor receptor (EGFR, also known as ErbB1) and its family members HER2 (also known as HER2/neu and ErbB2), HER3 (also known as ErbB3), and HER4 (also known as ErbB4) are important targets in BC (Riecke and Witzel 2020). All EGFRs are transmembrane glycoproteins with an extracellular ligand binding domain and an intracellular Tyr kinase (TK) domain. The cellular effects of the EGFRs are cell proliferation, cell adhesion and motility, invasion, angiogenesis, and protection against apoptosis (Eccles 2011), events that they govern by the signaling pathway activation including PI3K, Ras-Raf-MAPK, JNK, and phospholipase C-$\gamma$ (PLC-$\gamma$), resulting in a plethora of biological functions (Schulze et al. 2005). In addition, the biological response of EGFRs is variable and will depend not only on their overexpression but also on how these family receptors interact since it has been described that they can form homo- or heterodimers. This interrelationship between EGFRs also influences the response to therapy (Zaczek et al. 2005).

EGFR expression is reported in non-transformed epithelial cells of breast tissue; however, when an overexpression or hyperactivation of its signaling pathways is present, it gives rise to transformed or malignant cell phenotype (Zajchowski et al. 1988). Thus, aberrant EGFR expression or signaling can promote tumor development, angiogenesis, and metastasis as well as poor prognosis in many human malignancies (Liu et al. 2020a, b, c). In BC, EGFR overexpression is associated with higher proliferation, genomic instability, poor differentiation, lower hormone receptor levels, HER2 overexpression, poor clinical outcome, and a higher risk of relapse in patients. Its aberrant expression has mainly been reported in inflammatory and TNBC (Guerin et al. 1989; Burness et al. 2010; Rimawi et al. 2010; Changavi et al. 2015).

Some of the described mechanisms by which EGFR acts are due to the activation of multiple signaling pathways. These findings have been documented from in vitro and in vivo experiments. In this regard, EGFR can promote the cell proliferation of TNBC cells (SUM149, MDA-MB-468, and BT-20) by the upregulation of MAPK-ERK and Akt phosphorylation, having the cyclin-dependent kinase 2 (CDK2) as one of the main effectors, which even determines the sensitivity to erlotinib, an EGFR inhibitor (Yamasaki et al. 2007). Of note, the EGFR presence in this type of cell lines seems to be predominant since it has been reported that when these cells were treated with erlotinib, ERK and Akt phosphorylation was downregulated (Yamasaki et al. 2007). In addition, the administration of erlotinib in a SUM149 TNBC xenograft model significantly inhibited tumor growth and lung metastasis compared with the control group. In addition, erlotinib inhibited ERK and Akt phosphorylation and different mesenchymal markers, including marker E-cadherin and vimentin, in mice tumors (Zhang et al. 2009; Ueno and Zhang 2011).

One of the pathways that has recently been associated with promoting proliferation due to EGFR in TNBC cells is the JAK/STAT3 pathway. The overexpression of EGFR and the phosphorylated state of STAT3, as well as MCL1 and PIM1, important downstream proteins of JAK/STAT3 signaling, has been reported in patient samples, in cell lines and murine models with xenotransplants carried out with the TN subtype of BC (Song et al. 2020). In addition, it has been highlighted that EGFR promotes an anti-apoptotic effect via the EGFR-STAT3-Bcl-XL signaling axis in TNBC, contributing to the aberrant survival of malignant cells (Song et al. 2020). Of note, EGFR silencing inhibits tumor growth and MCL1, PIM1, and pSTAT3 expression. In vivo, experiments confirmed that EGFR silencing, STAT3 inhibitor III, and JAK/STAT3 inhibition by ruxolitinib inhibited the proliferation of TNBC cells. In addition, EGFR silencing significantly induced apoptosis (Song et al. 2007). This finding is important since the STAT family, particularly STAT3, maintains a pro-carcinogenic inflammatory microenvironment during malignant transformation and cancer progression (Mantovani et al. 2008).

On the other hand, EGFR expression has also been reported in $ER^+$ BC but to a lesser extent in $HER2^+$ and TNBC cells (Rimawi et al. 2010). Interestingly, although its expression does not significantly govern the cell proliferation in $ER^+$ BC cells (parental MCF-7 cell line); the activation of its signaling, specifically the PI3K/Akt pathway, can induce the expression of proliferative markers (RB1, CyclinD/CdK4/6), downregulated the pro-apoptotic factors (SGPL1, BIK, BMF, APAF1) and deregulate the inhibitory cell proliferation effect evoked by antihormonal therapies such as tamoxifen or fulvestrant (ICI 182,780) in MCF-7 cell line in which EGFR expression was induced with a retroviral vector (MCF-7/EGFR), concluding that EGFR-driven signaling is sufficient to overrule the tamoxifen inhibition of $E2^+$ cell proliferation (Moerkens et al. 2014). In addition, in vitro experiments carried out on MCF-7 cells resistant to tamoxifen (MVLN cells/TAM-R) have revealed that constitutive PI3K/Akt and MAPK pathway activation and EGFR aberrant expression and activation were reported (Knowlden et al. 2003; Ghayad et al. 2010). These works suggest that BC cell growth is mediated by the autocrine release and action of EGFR-specific ligands inducing preferential EGFR/HER2 dimerization (Knowlden et al. 2003; Ghayad et al. 2010). Murine experiments carried out in MCF-7 xenograft tumors showed that EGFR/HER2 expression was significantly increased in tamoxifen-resistant tumors compared to non-tamoxifen-resistant tumors (Massarweh et al. 2008). Thus, the authors propose that EGFR/HER2 may mediate tamoxifen resistance in $ER^+$ BC despite continued suppression of ERα by tamoxifen (Massarweh et al. 2008). The above corroborates with clinical data, indicating that tamoxifen-resistant BCs often have an increased expression of RTKs, specifically EGFR and its family members (Nicholson et al. 1993; Mansouri et al. 2018a, b). Of note, co-expression of ERα and EGFR has been observed in around 50% of the patients with luminal BC tamoxifen resistance (Gutteridge et al. 2010). These studies provide a rationale for combining EGFR/HER inhibitors in tamoxifen-resistant breast tumors, even in tumors that do not initially overexpress them, to solve an important problem in the clinic (Kim et al. 2015).

In addition to the above, despite research reports, the mechanism by which overexpression of RTKs induces anti-estrogen resistance is still unclear. Novel studies suggest that increased crosstalk between ERα, or different GPCRs, and RTKs might be responsible for drug resistance development (Hutcheson et al. 2003; Fan et al. 2007; Pietras and Marquez-Garban 2007; Kilpatrick and Hill 2021; Moody et al. 2023). In addition to the above, there are clinical reports where the correlation of p53, EGFR, and HER2 expression has been studied. Interestingly, the presence of these three targets is strongly associated with poor clinical outcomes (Visscher et al. 1993). Therefore, the study of EGFR expression with different markers could generate specific or personalized diagnostic panels in patients with BC that could even help in the response to therapy.

## 4.3 Role of HER2 in BC

As we described previously, the activation of EGFR family members promotes scattering and invasion of BC cells (Segovia-Mendoza et al. 2015). Specifically, HER2 is amplified and overexpressed in about 15–20% of all BCs; it controls an oncogenic signaling network that inhibits tumor cell death (Pupa et al. 2021). Like

EGFR, HER2 can also impact the activation of MAPK and PI3K/Akt pathways (Weigelt et al. 2010), however, HER2 lacks specific growth factor ligands, the reason why HER2 is the preferred heterodimerization partner for other EGFRs (Yarden and Pines 2012; Segovia-Mendoza et al. 2015). Interestingly, it has been reported that the overexpression of both EGFR and HER2 in non-transformed epithelial cells of the mammary gland is associated with the epithelial-mesenchymal transition and the expression of various markers involved in this process including cytokeratin (CK)14, CK19, E-cadherin, p63, P-cadherin, N-cadherin, and Axl (Ingthorsson et al. 2016). In addition, in the clinical context, HER2 overexpression in BC cells significantly correlated with a high cellular proliferation rate (S-phase), absence of ER and PR negative, positive axillary nodes, high tumor grade, and DNA aneuploidy (Anbazhagan et al. 1991). Nevertheless, different scientific evidence have reported that there may be co-expression of HER2 with the presence of ERα. These cases try to visualize a therapeutic panorama that covers treatment options for the inhibition of both receptors (Giuliano et al. 2013; Montemurro et al. 2013).

Interestingly, HER2 overexpression in BC cells has not only been linked to cellular proliferation, but different splice variants contribute to BC cell survival (Castagnoli et al. 2019). Specifically, HER2 splice and truncated variants (HER2 lacking exon 16, d16HER2) and (p95HER2), respectively, have recently been involved in the BC stem cell (BCSCs) regulation and maintenance (Pupa et al. 2021). In this regard, d16HER2 has been identified as one of the most oncogenic HER2 isoforms implicated in tumorigenesis, which is expressed in a proportion from 4% to 10% in BC samples (Mitra et al. 2009; Volpi et al. 2019) d16HER2 is involved in EMT/stemness and the response to trastuzumab targeted therapy, events linked with the crosstalk with the Notch signaling pathway. Meanwhile, the p95HER2 variant can regulate BCSC features (Castagnoli et al. 2017; Pupa et al. 2021). In addition, the d16HER2 variant is linked with Src activation and outstanding tumorigenesis of BC cells, but also with a greater response to trastuzumab; only when the d16HER2 variant is found alone since the co-expression of d16HER2 with the wildtype of HER2 favors the trastuzumab resistance (Castagnoli et al. 2014).

Different downstream signaling pathways have been identified after activation derived from mouse models of breast tumors that express both the d16HER2 variant and wildtype HER2 receptor. In this sense, the d16HER2 variant was able to activate MAPK, Akt, Src, and STAT3 pathways in a similar manner to the HER2 wildtype. However, the activation of d16HER2 significantly increases Src phosphorylation levels compared to HER2 wild-type tumors (Castagnoli et al. 2014). In addition, the d16HER2 splice variant presence in BC cells induces the formation of stable and constitutively active HER2 homodimers on the tumor cell surface, favoring cell survival (Castagnoli et al. 2019). The findings above suggest multiple heterogeneity within the HER2 BC subtype. Therefore, more in-depth studies on molecular classification should be conducted to improve the patient's prognosis, as has been previously proposed (Jackson et al. 2013).

Interestingly, crosstalk between HER2 and ERα or membrane estrogen receptor (GPER) has been documented (Schettini et al. 2016; Clusan et al. 2023; Pegram et al. 2023). ERα and its different isoforms are involved in nonnuclear/non-genomic pathways and can interact and activate several RTKs including HER2 and their downstream intermediates (MAPK and Akt) (Nemere et al. 2003). These interactions are involved in endocrine therapy resistance, and that is why it is important to understand their mechanisms and expand their study (Arpino et al. 2008).

## 4.4 Role of Insulin Receptor (IR) in BC

IR is also a transmembrane RTK whose conformation is mainly as dimers in the cell membrane (Ward and Lawrence 2009). It is known that this receptor can be activated by insulin or by different insulin growth factors (IGFs) and has significant roles in metabolism, cell growth, and development, among others (Haeusler et al. 2018). Like other RTKs, IR activates MAPK and PI3K pathways, the latter being the one most associated with the metabolic effects of the receptor (Zhang et al. 2022). In humans, IR plays a crucial role in nutrient homeostasis. Of note, dysregulation of the receptor expression and ligands function (insulin, IGF-1, and IGF-2) or its downstream signaling effectors drives several diseases, including diabetes (Boucher et al. 2014), obesity (Wondmkun 2020), and multiple cancers (Vella et al. 2018). Nevertheless, the relationship between IR activation in BC has been studied for a short time compared to other RTKs (Nguyen et al. 2022).

In the normal breast, IR and insulin-like growth factor receptor type 1 (IGF-1R) have a significant role during development and in the mature adult gland since they are involved in terminal end bud development and ductal morphogenesis in the mammary gland development and are also required for alveolar differentiation during pregnancy and lactation (Ruan and Kleinberg 1999; Neville et al. 2013). Interestingly, IR expression significantly increases in human BC samples compared to normal breast tissues (Aljada et al. 2015). In addition, IR overexpression correlated with tumor size, cancer staging, and ER status (Vella et al. 2018). In contrast, decreased IR expression in TNBC has been linked with greater tumor aggressiveness (Yang et al. 2020). Of note, other studies have made important contributions to the expression of IR in BC. Regarding the above, preclinical mouse mammary carcinoma models have exposed that IR expression correlates with CD24 stem cell markers (Rostoker et al. 2015, 2016). This fact opens new therapeutic guidelines that allow CD24 to be considered a predictive factor for BC responding to therapies directed against IGF-1R.

As mentioned, IGF-2 is also an important ligand for the insulin receptor, and through its binding to IR as well as the downstream activation of the PI3K/Akt pathway, it can promote the stem cell phenotype in BC (Belfiore et al. 2023). In addition, recent reports have evaluated the differential IGF-2 methylation gene pattern in BC samples and normal breast tissue. The methylation analyses showed that a unique site exists in the *IGF-2* gene that is hypermethylated in normal breasts and hypomethylated in BC. The above is relevant since IGF-2 promotes metastasis and chemoresistance and contributes to BC aggressiveness. Thus, the IGF-2 and IR evaluation might serve as biomarkers that improve BC diagnosis and treatment (Radhakrishnan et al. 2020).

On the other hand, it has been described that IR has several isoforms due to alternative splicing, thus we have IR-A and IR-B, which are present in the cells as homodimers or as heterodimers (Vella et al. 2018). Moreover, IR can also form a hybrid receptor with IGF-1R. Related to the above, the IR-A form has been associated with mitogenic signals, and it is known as oncofetal isoform, while the IR-B isoform has been related to metabolic effects in adult tissues. In this context, IR-A activation has been mostly related to cancer cell proliferation and therapy resistance (Belfiore et al. 2017; Galal et al. 2023).

Regarding the above, recent reports have documented that IR-A drives cell migration and invasion in vitro and in vivo models employing 4T1 TNBC cells overexpressing IR-A or IR-B. First, the authors evaluated transcriptome analysis in vitro after the exposure to insulin, which evidenced that approximately 89% of insulin-stimulated transcripts depended on IR-A, the tumor progression and immune evasion genes being the most modulated. In contrast, in 4T1 IR-B overexpressing cells, insulin induced the expression of genes primarily involved in metabolic regulation (Vella et al. 2021). In addition, IR-A was associated with metastasis and tumor growth promotion in vivo models of mice injected with 4T1/IR-A-overexpressing cells (Vella et al. 2021). Finally, the participation of IR-A was also related to angiogenesis in a xenograft zebrafish embryo tumor model carried out with 4T1/IR-A and 4T1/IR-B cell lines (Vella et al. 2021). It is important to mention that the IR-A response might depend on the ligand and the BC type; in this sense, the IR-A stimulation with IGF-2 has been related to metabolic cell reprogramming through an increase in aerobic glycolysis activity and mitochondrial biogenesis in IGF-2R overexpressing/IGF-1R-ablated MCF-7 cells, resulting in proliferative responses and invasion (Vella et al. 2019). However, the differential activation of IR-A in different types of BC has not been thoroughly elucidated.

On the other hand, the relationship between obesity and BC through the activation of IR has been addressed in different reports (Glassman et al. 2023). Generally, in BC, obesity is associated with a twofold increased relative risk of death, which has even been linked with increased estrogen production (Cleary and Grossmann 2009), oxidative stress (Glassman et al. 2023), altered adipokine expression, and enhanced inflammatory status (Devericks et al. 2022; Tewari et al. 2022).

In summary, IR research and its biological effects on BC remain under investigation. In the following section, the effects of IGF-1R on the pathophysiology of BC is described.

## 4.5 Role of Insulin-Like Growth Factor Receptor Type 1 (IGF-1R) in BC

IGF-1R is another RTK with important actions in BC. Like the receptors described above, IGF-1R also has catalytic activity on Tyr residues, which occurs after binding to its main ligand, insulin growth factor (IGF-1), but also after binding to other ligands such as insulin and IGF-2. Activation of IGF-1R is achieved in the extracellular α subunits, which leads to autophosphorylation of three Tyr residues within the intracellular kinase domain. Although IGF-1R is homologous to the insulin receptor (IR), it has been reported that they are constituted and have different physiological actions. Nevertheless, IGF-1R can form homodimers with insulin receptor (IR), which also contribute to cancer development (Arcaro 2013).

In non-tumorigenic breast cells, IGF-1R activates matrix metalloproteinases (MMPs) that release heparin-binding epidermal growth factor (HB-EGF) and activates the EGFR-dependent MAPK pathway (Song et al. 2007). In BC cells, IGF-1R can also activate the same mitogenic and metabolic pathways as EGFRs, MAPK, and Akt (Voudouri et al. 2015). Nevertheless, IGF-1R activation can induce the phosphorylation of other substrates such as Src homology 2/collagen-*α* proteins (Shc) (Giorgetti et al. 1994), growth factor, focal adhesion kinase (FAK), and carboxyl-terminal Src kinase (CSK) (Baron et al. 1998). Of note, IGF-1R signaling induces the expression of oncogenic miRNAs (103/107, 1826, 191, 93) and downregulates the miR-16 and let-7 families, which are classified as tumor suppressors in luminal BC cells (MCF-7); these actions are driven mainly through the PI3K/Akt pathway (Martin et al. 2012).

In addition, IGF-1R can be co-expressed with EGFR and HER2, conferring greater resistance to antiestrogenic therapies (Gee et al. 2005). The above has been linked with the activation of the MAPK pathway; besides, this triple-receptor interaction activates the ERα to promote the transcriptional activity of activation functions (AF-1), specifically in Ser residues associated with acquired tamoxifen resistance. Clinical data have demonstrated that IGF-1R promotes the phosphorylation of $Tyr^{1316}$, and $Ser^{118}$ of ERα, which are present in $ER^{+}$ tamoxifen-responsive disease and $ER^{+}$ de novo acquired tamoxifen-resistant BC (Gee et al. 2005; Peiró et al. 2011). Another characteristic that has been documented about the IGF-1R expression in BC is that it is mainly associated with luminal tumors; its expression in hormone receptor–negative tumors is less frequent. Furthermore, IGF-1R predominantly activates the PI3K/Akt pathway and to a lesser extent the MAPK pathway (Ueda et al. 2006; Peiró et al. 2011). It should be noted that IGR-1R expression has been linked with poor prognosis, considering its expression analysis to make treatment decisions in patients (Peiró et al. 2011). Experimental results have shown that E2 (20 ng/ml) can activate IGF-1R in MCF-7 cells. E2 induces a rapid MAPK activation through IGF-1R–MMP–HB-EGF–EGFR, and MAPK; these effects were corroborated by the employment of IGF-1R and MMP selective inhibitors or knockdown of IGF-1R (Song et al. 2007).

In vitro experiments conducted in ERα/HER2/IGF-1R-positive BC cells have also demonstrated the biological role of IGF-1R. The dual employment of tamoxifen and NVP-AEW541 and BMS-536924IGF-1R, ERα, and IFG-1R inhibitors enhanced growth inhibition in different $HER2^{+}$ BC cell lines. In addition, combined treatment with trastuzumab (a humanized monoclonal antibody against HER2), tamoxifen, and either of the IGF-1R inhibitors enhanced response compared to dual targeting strategies. Furthermore, the benefits of the IGF-1R blockade in a trastuzumab-resistant cell line model result in a similar enhanced response as observed in the non-trastuzumab-resistant cells (McDermott et al. 2017). Interestingly, clinical data reported different phosphorylation statuses of the chimeric IGF-1R/IR association in HER2$^+$ BC cell lines and tumors, again, the dual co-targeting of HER2 and IGF-1R may be beneficial in some HER2-amplified BCs (Browne et al. 2012).

On the other hand, metastatic TNBC also presents IGF-1R amplification that correlates with high ERβ protein expression, which together activates metabolic pathways that favor the progression of BC. In addition, IGF-1R signaling functions as positive feedback for *ERβ* gene amplification, which is also involved in cancer progression. In this sense, it has been suggested that the use of metformin can serve as an adjuvant treatment in TNBC (Hamilton et al. 2015). Recently, it has been documented that not only the expression of IGF-1R in BC cells is important for disease progression but also its localization is important. Luminal tumors preferentially exhibit a membrane localization, while in HER2$^+$ tumors, the IGF-1R is mainly found in the cell cytoplasm (Alkhayyal et al. 2020). In addition, recent reports about genomic profiles induced by the IGF-1R pathway by insulin action have been associated with the prognosis of BC as well as its treatments, where the most regulated genes correspond to three main processes, namely insulin pathway, chronic inflammation pathway, and IGF pathway (Gennari et al. 2021).

With the above evidence, IGF-1R can be considered a new marker in cases of resistance to treatment and in advanced stages of BC.

## 4.6 Role of Vascular Endothelial Growth Factor Receptor (VEGR) in BC

VEGFR is another RTK known as the main biological mediator involved in the growth of new blood vessels (Shibuya 2011). Three different isoforms of this receptor bind to different ligands, such as vascular endothelial growth factors and placental growth factors (VEGF-A, VEGF-B, and PIGF). VEGFR1 and VEGFR2 are directly related to angiogenesis and metastasis processes, while VEGR3 is associated with lymphangiogenesis (Shaik et al. 2020). VEGFR1 and VEGFR2 have a higher affinity for most of the known ligands for these receptors (Shibuya 2011; Wang et al. 2020).

In general terms, VEGFRs are membrane glycoproteins expressed in the endothelium, immune cells, and epithelial and neural tissues (Boussat et al. 2000; Al-Nedawi et al. 2009; Hoeres et al. 2018). After the binding of VEGFs to VEGFRs, they recruit and phosphorylate enzymes and effectors that affect a wide variety of responses, including cell proliferation, migration, and vascular permeability in different cells and tissues, but also in cancer cells that together allow the triggering of metastatic disease. VEGFs binding to the VEGFR extracellular domain promote receptor dimerization and tyrosine transphosphorylation of the intracellular domains, initiating a signaling cascade that culminates in the phospholipase C γ-1 (PLC-γ1) activation, diacylglycerol formation (PKC activation), calcium flux, PI3K/Akt, and MAPK pathways (Wang et al. 2020). Importantly, in a cancer context, VEGFR expression is upregulated by HIF-1α (Brito et al. 2011). In addition, recent evidence has reported that VEGF-A (VEGF), the most potent angiogenic factor, can stimulate in a paracrine manner endothelial progenitor cell angiogenesis and in an autocrine cancer cell manner, the survival for several tumors (Liang et al. 2017). It has been known that VEGF can also act in conjunction with other cytokines and thus form a pro-inflammatory microenvironment in BC (Mendez-Garcia et al. 2019). Moreover, high levels of VEGF have been found in serum samples of BC patients. VEGF expression also correlates with higher frequency in luminal B, HER2$^{+}$, and TN types versus luminal A BC. In addition, it is significantly associated with increased risk for BC mortality and inversely correlated with tumor grade (Adams et al. 2000; Dent 2009; Liu et al. 2011a; Srabovic et al. 2013; Zhu and Zhou 2015).

Interestingly, angiogenesis can be hormonally regulated in BC cells; it has been reported that 17β-estradiol (E2) induces VEGF and VEGFR2 expression (Sengupta et al. 2003; Higgins et al. 2006); meanwhile, the antihormonal therapy based in tamoxifen reverses these effects (Garvin and Dabrosin 2003; Garvin et al. 2005). Notably, VEGF and VEGFRs have been associated with breast tamoxifen-resistant cancer cells by MAPK overactivation (Mansouri et al. 2018a, b). Regarding the above, although luminal BC is the best response to therapies within BC pathology, the reduced or lost expression of full-length and membrane VEGFR1 results in adverse tumor features and shortened patient survival, resulting in a not satisfactory response to endocrine therapy (Lebok et al. 2016). In addition, it is highlighted that the dual VEGFR2 and the epithelial-mesenchymal transition marker expression, including Twist1 and Vimentin, in BC cells is linked with poor prognosis in the different BC subtypes and correlates with lymph node metastasis (Yan et al. 2015).

It has recently been documented that the expression of VEGF-A splicing variants influences the pathogenicity of BC, expanding the horizon of study, as well as therapy, of how VEGF-VEGFR complexes would have to be studied in the different subtypes of BC (Al Kawas et al. 2022). In addition, it is highlighted that the dual VEGFR2 and the epithelial-mesenchymal transition marker expression, including Twist1 and Vimentin, in BC cells is linked with poor prognosis in the different BC subtypes and correlates with lymph node metastasis (Yan et al. 2015). In particular, the $VEGF_{165b}$ isoform, an antiangiogenic splice variant, has been associated to higher invasive BC tumor cells. Interestingly, $VEGF_{165b}$ plasma levels are significantly elevated in BC patients, and they reduce in women after chemotherapy administration (Al Kawas et al. 2022). Although it has been mentioned that the expression of VEGF and its activation pathways is important in all BC subtypes, clinical evidence with immunohistochemistry analysis about VEGF expression in BC samples showed a strong correlation with ER-negative (ER$^{-}$) and HER2-positive BC types (Fuckar et al. 2006) so much so that preclinical studies show that double blockade against HER2 and VEGFR presents a good inhibitory response in BC xenografts mouse models (Le et al. 2008).

Therefore, we consider it important to mention that the biological role of the different VEGFR ligands, as well as the associations that these receptors generate to promote angiogenesis in different types of BC, is not yet completely clarified. In addition to the above, the cellular balance of the different pro- and anti-angiogenic factors will depend on the cellular context in which the pathology is immersed.

Nevertheless, blocking VEGFR activation has been an important therapeutic strategy in BC disease (Brogowska et al. 2023).

## 4.7 Role of Platelet-Derived Growth Factor Receptor (PDGFR) in BC

The platelet-derived growth factor (PDGF) and its receptors (PDGFRs) trigger signaling associated with cell proliferation, cell migration, cellular differentiation, and angiogenesis, actions mainly mediated in an autocrine or paracrine manner. The PDGF signaling network includes two monomeric receptors, PDGFR-α and PDGFR-β, and four ligands, PDGF-A, PDGF-B, PDGF-C, and PDGF-D, which are synthesized from independent genes (Jansson et al. 2018). PDGF-A and PDGF-B are secreted as active homodimers (PDGF-AA, PDGF-BB) or heterodimers (PDGF-AB) after intracellular enzymatic processing by proprotein convertases, while PDGF-C and PDGF-D are secreted as inactive homodimers (PDGF-CC, PDGF-DD) and need further activation by extracellular proteases (Shah and Sizemore 2022; Zou et al. 2022). Of them, PDGF-A and PDGF-B are frequently overexpressed in human breast tumors and function as an adverse prognostic factor in patients with advanced BC (Yu et al. 2015; Shah and Sizemore 2022). All PDGF isoforms except PDGF-DD promote PDGFR-α dimerization, though PDGFR-α binds to PDGF AA with higher affinity, whereas PDGF-BB and PDGF-DD activate PDGFR-β dimers.

Most studies have focused on the role of PDGF-AA and PDGF-BB and receptor subtypes in BC's development, progression, and aggressiveness, finding a direct relationship with their high expression (Yu et al. 2015; Shah and Sizemore 2022). Furthermore, recent evidence also demonstrated that upregulating PDGF-AB via loss of miRNA let-7d promotes BC brain metastasis, showing the AB heterodimer's relevance in BC's progression (Wyss et al. 2021).

However, recent studies have revealed PDGF-CC and PDGF-DD's role in BC. Jansson et al. reported that high PDGF-CC and PDGFR-α expression is linked with the aggressive biological phenotype of the TNBC subtype, and high PDGF-CC increased the risk of 5-year distant recurrence. Even more, tumor cell PDGFR-α was significantly upregulated in lymph node metastases and asynchronous recurrences, which supports an active role of the PDGF signaling pathway in tumor progression (Jansson et al. 2018). In the case of PDGF-D, its role in tumor progression or treatment is largely unknown; however, recent evidence suggests that PDGF-D, which only binds and activates PDGFR-β, can be associated with BC, triggering cell proliferation, apoptosis, migration, and angiogenesis (Lu et al. 2022).

Once activated by their ligand, three dimeric receptor combinations have been reported, PDGFR-αα, PDGFR-ββ, and PDGFR-αβ, which activate similar cellular signals associated with mitogenic signals. However, most available information comes from studies on the functional role of PDGFR-αα and PDFGR-ββ. PDGF binding to their PDGFRs promotes receptor dimerization and activation, triggering distinct signaling pathways such as PI3K/Akt, MAPK/ERK, JAK/STAT, and Notch.

Therefore, the PDGF/PDGFR signaling pathway is a critical promotor of tumor cell proliferation, invasion, metastasis, and angiogenesis (Zou et al. 2022). Thus, PDGF/PDGFR dysregulation is often present in cancer. In this regard, their expression has been identified in a variety of cancers, such as non-small cell lung cancer, pancreatic cancer, BC, ovarian cancer, hepatocellular carcinoma, colorectal cancer, and neuroendocrine tumors (Zou et al. 2022).

Enough evidence implicates anomalous PDGF/PDGFR signaling in BC development and metastases (REFs). In BC, a high PDGFR-α/β expression has been associated with an unfavorable prognostic factor, which correlates with tumor progression and metastasis (Pandey et al. 2023). For instance, PDGFR-α has been identified in human HER2-positive (Carvalho et al. 2005), ER- and PR-negative, and triple-negative BC subtypes (Jansson et al. 2014; Park et al. 2015). Moreover, a high stromal PDGFR-α expression was also detected in HER2-positive and high Ki67 BC (Park et al. 2015). Regarding PDGFR-β, expression in BC has only been reported in stroma, and increased expression has been associated with HER2-positive and high Ki67 (Park et al. 2015; Jansson et al. 2018).

Among the signaling pathways activated by PDGF/PDGFR associated with BC are the MAPK/ERK1/2 pathway, which promotes cancer cell growth and proliferation; the long noncoding RNA (lncRNA) LURAP1L-AS1-mediated activation of canonical NF-κB pathway (PDGF-BB) associated to the transformation of human fibroblast into cancer-associated fibroblasts (CAFs) (Ren et al. 2021), which are abundantly present in the tumor stroma and promote BC proliferation, migration, invasion, and angiogenesis (Hu et al. 2022).

Recently, it was shown that PDGF-B/PDGFR plays a critical role in the proliferation and metastatic phenotype of TNBC cells via the SHP-2/Akt pathway, which regulates the expression of the transcription factor ZEB1 (a well-known regulator of EMT) and miR-200a, which negatively regulates ZEB1 (Zhang et al. 2020). The above is relevant since ZEB1 is a transcriptional factor that regulates the expression of several genes involved in the migration and invasion of various cancers, including BC (Zhang et al. 2020).

Interestingly, it was also described that there is a crosstalk between platelets and BC cells where PDGF-B/PDGFR plays a critical role. It was shown that BC cells activate platelets by secreting inflammatory factors, increasing the expression of CD36, and promoting the release of many growth factors and cytokines, including high levels of PDGF-B. Acting in a paracrine form, PDGF-B activates the PDGFR-β/COX-2 signaling cascade in BC cells, which promotes elevated levels of inflammatory factors, leading to the increased hematogenous metastasis of BC (Tang et al. 2023).

## 4.8 Role of Fibroblast Growth Factor Receptor (FGFR) in BC

The fibroblast growth factor (FGF) family consists of 18 structurally related members, which mediate their cellular responses by activating four monomeric transmembrane RTKs, designated FGFR1-4 and one membrane-associated receptor lacking the intracellular domain (FGFR5 or FGFRL1) (Sobhani et al. 2020; Servetto et al. 2021). All 18 members have critical functions in embryonic development and regulation of cell proliferation, differentiation, and migration in adults.

Similar to most RTKs, after ligand binding, FGFRs dimerize and become activated, which leads to the activation of several downstream proteins such as the FGFR substrate 2 (FRS2), Shc, PLC-γ, STAT1 and Gab1, as well as distinct signaling pathways such as MAPK/ERK1/2, PI3K/Akt, IP3/$Ca^{2+}$, DAG/PKC, and JAK-STAT, which regulate target gene expression involved in cell proliferation, migration, invasion, and angiogenesis (Eswarakumar et al. 2005; Ornitz and Itoh 2015; Santolla and Maggiolini 2020).

Different genetic alterations associated with the activation of FGFR signaling have been identified in BC, mainly attributed to modifications in the *FGF* and *FGFR* genes, such as gene amplification, overexpression of nuclear FGFRs, chromosomal translocation, and point mutations (Servetto et al. 2021). The most frequent genetic alteration detected in BC is *FGFR-1-4* gene amplification (Santolla and Maggiolini 2020). However, amplification of *FGFR-1* represents the more frequent genomic alteration in BC, particularly in $ER^+$ BC, where it has been associated with increased *FGFR-1* overexpression and signaling and described as a strong predictor of poor prognosis (Elbauomy Elsheikh et al. 2007; André and Cortés 2015; Santolla and Maggiolini 2020).

The amplification of other FGFRs has also been identified in BC, such as that of *FGFR-2*, which was observed in 4% of TNBCs and detected in TNBC cell lines where *FGFR-2* is constitutively active and highly sensitive to inhibition, which suggests that *FGFR-2* might be a potential therapeutic target (Turner et al. 2010; André and Cortés 2015).

On the other hand, *FGFR-4*, initially considered to have a lower impact on BC, has aroused significant interest in recent years due to its association with the development of BC metastasis. For instance, *FGFR-4* amplification has also been detected in ER/$PR^+$ BC, and an elevated FGFR-4 mRNA level was also observed in BC samples, although the genetic amplification of this receptor is less common in patients with BC, suggesting that alternative mechanisms are involved (André and Cortés 2015; Levine et al. 2020; Santolla and Maggiolini 2020). The abnormal activation of FGFR-4, associated with receptor and specific ligand (FGF-19) overexpression, FGFR-4 somatic mutations, and FGFR-4 single-nucleotide polymorphism, has been related to cancer progression and resistance to cancer therapy. Furthermore, FGFR-4 pharmacological inhibition or knockdown can inhibit tumor growth and metastasis both in vitro and in vivo. A relevant aspect of FGFR-4 is its lowest homology among the FGFR family, allowing the design of specific FGFR-4 inhibitors (Levine et al. 2020). In addition, a recent study indicates that FGFR-4 promotes the transition from a luminal A primary breast tumor to a highly proliferative and metastatic, HER2-enriched subtype (Garcia-Recio et al. 2020).

## 4.9 Role of Hepatocyte Growth Factor (HGF)/MET Receptor in BC

Hepatocyte growth factor (HGF) is generated by stromal and mesenchymal cells in the liver, which activates its RTK, the HGF receptor, also known as the mesenchymal-epithelial transition receptors (MET), in a paracrine manner. The MET, which belongs to the MET proto-oncogene family, is a 190 kDa heterodimer consisting of an extracellular α-subunit of 50 kDa, with the N-terminal and a transmembrane β-subunit (140 kDa), linked by disulfide bonds (Parikh and Ghate 2018). The binding of HGF to MET initiates its activation, dimerization, and autophosphorylation of intracellular tyrosine residues, which allows the recruitment of signaling effectors, including GAB1 and GRB2, which in turn activates downstream signaling pathways such as Ras/MAPK, Ras/PAK, Src, PI3K/Akt/mTOR, JAK/STATs, Wnt/β-catenin, and NF-κB (Parikh and Ghate 2018). Under normal physiology conditions, activating these signaling pathways promotes motility, immune response, morphogenesis, EMT, angiogenesis, and proliferation of epithelial cells in various organs (Kato 2017). HGF/MET also plays a role in controlling ductal branching and expansion throughout the development of the mammary gland (Soriano et al. 1998; Acevedo et al. 2022).

Numerous studies have shown that HGF/MET is aberrantly upregulated, and its signaling promotes aggressive cellular invasiveness, which is closely associated with tumor metastasis and plays a critical role in oncogenesis and tumor progression in various malignancies, including BC (Tuck et al. 1996). For instance, MET is overexpressed in invasive ductal carcinomas (IDC), the most prevalent form of early-stage BC, which controls tumor development, metastasis, and chemoresistance. $Tyr^{1234/1235}$ residues in the activation loop of the kinase domain and $Tyr^{1349/1356}$ in the C-terminal tail of MET undergo autophosphorylation upon binding to HGF. This process creates docking sites for downstream effectors that promote BC cell survival, proliferation, invasion, and dispersion (Ho-Yen et al. 2015; Acevedo et al. 2022). Interestingly, inhibition of MET autophosphorylation has emerged as a strategy for BC therapy, promoting the design of specific MET kinase activity inhibitors (Puccini et al. 2019).

An interesting finding is the association of MET with chemoresistance. For instance, hyperactivation of MET and ephrin type-A receptor 2 (EphA2) in TNBC patients treated with 5-fluorouracil driven to chemotherapy correlates with lower disease-free survival (Mitra et al. 2023).

# 5 RTK Transactivation by GPCRS and Nuclear Receptors

## 5.1 GPCRs

G-protein-coupled receptors (GPCRs) represent the most prominent family of transmembrane receptors, which regulate a wide variety of intracellular signaling pathways in response to a great diversity of ligands, ranging from amino acids, photons, and ions to peptides, proteins, and fatty acids, thus playing a critical role in cellular pathophysiology and the therapy of various diseases (Hilger et al. 2018; Di Liberto et al. 2019; Odoemelam et al. 2020). Structurally, all GPCRs comprise an amino-terminal extracellular domain, seven transmembrane-spanning α-helix domains (7TM) connected by three intracellular and three extracellular loops, and finally, an intracellular carboxyl tail (Lefkowitz 2004; Kobilka 2007; Di Liberto et al. 2019). GPCRs have been classified according to their pharmacological properties and sequence homology into six main groups: class A rhodopsin-like GPCRs; class B of human secretin-like and adhesion GPCRs; class C consists of metabotropic glutamate-like GPCRs; class D pheromone GPCRs; class E the cAMP receptors; and class F, which includes frizzle/smoothened GPCRs. Class A is the most prominent and best-studied family of GPCRs, which include receptors with an important role in tumor biology (Bar-Shavit et al. 2016).

Ligand-mediated GPCR activation promotes specific conformational changes, allowing it to bind to its cognate G protein, composed of the Gα subunit and the Gβγ dimer (Jean-Charles et al. 2017; Di Liberto et al. 2019). In the classical G protein activation cycle, Gα associates with the Gβγ dimer when bound to GDP and forms the inactive heterotrimer. GPCR activation by ligand binding promotes GDP dissociation from Gα, the exchange of GDP by GTP, and a conformational change that favors the dissociation of the Gα and Gβγ subunits. The Gα subunit has been classified into four subfamilies based on sequence identity: G$\alpha_s$, G$\alpha_{i/o}$, G$\alpha_{q/11}$, and G$\alpha_{12/13}$ (Neves et al. 2002). Both Gα subunits and Gβγ dimers can signal through multiple downstream effectors such as AC, PI3K, Src, and MAPKs, leading to a wide range of cellular and physiological responses (Gurevich and Gurevich 2019).

The cellular response stops by several mechanisms, which happen at the G protein and GPCR levels. One of these mechanisms begins when the Gα subunit hydrolyzes GTP to GDP through its intrinsic GTPase activity and reassociates with Gβγ, ending the G protein activation cycle (Pierce et al. 2002; Di Liberto et al. 2019). Moreover, after receptor activation, most GPCRs undergo regulatory mechanisms such as desensitization and internalization triggered by ligands or agonists. These mechanisms are facilitated by agonist-induced receptor phosphorylation by G-protein-coupled receptor kinases (GRKs). Phosphorylation by GRKs enables the cytosolic recruitment of β-arrestins to the GPCR. β-arrestin then physically uncouples the interaction between the receptor and the G protein, ending the G--protein-mediated signaling. Furthermore, β-arrestins were subsequently found to target the receptor to clathrin-coated vesicle-mediated endocytosis by directly interacting with various components of the clathrin-mediated endocytosis machinery (Ranjan et al. 2016; Nieto Gutierrez and McDonald 2018; Rajagopal and Shenoy 2018).

In addition to their important roles in GPCR desensitization/internalization, β-arrestins are transducer proteins that mediate G-protein-independent signaling pathways by acting as multifunctional scaffolding proteins and signaling activators for intracellular signaling propagation and amplification (Shenoy and Lefkowitz 2011; Bagnato and Rosanò 2019).

## 5.2 GPCR-Mediated Transactivation of RTKs

Along with the canonical mechanisms regulated by G protein activation and generation of second messengers, GPCRs stimulate growth-promoting actions by activating RTKs and downstream signaling pathways triggered by these receptors, a process named transactivation (Daub et al. 1996; Di Liberto et al. 2019). It has been identified that RTKs can be activated by GPCRs in a ligand-dependent or -independent manner. Ligand-dependent transactivation mechanisms commonly involve the activation of membrane-bound MMPs or "a disintegrin and metalloproteinases" (ADAMs), a process well characterized for the EGFR (Prenzel et al. 1999; Cattaneo et al. 2014). MMPs or ADAMs cleave RTK pro-ligands bound to extracellular matrix components to generate a mature ligand that can bind and transactivate specific RTKs, triggering downstream signaling. Ligand-dependent transactivation mechanisms involve the activation of several MMPs, including MMP-1, MMP-2, MMP-7, MMP-9, and MMP-14. MMP-14 has been shown to be involved in GPCR-mediated transactivation of multiple RTKs, such as EGFR, VEGFR, and PDGFR (Chaplin et al. 2017). The activation of MMPs or ADAMs happens as a consequence of GPCR activation, although specific mechanisms, which involve Gβγ subunits (Overland and Insel 2015), β-arrestins (Tilley et al. 2009; Oligny-Longpré et al. 2012), or Src (Almendro et al. 2010) are not fully understood (Kilpatrick and Hill 2021).

Transactivation can also occur through ligand-independent mechanisms, in which RTK is activated without RTK ligands. Signaling kinases activated after GPCR activation, such as Src, can directly activate RTKs by phosphorylating tyrosine residues in the C terminus. Furthermore, second messengers generated through effector protein-mediated signaling (e.g., reactive oxygen species (ROS) produced by NADPH oxidase) can also facilitate the direct activation of RTKs (Kilpatrick and Hill 2021).

The best-studied and characterized GPCR-mediated RTK transactivation is that of the EGF receptor family, which has been demonstrated with class A and class B GPCRs, such as adrenergic receptors (β1-, β2-, and α-ARs), adenosine receptors, lysophosphatidic acid (LPA) receptors (LPARs), μ opioid receptors, muscarinic M1 receptors, endothelin receptors, and angiotensin II (Ang II) type 1 receptor ($AT_1R$) (Prenzel et al. 1999; Liebmann 2011). In addition, evidence of transactivation has been observed in a range of cell types for other RTK families, which include the VEGFRs and PDGFRs, as mentioned above, but also FGFRs, MET, IGF-1R-1, and IR (Kilpatrick and Hill 2021) (Fig. 2).

It is well recognized that the crosstalk between GPCRs and RTKs is a crucial signaling mechanism implicated in cancer growth, angiogenesis, and metastasis. In this context, GPCR-mediated EGFR transactivation through MMP-mediated release of EGF-like ligands has been identified as a significant contributor to cancer progression and metastasis in various malignancies, including lung, breast, prostate, and ovarian cancers (Lappano and Maggiolini 2011; Lappano et al. 2013).

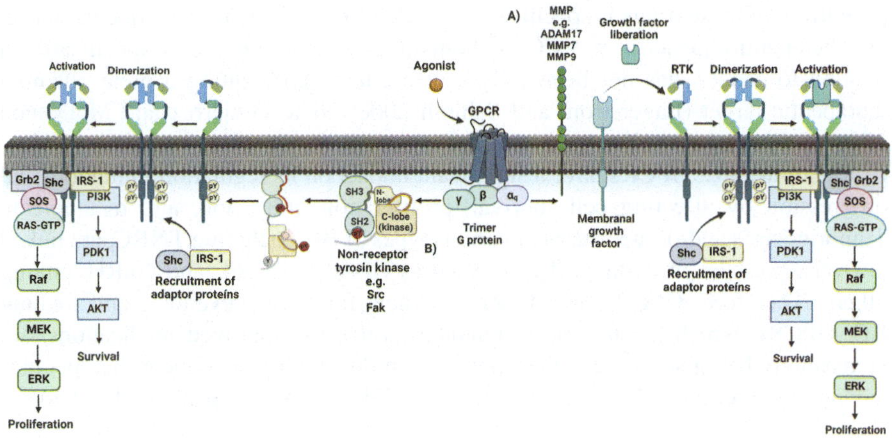


**Fig. 2** Mechanism of transactivation of RTKs with GPCRs. There are two mechanisms: (1) direct: it involves an interaction of an agonist toward a specific GPCR, after which different matrix metalloproteases (MMPs) are activated, which give rise to the release of ligands such as EFG, which can activate different RTKs. Regarding the above, many GPCR agonists, such as angiotensin and endothelin, among others, induce RTK transactivation. (2) indirect: this mechanism is carried out independent of ligand-binding, triggered by the intercellular tyrosine kinase activity induced by GPCR signaling

## 5.3 GPCRS in Cancer and BC

The extensive diversity of functions regulated by GPCRs has been linked to the development of various diseases, including cancer. In this context, GPCRs have been associated with angiogenesis, metastasis, tumor development, invasion, migration, and survival of several types of cancer, such as colorectal, lung, prostate, brain, and BC (Lappano et al. 2018). Among them, chemokine receptors, protease-activated receptors, ERs, angiotensin receptors, prostaglandin receptors, and receptors for bioactive lipids such as LPA and sphingosine-1-phosphate (S1P) are implicated in aberrant cell proliferation in a wide range of cancer cells, including BC (Dorsam and Gutkind 2007; Lappano et al. 2018).

In addition to normal physiological processes in which GPCRs are involved, aberrant expression of GPCRs and their ligands has been demonstrated to be tightly associated with various pathophysiological processes and disease conditions. For instance, many of the GPCR-regulated signal transduction pathways are critical for tumor initiation and progression, which includes cellular processes such as proliferation, apoptosis, stress signaling, immune evasion, invasion, angiogenesis, and metastasis (Segovia-Mendoza et al. 2022; O'Hayre et al. 2014; Nieto Gutierrez and McDonald 2018).

Because GPCRs regulate a wide range of physiological and pathophysiological responses and are considered valuable targets for therapeutic treatment, they have become an effective protein class of pharmaceutical targets. While approximately 30–40% of existing medications are designed to target GPCRs, only a small proportion of these drugs is specifically intended for use in cancer treatment. Hence, the therapeutic capacity of GPCRs, both in a general sense and specifically in relation to cancer, has not been fully utilized, thus representing a great potential therapeutic target (Lagerstrom and Schioth 2008; Nieto Gutierrez and McDonald 2018).

Regarding BC, GPCRs have a significant impact on its regulation, initiation, and progression by affecting cell survival, proliferation, migration, and metabolism. Tumorigenesis and/or metastasis in various types of BC, including TNBC, are linked to overexpression of certain GPCRs, such as GPER, CXCR4, and GnRH, among others. Therefore, GPCRs are attractive targets for both preventing and treating BC. GPCRs, which are the most abundant targets for approved medications, are increasingly being studied for their potential in developing anticancer therapeutics, having the potential to be useful targets for BC treatment (Lappano et al. 2018).

## 5.4 RTK Transactivation by Specific GPCRs in BC

Although many GPCRs have been identified in BC with important implications in its development, in the present review, we focus on those in which the transactivation of RTKs is a critical mechanism for their effects in this type of cancer.

### 5.4.1 Estrogens/GPER1

17β-estradiol (estrogen, E2), through receptor binding and activation, is necessary for the development of the mammary gland, promoting the growth of epithelial cells, ductal elongation, and morphogenesis. In addition to its role in normal breast development, E2 stimulates cell proliferation in tumorigenic environments, particularly BC. Classically, E2 binds to and activates ERα and ERβ, which regulates the transcription of genes involved in numerous physiological functions (Lappano and Maggiolini 2011). In this context, the proliferative effects of E2 in normal breasts and breast tumors are particularly linked to the ERα. Nevertheless, diverse in vitro studies have shown that the G-protein-coupled estrogen receptor 1 (GPER1), previously known as GPR30, can modulate proliferation in BC cells in different ways depending on the specific cell conditions (Scaling et al. 2014; Prossnitz and Barton 2023). GPER1 is a member of the GPCR superfamily, present in the plasma membrane, endoplasmic reticulum, and nuclear membranes (Leiber et al. 2015), which participates in mediating rapid non-genomic effects of E2 to regulate $Ca^{2+}$ mobilization, cAMP production, and ERK1/2/MAPKs and PI3K/Akt signaling pathways and modulate the progression of various hormone-responsive tumors such as endometrial, ovarian, and BC (Lappano et al. 2013; Arterburn and Prossnitz 2023).

Different studies have consistently shown that GPER1 effects in various cancer types, including BC, depend on distinct RTK transactivation. It has been shown that E2/GPER1 triggers rapid effects by activating numerous transduction signaling, including those that rely on EGFR transactivation. For instance, Filardo et al. demonstrated that in MDA-MB-231 BC cells ($ER\alpha^{-}$, $ER\beta^{-}$), E2/GPER activates the ERK1/2/MAPK pathway through EGFR transactivation. In the proposed mechanism, the βγ subunit of the Gαi-protein activates the Src/Shc proteins, which in turn triggers MMPs to cleave the pro-HB-EGF to generate HB-EGF, which then promotes the EGFR activation, resulting in the MAPK/ERK1/2 signaling pathway activation (Filardo et al. 2000). Additionally, it was demonstrated that in SKBr3 cells ($ER\alpha^-$, $ER\beta^-$, $GPER1^+$), the GPER1-mediated EGFR transactivation causes a rapid $Ca^{2+}$ mobilization and PI3K activation upon E2 stimulation (Revankar et al. 2005; Ariazi et al. 2010; Vajaria and Vasudevan 2018). These findings indicate that GPER1 may be necessary for the growth and survival of $ER^-$ breast tumors that still respond to estrogen through GPER-dependent activation of EGFR.

An interesting finding is the fact that EGFR transactivation mediated by E2/GPER1 is carried out indirectly through sphingosine kinase (SK) and integrin-α5β1, which act as mediators of transactivation in BC cells (Sukocheva et al. 2006; Quinn et al. 2009). The first study demonstrated that E2/GPER-induced EGFR transactivation in human BC cells is mediated by the lipid kinase sphingosine kinase-1 (Sph1K). Activation of Sph1K by E2 promotes S1P release, which, acting in a paracrine and autocrine manner, activates the S1P3 GPCR receptor (Edg3), resulting in the transactivation of EGFR in an MMP-dependent fashion (Sukocheva et al. 2006). The second study showed that when E2/GPER1 activates Src kinase, it causes the formation of Shc/integrin-α5β1 complex. These complexes help assemble the fibronectin matrix, trigger MMP-3, and potentially release EGF-like ligands, such as HB-EGF, to activate EGFRs (Quinn et al. 2009).

On the other hand, it has been shown that there is a reciprocal relationship between EGFR and GPER1 that involves the upregulation of GPER1 expression (Albanito et al. 2008; Ruan et al. 2012). EGF/EGFR stimulates GPER expression by rapidly activating ERK1/2 phosphorylation and inducing c-fos expression, which binds to the AP-1 site within the GPER1 promoter sequence. The EGFR-MAPK pathway activation creates a loop that involves estrogen through GPER1, enhancing the proliferation of breast tumor cells. This upregulation of GPER by EGF signaling sheds light on the known interaction between estrogen and EGF, which is linked to BC progression. Therefore, EGF may boost E2's effects by increasing GPER1 expression in $ER^-$ and $ER^+$ BC cells (Albanito et al. 2008; Vivacqua et al. 2009). Likewise, EGFR has been implicated in the upregulation of GPER1 expression under hypoxic conditions in BC cells and CAFs, indicating a potential role for GPER1 in hypoxia-induced angiogenesis (Recchia et al. 2011; De Francesco et al. 2013).

It was also reported that GPER1 activation promotes the upregulation of β1-integrin in MCF-7 cancer cells, which plays a critical role in tumor progression and survival. The effect was mediated by GPER1-mediated EGFR transactivation and activation of the ERK pathway, which leads to transcriptional regulation, cell migration, and EMT program activation (Yuan et al. 2015). Abnormal EMT activation is known to drive metastasis initiation through various mechanisms, including acquiring stemness properties.

The above evidence highlights the capacity of GPER1 to activate EGFR signaling, which falls within the general concept of GPCR-mediated RTK transactivation

(Kilpatrick and Hill 2021). However, diverse studies have revealed a connection between GPER1 and estrogenic signaling with IGF-1R-mediated action in the BC. De Marco et al. (2013) showed that IGF-1/IGF-1R upregulates GPER1 mRNA and protein levels through the PKCδ/ERK/c-fos/AP1 pathway, resulting in migratory effects in MCF7 BC cells and demonstrating reciprocal regulation of GPER1 by RTKs (De Marco et al. 2013). Furthermore, zinc chloride (ZnCl2) effects involve GPER1 in the transactivation of IGF-1R/EGFR-mediated signaling, which in turn triggers downstream pathways like ERK and Akt in CAFs and BC cell migration (Pisano et al. 2017). It is important to note that the interaction and crosstalk between GPER1 and IGF-1R have been mainly detected in cancer cells. A notable example of this difference is seen in the role of estradiol in neuroprotection. While the significance of the interaction between ERα and IGFR for E2-mediated neuroprotection has been established (Garcia-Segura et al. 2007; Sohrabji 2015; Vajaria and Vasudevan 2018), there is currently no evidence of such crosstalk between GPER1 and IGFR in the brain, although the role of E2/GPER1 in neuroprotection has been well documented (Liu et al. 2012a, b; Vajaria and Vasudevan 2018).

In BC-associated CAFs, an interesting relationship between GPER1, growth factors, and RTKs has been observed, which seems important for creating a pro-tumorigenic environment and breast tumor progression. For instance, both E2 and G-1 have been found to trigger heightened expression of HIF-1α and the angiogenic VEGF in BC CAFs, with the latter process being reliant on HIF-1α (De Francesco et al. 2014). This increased VEGF expression, and its action through its receptor supports tumor angiogenesis, indicating that GPER might play a role in developing a pro-tumorigenic environment (Shaw et al. 2024). Recent research has demonstrated that IGF-1/IGF-1R induces the upregulation of GPER1, HIF-1α, and VEGF in breast CAFs. Notably, the suppression of GPER1 or HIF-1α eliminated the observed increase in VEGF caused by IGF-1, providing further evidence of GPER's involvement in a CAF-induced tumor microenvironment that promotes blood vessel growth (De Francesco et al. 2017). Additionally, in $ER^{-}$ primary CAFs, both E2 and the GPER agonist G-1 increased the production and release of FGF2 through GPER1, mediated by the EGFR/ERK/c-fos/AP-1 signaling pathway. Notably, the FGF2 released from E2/GPER-stimulated CAF cells, and acting through FGFR1, triggered the expression of CTGF, which led to the migration and invasion of MDA-MB-231 cells (Santolla et al. 2019; Santolla and Maggiolini 2020). It is important to highlight that elevated FGF2 levels have been detected in the blood of patients with BC, particularly when metastasis is present (Brown et al. 2016; Santolla et al. 2019).

### 5.4.2 Chemokine Receptors

Chemokines are small signaling proteins (10–8 kDa) that belong to the superfamily of cytokine-like proteins. They interact with the class A rhodopsin family of GPCRs and play crucial roles in immune cell attraction and migration during wound healing, infections, and cancer progression. Based on their cysteine motif composition, the chemokine family is categorized into several groups: C–C, CXC, CX3C, or XC. Of these, the C–C chemokines specifically impact the functioning of T cells and macrophages (Acevedo et al. 2022). Currently, over 20 human chemokine receptors are categorized into four families based on the ligands they bind: CCR, CXCR, CX3CR, and XCR (Liu et al. 2020a, b, c).

The majority of chemokine-mediated responses are prevented by pertussis toxin, indicating that the Gi family plays a role in initiating signal transduction downstream of chemokine receptors, which classically leads to a decreased activity of the PKA as a consequence of the AC inactivation. However, chemokine receptors can also couple to Gq/11 in a cell- and stimulus-dependent manner, which activates PLC-β to produce IP3, essential for intracellular $Ca^{2+}$ release. This G-protein-signaling module is critical for cell chemotaxis and activation of PI3K-mediated pathways (Lacalle et al. 2017).

Chemokines and chemokine receptors are involved in the growth, invasion, and proliferation of tumor cells by inducing the MAPK/ERK signaling pathway, promoting the expression of genes that stimulate growth, such as cyclin D1, oncogene FOS, and human HB-EGF, and increasing the expression of the antiapoptotic gene MDM2. Furthermore, chemokines and chemokine receptors negatively regulate the expression of Bcl-2 and the activation of caspase-3 and caspase-9, further promoting tumor cell proliferation. Moreover, they raise leukocyte activation in tumor tissue to secrete factors such as IL-10, TGF-β, MMPs, growth factors, and angiogenic factors, which are vital in promoting tumor growth, invasion, and immunosuppression (Lacalle et al. 2017; Liu et al. 2020a, b, c).

Next, we address the main chemokines and their receptors related to the development of breast cancer, although different studies have revealed that CXCR4 and CXCR7 are the most commonly upregulated chemokine receptors in breast cancer cells compared with normal breast epithelial cells.

CXCR4 is a GPCR that specifically binds to the chemokine CXCL12, also known as stromal-derived factor 1α (SDF-1α). CXCR4 activation has been linked to the regulation and activation of the immune system and other processes such as morphogenesis, angiogenesis, and cell survival (Pozzobon et al. 2016). Ligand-mediated CXCR4 activation leads to the coupling and activation of Gi proteins, which inhibit adenylyl cyclase, reducing cAMP production. This, in turn, stimulates the activity of Src family tyrosine kinases, which activate the Ras/Raf/MEK/ERK pathway. This pathway modulates cell cycle progression by phosphorylating the adaptor protein Shc. CXCR4-dependent cell migration is mediated by the PI3Ks, which are activated by dimer Gβγ. Once active, PI3Ks control gene transcription, cell migration, and cell adhesion by activating Akt and multiple focal adhesion components (Pozzobon et al. 2016). Furthermore, the Gβγ dimer can trigger PLC activation, which catalyzes the breakdown of PIP2 into two second messengers, IP3 and DAG. IP3 triggers $Ca^{2+}$ release from intracellular stores, while DAG activates PKC and MAPK, which are involved in various cellular signaling pathways (Pozzobon et al. 2016).

CXCR4 is one of the most studied chemokine receptors due to its role in mediating metastasis, tumor growth, cell survival, and chemotaxis in various cancers, including BC (Balkwill 2004). Research reveals that BC tissues and cell lines overexpress CXCR4 and CXCL12, while normal breast tissue does not exhibit these expressions (Müller et al. 2001). Additionally, CXCL12 is present in organs where BC commonly spreads (Müller et al. 2001). In mouse models, treating with neutralizing antibodies against CXCR4 inhibits metastasis, highlighting the significance of CXCR4 in cancer progression (Müller et al. 2001; Smith et al. 2004).

CXCR4 can coordinate multiple signaling network changes to facilitate the transformation of mammary epithelial cells. These changes include transactivation and upregulation of several RTKs, disruption of the p53/MDM2 axis, increased expression of E-cadherin and c-myc, and modulation of cell cycle molecules (Su et al. 2011). Overexpression of CXCR4 in MCF10A cells led to overexpression and activation of MET, IGF-1R, ErbB2, and EGFR, which are associated with a more invasive and malignant phenotype (Su et al. 2011). An increase in c-MET due to CXCR4 overactivity could indicate a potential mechanism that triggers key cancer-causing pathways. This might also be linked to CXCR4's role in metastatic BC, as the MET pathways encourage cell separation through metalloprotease production, often leading to metastasis (Camp et al. 1999; Su et al. 2011).

In addition to promoting the overexpression of some RTKs in cancer cells, initial research indicates the existence of crosstalk between CXCR4 and RTKs. The presence of CXCR4 plays a significant role in the dissemination of cancer through ErbB-receptor pathways, particularly in ErbB2-related BC, whose strong connection with CXCR4 in human breast tumors has been reported. Early research by Cabioglu et al. showed that CXCR4 activation by CXCL12 increases ErbB2 signaling in BC cells through an Src-dependent transactivation mechanism (Cabioglu et al. 2005). It has also been found that, reciprocally, ErbB2 increases the expression of CXCR4, which is a significant mechanism in the progression of BC, as this increase in CXCR4 is necessary for ErbB2-mediated metastatic colonization (Li et al. 2004). Furthermore, ErbB2 inhibits the rate of CXCR4 degradation induced by its ligand. Thus, cancer cells that overexpress ErbB2, such as BC cells, also maintain a high level of CXCR4 expression, contributing to invasive processes, particularly in the lung (Li et al. 2004).

An interesting finding demonstrates the importance of ErbB2 transactivation mediated by CXCL12/CXCR4. It shows that cell proliferation promoted by E2 in BC cells relies on the production of CXCL12 and its autocrine/paracrine actions through CXCR4, leading to ErbB2 transactivation. Thus, CXCR4 activation is an intermediate signal that transduces signals from E2 to ErbB2 (Pattarozzi et al. 2008).

Another study demonstrated that the ErbB3 receptor, when activated by heregulin, also induces the upregulation of CXCR4 through transcriptional regulation by HIF-1$\alpha$ in BC cells, promoting increased cell motility via the P-Rex1/Rac1 signaling pathway when activated by CXCL12 (Lopez-Haber et al. 2016). Previous studies have shown that heregulin stimulation of ErbB3 can transactivate CXCR4, independently from its ligand, increasing its phosphorylation on Tyr and Ser residues and rapid recruitment of $\beta$-arrestin to the receptor (Sosa et al. 2010). These findings reveal a bidirectional relationship between specific transmembrane receptors involved in tumorigenesis and cancer spread, such as GPCRs and RTKs (Sosa et al. 2010; Lopez-Haber et al. 2016).

Other reported crosstalks between CXCR4 and RTKs, important for the growth and metastasis of BC, involve MET, IGF-1Rs, PDGFRs, and VEGFRs. In this context, it was demonstrated that HGF/MET induces CXCR4 expression in MCF7 cells through a mechanism that involves the transcription factors Ets1 and NF-κB (Maroni et al. 2007). Moreover, a different mechanism was proposed to mediate the expression of CXCR4 by HGF/MET, which involves PI3K and PKCζ (Huang et al. 2012). However, in both cases, the CXCR4 upregulation contributes to the tumor invasiveness/metastatic phenotype of breast carcinoma (Maroni et al. 2007; Huang et al. 2012).

On the other hand, it was proposed that CXCR4 constitutively forms a complex with the IGF-1R in MDA-MB-231 cells, an interaction that allows IGF-1 to activate migrational signaling pathways through CXCR4, Giα2, and Gβγ in the absence of CXCL12 (Akekawatchai et al. 2005; Pyne and Pyne 2011).

Liu et al. demonstrated that PDGFR-β-activation by PDGF-D, which is highly expressed in human BC, promotes lymphatic metastasis through increased CXCR4 expression and activation (Liu et al. 2011b). Additionally, the CXCR4/CXCL12 signaling axis can induce angiogenesis and tumor progression by increasing VEGF expression via the PI3K/Akt pathway (Liang et al. 2007).

The CXC chemokine receptor 7 (CXCR7), also known as atypical chemokine receptor 3 (ACKR3), is expressed in $ER^+$ and $ER^-$ BC cell lines. CXCR7 is also differentially expressed in normal breast tissues but in higher abundance in $ER^+$ cancer tissues. Knocking down CXCR7 in MCF7 cells reduced proliferation and cell cycle arrest. Through proximity ligation assays, it was found that CXCR7 and EGFR colocalize in both tissues and cancer cell lines. Interestingly, depletion of CXCR7 reduced levels of tyrosine-phosphorylated EGFR after EGF stimulation and phosphorylation of ERK1/2, indicating potential transactivation of EGFR by CXCR7s, which impacts mitogenic signaling in MCF7 cells. Depletion of β-arrestin 2 in EGFR-overexpressing cells led to decreased CXCR7-EGFR colocalization, suggesting that β-arrestin 2 acts as a scaffold to enhance CXCR7-dependent activation of EGFR after EGF stimulation (Salazar et al. 2014). A recent discovery revealed that CXCR7, together with CXCR4 and EGFR/ErbB2, carry out crosstalk in MDA-MB-361 BC cells (a luminal B subtype model with ErbB2 overexpression) to simultaneously promote ERK1/2 activation. This mechanism may be relevant to integrate concurrent signals from the tumor microenvironment and forward cancer progression (Neves et al. 2022).

Chemokine C-C ligand 2 (CCL2), also known as monocyte chemoattractant protein (MCP-1), plays a crucial role in breast cancer progression. It is involved in attracting macrophages and endothelial cells in cases of acute inflammation and cancer progression. CCL2 mainly binds CCR2 to perform this function. Interestingly, BCs that exhibit higher levels of CCL2 are also associated with elevated macrophages. High stromal expression of CCL2 is an unfavorable prognosis indicator for patients with BC. Inhibiting the activity of CCL2 has been observed to prevent the development, survival, and invasion of breast tumor xenografts by decreasing the recruitment of macrophages (Hao et al. 2020). Recent studies have shown that enhanced CCL2/CCR2 signaling in breast epithelial cells is often associated with forming DCIS, the most common early-stage BC (Acevedo et al. 2022). It is interesting to note that CCL2/CCR2 induces MET receptor tyrosine phosphorylation and activation in BC cells. The activity of MET receptors depends on interactions between Src, CCR2, and MET. Thus, CCL2/CCR2 MET-mediated transactivation regulates ductal BC's development, invasion, and metabolism (Acevedo et al. 2022).

Another receptor, CX3CR, which responds to the chemokine CX3CL, has also been implicated in the diversity of breast tumors and EMT. In part, these effects have been attributed to the transactivation of EGFRs through the proteolytic shedding by MMPs of an EGFR ligand, such as amphiregulin, HB-EGF, or TGF-α, leading to ERK pathway activation and cell proliferation (Tardáguila et al. 2013).

### 5.4.3 Cannabinoid Receptor 2

Cannabinoids are active components of cannabis, which have been found to promote antitumor effects in $HER2^+$ BC. These effects occur when cannabinoid receptors, $CB_1$ and $CB_2$ receptors ($CB_1R$ and $CB_2R$), belonging to the GPCR superfamily, are activated. When both receptors are activated by ligands, they are primarily coupled to the Gi proteins, which leads to the inhibition of AC and a decrease in cAMP levels (Dhopeshwarkar and Mackie 2014). While $CB_1R$ is responsible for the psychoactive effects of cannabis and is widely distributed throughout the central nervous system and the body, $CB_2R$ is mainly found in the immune system of healthy individuals. However, studies suggest increased $CB_2R$ expression in various diseases, including BC (Blasco-Benito et al. 2019; Kisková et al. 2019).

$CB_2R$ expression was found to be highly expressed in tumor breasts, correlating with the tumors' histologic grade and aggressiveness. In contrast, $CB_1R$ mRNA levels were lower in low-medium and high histologic-grade tumors than in normal, noncancerous breast tissue. Moreover, $CB_2R$ expression levels were almost imperceptible in healthy breast tissue (Caffarel et al. 2006; Kisková et al. 2019).

It is worth noting that tumors that are positive for ErbB2/HER2 express higher levels of $CB_2R$ mRNA compared to tumors that are negative for ErbB2/HER2 (Caffarel et al. 2006; Kisková et al. 2019). The increased $CB_2R$ expression in $HER2^+$ breast tumors is linked to a higher likelihood of local recurrence and development of metastases, which leads to a poor prognosis for patients. The reason behind the elevated $CB_2R$ expression is HER2 upregulates $CB_2R$ expression by activating ELK1 through the ERK signaling pathway (Pérez-Gómez et al. 2015). Interestingly, the subsequent increase in $CB_2R$ promotes the formation of the heterodimer HER2/$CB_2Rs$ that activate Src, which mediates pro-oncogenic signaling that also correlates with poor patient prognosis. Disrupting the HER2/$CB_2R$ heterodimer can promote antitumor responses (Pérez-Gómez et al. 2015; Blasco-Benito et al. 2019). The above findings suggest that simultaneous targeting of the cannabinoid receptor $CB_2$ and HER2 may be an effective therapeutic strategy.

### 5.4.4 Lysophosphatidic Acid (LPA) Receptors (LPARs)

LPA controls several biological processes, including cell shape, motility, and proliferation, when they bind to six possible LPA receptors (LPARs), designated $LPA_{1\text{-}6}$, which belong to the GPCR superfamily. LPARs are coupled to multiple Gα protein subunits, particularly Gαi/o, Gαq, and Gα12/13, to regulate cell proliferation, migration, and survival associated with cancer progression (Balijepalli et al. 2021).

Hopkins et al. demonstrated that LPA and EGF stimulate proliferation, ERK and Akt activation, and CCN1 induction in human BC cell lines. CCN1 is an inducible matricellular protein whose expression positively correlates with breast cancer progression. Moreover, LPA antagonists were shown to inhibit the effects of both LPA and EGF on MCF-7 and MDA-MB-231 cell migration and proliferation, indicating a possible crosstalk between EGF and LPA that mediates the actions of EGF in BC cells (Hopkins et al. 2016). By activating $LPA_1R$, LPA causes an increase in the expression of HB-EGF in BC cellular models. This is similar to what was found in patients, showing high $LPA_1R$ and HB-EGF expression levels in the breast tumors, so it has been proposed that HB-EGF be a biomarker for $LPA_1R$ activation in human BC (David et al. 2014). These results may suggest a hyperactivation of EGFR-mediated signaling pathways through LPA/$LPA_1R$ in BC. Previous research has shown that LPA can also transactivate EGFR through a direct mechanism involving Gβγ and Src-mediated EGFR phosphorylation on Tyr residues (Prenzel et al. 2000). It has been shown that LPA increases the expression of VEGF-C mRNA in vascular endothelial cells, leading to the formation of endothelial cell tubes, effects that are $LPA_1R$- and $LPA_3R$-dependent and requires COX-2, EGFR transactivation, and activation of NF-κB (Lin et al. 2008). Interestingly, COX-2 activation is highly correlated with VEGF-C expression and secretion in human BC, contributing to the formation of new lymph vessels, a relevant process for cancer development (Timoshenko et al. 2006).

### 5.4.5 Protease-Activated Receptors

Protease-activated receptors (PARs) are a family of GPCRs activated by serine proteases such as thrombin, trypsin, tryptase, and plasmin, which cleavage part of their amino-terminal region to act as ligands on the receptor. Four PARs, named PAR1-4, have been identified as critical regulators of vascular physiology and inflammation (Peach et al. 2023).

It has been shown that overexpression of PAR1 is associated with growth, cell invasion, tumor progression, and the development of metastases in breast cancer (Dorsam and Gutkind 2007). Regarding the molecular mechanism related to PAR1 involvement in BC, it was demonstrated that regulation of endogenous PAR1s is altered in metastatic breast carcinoma cells, such that during trafficking of the activated receptor, it is not sorted to lysosomes and degraded (Booden et al. 2004).

In fact, it has been described that protease-activated receptor 1 (PAR1) activation by thrombin induces a persistent EGFR/HER2 transactivation through ERK1/2 phosphorylation in triple-negative invasive breast cancer cells. The mechanisms involved in this transactivation are the Gαi/o and metalloproteases protein activation (Arora et al. 2008). Interestingly, it is highlighted that the authors also found a direct participation of PAR1 with fibroblast function; this gives rise to thinking about a paracrine regulation between breast cancer cells and fibroblasts.

Interactions between GPCRs and RTKs serve as alternative pathways involved in proliferative processes or as resistance mechanisms that are worth exploring or contemplating in different types of breast cancer. However, the area of study of transactivation between these receptors has not yet been explored in depth.

## 6 Blocking GPCRS-RTKs Transactivation, Therapeutic Strategies with a Clinical Approach

It is widely known that different target therapies for RTKs, including monoclonal antibodies and small molecules known as tyrosine kinase inhibitors, have a significant response in BC patients (Segovia-Mendoza et al. 2015). Currently, at the clinical level, no combined therapies block the transactivation between RTKs and GPCRs. Nevertheless, recent articles evaluate dual receptor inhibition due to the cellular importance and pro-tumor effects of this protein interaction in BC. Below, some strategies that have been proposed at a basic level and their significance in the clinic are described.

In general, transactivation between GPCRs and RTKs occurs in lipid domains of the plasma membrane, termed lipid rafts, and there are two pathways in which transactivation can occur, one from GPCRs to RTKs and vice versa (Ostrom and Insel 2004; Kilpatrick and Hill 2021).

The first mechanism involves an interaction of an agonist toward a specific GPCR, after which different matrix metalloproteases (MMPs) are activated, which gives rise to the release of ligands such as HB-EGF, which can activate different RTKs. Regarding the above, many GPCR agonists, such as angiotensin and endothelin, among others, induce RTK transactivation (Heeneman et al. 2000). A second mechanism is the RTK transactivation independent of ligand-binding, triggered by the intercellular tyrosine kinase activity induced by GPCR signaling. Similarly, RTK-GPCR transactivation occurs in a ligand-dependent and -independent manner, where the ligand/RTK-induced transcription enhancement for a specific GPCRs can occur, but also the transactivation can promote the recruitment of a particular G protein (Gs, Gi, or Gq/11), giving a hybrid activation (Crudden et al. 2018).

Regarding the above, some reports have exposed that the transactivation between GPCRs and RTKs needs different adapter proteins, such as GTPases or β-arrestins, which lead to receptor internalization and their redistribution into the cell membrane (Mores et al. 2019). Recent studies have focused on testing a dual inhibitor of the Ras and ARF6 GTPases, known as Rasarfin, shown to block GPCR internalization (taking the angiotensin II type 1 receptor ($AT_1R$) as a prototype) through the inhibition of recruitment of clathrin adaptor AP-2 to β-arrestin 2 as well as the subsequent MAPK activation. These effects were carried out by the binding of Rasarfin to the Ras-binding pocket in a specific hydrophobic groove, which resulted in the conformational modifications of the Ras protein and avoided both GPCR internalization and the subsequent ERK1/2 activation in human embryonic kidney 293T (HEK293T) cells (Giubilaro et al. 2021).

It is worth mentioning that some strategies for arrestins blocking could be considered strategic targets for the joint block of GPCRs and RTKs since they are regarded as important signaling scaffolds and adaptors proteins that promote the desensitization, sequestration, and recycling of different receptors (Suleymanova et al. 2017). The above is based on research evidence that exposes that arrestins impact BC development (Shukla and Dwivedi-Agnihotri 2020). In relation to the above, chemokine receptor-7 (CXCR7)-mediated EGFR transactivation have been reported in BC cells; this interaction leads to the promotion of cell proliferation and cell-cell adhesion, as well as the activation of the anti-apoptotic signal, with a more significant predominance in $ER^+$ BC cell lines. Importantly, CXCR7 depletion reduces the levels of phospho-EGFR and phospho-ERK1/2 after EGF stimulation, suggesting a potentially direct impact on mitogenic signaling. Interestingly, using siRNA to knockdown β-arrestin 2 in BC cells with EGFR overexpression, the CXCR7-EGFR colocalization was also inhibited, suggesting that β-arrestin 2 acts as a scaffold protein to enhance CXCR7-dependent activation of EGFR after EGF-stimulation (Salazar et al. 2014). The previous results indicate that the blockade of β-arrestin 2 would be important to inhibit CXCR7-EGFR transactivation, presenting a particular focus on CXCR7 signaling (Fig. 3).

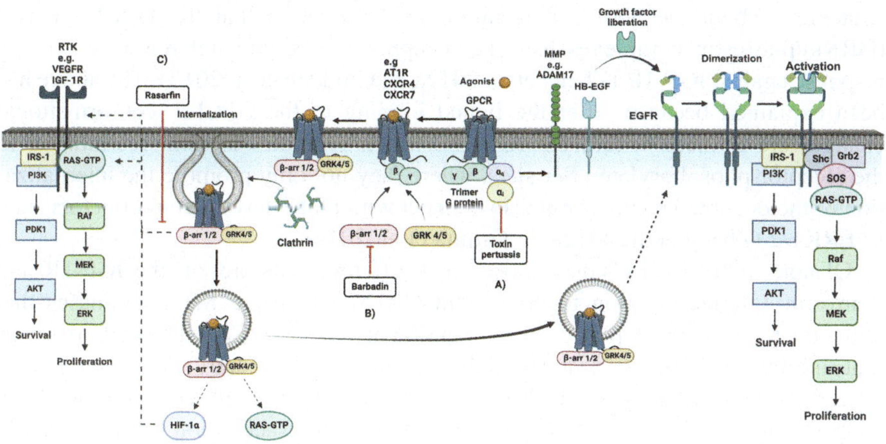


**Fig. 3** Therapeutic strategies to inhibit transactivation between GPCRs-RTKs. Therapeutic proposals for GPCRs-RTKs transactivation inhibition can be included in the following mechanisms: (a) Blocking the arrestins: this strategy is since arrestins are responsible for promoting receptor desensitization and recycling. Many GPCRs can interact with different RTKs through arrestins, which promote RTK phosphorylation. Furthermore, arrestins can activate different transcription factors, such as the HIF/VEGF axis, inducing cell metastasis. However, it is important to mention that arrestins have differential effects, therefore, it has been suggested to explore the affinity of arrestins for different types of RTKs. (b) Blocking GRK activity: GRKs are serine-threonine kinases that allow β-arrestin recruitment and inhibition of activated GPCRs. In addition, GRK inhibition has been significantly associated with PI3K/Akt and MAPK blocking activation by different RTKs such as IGF-1R, EGFR, or HER2

Likewise, it was reported that β-arrestin 1 is involved in the growth of metastatic BC due to its interaction with HIF-1α through the activation of VEGFR (Shenoy et al. 2012). Thus, the authors suggest that β-arrestin 1 regulates nuclear signaling during hypoxia to promote the survival of BC cells and breast tumor xenografts via VEGF signaling. The above correlates with elevated expression levels of β-arrestin 1 and VEGF-A in metastatic human BC tissues. However, the authors did not demonstrate what GPCR would be directly involved in the transactivation of β-arrestin 1 with HIF-1α and VEGFR. Interestingly, the authors found that VEGF-A inhibition by anti-angiogenic compounds results in β-arrestin 1/HIF-1α signaling disruption (Shenoy et al. 2012).

In addition, different studies have focused on the β-arrestin 1 role and its IGF-1R interaction, where there is a strong ERK1/2 activation and the subsequent receptor degradation. In contrast, the role of IGF-1R and its interaction with β-arrestin 2 has not been widely explored. Nevertheless, it is suggested that the interaction of IGFR-1 with β-arrestin 2 protects the IGF-1R degradation, induces weak ERK1/2 activation, and promotes cell proliferation after ligand binding (Suleymanova et al. 2017). The affinity of IGF-1R toward the different β-arrestin isoforms has been studied through co-immunoprecipitation studies; β-arrestin 1 presents more affinity when the receptor is occupied by its ligand, while β-arrestin 2 has more affinity when IGF-1R is not bound to IGF (Suleymanova et al. 2017). The above suggests a new way to explore the affinities of arrestins to different RTKs and their behavior in therapeutic strategies. About the above, it is important to mention that the GPCR kinases (GRKs)/β-arrestin complexes have been reported as significant regulators of the response against IGF-1R (Zheng et al. 2012a, b; Crudden et al. 2015). The above has been explained because, after the ligand binding to the GPCRs, conformational changes are induced within the IGF-1R, which promotes arrestin recruitment to the receptor-phosphorylated Ser sites, where they not only promote the internalization of the occupied receptor but also connect with other downstream effectors such as ERK1/2 (Zheng et al. 2012a, b; Girnita et al. 2016).

Of note, other works have exposed that different cells present the IGF-1R and β-arrestin 1 transactivation together with GRKs, which are Ser-Thr kinases that allow the recruitment of β-arrestins and inhibit activated GPCRs. The authors report that inhibiting GRK5/6 reduces IGF-1-mediated ERK and Akt activation in HEK293T cells (Zheng et al. 2012a, b). Although these effects have not been analyzed in BC cells, it could be assumed that this transactivation occurs similarly in BC cells since IGF-1R is significantly associated with cell survival processes in this type of neoplasia. Concerning the above, GRKs have been considered as negative regulators of GPCR activity, pointing out their role as therapeutic targets in cancer and different pathologies (Homan and Tesmer 2015; Yu et al. 2018). In addition, there is evidence that GRKs are overexpressed in BC (Nogues et al. 2016). In this context, Nogues et al. found that GRK2 expression up-regulation correlates with the activation of EGFR/HER2 through the PI3K/Akt signaling pathway (Nogues et al. 2016). Furthermore, the up-regulation of GRK2 increased its functionality, leading to the activation of Histone Deacetylase 6 (HDAC6) and Prolyl

Isomerase Pin1, both critical modulators of cell proliferation and tumor growth (Nogues et al. 2016) (Fig. 3).

Another piece of evidence proposes the GRK2 role in the CXCR4/EGFR transactivation in BC, which is related to metastasis (Mukherjee and Zhao 2013). It has also been shown that the CXCR4/IGF-1R transactivation impacts BC cells' migratory capacity. Interestingly, metastatic characteristics after IGF-1 stimulation were inhibited by the pretreatment of the cells with pertussis toxin and RNAi-mediated knockdown of CXCR4. Moreover, coprecipitation studies revealed the presence of a constitutive complex containing IGF-1R, CXCR4, and G protein subunits, Giα2 and Gβ, after the stimulation with IGF-1 in BC cells (Akekawatchai et al. 2005). In addition, the transactivation of EGFR by CXCR4 has also been demonstrated in lung cancer cells, where EGFR activation by EGF under hypoxic conditions causes a dramatic upregulation of CXCR4 expression and PI3K/Akt/PTEN/mTOR activation, promoting the migratory capacity of lung cancer cells. The above was evidenced by employing the PI3K inhibitors, wortmannin, and LY294002 (Phillips et al. 2005).

Less studied in BC is the transactivation of the vasoactive intestinal peptide (VIP) receptors, VPAC1, and VPAC2, with EGFR/HER2, which impacts both RTK protein level induction and their phosphorylation via Src activation. Interestingly, the employment of VIP receptor antagonists or Src kinase inhibitors blocked EGFR/HER2 phosphorylation (Valdehita et al. 2009). With the previous results, the authors propose using VIP antagonists in BC. More recently, scientific literature has also proven that VIP-mediated EGFR/HER2 transactivation promotes proliferation in different BC cell lines (Asano et al. 2023). The above data are relevant since there is no robust evidence about the effects of VIP on BC, and its functions as a neuropeptide, neuromodulator, and neurotransmitter have been largely explored over the years.

Besides, a new therapeutic strategy proposed against BC is the modulation of GPCR-mediated RTK transactivation through the use of biased ligands for the GPCRs. A biased ligand is defined as a chemical ligand that can induce a distinct and specific "active" receptor conformation for the receptor, and this specific conformation entails the activation of one specific subset of signal transduction (Wisler et al. 2018).

Some reports identified that different ligands exhibit bias on a functional selectivity for GPCR activation through specific G proteins or other signal transducers such as GRK and β-arrestin, causing the activation of different processes. The study of biased signaling by diverse pharmaceutical drugs involved in therapies against BC is new ongoing research. However, biased ligands have been investigated in other pathologies, like cardiovascular disease (Roy et al. 2017) (Fig. 4).

For example, the chemical compound TRV120027, which is a biased agonist for the $AT_1R$, has been used in acute decompensated heart failure, but it has also been implicated in promoting cardiac contractility, like in hypertension and chronic systolic heart failure; both processes are induced by the signaling of β-arrestin caused by TRV120027, which has caused its discontinuation in clinical trails. Meanwhile, Carvedilol, a drug that acts as an antagonism for the β-adrenergic, can

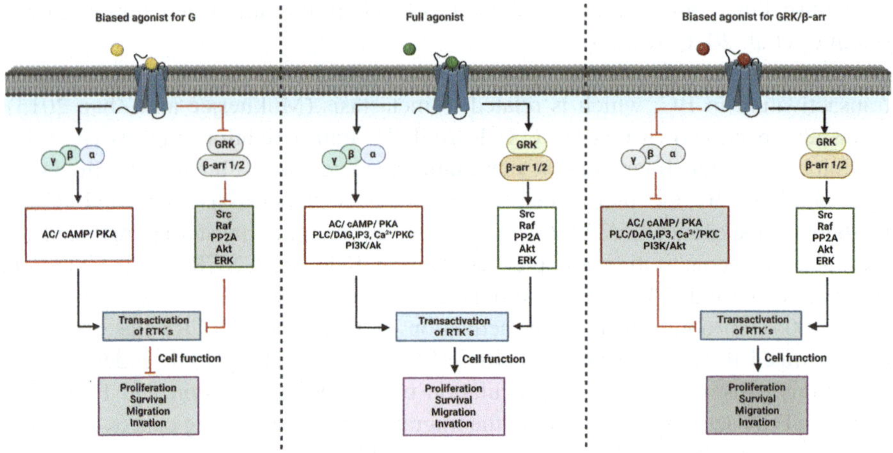


**Fig. 4** Molecular mechanisms of biased ligands. The biased ligands (left and right panels) are chemical compounds that modulate and activate a specific GPCR-dependent signaling pathway, in contrast to the balanced or full agonist actions (middle panel), which activate both G-protein- and GRK/β-arrestin-dependent signaling. Agonist activation of both cascades can modulate different responses. In the case of G protein signaling, the activation of a specific subtype of Gα protein, such as Gαs or Gαq, can upregulate the activity of effector proteins, such as adenylyl cyclase (AC) or phospholipase C (PLC); once effector proteins are activated, they produce second messengers, which in turn modulate the activity of second messenger activated protein kinases, such as PKA and PKC, which can phosphorylate proteins and regulate the proliferation and survivance of some tumors. On the other hand, signaling mediated by G-protein-independent signaling is mediated by GRKs/β-arrestins, in which β-arrestins participate in modulating different signaling pathways, thanks to its ability to act as a scaffolding protein, which recruits different signaling proteins that regulate distinct physiological outcomes. It is important to note that each independent signaling pathway (dependent on G proteins or GRKs/β-arrestins) is able to modulate the transactivation of RTKs, which represents a challenge for the design of biased agonists that can modulate the activity of specific RTKs involved in different types of cancers

mediate signaling by β-arrestin but not by signaling G protein; nevertheless, the clinical implications of this drug are not fully understood, but research on its effects is ongoing (Kim et al. 2008; Chaudhary and Kim 2021) (Fig. 4).

In the cancer field, research has been developed on biased molecule agonists for GPCR, like β adrenergic receptors, opioid receptors, angiotensin receptors, dopaminergic, and serotonergic receptors, and identifying the signaling pathway by either G proteins or arrestins. One of the approaches has been the development of recombinant proteins, like the recombinant wild-type chemokine CXCL12 in a monomeric and dimeric form can interact with the CXCR4 and inhibit the metastasis in a model of colorectal cancer, which can be explored in a BC model (Drury et al. 2011; Takekoshi et al. 2012).

In summary, the GPCR-RTK interaction could be assumed to be a promising target therapy in BC. However, personalized studies would have to be carried out to achieve maximum effectiveness, and we would consider arrestins or GRKs regulation by biased agonists as candidate strategies to inhibit GPCR-RTK transactivation induced by GPCRs.

**Acknowledgments** The authors thank the Center for Research and Advanced Studies of the National Polytechnic Institute (Cinvestav-IPN) for the support received (to JAO-R) and CONAHCYT for scholarship grant No: 1073659 (to EA-R). MS-M thanks to the Faculty of Medicine, UNAM for the support provided.

## References

Acevedo DS, Fang WB, Rao V, Penmetcha V, Leyva H, Acosta G, Cote P, Brodine R, Swerdlow R, Tan L, Lorenzi PL, Cheng N (2022) Regulation of growth, invasion and metabolism of breast ductal carcinoma through CCL2/CCR2 signaling interactions with MET receptor tyrosine kinases. Neoplasia 28:100791

Acuner Ozbabacan SE, Gursoy A, Nussinov R, Keskin O (2014) The structural pathway of interleukin 1 (IL-1) initiated signaling reveals mechanisms of oncogenic mutations and SNPs in inflammation and cancer. PLoS Comput Biol 10:e1003470

Adams J, Carder PJ, Downey S, Forbes MA, MacLennan K, Allgar V, Kaufman S, Hallam S, Bicknell R, Walker JJ, Cairnduff F, Selby PJ, Perren TJ, Lansdown M, Banks RE (2000) Vascular endothelial growth factor (VEGF) in breast cancer: comparison of plasma, serum, and tissue VEGF and microvessel density and effects of Tamoxifen. Cancer Res 60:2898–2905

Akekawatchai C, Holland JD, Kochetkova M, Wallace JC, McColl SR (2005) Transactivation of CXCR4 by the insulin-like growth factor-1 receptor (IGF-1R) in human MDA-MB-231 breast cancer epithelial cells. J Biol Chem 280:39701–39708

Al Kawas H, Saaid I, Jank P, Westhoff CC, Denkert C, Pross T, Weiler KBS, Karsten MM (2022) How VEGF-A and its splice variants affect breast cancer development – clinical implications. Cell Oncol 45:227–239

Al-Nedawi K, Meehan B, Kerbel RS, Allison AC, Rak J (2009) Endothelial expression of autocrine VEGF upon the uptake of tumor-derived microvesicles containing oncogenic EGFR. Proc Natl Acad Sci USA 106:3794–3799

Albanito L, Sisci D, Aquila S, Brunelli E, Vivacqua A, Madeo A, Lappano R, Pandey DP, Picard D, Mauro L, Andò S, Maggiolini M (2008) Epidermal growth factor induces G protein-coupled receptor 30 expression in estrogen receptor-negative breast cancer cells. Endocrinology 149: 3799–3808

Aljada A, Saleh AM, Al-Aqeel SM, Shamsa HB, Al-Bawab A, Al Dubayee M, Ahmed AA (2015) Quantification of insulin receptor mRNA splice variants as a diagnostic tumor marker in breast cancer. Cancer Biomark 15:653–661

Alkhayyal N, Talaat I, Vinodnadat A, Maghazachi A, Abusnana S, Syrjanen K, Bendardaf R (2020) Correlation of insulin-like growth factor 1 receptor expression with different molecular subtypes of breast cancer in the UAE. Anticancer Res 40:1555–1561

Allen F, Maillard I (2021) Therapeutic targeting of notch signaling: from cancer to inflammatory disorders. Front Cell Dev Biol 9. https://doi.org/10.3389/fcell.2021.649205

Almendro V, Garcia-Recio S, Gascon P (2010) Tyrosine kinase receptor transactivation associated to G protein-coupled receptors. Curr Drug Targets 11:1169–1180

American Cancer Society (2022) Breast Cancer Facts & Figures 2022-2024. Atlanta, American Cancer Society, Inc 2022:1–3

Anbazhagan R, Gelber RD, Bettelheim R, Goldhirsch A, Gusterson BA (1991) Association of c-erbB-2 expression and S-phase fraction in the prognosis of node positive breast cancer. Ann Oncol 2:47–53

André F, Cortés J (2015) Rationale for targeting fibroblast growth factor receptor signaling in breast cancer. Breast Cancer Res Treat 150:1–8

Anstead GM, Carlson KE, Katzenellenbogen JA (1997) The estradiol pharmacophore: ligand structure-estrogen receptor binding affinity relationships and a model for the receptor binding site. Steroids 62:268–303

Arcaro A (2013) Targeting the insulin-like growth factor-1 receptor in human cancer. Front Pharmacol 4:30

Ariazi EA, Brailoiu E, Yerrum S, Shupp HA, Slifker MJ, Cunliffe HE, Black MA, Donato AL, Arterburn JB, Oprea TI, Prossnitz ER, Dun NJ, Jordan VC (2010) The G protein–coupled receptor GPR30 inhibits proliferation of estrogen receptor–positive breast cancer cells. Cancer Res 70:1184–1194

Arora P, Cuevas BD, Russo A, Johnson GL, Trejo J (2008) Persistent transactivation of EGFR and ErbB2/HER2 by protease-activated receptor-1 promotes breast carcinoma cell invasion. Oncogene 27:4434–4445

Arpino G, Wiechmann L, Osborne CK, Schiff R (2008) Crosstalk between the estrogen receptor and the HER tyrosine kinase receptor family: molecular mechanism and clinical implications for endocrine therapy resistance. Endocr Rev 29:217–233

Arterburn JB, Prossnitz ER (2023) G protein–coupled estrogen receptor GPER: molecular pharmacology and therapeutic applications. Annu Rev Pharmacol Toxicol 63:295–320

Asano S, Ono A, Sakamoto K, Hayata-Takano A, Nakazawa T, Tanimoto K, Hashimoto H, Ago Y (2023) Vasoactive intestinal peptide receptor 2 signaling promotes breast cancer cell proliferation by enhancing the ERK pathway. Peptides 161:170940

Bagnato A, Rosanò L (2019) New routes in GPCR/β-arrestin-driven signaling in cancer progression and metastasis. Front Pharmacol 10:114

Bai J-W, Wei M, Li J-W, Zhang G-J (2020) Notch signaling pathway and endocrine resistance in breast cancer. Front Pharmacol 11:924

Baker C, Mansfield Z (2023) Cancer statistics for England, Commons Library research briefing. House of Commons Library, London, pp 1–29

Balijepalli P, Sitton CC, Meier KE (2021) Lysophosphatidic acid signaling in cancer cells: what makes LPA so special? Cells 10:2059

Balkwill F (2004) Cancer and the chemokine network. Nat Rev Cancer 4:540–550

Balmaña J, Díez O, Rubio IT, Cardoso F (2011) BRCA in breast cancer: ESMO clinical practice guidelines. Ann Oncol 22:vi31–vi34

Bar-Shavit R, Maoz M, Kancharla A, Nag J, Agranovich D, Grisaru-Granovsky S, Uziely B (2016) G protein-coupled receptors in cancer. Int J Mol Sci 17:1320

Baron V, Calleja V, Ferrari P, Alengrin F, Van Obberghen E (1998) p125Fak focal adhesion kinase is a substrate for the insulin and insulin-like growth factor-I tyrosine kinase receptors. J Biol Chem 273:7162–7168

Batool Z, Azfal A, Liaquat L, Sadir S, Nisar R, Inamullah A, Faiz Ghalib AU, Haider S (2023) Receptor tyrosine kinases (RTKs): from biology to pathophysiology. In: Mansour HM, Khattab MM, El-khatib AS (eds) Receptor tyrosine kinases in neurodegenerative and psychiatric disorders. Academic Press, Amsterdam, pp 117–185

Belfiore A, Malaguarnera R, Vella V, Lawrence MC, Sciacca L, Frasca F, Morrione A, Vigneri R (2017) Insulin receptor isoforms in physiology and disease: an updated view. Endocr Rev 38: 379–431

Belfiore A, Rapicavoli RV, Le Moli R, Lappano R, Morrione A, De Francesco EM, Vella V (2023) IGF2: a role in metastasis and tumor evasion from immune surveillance? Biomedicines 11:229

Blasco-Benito S, Moreno E, Seijo-Vila M, Tundidor I, Andradas C, Caffarel MM, Caro-Villalobos-M, Urigüen L, Diez-Alarcia R, Moreno-Bueno G, Hernández L, Manso L, Homar-Ruano P, Mccormick PJ, Bibic L, Bernadó-Morales C, Arribas J, Canals M, Casadó V, Canela EI, Guzmán M, Pérez-Gómez E, Sánchez C (2019) Therapeutic targeting of HER2–CB2R heteromers in HER2-positive breast cancer. Proc Natl Acad Sci 116:3863–3872

Bogoyevitch MA (2006) The isoform-specific functions of the c-Jun N-terminal Kinases (JNKs): differences revealed by gene targeting. BioEssays 28:923–934

Booden MA, Eckert LB, Der CJ, Trejo J (2004) Persistent signaling by dysregulated thrombin receptor trafficking promotes breast carcinoma cell invasion. Mol Cell Biol 24:1990–1999

Boucher J, Kleinridders A, Kahn CR (2014) Insulin receptor signaling in normal and insulin-resistant states. Cold Spring Harb Perspect Biol 6:a009191

Boussat S, Eddahibi S, Coste A, Fataccioli V, Gouge M, Housset B, Adnot S, Maitre B (2000) Expression and regulation of vascular endothelial growth factor in human pulmonary epithelial cells. Am J Physiol Lung Cell Mol Physiol 279:L371–L378

Brito LG, Schiavon VF, Andrade JM, Tiezzi DG, Peria FM, Marana HR (2011) Expression of hypoxia-inducible factor 1-alpha and vascular endothelial growth factor-C in locally advanced breast cancer patients. Clinics (Sao Paulo) 66:1313–1320

Brogowska KK, Zajkowska M, Mroczko B (2023) Vascular endothelial growth factor ligands and receptors in breast cancer. J Clin Med 12:2412

Bromberg J, Darnell JE (2000) The role of STATs in transcriptional control and their impact on cellular function. Oncogene 19:2468–2473

Brown WS, Tan L, Smith A, Gray NS, Wendt MK (2016) Covalent targeting of fibroblast growth factor receptor inhibits metastatic breast cancer. Mol Cancer Ther 15:2096–2106

Browne BC, Eustace AJ, Kennedy S, O'Brien NA, Pedersen K, McDermott MS, Larkin A, Ballot J, Mahgoub T, Sclafani F, Madden S, Kennedy J, Duffy MJ, Crown J, O'Donovan N (2012) Evaluation of IGF1R and phosphorylated IGF1R as targets in HER2-positive breast cancer cell lines and tumours. Breast Cancer Res Treat 136:717–727

Burness ML, Grushko TA, Olopade OI (2010) Epidermal growth factor receptor in triple-negative and basal-like breast cancer: promising clinical target or only a marker? Cancer J 16:23–32

Butti R, Das S, Gunasekaran VP, Yadav AS, Kumar D, Kundu GC (2018) Receptor tyrosine kinases (RTKs) in breast cancer: signaling, therapeutic implications and challenges. Mol Cancer 17:34

Cabioglu N, Summy J, Miller C, Parikh NU, Sahin AA, Tuzlali S, Pumiglia K, Gallick GE, Price JE (2005) CXCL-12/stromal cell–derived factor-1α Transactivates HER2-neu in breast cancer cells by a novel pathway involving Src kinase activation. Cancer Res 65:6493–6497

Caffarel MAM, Sarrió D, Palacios J, Guzmán M, SáNchez C (2006) Δ9-tetrahydrocannabinol inhibits cell cycle progression in human breast cancer cells through Cdc2 regulation. Cancer Res 66:6615–6621

Camp RL, Rimm EB, Rimm DL (1999) Met expression is associated with poor outcome in patients with axillary lymph node negative breast carcinoma. Cancer 86:2259–2265

Carvalho I, Milanezi F, Martins A, Reis RM, Schmitt F (2005) Overexpression of platelet-derived growth factor receptor α in breast cancer is associated with tumour progression. Breast Cancer Res 7:R788

Castagnoli L, Iezzi M, Ghedini GC, Ciravolo V, Marzano G, Lamolinara A, Zappasodi R, Gasparini P, Campiglio M, Amici A, Chiodoni C, Palladini A, Lollini PL, Triulzi T, Menard S, Nanni P, Tagliabue E, Pupa SM (2014) Activated d16HER2 homodimers and SRC kinase mediate optimal efficacy for trastuzumab. Cancer Res 74:6248–6259

Castagnoli L, Ghedini GC, Koschorke A, Triulzi T, Dugo M, Gasparini P, Casalini P, Palladini A, Iezzi M, Lamolinara A, Lollini PL, Nanni P, Chiodoni C, Tagliabue E, Pupa SM (2017) Pathobiological implications of the d16HER2 splice variant for stemness and aggressiveness of HER2-positive breast cancer. Oncogene 36:1721–1732

Castagnoli L, Ladomery M, Tagliabue E, Pupa SM (2019) The d16HER2 splice variant: a friend or foe of HER2-positive cancers? Cancers (Basel) 11:902

Cathcart-Rake EJ, Ruddy KJ, Bleyer A, Johnson RH (2021) Breast cancer in adolescent and young adult women under the age of 40 years. JCO Oncol Pract 17:305–313

Cattaneo F, Guerra G, Parisi M, De Marinis M, Tafuri D, Cinelli M, Ammendola R (2014) Cell-surface receptors transactivation mediated by g protein-coupled receptors. Int J Mol Sci 15: 19700–19728

Cejudo-Arteaga S, Guerrero-Ramos MA, Kuri-Exome R, Martinez-Cordero E, Farias-Serratos F, Maldonado-Vega M (2022) Epidemiology of breast cancer in Mexican women with obesity as a risk factor. Int J Mol Sci 23:10742

Cerma K, Piacentini F, Moscetti L, Barbolini M, Canino F, Tornincasa A, Caggia F, Cerri S, Molinaro A, Dominici M, Omarini C (2023) Targeting PI3K/AKT/mTOR pathway in breast cancer: from biology to clinical challenges. Biomedicines 11:109

Chan YT, Lai ACY, Lin RJ, Wang YH, Wang YT, Chang WW, Wu HY, Lin YJ, Chang WY, Wu JC, Yu JC, Chen YJ, Yu AL (2020) GPER-induced signaling is essential for the survival of breast cancer stem cells. Int J Cancer 146:1674–1685

Changavi AA, Shashikala A, Ramji AS (2015) Epidermal growth factor receptor expression in triple negative and nontriple negative breast carcinomas. J Lab Phys 7:79–83

Chaplin R, Thach L, Hollenberg MD, Cao Y, Little PJ, Kamato D (2017) Insights into cellular signalling by G protein coupled receptor transactivation of cell surface protein kinase receptors. J Cell Commun Signal 11:117–125

Chaudhary PK, Kim S (2021) An insight into GPCR and G-proteins as cancer drivers. Cells 10: 3288

Chen P, Li B, Ou-Yang L (2022) Role of estrogen receptors in health and disease. Front Endocrinol (Lausanne) 13:839005

Cleary MP, Grossmann ME (2009) Minireview: obesity and breast cancer: the estrogen connection. Endocrinology 150:2537–2542

Clevenger CV (2004) Roles and regulation of Stat family transcription factors in human breast cancer. Am J Pathol 165:1449–1460

Clusan L, Ferriere F, Flouriot G, Pakdel F (2023) A basic review on estrogen receptor signaling pathways in breast cancer. Int J Mol Sci 24:6834

Cortés I, Sánchez-Ruíz J, Zuluaga S, Calvanese V, Marqués M, Hernández C, Rivera T, Kremer L, González-García A, Carrera AC (2012) p85β phosphoinositide 3-kinase subunit regulates tumor progression. Proc Natl Acad Sci 109:11318–11323

Critchley WR, Pellet-Many C, Ringham-Terry B, Harrison MA, Zachary IC, Ponnambalam S (2018) Receptor tyrosine kinase ubiquitination and de-ubiquitination in signal transduction and receptor trafficking. Cells 7:22

Crudden C, Ilic M, Suleymanova N, Worrall C, Girnita A, Girnita L (2015) The dichotomy of the insulin-like growth factor 1 receptor: RTK and GPCR: friend or foe for cancer treatment? Growth Hormon IGF Res 25:2–12

Crudden C, Shibano T, Song D, Suleymanova N, Girnita A, Girnita L (2018) Blurring boundaries: receptor tyrosine kinases as functional G protein-coupled receptors. Int Rev Cell Mol Biol 339: 1–40

Daub H, Weiss FU, Wallasch C, Ullrich A (1996) Role of transactivation of the EGF receptor in signalling by G-protein-coupled receptors. Nature 379:557–560

David M, Sahay D, Mege F, Descotes F, Leblanc R, Ribeiro J, Clézardin P, Peyruchaud O (2014) Identification of heparin-binding EGF-like growth factor (HB-EGF) as a biomarker for lysophosphatidic acid receptor type 1 (LPA1) activation in human breast and prostate cancers. PLoS One 9:e97771

De Francesco EM, Lappano R, Santolla MF, Marsico S, Caruso A, Maggiolini M (2013) HIF-1α/GPER signaling mediates the expression of VEGF induced by hypoxia in breast cancer associated fibroblasts (CAFs). Breast Cancer Res 15:R64

De Francesco EM, Pellegrino M, Santolla MF, Lappano R, Ricchio E, Abonante S, Maggiolini M (2014) GPER mediates activation of HIF1α/VEGF signaling by estrogens. Cancer Res 74: 4053–4064

De Francesco EM, Sims AH, Maggiolini M, Sotgia F, Lisanti MP, Clarke RB (2017) GPER mediates the angiocrine actions induced by IGF1 through the HIF-1α/VEGF pathway in the breast tumor microenvironment. Breast Cancer Res 19:129

De Francesco E, Maggiolini M, Musti A (2018) Crosstalk between notch, HIF-1α and GPER in breast cancer EMT. Int J Mol Sci 19:2011

De Marco P, Bartella V, Vivacqua A, Lappano R, Santolla MF, Morcavallo A, Pezzi V, Belfiore A, Maggiolini M (2013) Insulin-like growth factor-I regulates GPER expression and function in cancer cells. Oncogene 32:678–688

Dent SF (2009) The role of VEGF in triple-negative breast cancer: where do we go from here? Ann Oncol 20:1615–1617

Devanaboyina M, Kaur J, Whiteley E, Lin L, Einloth K, Morand S, Stanbery L, Hamouda D, Nemunaitis J (2022) NF-κB signaling in tumor pathways focusing on breast and ovarian cancer. Oncol Rev 16:10568

Devericks EN, Carson MS, McCullough LE, Coleman MF, Hursting SD (2022) The obesity-breast cancer link: a multidisciplinary perspective. Cancer Metastasis Rev 41:607–625

Dhopeshwarkar A, Mackie K (2014) CB2 cannabinoid receptors as a therapeutic target—what does the future hold? Mol Pharmacol 86:430–437

Di Liberto V, Mudo G, Belluardo N (2019) Crosstalk between receptor tyrosine kinases (RTKs) and G protein-coupled receptors (GPCR) in the brain: focus on heteroreceptor complexes and related functional neurotrophic effects. Neuropharmacology 152:67–77

do Nascimento RG, Otoni KM (2020) Histological and molecular classification of breast cancer: what do we know? Mastology 30:1–8

Dorsam RT, Gutkind JS (2007) G-protein-coupled receptors and cancer. Nat Rev Cancer 7:79–94

Drury LJ, Ziarek JJ, Gravel S, Veldkamp CT, Takekoshi T, Hwang ST, Heveker N, Volkman BF, Dwinell MB (2011) Monomeric and dimeric CXCL12 inhibit metastasis through distinct CXCR4 interactions and signaling pathways. Proc Natl Acad Sci 108:17655–17660

Du Z, Lovly CM (2018) Mechanisms of receptor tyrosine kinase activation in cancer. Mol Cancer 17:58

Eccles SA (2011) The epidermal growth factor receptor/Erb-B/HER family in normal and malignant breast biology. Int J Dev Biol 55:685–696

Elbauomy Elsheikh S, Green AR, Lambros MB, Turner NC, Grainge MJ, Powe D, Ellis IO, Reis-Filho JS (2007) FGFR1 amplification in breast carcinomas: a chromogenic in situhybridisation analysis. Breast Cancer Res 9:R23

Eswarakumar VP, Lax I, Schlessinger J (2005) Cellular signaling by fibroblast growth factor receptors. Cytokine Growth Factor Rev 16:139–149

Fan P, Wang J, Santen RJ, Yue W (2007) Long-term treatment with tamoxifen facilitates translocation of estrogen receptor alpha out of the nucleus and enhances its interaction with EGFR in MCF-7 breast cancer cells. Cancer Res 67:1352–1360

Ferlay J, Colombet M, Soerjomataram I, Parkin DM, Piñeros M, Znaor A, Bray F (2021) Cancer statistics for the year 2020: an overview. Int J Cancer 149:778–789

Ferlay J, Ervik M, Lam F, Laversanne M, Colombet M, Mery L, Piñeros M, Znaor A, Soerjomataram I, Bray F (2024) Global cancer observatory: cancer today. International Agency for Research on Cancer, Lyon. From https://gco.iarc.who.int/today

Filardo EJ, Quinn JA, Bland KI, Frackelton AR (2000) Estrogen-induced activation of Erk-1 and Erk-2 requires the G protein-coupled receptor homolog, GPR30, and occurs via trans-activation of the epidermal growth factor receptor through release of HB-EGF. Mol Endocrinol 14: 1649–1660

Forbes C, Fayter D, de Kock S, Quek RGW (2019) A systematic review of international guidelines and recommendations for the genetic screening, diagnosis, genetic counseling, and treatment of BRCA-mutated breast cancer. Cancer Manag Res 11:2321–2337

Fuckar D, Dekanic A, Stifter S, Mustac E, Krstulja M, Dobrila F, Jonjic N (2006) VEGF expression is associated with negative estrogen receptor status in patients with breast cancer. Int J Surg Pathol 14:49–55

Fuentes N, Silveyra P (2019) Estrogen receptor signaling mechanisms. Adv Protein Chem Struct Biol 116:135–170

Galal MA, Alouch SS, Alsultan BS, Dahman H, Alyabis NA, Alammar SA, Aljada A (2023) Insulin receptor isoforms and insulin growth factor-like receptors: implications in cell signaling, carcinogenesis, and chemoresistance. Int J Mol Sci 24:15006

Gamble P, Jaroensri R, Wang H, Tan F, Moran M, Brown T, Flament-Auvigne I, Rakha EA, Toss M, Dabbs DJ, Regitnig P, Olson N, Wren JH, Robinson C, Corrado GS, Peng LH, Liu Y, Mermel CH, Steiner DF, Chen PC (2021) Determining breast cancer biomarker status and associated morphological features using deep learning. Commun Med (Lond) 1:14

Garcia-Recio S, Thennavan A, East MP, Parker JS, Cejalvo JM, Garay JP, Hollern DP, He X, Mott KR, Galván P, Fan C, Selitsky SR, Coffey AR, Marron D, Brasó-Maristany F, Burgués O, Albanell J, Rojo F, Lluch A, De Dueñas EM, Rosen JM, Johnson GL, Carey LA, Prat A, Perou CM (2020) FGFR4 regulates tumor subtype differentiation in luminal breast cancer and metastatic disease. J Clin Invest 130:4871–4887

Garcia-Segura LM, Diz-Chaves Y, Perez-Martin M, Darnaudéry M (2007) Estradiol, insulin-like growth factor-I and brain aging. Psychoneuroendocrinology 32:S57–S61

Garvin S, Dabrosin C (2003) Tamoxifen inhibits secretion of vascular endothelial growth factor in breast cancer in vivo. Cancer Res 63:8742–8748

Garvin S, Nilsson UW, Dabrosin C (2005) Effects of oestradiol and tamoxifen on VEGF, soluble VEGFR-1, and VEGFR-2 in breast cancer and endothelial cells. Br J Cancer 93:1005–1010

Gee JM, Robertson JF, Gutteridge E, Ellis IO, Pinder SE, Rubini M, Nicholson RI (2005) Epidermal growth factor receptor/HER2/insulin-like growth factor receptor signalling and oestrogen receptor activity in clinical breast cancer. Endocr Relat Cancer 12(Suppl 1):S99–S111

Gennari A, Sormani M, Puntoni M, Martini V, Amaro A, Bruzzi P, Pfeffer U (2021) Identification of a prognostic signature based on the expression of genes related to the insulin pathway in early breast cancer. Breast Care (Basel) 16:299–306

Ghayad SE, Vendrell JA, Ben Larbi S, Dumontet C, Bieche I, Cohen PA (2010) Endocrine resistance associated with activated ErbB system in breast cancer cells is reversed by inhibiting MAPK or PI3K/Akt signaling pathways. Int J Cancer 126:545–562

Giaquinto AN, Sung H, Miller KD, Kramer JL, Newman LA, Minihan A, Jemal A, Siegel RL (2022) Breast cancer statistics, 2022. CA Cancer J Clin 72:524–541

Giorgetti S, Pelicci PG, Pelicci G, Van Obberghen E (1994) Involvement of Src-homology/collagen (SHC) proteins in signaling through the insulin receptor and the insulin-like-growth-factor-I-receptor. Eur J Biochem 223:195–202

Girnita L, Takahashi SI, Crudden C, Fukushima T, Worrall C, Furuta H, Yoshihara H, Hakuno F, Girnita A (2016) Chapter seven – When phosphorylation encounters ubiquitination: a balanced perspective on IGF-1R signaling. Prog Mol Biol Transl Sci 141:277–311

Girnius N, Edwards YJ, Garlick DS, Davis RJ (2018) The cJUN NH(2)-terminal kinase (JNK) signaling pathway promotes genome stability and prevents tumor initiation. elife 7:e36389

Giubilaro J, Schuetz DA, Stepniewski TM, Namkung Y, Khoury E, Lara-Marquez M, Campbell S, Beautrait A, Armando S, Radresa O, Duchaine J, Lamarche-Vane N, Claing A, Selent J, Bouvier M, Marinier A, Laporte SA (2021) Discovery of a dual Ras and ARF6 inhibitor from a GPCR endocytosis screen. Nat Commun 12:4688

Giuliano M, Trivedi MV, Schiff R (2013) Bidirectional crosstalk between the estrogen receptor and human epidermal growth factor receptor 2 signaling pathways in breast cancer: molecular basis and clinical implications. Breast Care 8:256–262

Gkouveris I, Nikitakis N, Karanikou M, Rassidakis G, Sklavounou A (2014) Erk1/2 activation and modulation of STAT3 signaling in oral cancer. Oncol Rep 32:2175–2182

Glassman I, Le N, Asif A, Goulding A, Alcantara CA, Vu A, Chorbajian A, Mirhosseini M, Singh M, Venketaraman V (2023) The role of obesity in breast cancer pathogenesis. Cells 12:2061

Gross C, Bassell GJ (2014) Neuron-specific regulation of class I PI3K catalytic subunits and their dysfunction in brain disorders. Front Mol Neurosci 7:12

Guerin M, Gabillot M, Mathieu MC, Travagli JP, Spielmann M, Andrieu N, Riou G (1989) Structure and expression of c-erbB-2 and EGF receptor genes in inflammatory and non-inflammatory breast cancer: prognostic significance. Int J Cancer 43:201–208

Gurevich VV, Gurevich EV (2019) GPCR signaling regulation: the role of GRKs and Arrestins. Front Pharmacol 10:125

Gutteridge E, Agrawal A, Nicholson R, Cheung KL, Robertson J, Gee J (2010) The effects of gefitinib in tamoxifen-resistant and hormone-insensitive breast cancer: a phase II study. Int J Cancer 126:1806–1816

Ha J, Kang E, Seo J, Cho S (2019) Phosphorylation dynamics of JNK signaling: effects of dual-specificity phosphatases (DUSPs) on the JNK pathway. Int J Mol Sci 20:6157

Haeusler RA, McGraw TE, Accili D (2018) Biochemical and cellular properties of insulin receptor signalling. Nat Rev Mol Cell Biol 19:31–44

Hamilton N, Marquez-Garban D, Mah VH, Elshimali Y, Elashoff D, Garon EB, Vadgama J, Pietras R (2015) Estrogen receptor-beta and the insulin-like growth factor axis as potential therapeutic targets for triple-negative breast cancer. Crit Rev Oncog 20:373–390

Hao Q, Vadgama JV, Wang P (2020) CCL2/CCR2 signaling in cancer pathogenesis. Cell Commun Signal 18:1–13

Harbeck N, Penault-Llorca F, Cortes J, Gnant M, Houssami N, Poortmans P, Ruddy K, Tsang J, Cardoso F (2019) Breast cancer Nat Rev Dis Primers 5:66

He Y, Sun MM, Zhang GG, Yang J, Chen KS, Xu WW, Li B (2021) Targeting PI3K/Akt signal transduction for cancer therapy. Signal Transduct Target Ther 6:425

Heeneman S, Haendeler J, Saito Y, Ishida M, Berk BC (2000) Angiotensin II induces transactivation of two different populations of THE platelet-derived growth factor β receptor: key role for the p66 adaptor protein Shc. J Biol Chem 275:15926–15932

Hemmings BA, Restuccia DF (2012) PI3K-PKB/Akt Pathway. Cold Spring Harb Perspect Biol 4: a011189

Higgins KJ, Liu S, Abdelrahim M, Yoon K, Vanderlaag K, Porter W, Metz RP, Safe S (2006) Vascular endothelial growth factor receptor-2 expression is induced by 17beta-estradiol in ZR-75 breast cancer cells by estrogen receptor alpha/Sp proteins. Endocrinology 147: 3285–3295

Hilger D, Masureel M, Kobilka BK (2018) Structure and dynamics of GPCR signaling complexes. Nat Struct Mol Biol 25:4–12

Hirtz A, Bailly Y, Rech F, Pierson J, Dumond H, Dubois-Pot-Schneider H (2022) Molecular characterization of the dual effect of the GPER agonist G-1 in glioblastoma. Int J Mol Sci 23: 14309

Ho-Yen CM, Jones JL, Kermorgant S (2015) The clinical and functional significance of c-Met in breast cancer: a review. Breast Cancer Res 17:52

Hoeres T, Wilhelm M, Smetak M, Holzmann E, Schulze-Tanzil G, Birkmann J (2018) Immune cells regulate VEGF signalling via release of VEGF and antagonistic soluble VEGF receptor-1. Clin Exp Immunol 192:54–67

Homan KT, Tesmer JJ (2015) Molecular basis for small molecule inhibition of G protein-coupled receptor kinases. ACS Chem Biol 10:246–256

Hopkins M, Zhang Z, Liu Z, Meier K (2016) Eicosopentaneoic acid and other free fatty acid receptor agonists inhibit lysophosphatidic acid- and epidermal growth factor-induced proliferation of human breast cancer cells. J Clin Med 5:16

Horvath CM, Wen Z, Darnell JE (1995) A STAT protein domain that determines DNA sequence recognition suggests a novel DNA-binding domain. Genes Dev 9:984–994

Hoxhaj G, Manning BD (2020) The PI3K–AKT network at the interface of oncogenic signalling and cancer metabolism. Nat Rev Cancer 20:74–88

Hu X, Li J, Fu M, Zhao X, Wang W (2021) The JAK/STAT signaling pathway: from bench to clinic. Signal Transduct Target Ther 6:402

Hu D, Li Z, Zheng B, Lin X, Pan Y, Gong P, Zhuo W, Hu Y, Chen C, Chen L, Zhou J, Wang L (2022) Cancer-associated fibroblasts in breast cancer: challenges and opportunities. Cancer Commun 42:401–434

Huang S, Ouyang N, Lin L, Chen L, Wu W, Su F, Yao Y, Yao H (2012) HGF-induced PKCζ activation increases functional CXCR4 expression in human breast cancer cells. PLoS One 7: e29124

Hutcheson IR, Knowlden JM, Madden TA, Barrow D, Gee JM, Wakeling AE, Nicholson RI (2003) Oestrogen receptor-mediated modulation of the EGFR/MAPK pathway in tamoxifen-resistant MCF-7 cells. Breast Cancer Res Treat 81:81–93

Ihle JN, Hawkins PT, Parker PJ (1996) Janus kinases in cytokine signalling. Philos Trans R Soc Lond B Biol Sci 351:159–166

Ingthorsson S, Andersen K, Hilmarsdottir B, Maelandsmo GM, Magnusson MK, Gudjonsson T (2016) HER2 induced EMT and tumorigenicity in breast epithelial progenitor cells is inhibited by coexpression of EGFR. Oncogene 35:4244–4255

Irtegun S, Wood RJ, Ormsby AR, Mulhern TD, Hatters DM (2013) Tyrosine 416 is phosphorylated in the closed, repressed conformation of c-Src. PLoS ONE 8:e71035

Jackson C, Browell D, Gautrey H, Tyson-Capper A (2013) Clinical significance of HER-2 splice variants in breast cancer progression and drug resistance. Int J Cell Biol 2013:973584

Jansson S, Bendahl P-O, Grabau DA, Falck A-K, Fernö M, Aaltonen K, Rydén L (2014) The three receptor tyrosine kinases c-KIT, VEGFR2 and PDGFRα, closely spaced at 4q12, show increased protein expression in triple-negative breast cancer. PLoS One 9:e102176

Jansson S, Aaltonen K, Bendahl P-O, Falck A-K, Karlsson M, Pietras K, Rydén L (2018) The PDGF pathway in breast cancer is linked to tumour aggressiveness, triple-negative subtype and early recurrence. Breast Cancer Res Treat 169:231–241

Jean-Charles PY, Kaur S, Shenoy SK (2017) G protein-coupled receptor signaling through beta-arrestin-dependent mechanisms. J Cardiovasc Pharmacol 70:142–158

Jiang N, Dai Q, Su X, Fu J, Feng X, Peng J (2020) Role of PI3K/AKT pathway in cancer: the framework of malignant behavior. Mol Biol Rep 47:4587–4629

Johnson KS, Conant EF, Soo MS (2020) Molecular subtypes of breast cancer: a review for breast radiologists. J Breast Imaging 3:12–24

Kato T (2017) Biological roles of hepatocyte growth factor-Met signaling from genetically modified animals (review). Biomed Rep 7:495

Katzenellenbogen JA, Mayne CG, Katzenellenbogen BS, Greene GL, Chandarlapaty S (2018) Structural underpinnings of oestrogen receptor mutations in endocrine therapy resistance. Nat Rev Cancer 18:377–388

Ke X, Shen L (2017) Molecular targeted therapy of cancer: the progress and future prospect. Front Lab Med 1:69–75

Kilpatrick LE, Hill SJ (2021) Transactivation of G protein-coupled receptors (GPCRs) and receptor tyrosine kinases (RTKs): recent insights using luminescence and fluorescence technologies. Curr Opin Endocr Metab Res 16:102–112

Kim I-M, Tilley DG, Chen J, Salazar NC, Whalen EJ, Violin JD, Rockman HA (2008) β-Blockers alprenolol and carvedilol stimulate β-arrestin-mediated EGFR transactivation. Proc Natl Acad Sci USA 105:14555–14560

Kim S, Lee J, Oh SJ, Nam SJ, Lee JE (2015) Differential effect of EGFR inhibitors on tamoxifen-resistant breast cancer cells. Oncol Rep 34:1613–1619

Kisková T, Mungenast F, Suváková M, Jäger W, Thalhammer T (2019) Future aspects for cannabinoids in breast cancer therapy. Int J Mol Sci 20:1673

Knowlden JM, Hutcheson IR, Jones HE, Madden T, Gee JM, Harper ME, Barrow D, Wakeling AE, Nicholson RI (2003) Elevated levels of epidermal growth factor receptor/c-erbB2 heterodimers mediate an autocrine growth regulatory pathway in tamoxifen-resistant MCF-7 cells. Endocrinology 144:1032–1044

Kobilka BK (2007) G protein coupled receptor structure and activation. Biochim Biophys Acta 1768:794–807

Kohale IN, Yu J, Zhuang Y, Fan X, Reddy RJ, Sinnwell J, Kalari KR, Boughey JC, Carter JM, Goetz MP, Wang L, White FM (2022) Identification of Src family kinases as potential therapeutic targets for chemotherapy-resistant triple negative breast cancer. Cancers 14:4220

Kudaravalli S, den Hollander P, Mani SA (2022) Role of p38 MAP kinase in cancer stem cells and metastasis. Oncogene 41:3177–3185

Kumar R, Zakharov MN, Khan SH, Miki R, Jang H, Toraldo G, Singh R, Bhasin S, Jasuja R (2011) The dynamic structure of the estrogen receptor. J Amino Acids 2011:812540

Lacalle RA, Blanco R, Carmona-Rodríguez L, Martín-Leal A, Mira E, Mañes S (2017) Chemokine receptor signaling and the hallmarks of cancer. In: Galluzzi L (ed) International review of cell and molecular biology, vol 331. Academic Press, Amsterdam, pp 181–244

Lagerstrom MC, Schioth HB (2008) Structural diversity of G protein-coupled receptors and significance for drug discovery. Nat Rev Drug Discov 7:339–357

Lappano R, Maggiolini M (2011) G protein-coupled receptors: novel targets for drug discovery in cancer. Nat Rev Drug Discov 10:47–60

Lappano R, De Marco P, De Francesco EM, Chimento A, Pezzi V, Maggiolini M (2013) Cross-talk between GPER and growth factor signaling. J Steroid Biochem Mol Biol 137:50–56

Lappano R, Jacquot Y, Maggiolini M (2018) GPCR modulation in breast cancer. Int J Mol Sci 19:3840

Le XF, Mao W, Lu C, Thornton A, Heymach JV, Sood AK, Bast RC Jr (2008) Specific blockade of VEGF and HER2 pathways results in greater growth inhibition of breast cancer xenografts that overexpress HER2. Cell Cycle 7:3747–3758

Lebok P, Huber J, Burandt EC, Lebeau A, Marx AH, Terracciano L, Heilenkotter U, Janicke F, Muller V, Paluchowski P, Geist S, Wilke C, Simon R, Sauter G, Quaas A (2016) Loss of membranous VEGFR1 expression is associated with an adverse phenotype and shortened survival in breast cancer. Mol Med Rep 14:1443–1450

Lefkowitz RJ (2004) Historical review: a brief history and personal retrospective of seven-transmembrane receptors. Trends Pharmacol Sci 25:413–422

Leiber D, Burlina F, Byrne C, Robin P, Piesse C, Gonzalez L, Leclercq G, Tanfin Z, Jacquot Y (2015) The sequence Pro295–Thr311 of the hinge region of oestrogen receptor α is involved in ERK1/2 activation via GPR30 in leiomyoma cells. Biochem J 472:97–109

Lemmon MA, Schlessinger J (2010) Cell signaling by receptor tyrosine kinases. Cell 141: 1117–1134

Levine KM, Ding K, Chen L, Oesterreich S (2020) FGFR4: a promising therapeutic target for breast cancer and other solid tumors. Pharmacol Ther 214:107590

Li YM, Pan Y, Wei Y, Cheng X, Zhou BP, Tan M, Zhou X, Xia W, Hortobagyi GN, Yu D, Hung M-C (2004) Upregulation of CXCR4 is essential for HER2-mediated tumor metastasis. Cancer Cell 6:459–469

Liang Z, Brooks J, Willard M, Liang K, Yoon Y, Kang S, Shim H (2007) CXCR4/CXCL12 axis promotes VEGF-mediated tumor angiogenesis through Akt signaling pathway. Biochem Biophys Res Commun 359:716–722

Liang L, Yue Z, Du W, Li Y, Tao H, Wang D, Wang R, Huang Z, He N, Xie X, Han Z, Liu N, Li Z (2017) Molecular imaging of inducible VEGF expression and tumor progression in a breast cancer model. Cell Physiol Biochem 42:407–415

Liebmann C (2011) EGF receptor activation by GPCRs: an universal pathway reveals different versions. Mol Cell Endocrinol 331:222–231

Lin C-I, Chen C-N, Huang M-T, Lee S-J, Lin C-H, Chang C-C, Lee H (2008) Lysophosphatidic acid upregulates vascular endothelial growth factor-C and tube formation in human endothelial cells through LPA1/3, COX-2, and NF-κB activation- and EGFR transactivation-dependent mechanisms. Cell Signal 20:1804–1814

Lin AH, Li RW, Ho EY, Leung GP, Leung SW, Vanhoutte PM, Man RY (2013) Differential ligand binding affinities of human estrogen receptor-α isoforms. PLoS One 8:e63199

Lin C-C, Wieteska L, Poncet-Montange G, Suen KM, Arold ST, Ahmed Z, Ladbury JE (2020) Regulation of kinase activity by combined action of juxtamembrane and C-terminal regions of receptors. BioRxiv. https://doi.org/10.1101/2020.10.01.322123

Liu Y, Tamimi RM, Collins LC, Schnitt SJ, Gilmore HL, Connolly JL, Colditz GA (2011a) The association between vascular endothelial growth factor expression in invasive breast cancer and survival varies with intrinsic subtypes and use of adjuvant systemic therapy: results from the Nurses' Health Study. Breast Cancer Res Treat 129:175–184

Liu J, Liao S, Huang Y, Samuel R, Shi T, Naxerova K, Huang P, Kamoun W, Jain RK, Fukumura D, Xu L (2011b) PDGF-D improves drug delivery and efficacy via vascular normalization, but promotes lymphatic metastasis by activating CXCR4 in breast cancer. Clin Cancer Res 17:3638–3648

Lopez-Haber C, Barrio-Real L, Casado-Medrano V, Kazanietz MG (2016) Heregulin/ErbB3 signaling enhances CXCR4-driven Rac1 activation and breast cancer cell motility via hypoxia-inducible factor 1α. Molecular and Cell Biol 36:2011–2026

Liu F, Bardhan K, Yang D, Thangaraju M, Ganapathy V, Waller JL, Liles GB, Lee JR, Liu K (2012a) NF-kB directly regulates Fas transcription to modulate Fas-mediated apoptosis and tumor suppression. J Biol Chem 287:25530–25540

Liu S-B, Zhang N, Guo Y-Y, Zhao R, Shi T-Y, Feng S-F, Wang S-Q, Yang Q, Li X-Q, Wu Y-M, Ma L, Hou Y, Xiong L-Z, Zhang W, Zhao M-G (2012b) G-protein-coupled receptor 30 mediates rapid neuroprotective effects of estrogen via depression of NR2B-containing NMDA receptors. J Neurosci 32:4887–4900

Liu S, Meng X, Chen H, Liu W, Miller T, Lu MMY, Zhang F, Gagea M, Arteaga CL, Mills GB, Meric-Bernstam F, González-Angulo AM (2014) Targeting tyrosine-kinases and estrogen receptor abrogates resistance to endocrine therapy in breast cancer. Oncotarget 5(19):9049

Liu H, Yang Z, Lu W, Chen Z, Chen L, Han S, Wu X, Cai T, Cai Y (2020a) Chemokines and chemokine receptors: a new strategy for breast cancer therapy. Cancer Med 9:3786–3799

Liu H, Zhang B, Sun Z (2020b) Spectrum of EGFR aberrations and potential clinical implications: insights from integrative pan-cancer analysis. Cancer Commun (Lond) 40:43–59

Liu Y, Ma H, Yao J (2020c) ERα, a key target for cancer therapy: a review. Onco Targets Ther 13: 2183–2191

Lobry C, Oh P, Mansour MR, Look AT, Aifantis I (2014) Notch signaling: switching an oncogene to a tumor suppressor. Blood 123:2451–2459

Lu Z, Xu S (2006) ERK1/2 MAP kinases in cell survival and apoptosis. IUBMB Life 58:621–631

Lu J-F, Hu Z-Q, Yang M-X, Liu W-Y, Pan G-F, Ding J-B, Liu J-Z, Tang L, Hu B, Li H-C (2022) Downregulation of PDGF-D inhibits proliferation and invasion in breast cancer MDA-MB-231 cells. Clin Breast Cancer 22:e173–e183

Luo J, Liu D (2020) Does GPER really function as a G protein-coupled estrogen receptor in vivo? Front Endocrinol 11:148

Luo J, Zou H, Guo Y, Tong T, Ye L, Zhu C, Deng L, Wang B, Pan Y, Li P (2022) SRC kinase-mediated signaling pathways and targeted therapies in breast cancer. Breast Cancer Res 24:99

Ma Y, Nicolet J (2023) Specificity models in MAPK cascade signaling. FEBS Open Bio 13: 1177–1192

Makki J (2015) Diversity of breast carcinoma: histological subtypes and clinical relevance. Clinical medicine insights. Pathology 8:CPath.S31563

Mansouri S, Farahmand L, Teymourzadeh A, Majidzadeh AK (2018a) Clinical evidence on the magnitude of change in growth pathway activity in relation to tamoxifen resistance is required. Curr Cancer Drug Targets 18:668–676

Mansouri S, Feizi N, Mahdi A, Majidzadeh AK, Farahmand L (2018b) A review on the role of VEGF in tamoxifen resistance. Anti Cancer Agents Med Chem 18:2006–2009

Mantovani A, Allavena P, Sica A, Balkwill F (2008) Cancer-related inflammation. Nature 454: 436–444

Maroni P, Bendinelli P, Matteucci E, Desiderio MA (2007) HGF induces CXCR4 and CXCL12-mediated tumor invasion through Ets1 and NF-κB. Carcinogenesis 28:267–279

Martellucci S, Clementi L, Sabetta S, Mattei V, Botta L, Angelucci A (2020) Src family kinases as therapeutic targets in advanced solid tumors: what we have learned so far. Cancers 12:1448

Martin EC, Bratton MR, Zhu Y, Rhodes LV, Tilghman SL, Collins-Burow BM, Burow ME (2012) Insulin-like growth factor-1 signaling regulates miRNA expression in MCF-7 breast cancer cell line. PLoS One 7:e49067

Martin-Vega A, Cobb MH (2023) Navigating the ERK1/2 MAPK Cascade. Biomol Ther 13:1555

Martínez-Limón A, Joaquin M, Caballero M, Posas F, De Nadal E (2020) The p38 pathway: from biology to cancer therapy. Int J Mol Sci 21:1913

Massarweh S, Osborne CK, Creighton CJ, Qin L, Tsimelzon A, Huang S, Weiss H, Rimawi M, Schiff R (2008) Tamoxifen resistance in breast tumors is driven by growth factor receptor signaling with repression of classic estrogen receptor genomic function. Cancer Res 68:826–833

McDermott MSJ, Canonici A, Ivers L, Browne BC, Madden SF, O'Brien NA, Crown J, O'Donovan N (2017) Dual inhibition of IGF1R and ER enhances response to trastuzumab in HER2 positive breast cancer cells. Int J Oncol 50:2221–2228

Mendez-Garcia LA, Nava-Castro KE, Ochoa-Mercado TL, Palacios-Arreola MI, Ruiz-Manzano RA, Segovia-Mendoza M, Solleiro-Villavicencio H, Cazarez-Martinez C, Morales-Montor J (2019) Breast cancer metastasis: are cytokines important players during its development and progression? J Interf Cytokine Res 39:39–55

Mitra D, Brumlik MJ, Okamgba SU, Zhu Y, Duplessis TT, Parvani JG, Lesko SM, Brogi E, Jones FE (2009) An oncogenic isoform of HER2 associated with locally disseminated breast cancer and trastuzumab resistance. Mol Cancer Ther 8:2152–2162

Mitra D, Saha D, Das G, Mukherjee R, Banerjee S, Alam N, Mustafi SM, Nath P, Majumder A, Majumder B, Murmu N (2023) Lupeol synergizes with 5-fluorouracil to combat c-MET/EphA2 mediated chemoresistance in triple negative breast cancer. iScience 26:108395

Moerkens M, Zhang Y, Wester L, van de Water B, Meerman JHN (2014) Epidermal growth factor receptor signalling in human breast cancer cells operates parallel to estrogen receptor α signalling and results in tamoxifen insensitive proliferation. BMC Cancer 14:283

Montemurro F, Di Cosimo S, Arpino G (2013) Human epidermal growth factor receptor 2 (HER2)-positive and hormone receptor-positive breast cancer: new insights into molecular interactions and clinical implications. Ann Oncol 24:2715–2724

Moody TW, Ramos-Alvarez I, Jensen RT (2023) Peptide G-protein-coupled receptors and ErbB receptor tyrosine kinases in cancer. Biology (Basel) 12:957

Mores KL, Cassell RJ, van Rijn RM (2019) Arrestin recruitment and signaling by G protein-coupled receptor heteromers. Neuropharmacology 152:15–21

Morrison DK (2012) MAP kinase pathways. Cold Spring Harb Perspect Biol 4:a011254

Mukherjee D, Zhao J (2013) The role of chemokine receptor CXCR4 in breast cancer metastasis. Am J Cancer Res 3:46–57

Müller A, Homey B, Soto H, Ge N, Catron D, Buchanan ME, Mcclanahan T, Murphy E, Yuan W, Wagner SN, Barrera JL, Mohar A, Verástegui E, Zlotnik A (2001) Involvement of chemokine receptors in breast cancer metastasis. Nature 410:50–56

Nemere I, Pietras RJ, Blackmore PF (2003) Membrane receptors for steroid hormones: signal transduction and physiological significance. J Cell Biochem 88:438–445

Neves SR, Ram PT, Iyengar R (2002) G protein pathways. Science 296:1636–1639

Neves M, Marolda V, Mayor F, Penela P (2022) Crosstalk between CXCR4/ACKR3 and EGFR signaling in breast cancer cells. Int J Mol Sci 23:11887

Neville MC, Webb P, Ramanathan P, Mannino MP, Pecorini C, Monks J, Anderson SM, MacLean P (2013) The insulin receptor plays an important role in secretory differentiation in the mammary gland. Am J Physiol Endocrinol Metab 305:E1103–E1114

Nguyen TNQ, Jung S, Nguyen HA, Lee B, Vu SH, Myagmarjav D, Eum HH, Lee H-O, Jo T, Choi Y, Lee M-S (2022) The regulation of insulin receptor/insulin-like growth factor 1 receptor ratio, an important factor for breast cancer prognosis, by TRIP-Br1. J Hematol Oncol 15:82

Nicholson RI, McClelland RA, Finlay P, Eaton CL, Gullick WJ, Dixon AR, Robertson JF, Ellis IO, Blamey RW (1993) Relationship between EGF-R, c-erbB-2 protein expression and Ki67 immunostaining in breast cancer and hormone sensitivity. Eur J Cancer 29A:1018–1023

Nieto Gutierrez A, McDonald PH (2018) GPCRs: emerging anti-cancer drug targets. Cell Signal 41:65–74

Nogues L, Reglero C, Rivas V, Salcedo A, Lafarga V, Neves M, Ramos P, Mendiola M, Berjon A, Stamatakis K, Zhou XZ, Lu KP, Hardisson D, Mayor F Jr, Penela P (2016) G protein-coupled receptor kinase 2 (GRK2) promotes breast tumorigenesis through a HDAC6-Pin1 axis. EBioMedicine 13:132–145

O'Hayre M, Degese MS, Gutkind JS (2014) Novel insights into G protein and G protein-coupled receptor signaling in cancer. Curr Opin Cell Biol 27:126–135

Odoemelam CS, Percival B, Wallis H, Chang M-W, Ahmad Z, Scholey D, Burton E, Williams IH, Kamerlin CL, Wilson PB (2020) G-protein coupled receptors: structure and function in drug discovery. RSC Adv 10:36337–36348

Oh A, Pardo M, Rodriguez A, Yu C, Nguyen L, Liang O, Chorzalska A, Dubielecka PM (2023) NF-κB signaling in neoplastic transition from epithelial to mesenchymal phenotype. Cell Commun Signal 21:291

Oligny-Longpré G, Corbani M, Zhou J, Hogue M, Guillon G, Bouvier M (2012) Engagement of β-arrestin by transactivated insulin-like growth factor receptor is needed for V2 vasopressin receptor-stimulated ERK1/2 activation. Proc Natl Acad Sci 109:E1028–E1037

Ornitz DM, Itoh N (2015) The fibroblast growth factor signaling pathway. Wiley Interdiscip Rev Dev Biol 4:215–266

Ostrom RS, Insel PA (2004) The evolving role of lipid rafts and caveolae in G protein-coupled receptor signaling: implications for molecular pharmacology. Br J Pharmacol 143:235–245

Overland AC, Insel PA (2015) Heterotrimeric G proteins directly regulate MMP14/membrane type-1 matrix metalloprotease: a novel mechanism for GPCR-EGFR transactivation. J Biol Chem 290:9941–9947

Pandey P, Khan F, Upadhyay TK, Seungjoon M, Park MN, Kim B (2023) New insights about the PDGF/PDGFR signaling pathway as a promising target to develop cancer therapeutic strategies. Biomed Pharmacother 161:114491

Parikh PK, Ghate MD (2018) Recent advances in the discovery of small molecule c-met kinase inhibitors. Eur J Med Chem 143:1103–1138

Park SY, Kim HM, Koo JS (2015) Differential expression of cancer-associated fibroblast-related proteins according to molecular subtype and stromal histology in breast cancer. Breast Cancer Res Treat 149:727–741

Pattarozzi A, Gatti M, Barbieri F, Würth R, Porcile C, Lunardi G, Ratto A, Favoni R, Bajetto A, Ferrari A, Florio T (2008) 17β-Estradiol promotes breast cancer cell proliferation-inducing stromal cell-derived factor-1-mediated epidermal growth factor receptor transactivation: reversal by Gefitinib Pretreatment. Mol Pharmacol 73:191–202

Pavitra E, Kancharla J, Gupta VK, Prasad K, Sung JY, Kim J, Tej MB, Choi R, Lee J-H, Han Y-K, Raju GSR, Bhaskar L, Huh YS (2023) The role of NF-κB in breast cancer initiation, growth, metastasis, and resistance to chemotherapy. Biomed Pharmacother 163:114822

Peach CJ, Edgington-Mitchell LE, Bunnett NW, Schmidt BL (2023) Protease-activated receptors in health and disease. Physiol Rev 103:717–785

Pegram M, Jackisch C, Johnston SRD (2023) Estrogen/HER2 receptor crosstalk in breast cancer: combination therapies to improve outcomes for patients with hormone receptor-positive/HER2-positive breast cancer. NPJ Breast Cancer 9:45

Peiró G, Adrover E, Sánchez-Tejada L, Lerma E, Planelles M, Sánchez-Payá J, Aranda FI, Giner D, Gutiérrez-Aviñó FJ (2011) Increased insulin-like growth factor-1 receptor mRNA expression predicts poor survival in immunophenotypes of early breast carcinoma. Mod Pathol 24:201–208

Pérez-Gómez E, Andradas C, Blasco-Benito S, Caffarel MM, García-Taboada E, Villa-Morales M, Moreno E, Hamann S, Martín-Villar E, Flores JM, Wenners A, Alkatout I, Klapper W, Röcken C, Bronsert P, Stickeler E, Staebler A, Bauer M, Arnold N, Soriano J, Pérez-Martínez M, Megías D, Moreno-Bueno G, Ortega-Gutiérrez S, Artola M, Vázquez-Villa H, Quintanilla M, Fernández-Piqueras J, Canela EI, Mccormick PJ, Guzmán M, Sánchez C (2015) Role of cannabinoid receptor CB2 in HER2 pro-oncogenic signaling in breast cancer. JNCI J Natl Cancer Inst 107:djv077

Phillips RJ, Mestas J, Gharaee-Kermani M, Burdick MD, Sica A, Belperio JA, Keane MP, Strieter RM (2005) Epidermal growth factor and hypoxia-induced expression of CXC chemokine receptor 4 on non-small cell lung cancer cells is regulated by the phosphatidylinositol 3-kinase/PTEN/AKT/mammalian target of rapamycin signaling pathway and activation of hypoxia inducible factor-1alpha. J Biol Chem 280:22473–22481

Pierce KL, Premont RT, Lefkowitz RJ (2002) Seven-transmembrane receptors. Nat Rev Mol Cell Biol 3:639–650

Pietras RJ, Marquez-Garban DC (2007) Membrane-associated estrogen receptor signaling pathways in human cancers. Clin Cancer Res 13:4672–4676

Pinzone JJ, Stevenson H, Strobl JS, Berg PE (2004) Molecular and cellular determinants of estrogen receptor alpha expression. Mol Cell Biol 24:4605–4612

Pircher TJ, Petersen H, Gustafsson J-A, Haldosén L-A (1999) Extracellular signal-regulated kinase (ERK) interacts with signal transducer and activator of transcription (STAT) 5a. Mol Endocrinol 13:555–565

Pisano A, Santolla MF, De Francesco EM, De Marco P, Rigiracciolo DC, Perri MG, Vivacqua A, Abonante S, Cappello AR, Dolce V, Belfiore A, Maggiolini M, Lappano R (2017) GPER, IGF-IR, and EGFR transduction signaling are involved in stimulatory effects of zinc in breast cancer cells and cancer-associated fibroblasts. Mol Carcinog 56:580–593

Pozzobon T, Goldoni G, Viola A, Molon B (2016) CXCR4 signaling in health and disease. Immunol Lett 177:6–15

Prenzel N, Zwick E, Daub H, Leserer M, Abraham R, Wallasch C, Ullrich A (1999) EGF receptor transactivation by G-protein-coupled receptors requires metalloproteinase cleavage of proHB-EGF. Nature 402:884–888

Prenzel N, Zwick E, Leserer M, Ullrich A (2000) Tyrosine kinase signalling in breast cancer. Epidermal growth factor receptor: convergence point for signal integration and diversification. Breast Cancer Res 2:184–190

Prossnitz ER, Barton M (2023) The G protein-coupled oestrogen receptor GPER in health and disease: an update. Nat Rev Endocrinol 19:407–424

Pua LJW, Mai CW, Chung FF, Khoo AS, Leong CO, Lim WM, Hii LW (2022) Functional roles of JNK and p38 MAPK signaling in nasopharyngeal carcinoma. Int J Mol Sci 23:1108

Puccini A, Marín-Ramos NI, Bergamo F, Schirripa M, Lonardi S, Lenz H-J, Loupakis F, Battaglin F (2019) Safety and tolerability of c-MET inhibitors in cancer. Drug Saf 42:211–233

Pupa SM, Ligorio F, Cancila V, Franceschini A, Tripodo C, Vernieri C, Castagnoli L (2021) HER2 signaling and breast cancer stem cells: the bridge behind HER2-positive breast cancer aggressiveness and therapy refractoriness. Cancers (Basel) 13:4778

Pupo M, Pisano A, Abonante S, Maggiolini M, Musti AM (2014) GPER activates notch signaling in breast cancer cells and cancer-associated fibroblasts (CAFs). Int J Biochem Cell Biol 46: 56–67

Pyne NJ, Pyne S (2011) Receptor tyrosine kinase-G-protein-coupled receptor signalling platforms: out of the shadow? Trends Pharmacol Sci 32:443–450

Quesnelle KM, Boehm AL, Grandis JR (2007) STAT-mediated EGFR signaling in cancer. J Cell Biochem 102:311–319

Quinn JA, Graeber CT, Frackelton AR, Kim M, Schwarzbauer JE, Filardo EJ (2009) Coordinate regulation of estrogen-mediated fibronectin matrix assembly and epidermal growth factor receptor transactivation by the G protein-coupled receptor, GPR30. Mol Endocrinol 23: 1052–1064

Radhakrishnan VK, Ravichandran K, Eke C, Ortiz-Vicil A, Tan Q, Leon M, Leon DD (2020) Methylation of a newly identified region of the INS-IGF2 gene determines IGF2 expression in breast cancer tumors and in breast cancer cells. Oncotarget 11:3904–3920

Rajagopal S, Shenoy SK (2018) GPCR desensitization: acute and prolonged phases. Cell Signal 41: 9–16

Ranjan R, Gupta P, Shukla Arun K (2016) GPCR signaling: β-arrestins kiss and remember. Curr Biol 26:R285–R288

Recchia AG, De Francesco EM, Vivacqua A, Sisci D, Panno ML, Andò S, Maggiolini M (2011) The G protein-coupled receptor 30 is up-regulated by hypoxia-inducible factor-1α (HIF-1α) in breast cancer cells and cardiomyocytes. J Biol Chem 286:10773–10782

Ren X, Li L, Wu J, Lin K, He Y, Bian L (2021) PDGF-BB regulates the transformation of fibroblasts into cancer-associated fibroblasts via the lncRNA LURAP1L-AS1/LURAP1L/IKK/IκB/NF-κB signaling pathway. Oncol Lett 22:537

Revankar CM, Cimino DF, Sklar LA, Arterburn JB, Prossnitz ER (2005) A transmembrane intracellular estrogen receptor mediates rapid cell signaling. Science 307:1625–1630

Riecke K, Witzel I (2020) Targeting the human epidermal growth factor receptor family in breast cancer beyond HER2. Breast Care (Basel) 15:579–585

Rimawi MF, Shetty PB, Weiss HL, Schiff R, Osborne CK, Chamness GC, Elledge RM (2010) Epidermal growth factor receptor expression in breast cancer association with biologic phenotype and clinical outcomes. Cancer 116:1234–1242

Roskoski R (2004) Src protein–tyrosine kinase structure and regulation. Biochem Biophys Res Commun 324:1155–1164

Rostoker R, Abelson S, Bitton-Worms K, Genkin I, Ben-Shmuel S, Dakwar M, Orr ZS, Caspi A, Tzukerman M, LeRoith D (2015) Highly specific role of the insulin receptor in breast cancer progression. Endocr Relat Cancer 22:145–157

Rostoker R, Ben-Shmuel S, Rashed R, Shen Orr Z, LeRoith D (2016) CD24 cell surface expression in Mvt1 mammary cancer cells serves as a biomarker for sensitivity to anti-IGF1R therapy. Breast Cancer Res 18:51

Roy I, Getschman AE, Volkman BF, Dwinell MB (2017) Exploiting agonist biased signaling of chemokines to target cancer. Mol Carcinog 56:804–813

Ruan W, Kleinberg DL (1999) Insulin-like growth factor I is essential for terminal end bud formation and ductal morphogenesis during mammary development. Endocrinology 140: 5075–5081

Ruan S-Q, Wang Z-H, Wang S-W, Fu Z-X, Xu K-L, Li D-B, Zhang S-Z (2012) Heregulin-β1-induced GPR30 upregulation promotes the migration and invasion potential of SkBr3 breast cancer cells via ErbB2/ErbB3–MAPK/ERK pathway. Biochem Biophys Res Commun 420: 385–390

Salazar N, Muñoz D, Kallifatidis G, Singh RK, Jordà M, Lokeshwar BL (2014) The chemokine receptor CXCR7 interacts with EGFR to promote breast cancer cell proliferation. Mol Cancer 13:198

Santolla MF, Maggiolini M (2020) The FGF/FGFR system in breast cancer: oncogenic features and therapeutic perspectives. Cancers 12:3029

Santolla MF, Vivacqua A, Lappano R, Rigiracciolo DC, Cirillo F, Galli GR, Talia M, Brunetti G, Miglietta AM, Belfiore A, Maggiolini M (2019) GPER mediates a feedforward FGF2/FGFR1 paracrine activation coupling CAFs to cancer cells toward breast tumor progression. Cells 8:223

Sanz-Ezquerro JJ, Cuenda A (2021) p38 Signalling Pathway. Int J Mol Sci 22:1003

Ségaliny AI, Tellez-Gabriel M, Heymann M-F, Heymann D (2015) Receptor tyrosine kinases: characterisation, mechanism of action and therapeutic interests for bone cancers. J Bone Oncol 4:1–12

Scaling AL, Prossnitz ER, Hathaway HJ (2014) GPER mediates estrogen-induced signaling and proliferation in human breast epithelial cells and Normal and malignant breast. Horm Cancer 5: 146–160

Schettini F, Buono G, Cardalesi C, Desideri I, De Placido S, Del Mastro L (2016) Hormone receptor/human epidermal growth factor receptor 2-positive breast cancer: where we are now and where we are going. Cancer Treat Rev 46:20–26

Schulze WX, Deng L, Mann M (2005) Phosphotyrosine interactome of the ErbB-receptor kinase family. Mol Syst Biol 1:2005 0008

Segovia-Mendoza M, Gonzalez-Gonzalez ME, Barrera D, Diaz L, Garcia-Becerra R (2015) Efficacy and mechanism of action of the tyrosine kinase inhibitors gefitinib, lapatinib and neratinib in the treatment of HER2-positive breast cancer: preclinical and clinical evidence. Am J Cancer Res 5:2531–2561

Segovia-Mendoza M, Fernández-López BE, Lemini C, Carlos-Reyes A, Prado-Garcia H (2022) Role of membrane estrogen receptor (GPER1) on the function of immune cells and its consequences on breast cancer pathophysiology. Springer, Cham, pp 1–20

Sengupta K, Banerjee S, Saxena N, Banerjee SK (2003) Estradiol-induced vascular endothelial growth factor-A expression in breast tumor cells is biphasic and regulated by estrogen receptor-alpha dependent pathway. Int J Oncol 22:609–614

Servetto A, Formisano L, Arteaga CL (2021) FGFR signaling and endocrine resistance in breast cancer: challenges for the clinical development of FGFR inhibitors. Biochim Biophys Acta Rev Cancer 1876:188595

Shah S, Sizemore GM (2022) Diverse roles of tumor-stromal PDGFB-to-PDGFRβ signaling in breast cancer growth and metastasis. In: Angel PM, Ostrowski MC (eds) Advances in cancer research, vol 154. Academic Press, Amsterdam, pp 93–140

Shaik F, Cuthbert GA, Homer-Vanniasinkam S, Muench SP, Ponnambalam S, Harrison MA (2020) Structural basis for vascular endothelial growth factor receptor activation and implications for disease therapy. Biomol Ther 10:1673

Shaul YD, Seger R (2007) The MEK/ERK cascade: from signaling specificity to diverse functions. Biochim Biophys Acta 1773:1213–1226

Shaw P, Dwivedi SKD, Bhattacharya R, Mukherjee P, Rao G (2024) VEGF signaling: role in angiogenesis and beyond. Biochim Biophys Acta Rev Cancer 1879:189079

Shembade N, Harhaj EW (2012) Regulation of NF-κB signaling by the A20 deubiquitinase. Cell Mol Immunol 9:123–130

Shenoy SK, Lefkowitz RJ (2011) β-Arrestin-mediated receptor trafficking and signal transduction. Trends Pharmacol Sci 32:521–533

Shenoy SK, Han S, Zhao YL, Hara MR, Oliver T, Cao Y, Dewhirst MW (2012) β-arrestin1 mediates metastatic growth of breast cancer cells by facilitating HIF-1-dependent VEGF expression. Oncogene 31:282–292

Shibuya M (2011) Vascular endothelial growth factor (VEGF) and its receptor (VEGFR) signaling in angiogenesis: a crucial target for anti- and pro-angiogenic therapies. Genes Cancer 2: 1097–1105

Shukla AK, Dwivedi-Agnihotri H (2020) Structure and function of beta-arrestins, their emerging role in breast cancer, and potential opportunities for therapeutic manipulation. Adv Cancer Res 145:139–156

Skolariki A, D'Costa J, Little M, Lord S (2022) Role of PI3K/Akt/mTOR pathway in mediating endocrine resistance: concept to clinic. Explor Target Antitumor Ther 3:172–199

Smith MCP, Luker KE, Garbow JR, Prior JL, Jackson E, Piwnica-Worms D, Luker GD (2004) CXCR4 regulates growth of both primary and metastatic breast cancer. Cancer Res 64:8604–8612

Sobhani N, Fan CO, Flores-Villanueva P, Generali D, Li Y (2020) The fibroblast growth factor receptors in breast cancer: from oncogenesis to better treatments. Int J Mol Sci 21:2011

Sohrabji F (2015) Estrogen-IGF-1 interactions in neuroprotection: ischemic stroke as a case study. Front Neuroendocrinol 36:1–14

Song RX-D, Zhang Z, Chen Y, Bao Y, Santen RJ (2007) Estrogen signaling via a linear pathway involving insulin-like growth factor I receptor, matrix metalloproteinases, and epidermal growth factor receptor to activate mitogen-activated protein kinase in MCF-7 breast cancer cells. Endocrinology 148:4091–4101

Song X, Liu Z, Yu Z (2020) EGFR promotes the development of triple negative breast cancer through JAK/STAT3 signaling. Cancer Manag Res 12:703–717

Soriano JV, Pepper MS, Orci L, Montesano R (1998) Roles of hepatocyte growth factor/scatter factor and transforming growth factor-β1 in mammary gland ductal morphogenesis. J Mammary Gland Biol Neoplasia 3:133–150

Sosa MS, Lopez-Haber C, Yang C, Wang H, Lemmon MA, Busillo JM, Luo J, Benovic JL, Klein-Szanto A, Yagi H, Gutkind JS, Parsons RE, Kazanietz MG (2010) Identification of the Rac-GEF P-Rex1 as an essential mediator of ErbB signaling in breast cancer. Mol Cell 40:877–892

Srabovic N, Mujagic Z, Mujanovic-Mustedanagic J, Softic A, Muminovic Z, Rifatbegovic A, Begic L (2013) Vascular endothelial growth factor Receptor-1 expression in breast cancer and its correlation to vascular endothelial growth factor a. Int J Breast Cancer 2013:746749

Stefkovich ML, Arao Y, Hamilton KJ, Korach KS (2018) Experimental models for evaluating non-genomic estrogen signaling. Steroids 133:34–37

Stronach B, Perrimon N (2002) Activation of the JNK pathway during dorsal closure in Drosophila requires the mixed lineage kinase, slipper. Genes Dev 16:377–387

Su H, Sobrino Najul EJ, Toth TA, Ng CM, Lelievre SA, Fred M, Tang CK (2011) Chemokine receptor CXCR4-mediated transformation of mammary epithelial cells by enhancing multiple RTKs expression and deregulation of the p53/MDM2 axis. Cancer Lett 307:132–140

Sukocheva O, Wadham C, Holmes A, Albanese N, Verrier E, Feng F, Bernal A, Derian CK, Ullrich A, Vadas MA, Xia P (2006) Estrogen transactivates EGFR via the sphingosine 1-phosphate receptor Edg-3: the role of sphingosine kinase-1. J Cell Biol 173:301–310

Suleymanova N, Crudden C, Shibano T, Worrall C, Oprea I, Tica A, Calin GA, Girnita A, Girnita L (2017) Functional antagonism of β-arrestin isoforms balance IGF-1R expression and signalling with distinct cancer-related biological outcomes. Oncogene 36:5734–5744

Sung H, Ferlay J, Siegel RL, Laversanne M, Soerjomataram I, Jemal A, Bray F (2021) Global cancer statistics 2020: GLOBOCAN estimates of incidence and mortality worldwide for 36 cancers in 185 countries. CA Cancer J Clin 71:209–249

Takekoshi T, Ziarek JJ, Volkman BF, Hwang ST (2012) A locked, dimeric CXCL12 variant effectively inhibits pulmonary metastasis of CXCR4-expressing melanoma cells due to enhanced serum stability. Mol Cancer Ther 11:2516–2525

Tang Y, Qian C, Zhou Y, Yu C, Song M, Zhang T, Min X, Wang A, Zhao Y, Lu Y (2023) Activated platelets facilitate hematogenous metastasis of breast cancer by modulating the PDGFR-β/COX-2 axis. iScience 26:107704

Tatosyan AG, Mizenina OA (2000) Kinases of the Src family: structure and functions. Biochemistry (Mosc) 65:49–58

Tardáguila M, Mira E, García-Cabezas MA, Feijoo AM, Quintela-Fandino M, Azcoitia I, Lira SA, Mañes S (2013) CX3CL1 promotes breast cancer via transactivation of the EGF pathway. Cancer Res 73:4461–4473

Taylor A, Brady AF, Frayling IM, Hanson H, Tischkowitz M, Turnbull C, Side L (2018) Consensus for genes to be included on cancer panel tests offered by UK genetics services: guidelines of the UK Cancer Genetics Group. J Med Genet 55:372–377

Tewari S, Vargas R, Reizes O (2022) The impact of obesity and adipokines on breast and gynecologic malignancies. Ann N Y Acad Sci 1518:131–150

Tilley DG, Kim IM, Patel PA, Violin JD, Rockman HA (2009) β-arrestin mediates β1-adrenergic receptor-epidermal growth factor receptor interaction and downstream signaling. J Biol Chem 284:20375–20386

Timoshenko AV, Chakraborty C, Wagner GF, Lala PK (2006) COX-2-mediated stimulation of the lymphangiogenic factor VEGF-C in human breast cancer. Br J Cancer 94:1154–1163

Torre LA, Siegel RL, Ward EM, Jemal A (2016) Global cancer incidence and mortality rates and trends—an update. Cancer Epidemiol Biomarkers Prev 25:16–27

Trenker R, Jura N (2020) Receptor tyrosine kinase activation: from the ligand perspective. Curr Opin Cell Biol 63:174–185

Tsang JYS, Tse GM (2020) Molecular classification of breast cancer. Adv Anat Pathol 27:27–35

Tuck AB, Park M, Sterns EE, Boag A, Elliott BE (1996) Coexpression of hepatocyte growth factor and receptor (Met) in human breast carcinoma. Am J Pathol 148:225–232

Turner N, Lambros MB, Horlings HM, Pearson A, Sharpe R, Natrajan R, Geyer FC, Van Kouwenhove M, Kreike B, Mackay A, Ashworth A, Van De Vijver MJ, Reis-Filho JS (2010) Integrative molecular profiling of triple negative breast cancers identifies amplicon drivers and potential therapeutic targets. Oncogene 29:2013–2023

Ueda S, Tsuda H, Sato K, Takeuchi H, Shigekawa T, Matsubara O, Hiraide H, Mochizuki H (2006) Alternative tyrosine phosphorylation of signaling kinases according to hormone receptor status in breast cancer overexpressing the insulin-like growth factor receptor type 1. Cancer Sci 97: 597–604

Ueno NT, Zhang D (2011) Targeting EGFR in triple negative breast cancer. J Cancer 2:324–328

Ullrich A, Schlessinger J (1990) Signal transduction by receptors with tyrosine kinase activity. Cell 61:203–212

Vainchenker W, Dusa A, Constantinescu SN (2008) JAKs in pathology: role of Janus kinases in hematopoietic malignancies and immunodeficiencies. Semin Cell Dev Biol 19:385–393
Vajaria R, Vasudevan N (2018) Is the membrane estrogen receptor, GPER1, a promiscuous receptor that modulates nuclear estrogen receptor-mediated functions in the brain? Horm Behav 104: 165–172
Valdehita A, Bajo AM, Schally AV, Varga JL, Carmena MJ, Prieto JC (2009) Vasoactive intestinal peptide (VIP) induces transactivation of EGFR and HER2 in human breast cancer cells. Mol Cell Endocrinol 302:41–48
Vella V, Milluzzo A, Scalisi NM, Vigneri P, Sciacca L (2018) Insulin receptor isoforms in cancer. Int J Mol Sci 19:3615
Vella V, Nicolosi ML, Giuliano M, Morrione A, Malaguarnera R, Belfiore A (2019) Insulin receptor isoform a modulates metabolic reprogramming of breast cancer cells in response to IGF2 and insulin stimulation. Cells 8:1017
Vella V, Giuliano M, La Ferlita A, Pellegrino M, Gaudenzi G, Alaimo S, Massimino M, Pulvirenti A, Dicitore A, Vigneri P, Vitale G, Malaguarnera R, Morrione A, Sims AH, Ferro A, Maggiolini M, Lappano R, De Francesco EM, Belfiore A (2021) Novel mechanisms of tumor promotion by the insulin receptor isoform a in triple-negative breast cancer cells. Cells 10:3145
Vignais M-L, Gilman M (1999) Distinct mechanisms of activation of Stat1 and Stat3 by platelet-derived growth factor receptor in a cell-free system. Mol Cell Biol 19:3727–3735
Villarreal-Garza C, Platas A, Miaja M, Fonseca A, Mesa-Chavez F, Garcia-Garcia M, Chapman JA, Lopez-Martinez EA, Pineda C, Mohar A, Galvez-Hernandez CL, Castro-Sanchez A, Martinez-Cannon BA, Barragan-Carrillo R, Munoz-Lozano JF, Goss P, Bargallo-Rocha JE, Aguilar D, Cardona S, Canavati M (2020) Young women with breast cancer in Mexico: results of the pilot phase of the Joven & Fuerte prospective cohort. JCO Glob Oncol 6:395–406
Visscher DW, Castellani R, Wykes SM, Sarkar FH, Hussain ME (1993) Concurrent abnormal expression of ERBB-2, EGFR, and p53 genes and clinical disease progression of breast carcinoma. Breast Cancer Res Treat 28:261–266
Vivacqua A, Lappano R, De Marco P, Sisci D, Aquila S, De Amicis F, Fuqua SAW, Andò S, Maggiolini M (2009) G protein-coupled receptor 30 expression is up-regulated by EGF and TGFα in estrogen receptor α-positive cancer cells. Mol Endocrinol 23:1815–1826
Volpi CC, Pietrantonio F, Gloghini A, Fuca G, Giordano S, Corso S, Pruneri G, Antista M, Cremolini C, Fasano E, Saggio S, Faraci S, Di Bartolomeo M, de Braud F, Di Nicola M, Tagliabue E, Pupa SM, Castagnoli L (2019) The landscape of d16HER2 splice variant expression across HER2-positive cancers. Sci Rep 9:3545
Voudouri K, Berdiaki A, Tzardi M, Tzanakakis GN, Nikitovic D (2015) Insulin-like growth factor and epidermal growth factor signaling in breast cancer cell growth: focus on endocrine resistant disease. Anal Cell Pathol (Amst) 2015:975495
Wang W, Nag SA, Zhang R (2015) Targeting the NFkappaB signaling pathways for breast cancer prevention and therapy. Curr Med Chem 22:264–289
Wang X, Bove AM, Simone G, Ma B (2020) Molecular bases of VEGFR-2-mediated physiological function and pathological role. Front Cell Dev Biol 8:599281
Ward CW, Lawrence MC (2009) Ligand-induced activation of the insulin receptor: a multi-step process involving structural changes in both the ligand and the receptor. BioEssays 31:422–434
Weigelt B, Lo AT, Park CC, Gray JW, Bissell MJ (2010) HER2 signaling pathway activation and response of breast cancer cells to HER2-targeting agents is dependent strongly on the 3D microenvironment. Breast Cancer Res Treat 122:35–43
Wheeler DL, Yarden Y (2015) Receptor tyrosine kinases: family and subfamilies. Springer, Cham
Wisler JW, Rockman HA, Lefkowitz RJ (2018) Biased G protein–coupled receptor signaling. Circulation 137:2315–2317
Wondmkun YT (2020) Obesity, insulin resistance, and type 2 diabetes: associations and therapeutic implications. Diabetes Metab Syndr Obes 13:3611–3616

Wyss CB, Duffey N, Peyvandi S, Barras D, Martinez Usatorre A, Coquoz O, Romero P, Delorenzi M, Lorusso G, Rüegg C (2021) Gain of HIF1 activity and loss of miRNA let-7d promote breast cancer metastasis to the brain via the PDGF/PDGFR axis. Cancer Res 81: 594–605

Xu W, Allbritton N, Lawrence DS (2012) Src kinase regulation in progressively invasive cancer. PLoS One 7:e48867

Yamasaki F, Zhang D, Bartholomeusz C, Sudo T, Hortobagyi GN, Kurisu K, Ueno NT (2007) Sensitivity of breast cancer cells to erlotinib depends on cyclin-dependent kinase 2 activity. Mol Cancer Ther 6:2168–2177

Yan JD, Liu Y, Zhang ZY, Liu GY, Xu JH, Liu LY, Hu YM (2015) Expression and prognostic significance of VEGFR-2 in breast cancer. Pathol Res Pract 211:539–543

Yang J, Nie J, Ma X, Wei Y, Peng Y, Wei X (2019) Targeting PI3K in cancer: mechanisms and advances in clinical trials. Mol Cancer 18:26

Yang Y, Chen X, Ma C (2020) Insulin receptor is implicated in triple-negative breast cancer by decreasing cell mobility. J Biomed Res 35:189–196

Yarden Y, Pines G (2012) The ERBB network: at last, cancer therapy meets systems biology. Nat Rev Cancer 12:553–563

Yarza R, Vela S, Solas M, Ramirez MJ (2016) c-Jun N-terminal kinase (JNK) signaling as a therapeutic target for Alzheimer's disease. Front Pharmacol 6:321

Yasar P, Ayaz G, User SD, Gupur G, Muyan M (2017) Molecular mechanism of estrogen-estrogen receptor signaling. Reprod Med Biol 16:4–20

Yu G, Zhou A, Xue J, Huang C, Zhang X, Kang S-H, Chiu W-T, Tan C, Xie K, Wang J, Huang S (2015) FoxM1 promotes breast tumorigenesis by activating PDGF-A and forming a positive feedback loop with the PDGF/AKT signaling pathway. Oncotarget 6:11281–11294

Yu S, Sun L, Jiao Y, Lee LTO (2018) The role of G protein-coupled receptor kinases in cancer. Int J Biol Sci 14:189–203

Yuan J, Liu M, Yang L, Tu G, Zhu Q, Chen M, Cheng H, Luo H, Fu W, Li Z, Yang G (2015) Acquisition of epithelial-mesenchymal transition phenotype in the tamoxifen-resistant breast cancer cell: a new role for G protein-coupled estrogen receptor in mediating tamoxifen resistance through cancer-associated fibroblast-derived fibronectin and β1-integrin signaling pathway in tumor cells. Breast Cancer Res 17:69

Zaczek A, Brandt B, Bielawski KP (2005) The diverse signaling network of EGFR, HER2, HER3 and HER4 tyrosine kinase receptors and the consequences for therapeutic approaches. Histol Histopathol 20:1005–1015

Zajchowski D, Band V, Pauzie N, Tager A, Stampfer M, Sager R (1988) Expression of growth factors and oncogenes in normal and tumor-derived human mammary epithelial cells. Cancer Res 48:7041–7047

Zardavas D, Phillips WA, Loi S (2014) PIK3CA mutations in breast cancer: reconciling findings from preclinical and clinical data. Breast Cancer Res 16:201

Zewdie ET, Tessema AW, Simegn GL (2021) Classification of breast cancer types, sub-types and grade from histopathological images using deep learning technique. Heal Technol 11: 1277–1290

Zhang D, LaFortune TA, Krishnamurthy S, Esteva FJ, Cristofanilli M, Liu P, Lucci A, Singh B, Hung MC, Hortobagyi GN, Ueno NT (2009) Epidermal growth factor receptor tyrosine kinase inhibitor reverses mesenchymal to epithelial phenotype and inhibits metastasis in inflammatory breast cancer. Clin Cancer Res 15:6639–6648

Zhang L, Yuan C, Peng J, Zhou L, Jiang Y, Lin Y, Yin W, Xu S, Ma J, Lu J (2020) SHP-2-mediated upregulation of ZEB1 is important for PDGF-B-induced cell proliferation and metastatic phenotype in triple negative breast cancer. Front Oncol 10:1230

Zhang X, Zhu X, Bi X, Huang J, Zhou L (2022) The insulin receptor: an important target for the development of novel medicines and pesticides. Int J Mol Sci 23:7793

Zheng H, Shen H, Oprea I, Worrall C, Stefanescu R, Girnita A, Girnita L (2012a) Beta-arrestin-biased agonism as the central mechanism of action for insulin-like growth factor 1 receptor-targeting antibodies in Ewing's sarcoma. Proc Natl Acad Sci USA 109:20620–20625

Zheng H, Worrall C, Shen H, Issad T, Seregard S, Girnita A, Girnita L (2012b) Selective recruitment of G protein-coupled receptor kinases (GRKs) controls signaling of the insulin-like growth factor 1 receptor. Proc Natl Acad Sci USA 109:7055–7060

Zhou B, Lin W, Long Y, Yang Y, Zhang H, Wu K, Chu Q (2022) Notch signaling pathway: architecture, disease, and therapeutics. Signal Transduct Target Ther 7:95

Zhu X, Zhou W (2015) The emerging regulation of VEGFR-2 in triple-negative breast cancer. Front Endocrinol (Lausanne) 6:159

Zou X, Tang X-Y, Qu Z-Y, Sun Z-W, Ji C-F, Li Y-J, Guo S-D (2022) Targeting the PDGF/PDGFR signaling pathway for cancer therapy: a review. Int J Biol Macromol 202:539–557

# Addressing *ESR1* Mutation: A Key Factor in Hormone Therapy Resistance in Breast Cancer

Takashi Takeshita and Hirotaka Iwase

**Abstract**

Breast cancer is a prevalent and heterogeneous disease, particularly the hormone receptor (HR)-positive/HER2-negative subtype. Historically, endocrine therapy (ET) was the primary approach for HR-positive/HER2-negative advanced breast cancer (ABC). Despite its efficacy, about 20% develop resistance. Biologically directed therapies, including cyclin-dependent kinase 4/6 inhibitors (CDK4/6i), have gained prominence. Understanding ET resistance mechanisms, particularly *ESR1* mutations, is crucial for effective HR-positive/HER2-negative ABC treatment.

*ESR1* mutations play a pivotal role in ET resistance in HR-positive/HER2-negative ABC. These mutations confer constitutive, neomorphic, and hypermorphic activity to *ESR1*, contributing to resistance mechanisms. Liquid biopsy, particularly circulating cell-free DNA analysis, has emerged as a noninvasive method for assessing *ESR1* mutations, overcoming challenges associated with tumor tissue analysis. *ESR1* mutations, detected in 20–40% of ABC cases undergoing aromatase inhibition, significantly impact prognosis and therapeutic responses. Strategic management of *ESR1* mutations involves monitoring and adjusting treatment approaches, such as transitioning to fulvestrant in the presence of mutations during aromatase inhibitors plus cycline-dependent kinase 4/6 inhibitors therapy. The importance of assessing other genetic mutations, like *PIK3CA*, in predicting response to fulvestrant is also highlighted.

This chapter explores *ESR1* mutations in HR-positive/HER2-negative ABC, focusing on their prevalence, clinical implications, and role in therapy resistance. Strategic management involves vigilant monitoring and tailored treatments, emphasizing the clinical potential of liquid biopsy.

T. Takeshita (✉) · H. Iwase
Department of Breast and Endocrine Surgery, Kumamoto City Hospital, Kumamoto, Japan
e-mail: takashi.takeshita@hyper.ocn.ne.jp

 
Interdisciplinary Cancer Research, https://doi.org/10.1007/16833_2024_251
Published online: 24 May 2024

**Keywords**

Advanced breast cancer · *ESR1* mutations · Endocrine therapy resistance · Liquid biopsy · *PIK3CA* mutations

## Abbreviations

| | |
|---|---|
| ABC | Advanced breast cancer |
| AI | Aromatase inhibitor |
| CDK4/6i | Cyclin-dependent kinase 4 and 6 inhibitors |
| cfDNA | Cell-free DNA |
| CI | Confidence interval |
| CNVs | Copy number variations |
| ER | Estrogen receptor |
| ET | Endocrine therapy |
| HER2 | Human epidermal growth factor receptor 2 |
| HR | Hormone receptor |
| LBD | Ligand binding domain |
| NGS | Next-generation sequencing |
| OS | Overall survival |
| PFS | Progression-free survival |
| WES | Whole-exome sequencing |
| WGS | Whole-genome sequencing |

## 1 Introduction

Breast cancer remains the most common cancer type affecting women worldwide, with one in nine Japanese women expected to be diagnosed within their lifetime. The Foundation for Promotion of Cancer Research projected that in 2022, 94,300 Japanese women would receive a breast cancer diagnosis, comprising 22% of all newly reported cancer cases in women. Additionally, the estimation indicated that 15,600 Japanese women would succumb to breast cancer in 2022, accounting for 10% of all female cancer–related deaths that year (The Japanese Breast Cancer Society Clinical Practice Guidlines for Breast Cancer 2022).

Breast cancer exhibits significant heterogeneity, even within its diverse subtypes. The most prevalent subtype remains hormone receptor (HR)-positive/human epidermal growth factor receptor 2 (HER2)-negative breast cancer, constituting approximately 80% of all breast cancer cases (Siegel et al. 2023). The estrogen signaling pathway plays a crucial role in estrogen receptor (ER)-positive breast cancer, and various strategies have been introduced to inhibit it. Historically, prior to 2010, the predominant approach for HR-positive/HER2-negative advanced breast cancer (ABC) in women involved endocrine therapy (ET). This included, for premenopausal women, the use of luteinizing hormone–releasing hormone agonists and selective ER modulators like tamoxifen and toremifene, and for postmenopausal women, aromatase inhibitors (AIs) such as exemestane (a steroidal AI), anastrozole, and letrozole (a nonsteroidal AI) due to their effectiveness, carefully balanced against associated side effects (Lumachi et al. 2015).

Despite the efficacy of these ETs in reducing recurrence and breast cancer-specific deaths, about 20% of all primary breast cancers develop resistance to ET (Pan et al. 2017). Nearly all ER-positive recurrent ABC patients acquire anti-estrogen resistance following repeated ETs (Osborne and Schiff 2011). Over the past decade and a half, there has been a notable shift towards the development of biologically targeted therapies in breast cancer management (Raheem et al. 2023). A significant finding from the FALCON study indicated improved progression-free survival (PFS) with fulvestrant 500 mg compared to anastrozole 1 mg as a first-line treatment for individuals with metastatic disease confined to the bones (Robertson et al. 2016). In the context of second-line metastatic treatment, the BOLERO-2 (everolimus (mTOR inhibitor) and exemestane versus exemestane and placebo in patients with HR-positive, HER2-negative aABC who has recurrence or progression while receiving previous therapy with a nonsteroidal AI in the adjuvant setting or to treat advanced disease (phase III trial)) study demonstrated enhanced PFS, albeit without a corresponding overall survival (OS) advantage, by incorporating the mammalian target of rapamycin inhibitor everolimus alongside exemestane, as opposed to exemestane alone (Baselga et al. 2012). Cyclin-dependent kinase 4 and 6 inhibitors (CDK4/6i) have emerged as pivotal agents in the landscape of breast cancer therapeutics (Fig. 1). Concerning palbociclib, the PALOMA-2 (fulvestrant

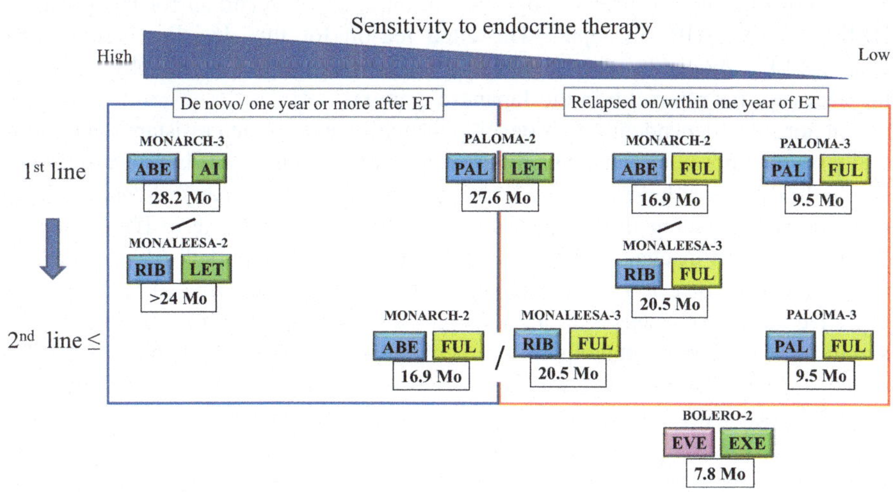


**Fig. 1** Correlation between key clinical trials and sensitivity to endocrine therapy in hormone receptor-positive/human epidermal growth receptor 2–negative breast Cancer therapeutics. *ET* endocrine therapy, *ABE* abemaciclib, *PAL* palbociclib, *RIB* ribociclib, *AI* aromatase inhibitor, *LET* letrozole, *FUL* fulvestrant, *EVE* everolimus, *EXE* exemestane

with or without palbociclib in premenopausal and postmenopausal women with HR-positive, HER2-negative ABC (phase III trial)) trial, which randomized 666 patients in a 2:1 ratio between the AI and palbociclib arm and the AI and placebo arm, was the initial trial to report its findings. The trial revealed a PFS hazard ratio of 0.58 with a 95% confidence interval (CI) of [0.46; 0.72] (Finn et al. 2016). The PALOMA-3 trial, which randomized 521 patients in a 2:1 ratio between fulvestrant and palbociclib and fulvestrant and placebo, was the first reported, showing a PFS hazard ratio of 0.46 with a 95% CI of [0.36; 0.59] (Cristofanilli et al. 2016). Concerning ribociclib, the MONALEESA-2 trial randomized 668 patients in a 1:1 fashion between AI and ribociclib, and AI and placebo, revealing a PFS hazard ratio of 0.56, 95% CI [0.43; 0.72] (Hortobagyi et al. 2016a). In the MONALEESA-3 trial, 726 patients were randomized in a 2:1 ratio between fulvestrant and ribociclib, and fulvestrant and placebo, disclosing a PFS hazard ratio of 0.59, 95% CI [0.48; 0.73] (Slamon et al. 2018). Regarding abemaciclib, the MONARCH-3 (abemaciclib (CDK4/6i) or placebo plus a nonsteroidal AI in postmenopausal women with HR-positive, HER2-negative ABC who had no prior systemic theory in the advanced setting (phase III trial)) trial included 493 patients randomized in a 2:1 ratio between AI and abemaciclib, and AI and placebo, with a PFS hazard ratio of 0.54, 95% CI [0.41; 0.72] (Goetz et al. 2017). The MONARCH-2 (abemaciclib monotherapy or in combination with tamoxifen in patients with endocrine-refractory HR-positive, HER2-negative ABC (phase II trial)) trial involved 669 patients randomized in a 2:1 ratio between AI and abemaciclib, and AI and placebo, with a PFS hazard ratio of 0.54, 95% CI [0.45; 0.65] (Sledge et al. 2020). In Japan, palbociclib obtained approval in 2018 as a first-line treatment for HR-positive/HER2-negative ABC in postmenopausal women, followed by the approval of abemaciclib in 2020. However, ribociclib did not receive approval for any line of treatment for HR-positive/HER2-negative ABC in Japan. The cited reason for this decision includes the approval of palbociclib, which addressed unmet medical needs, thereby resolving the necessity for ribociclib in the Japanese context.

All hormone-dependent ABCs develop primarily antiestrogen resistance and eventually become hormone-independent diseases, even resistant to the innovative combination therapies that have emerged in recent years. Therefore, many researchers have focused on elucidating the molecular mechanisms that contribute to ET resistance. Although several mechanisms of ET resistance have been reported, *ESR1* mutations are a strong candidate as a definitive biomarker of ET resistance (Brett et al. 2021). The objective of this chapter is to furnish oncologists with an understanding of the significance of *ESR1* mutations in the therapeutic landscape of HR-positive/HER2-negative ABC. Additionally, we offer perspectives from our clinical practice and present recommendations based on best practices where relevant.

## 2 Unraveling the Molecular Landscape of *ESR1* Mutations

Considerable insight into the molecular ramifications of *ESR1* mutations has been gained, benefiting from a robust understanding of ER structure and signaling, coupled with ongoing mechanism-based drug development (Brett et al. 2021). The fundamental mechanism by which *ESR1* mutations induce resistance is through constitutive activity. Unlike wild-type *ESR1*, which relies on estrogen ligand binding for coactivator recruitment, ligand binding domain (LBD)-mutated *ESR1* exhibits constitutive activity, rendering it impervious to AI-induced estrogen depletion (Robinson et al. 2013; Toy et al. 2013; Jeselsohn et al. 2014). Beyond constitutive activity, *ESR1* mutations acquire neomorphic and hypermorphic traits, transactivating modified sets of target genes that augment motility and metastasis (Martin et al. 2017; Jeselsohn et al. 2018; Arnesen et al. 2021). The Japan Breast Cancer Research Group-M07 (FUTURE—fulvestrant with additional palbociclib in advanced or metastatic HR-positive, HER2-negative breast cancer after progression to fulvestrant monotherapy (phase II trial)) trial investigated the use of palbociclib as a supplementary therapy following fulvestrant resistance in HR-positive/HER2-negative ABC. Recent findings from this study indicate a diminishing correlation between *ESR1* mutations and mutations in *PIK3CA*, *MAP3K1*, and *TP53* as the treatment with palbociclib in fulvestrant-resistant breast cancer advances (Takeshita et al. 2024). Moreover, *ESR1* mutations demonstrate slightly modified interactomes, displaying an enhanced association with *FOXA1* and *GREB1*, although entirely new partners have not been conclusively identified (Martin et al. 2017). Additionally, even in the presence of estrogen, mutated *ESR1* can exhibit significantly higher transactivation ability than wild-type *ESR1* (Li et al. 2013; Robinson et al. 2013; Jeselsohn et al. 2014).

## 3 Advancements in Liquid Biopsy: Overcoming Challenges in *ESR1* Mutation Assessment and Beyond

While these features underscore the importance of assessing the presence of *ESR1* mutations in ABC, there are three limitations associated with analyzing genetic abnormalities using tumor tissue material. Firstly, tumor biopsy poses inconvenience from the perspective of a monitoring tool, especially when the tumor site is not easily accessible. Secondly, genetic degradation resulting from formalin-fixed paraffin embedding and intra- or inter-tumor heterogeneity further hinders the utilization of tumor tissue material. Lastly, genetic mutation monitoring demands multiple accesses, always carrying the inherent risk of clinical complications (Overman et al. 2013).

Circulating cell-free DNA (cfDNA) analysis has been developed to overcome the limitations associated with tumor tissue material and has emerged as a noninvasive method for rapidly assessing mutational profiles and monitoring molecular changes during treatment, leveraging advancements in digital genomic technologies (Diaz and Bardelli 2014). To underscore the clinical applicability of cfDNA *ESR1* mutations, the ability to identify *ESR1* point mutations from limited cfDNA lengths and quantities is imperative (Dawson et al. 2013). Conventional sequencing approaches such as Sanger sequencing or pyrosequencing can only identify tumor-derived mutant fragments in patients with a substantial tumor burden and elevated levels of cfDNA. Droplet digital polymerase chain reaction technology addresses this limitation by offering superior accuracy (Hindson et al. 2013). This assay can detect *ESR1* LBD–variant molecules, specifically *ESR1* Y537S/N/C and D538G, at concentrations as low as 0.25 copies/μL (Takeshita and Iwase 2019), making it well suited for identifying cfDNA mutations in plasma (Takeshita et al. 2016, 2017, 2018). While droplet digital polymerase chain reaction methods offer sensitivity and cost-effectiveness, their screening capabilities are restricted to known variants, and they have limitations in terms of input and speed.

Next-generation sequencing (NGS) stands out for its high throughput and ability to screen for unknown variants, with the current capacity to detect minor allele frequencies <1% (Baer et al. 2016). Indeed, various methods, such as unique molecular identifiers (Hu et al. 2021) or unique barcodes (Phallen et al. 2017), can enhance sensitivity and reduce false negatives when utilizing NGS. NGS has been implemented in the sequencing of cfDNA in the realm of liquid biopsy. The utilization of targeted panels in NGS facilitates precise and highly sensitive detection of cfDNA mutations (Takeshita et al. 2024). Numerous methods leverage NGS for targeted panels, including Tagged-Amplicon deep sequencing, Safe-Sequencing System, CAncer Personalized Profiling by deep sequencing, and Ion Torrent (Chen and Zhao 2019). While targeted panels are favored for their high sensitivity and cost-effectiveness, they are limited to detecting point mutations and indels. NGS's unique advantage lies in its application to untargeted panels for genome-wide DNA variation discovery. Untargeted approaches, although lacking prior knowledge, can reveal genome-wide copy number patterns or assess mutation spectra. Whole-genome sequencing (WGS) is commonly employed to obtain the complete genomic profile of tumor DNA, encompassing point mutations, indels, rearrangements, and copy number variations (CNVs) (Chen and Zhao 2019). Despite offering comprehensive information, WGS is expensive and less sensitive. Whole-exome sequencing (WES) serves as a popular, cost-effective alternative by focusing solely on exonic sequencing. However, both WGS and WES necessitate high input sample volumes, limiting their application in scenarios with low cfDNA concentrations, such as screening and early diagnosis. Various genome-wide sequencing methods, tailored for different variation types, have been proposed, including personalized analysis of rearranged ends for rearrangement detection, digital karyotyping for DNA content quantification, and fast aneuploidy screening test-sequencing system for CNV detection (Chen and Zhao 2019). Liquid biopsy assays hold clinical promise; however, their technical intricacies present challenges. The growing adoption of liquid biopsies in clinical settings underscores their potential, yet the acquisition of further clinical validity and utility data remains imperative.

## 4 *ESR1* Mutations in Breast Cancer: Implications for Endocrine Therapy Resistance, Prognosis, and Therapeutic Approaches

Nearly two decades ago, the discovery of *ESR1* mutations as contributors to ET resistance in breast cancer was initially reported (Karnik et al. 1994; Roodi et al. 1995; Weis et al. 1996; Zhang et al. 1997). Recent breakthroughs in highly sensitive technologies have unequivocally substantiated the role of *ESR1* mutations as primary drivers of endocrine therapy resistance. (Merenbakh-Lamin et al. 2013; Robinson et al. 2013; Toy et al. 2013; Jeselsohn et al. 2014). In advanced ER-positive hormone–resistant breast cancer, various research groups (Garcia-Murillas et al. 2015; Schiavon et al. 2015; Chandarlapaty et al. 2016; Fribbens et al. 2016; Takeshita et al. 2016) have identified approximately 12 acquired *ESR1* point mutations, concentrated in a hot spot located at Tyr537 and Asp538 in helix 12 of the LBD of the *ESR1* gene. In a clinical setting, the prospective sequencing of 1,501 HR-positive breast cancers by Razavi et al. unveiled that approximately half of the instances of endocrine resistance are linked to *ESR1* mutations (Razavi et al. 2018). Moreover, there has been a growing revelation of other mechanisms, such as changes in the PI3K-AKT-mTORC1, RAS-MAPK, and CDK4/6-RB-E2F pathways, along with occurrences of *ESR1* loss, amplification, and translocation. Notably, *ESR1* mutations were found to be the most prevalent among these mechanisms. Hence, *ESR1*-activating hot spot mutations are considered key potential mechanisms contributing to the failure of ETs.

*ESR1* mutations are detected in approximately 20–40% of patients receiving aromatase inhibition for ABC, and the prevalence varies depending on the sites of metastatic disease (Takeshita et al. 2015; Zhang et al. 2018). Contrastingly, the incidence of *ESR1* mutations is markedly lower, ranging only between 4 and 5%, in cases of ABC following prior adjuvant AI, encompassing instances of recurrence during adjuvant AI treatment (Zhang et al. 2018). Therefore, *ESR1* mutations in HR-positive breast cancer predominantly arise following AI treatment in the metastatic setting. Even within the metastatic setting, our research indicates that the frequency of *ESR1* mutations increases with advancing treatment lines (Zhang et al. 2018). Furthermore, considering the current treatment approaches that combine ET with targeted therapies like CDK4/6, PI3K, or mTORC1 inhibitors, it is essential to assess their influence on *ESR1* mutations. In the MONARCH-3 trial, *ESR1* mutations were rare at baseline in both the placebo and abemaciclib groups but were most commonly acquired during the study, with 31.6% in the placebo group and 19.8% in the abemaciclib group at the end of treatment (Goetz et al. 2023) (Table 1). This suggests that combining AI with CDK4/6i does not necessarily suppress the development of *ESR1* mutations. The BOLERO-2 trial provides thorough evidence of the importance of *ESR1* mutations in the context of second-line therapy. In a cohort of 541 ER-positive ABC patients, 156 (28.8%) displayed the D538G and/or the Y537S *ESR1* mutation in plasma cfDNA. These mutations, indicative of *ESR1* LBD mutations, were each associated with a shortened PFS and OS (Chandarlapaty et al. 2016). The pooled analysis revealed a substantial correlation between plasma *ESR1* mutations and poorer PFS when compared with wild-type *ESR1* (hazard ratio: 1.40, 95% CI: 1.17–1.66; $P < 0.0001$) in patients with ER-positive ABC. The combined hazard ratio, utilizing a fixed-effects model for OS, was 1.65 (95% CI: 1.36–2.01; $P < 0.001$), signifying a prognostic association of *ESR1* mutations in individuals with ER-positive ABC (Zhang et al. 2018). Furthermore, the prognostic significance of *ESR1* mutations might be more pronounced in

**Table 1** Comprehensive analysis of the impact of ESR1 mutations on clinical outcomes

| Study | % Prior AI | % ESR1 mut | % Changes due to treatment of ESR1 mut | Drugs | Median months | | | | | | References |
|---|---|---|---|---|---|---|---|---|---|---|---|
| | | | | | PFS | | | OS | | | |
| | | | | | ESR1 mut | ESR1 WT | Significant difference | ESR1 mut | ESR1 WT | Significant difference | |
| FERGI | 100 | 37.3 | | FUL | 3.5–7.4 | 3.7 | | | | | Spoerke et al. (2016) |
| SoFEA/EFECT | 98.9 | 23.4 | | EXE | 2.4 | 4.8 | Yes | | | | Turner et al. (2020) |
| | | | | FUL | 3.9 | 4.1 | | | | | |
| BOLERO-2 | 100 | 28.8 | | EXE | 2.8 | 4 | Yes | 20.73 | 32.1 | Yes | Chandarlapaty et al. (2016) |
| | | | | EXE + EVE | 5.4 | 8.5 | Yes | | | | |
| MONARCH-3 | 28.7 | 5.1 | Acquired 31.6 | AI | 19.2* | 9.4* | | | | | Goetz et al. (2023) |
| | 26.4 | | Acquired 19.8 | AI+ABE | 14.2* | 16.7* | | | | | |
| nextMONARCH | NA | 40.3 | Acquired 6.3 | TAM + ABE | 7.2 | 9.01 | | | | | |
| PALOMA-3 | 25.7 | 21.8 | EOT 18.4 | FUL | 3.6 | 5.4 | | 20.2 | 28 | Yes | Fribbens et al. (2016), Cristofanilli et al. (2022) |
| | 51.7 | | | FUL + PAL | 11.3 | 11.1 | | 27.7 | 32.8 | | |
| FUTURE | 81.9 | 30 | D15 21, EOT 41 | FUL + PAL | 7.9 | 3.4 | | | | | Takeshita et al. (2024) |
| PADA-1 | 36.9 | 100 | | AI+PAL | 5.7 | | Yes | | | | Bidard et al. (2022) |
| | 34.1 | 100 | | (AI→FUL) +PAL | 11.9 | | | | | | |

*AI* aromatase inhibitor, *PFS* progression free survival, *OS* overall survival, *WT* wild-type, *FUL* fulvestrant, *EXE* exemestane, *EVE* everolimus, *ABE* abemaciclib, *NA* not applicable, *TAM* tamoxifen, *EOT* end of treatment, *PAL* palbociclib, *D15* day 15 of one cycle

*The analysis did not rely on the presence or absence of ESR1 mutations but rather on the presence or absence of acquired ESR1 mutations compared to baseline

patients undergoing second-line and subsequent ET due to the notably higher mutation rates, in contrast to patients undergoing first-line ET (Liao et al. 2020).

Breast cancer harboring *ESR1* mutations is anticipated to retain some sensitivity to fulvestrant. The FERGI (pictilisib (an oral inhibitor of class I phosphatidylinositol 3 kinase) for ER-positive, AI-resistant ABC (phase II trial)) (Spoerke et al. 2016), SoFEA (fulvestrant plus anastrozole or placebo versus exemestane alone after progression on nonsteroidal AIs in postmenopausal patients with HR-positive locally ABC (phase III trial)) and EFECT (evaluation of fulvestrant versus exemestane (phase III trial)) (Turner et al. 2020), PALOMA-3 (O'Leary et al. 2018; Cristofanilli et al. 2022), and FUTURE (Takeshita et al. 2024) trials collectively illustrate that the existence of cfDNA *ESR1* mutations correlates with reduced sensitivity to AI and diminished survival, while the overall *ESR1* mutation status does not forecast an inferior response to fulvestrant (Table 1). For instance, the SoFEA and EFECT trials, which evaluated fulvestrant against exemestane in ABC patients progressing on nonsteroidal AIs, underwent retrospective analysis for preexisting *ESR1* mutations in plasma cfDNA (Turner et al. 2020). Baseline *ESR1* mutations were detected in 39% of SoFEA patients and 23% of EFECT patients, indicating a significantly shorter PFS compared to *ESR1* wild-type counterparts: 2.4 months versus 4.8 months. Furthermore, patients with *ESR1* mutations exhibited a notably lower one-year OS rate, with rates of 62% in contrast to 79%. Recommendations for fulvestrant in the context of *ESR1* mutations may benefit from insights provided by databases like OncoKB, a platform offering information on genomic alterations and therapeutic agents in cancer. *ESR1* mutations could preferentially influence the effectiveness of the anti-estrogen element within a combined regimen, making the tumor response contingent on the sensitivity to the companion agent (such as CDK4/6i, PI3K inhibitors, or mTORC1 inhibitors). Therefore, effective modeling of combination therapies is crucial for translational research efforts.

## 5 Exploring the Significance of *ESR1* Mutations in Hormone Receptor–Positive Advanced Breast Cancer

According to the Japanese Breast Cancer Society Clinical Practice Guidelines, ABC is categorized into distinct forms: de novo, recurrence during adjuvant therapy (termed early recurrence), recurrence within one year post-adjuvant therapy (referred to as intermediate recurrence), and recurrence elapsed for more than one year post-adjuvant therapy (referred to as late recurrence). In HR-positive/HER2-negative ABC without life-threatening visceral metastases, both de novo and late recurrence are considered hormone-sensitive. For these cases, the recommended treatment involves an AI in combination with a CDK4/6i. Early and intermediate recurrence are deemed hormone resistant, and the guideline suggests fulvestrant in combination with a CDK4/6i (The Japanese Breast Cancer Society Clinical Practice Guidlines for Breast Cancer 2022).

The significance of *ESR1* mutations in the context of fulvestrant plus CDK4/6i was thoroughly investigated in the PALOMA-3 (Fribbens et al. 2016; Cristofanilli et al. 2022), FUTURE (Takeshita et al. 2024), and PADA-1 (switch to fulvestrant and palbociclib versus no switch in ABC with rising ESR1 mutation during AI and palbociclib therapy (phase III trial)) (Bidard et al. 2022) studies (Table 1). The PALOMA-3 study revealed that *ESR1* mutations were present in 25.3% (91/360) of plasma samples at baseline and/or the end of treatment, and the *ESR1* status did not impact the PFS benefit of palbociclib (Fribbens et al. 2016). However, when considering OS, palbociclib in combination with fulvestrant demonstrated a clear advantage over placebo combined with fulvestrant (Cristofanilli et al. 2022). In individuals harboring *ESR1* mutations, the median OS was 27.7 months (95% CI, 20.4–36.1) when treated with palbociclib plus fulvestrant, contrasting with 20.2 months (95% CI, 15.3–27.1) observed with placebo plus fulvestrant. The hazard ratio was 0.59 (95% CI, 0.37–0.94), indicating a significant improvement in survival with the use of palbociclib plus fulvestrant (Cristofanilli et al. 2022). The FUTURE study revealed that *ESR1* mutations in 30% of patients at fulvestrant resistance, increasing to 41% by the treatment's end, and the *ESR1* status did not influence the PFS benefit of palbociclib add-on therapy for fulvestrant-resistant breast cancer (Takeshita et al. 2024). The significance of *ESR1* mutations in the context of an AI and a CDK4/6i has been comprehensively explored in the PADA1 study. PADA1 is a multicenter, open-label, randomized controlled phase III trial conducted across 83 hospitals in France (Bidard et al. 2022). The study focused on investigating *ESR1* mutations during an AI plus palbociclib therapy in women aged 18 years or older with ER positive/HER2 negative, locally evaluated ABC who had not received prior systemic therapy for advanced disease. Patients with elevated *ESR1* mutations and no concurrent disease progression were monitored and randomly assigned to either continue treatment with the same AI or switch to fulvestrant. Both groups continued to receive palbociclib. The median PFS from random assignment was 11.9 months (95% CI 9.1–13.6) in the fulvestrant and palbociclib group, contrasting with 5.7 months (95% CI 3.9–7.5) in the AI and palbociclib group. The stratified hazard ratio was 0.61 (95% CI 0.43–0.86; $p = 0.004$), underscoring the significance of monitoring *ESR1* mutations in influencing treatment outcomes.

## 6 Strategic Management of *ESR1* Mutations in Hormone-Positive Breast Cancer

If *ESR1* mutations occur, they can impact OS (Chandarlapaty et al. 2016; Cristofanilli et al. 2022), influencing subsequent treatment. Therefore, it is crucial to implement treatment strategies while actively managing *ESR1* mutations. Treatment strategies for HR-positive/HER2-negative ABC based on *ESR1* mutations are easier to understand based on sensitivity to ETs (Fig. 2). ABC, which exhibits high sensitivity to ETs, is initially treated with AI plus CDK4/6i therapy. In this context, effective management of the *ESR1* mutation is crucial. Treatment approaches involving AI plus CDK4/6i and considering *ESR1* mutations were investigated in

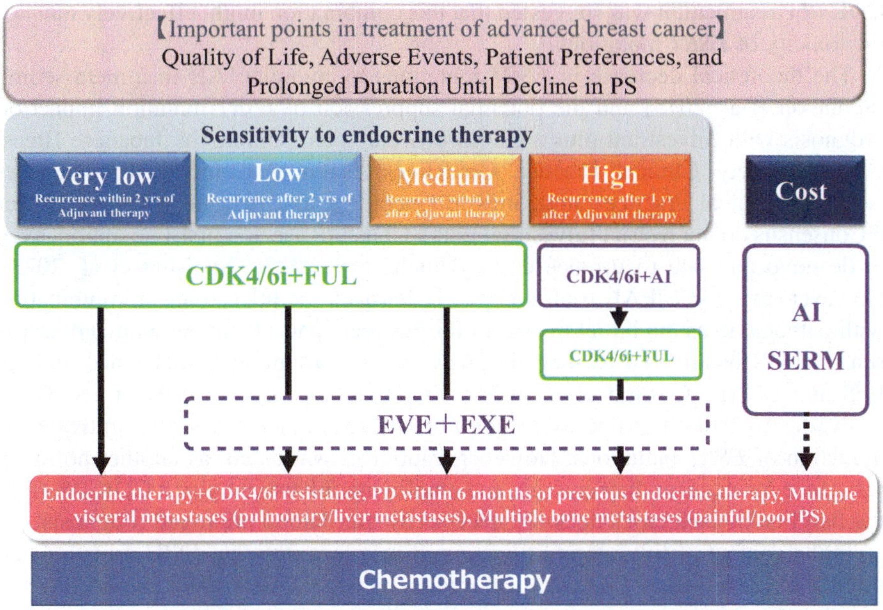


**Fig. 2** Tailoring treatment for hormone receptor-positive/human epidermal growth receptor 2–negative breast cancer based on endocrine therapy sensitivity. *PS* performance status, *CDK4/6i* cycline dependent kinase 4/6 inhibitor, *FUL* fulvestrant, *AI* aromatase inhibitor, *SERM* selective estrogen receptor modulator, *EVE* everolimus, *EXE* exemestane, *PD* progressive disease

the MONARCH-3 study, wherein genetic mutations were evaluated at the initiation and completion of treatment. The study revealed that 19.8% of *ESR1* mutations were acquired during the course of AI plus abemaciclib treatment (Goetz et al. 2023). Moreover, optimal utilization of an AI plus CDK4/6i strategy seems to require monitoring *ESR1* mutations, aligning with the approach outlined in the aforementioned PADA1 study (Bidard et al. 2022). Even if treatment is initiated with AI, transitioning the patient to fulvestrant upon the emergence of *ESR1* mutations allows for the full utilization of an AI plus a CDK4/6i treatment.

For the treatment of ET-low sensitivity (very low to medium) ABC, a regimen involving fulvestrant plus CDK4/6i is employed. ET-resistant cancers, driven by *ESR1* mutations, can be addressed through fulvestrant treatment. In the PALOMA-3 study, *ESR1* mutations were identified in 25.1% of patients at the treatment onset, with a slight increase to 31.3% at the conclusion of fulvestrant plus palbociclib therapy. Notably, 12.8% of patients initially lacking *ESR1* mutations exhibited new mutations upon disease progression (O'Leary et al. 2018). The FUTURE trial revealed *ESR1* mutations in 30% of patients at fulvestrant resistance, increasing to 41% by the treatment's end (Takeshita et al. 2024). *ESR1* mutation did not emerge as a prognostic factor for PFS in fulvestrant plus palbociclib in both PALOMA-3 and FUTURE studies. Despite the increase in *ESR1* mutations during fulvestrant plus a

CDK4/6i treatment, it was suggested that this combination might effectively manage the toxicity of *ESR1* mutations.

The theoretical decrease in *ESR1* mutations by avoiding AIs in a meta setting (Schiavon et al. 2015) and the potential suppression of *ESR1* mutation impact on prognosis with fulvestrant plus a CDK4/6i were noted. Unlike the Japanese Breast Cancer Society Clinical Practice Guidelines, National Comprehensive Cancer Network Clinical Practice Guidelines in Oncology, and Advanced Breast Cancer 6 Consensus do not restrict fulvestrant plus a CDK4/6i use as a first-line option, even in de novo and late recurrence cases (Untch et al. 2022; Gradishar et al. 2023). However, the PARSIFAL trial, comparing letrozole or fulvestrant in combination with palbociclib as the initial treatment for postmenopausal women with endocrine-sensitive HR-positive/HER2-negative ABC, found no significant differences in both PFS and OS (Di Cosimo et al. 2023). Consequently, utilizing fulvestrant plus a CDK4/6i from the first line in de novo or late recurrence cases may mitigate the induction of *ESR1* mutations. However, caution is warranted, as another hormone resistance mechanism may emerge, and employing fulvestrant plus a CDK4/6i from the first line in such cases does not necessarily translate to improved prognosis.

In the context of fulvestrant utilization, it is noteworthy that *ESR1* mutations do not represent the primary mechanism of resistance to ET. Therefore, the exploration of other genetic mutations becomes clinically imperative. The potential use of *PIK3CA* mutations as a biomarker for predicting clinical response to ET, including fulvestrant, is under investigation in large clinical trials analyzing *ESR1* mutations. Hotspot mutations, particularly in the helical (E542K and E545K) and kinase (H1047R) domains of *PIK3CA*, represent 80% to 90% of all observed *PIK3CA* mutations in human malignancies (Karakas et al. 2006). Digital PCR assays on plasma cfDNA in two phase III cohorts revealed the clinical characteristics of hotspot mutations in *PIK3CA*, a prevalent mutation in ABC. The BOLERO-2 cohort demonstrated varied frequencies of *PIK3CA* mutations between metastatic and primary breast tumors (Hortobagyi et al. 2016b).

Moynahan et al. also found that the median PFS was similar between the everolimus and placebo groups, irrespective of *PIK3CA* mutation status (hazard ratio of 0.43 for wild-type and 0.37 for mutant *PIK3CA*). Additionally, everolimus showed a comparable extension in median PFS for patients with specific *PIK3CA* mutations, including H1047R (hazard ratio: 0.37) and E545K/E542K mutations (hazard ratio: 0.30) (Moynahan et al. 2017). In the PALOMA-3 study, Cristofanilli et al. evaluated treatment resistance based on the *PIK3CA* mutational status in circulating cfDNA. The *PIK3CA* status did not significantly affect the treatment response to fulvestrant, with or without CDK4/6i (Cristofanilli et al. 2016). However, our previous research indicated that patients with *PIK3CA* mutations in cfDNA exhibit a shorter duration of ET response in the initial stages of treatment, compared to those without mutations (Takeshita et al. 2018). More recently, the FUTURE study demonstrated that *PIK3CA* mutations at the onset of fulvestrant resistance significantly contribute to poor prognosis. Moreover, these mutations at fulvestrant resistance continued to serve as strong prognostic factors, surpassing the impact of mutations that emerged early after the addition of palbociclib therapy (Takeshita et al. 2024). These findings suggest that *PIK3CA* mutations may serve as significant prognostic indicators in the context of fulvestrant utilization.

## 7 Conclusion

This chapter sheds light on the intricate landscape of *ESR1* mutations in HR-positive/HER2-negative ABC. The prevalence and clinical implications of *ESR1* mutations, particularly in the context of ET resistance, prognosis, and various therapeutic approaches, have been thoroughly explored. The emergence of liquid biopsy assays, with a focus on circulating cfDNA analysis, has provided valuable insights into overcoming the limitations associated with traditional tumor tissue material. The strategic management of *ESR1* mutations in HR-positive ABC involves vigilant monitoring and tailored treatment strategies, considering the impact on OS and response to various therapeutic combinations. Additionally, the exploration of other genetic mutations, such as *PIK3CA* mutations, is essential for a comprehensive understanding of treatment resistance and prognosis. While challenges persist, the growing adoption of liquid biopsy techniques underscores their clinical potential, emphasizing the need for further validation and utility data to enhance their role in managing HR-positive/HER2-negative ABC.

**Acknowledgements and Funding Information** Not applicable.

## References

Arnesen S, Blanchard Z, Williams MM, Berrett KC, Li Z, Oesterreich S, Richer JK, Gertz J (2021) Estrogen receptor alpha mutations in breast cancer cells cause gene expression changes through constant activity and secondary effects. Cancer Res 81:539–551

Baer C, Kern W, Koch S, Nadarajah N, Schindela S, Meggendorfer M, Haferlach C, Haferlach T (2016) Ultra-deep sequencing leads to earlier and more sensitive detection of the tyrosine kinase inhibitor resistance mutation T315I in chronic myeloid leukemia. Haematologica 101:830–838

Baselga J, Campone M, Piccart M, Burris HA 3rd, Rugo HS, Sahmoud T, Noguchi S, Gnant M, Pritchard KI, Lebrun F, Beck JT, Ito Y, Yardley D, Deleu I, Perez A, Bachelot T, Vittori L, Xu Z, Mukhopadhyay P, Lebwohl D, Hortobagyi GN (2012) Everolimus in postmenopausal hormone-receptor-positive advanced breast cancer. N Engl J Med 366:520–529

Bidard FC, Hardy-Bessard AC, Dalenc F, Bachelot T, Pierga JY, de la Motte Rouge T, Sabatier R, Dubot C, Frenel JS, Ferrero JM, Ladoire S, Levy C, Mouret-Reynier MA, Lortholary A, Grenier J, Chakiba C, Stefani L, Plaza JE, Clatot F, Teixeira L, D'Hondt V, Vegas H, Derbel O, Garnier-Tixidre C, Canon JL, Pistilli B, Andre F, Arnould L, Pradines A, Bieche I, Callens C, Lemonnier J, Berger F, Delaloge S, investigators P (2022) Switch to fulvestrant and palbociclib versus no switch in advanced breast cancer with rising ESR1 mutation during aromatase inhibitor and palbociclib therapy (PADA-1): a randomised, open-label, multicentre, phase 3 trial. Lancet Oncol 23:1367–1377

Brett JO, Spring LM, Bardia A, Wander SA (2021) ESR1 mutation as an emerging clinical biomarker in metastatic hormone receptor-positive breast cancer. Breast Cancer Res 23:85

Chandarlapaty S, Chen D, He W, Sung P, Samoila A, You D, Bhatt T, Patel P, Voi M, Gnant M, Hortobagyi G, Baselga J, Moynahan ME (2016) Prevalence of ESR1 mutations in cell-free

DNA and outcomes in metastatic breast cancer: a secondary analysis of the BOLERO-2 clinical trial. JAMA Oncol 2:1310–1315

Chen M, Zhao H (2019) Next-generation sequencing in liquid biopsy: cancer screening and early detection. Hum Genomics 13:34

Cristofanilli M, Turner NC, Bondarenko I, Ro J, Im SA, Masuda N, Colleoni M, DeMichele A, Loi S, Verma S, Iwata H, Harbeck N, Zhang K, Theall KP, Jiang Y, Bartlett CH, Koehler M, Slamon D (2016) Fulvestrant plus palbociclib versus fulvestrant plus placebo for treatment of hormone-receptor-positive, HER2-negative metastatic breast cancer that progressed on previous endocrine therapy (PALOMA-3): final analysis of the multicentre, double-blind, phase 3 randomised controlled trial. Lancet Oncol 17:425–439

Cristofanilli M, Rugo HS, Im SA, Slamon DJ, Harbeck N, Bondarenko I, Masuda N, Colleoni M, DeMichele A, Loi S, Iwata H, O'Leary B, Andre F, Loibl S, Bananis E, Liu Y, Huang X, Kim S, Lechuga Frean MJ, Turner NC (2022) Overall survival with Palbociclib and Fulvestrant in women with HR+/HER2- ABC: updated exploratory analyses of PALOMA-3, a double-blind, phase III randomized study. Clin Cancer Res 28:3433–3442

Dawson SJ, Tsui DW, Murtaza M, Biggs H, Rueda OM, Chin SF, Dunning MJ, Gale D, Forshew T, Mahler-Araujo B, Rajan S, Humphray S, Becq J, Halsall D, Wallis M, Bentley D, Caldas C, Rosenfeld N (2013) Analysis of circulating tumor DNA to monitor metastatic breast cancer. N Engl J Med 368:1199–1209

Di Cosimo S, Perez-Garcia JM, Bellet M, Dalenc F, Gil Gil MJ, Ruiz Borrego M, Gavila J, Sampayo-Cordero M, Aguirre E, Schmid P, Marme F, Gligorov J, Schneeweiss A, Albanell J, Zamora P, Wheatley D, Martinez-De Duenas E, Caranana V, Amillano K, Mina L, Malfettone A, Cortes J, Llombart-Cussac A (2023) Palbociclib with Fulvestrant or Letrozole in endocrine-sensitive patients with HR-positive/HER2-negative advanced breast cancer: a detailed safety analysis of the randomized PARSIFAL trial. Oncologist 28:23–32

Diaz LA Jr, Bardelli A (2014) Liquid biopsies: genotyping circulating tumor DNA. J Clin Oncol 32: 579–586

Finn RS, Martin M, Rugo HS, Jones S, Im SA, Gelmon K, Harbeck N, Lipatov ON, Walshe JM, Moulder S, Gauthier E, Lu DR, Randolph S, Dieras V, Slamon DJ (2016) Palbociclib and Letrozole in advanced breast cancer. N Engl J Med 375:1925–1936

Fribbens C, O'Leary B, Kilburn L, Hrebien S, Garcia-Murillas I, Beaney M, Cristofanilli M, Andre F, Loi S, Loibl S, Jiang J, Bartlett CH, Koehler M, Dowsett M, Bliss JM, Johnston SR, Turner NC (2016) Plasma ESR1 mutations and the treatment of Estrogen receptor-positive advanced breast cancer. J Clin Oncol 34:2961–2968

Garcia-Murillas I, Schiavon G, Weigelt B, Ng C, Hrebien S, Cutts RJ, Cheang M, Osin P, Nerurkar A, Kozarewa I, Garrido JA, Dowsett M, Reis-Filho JS, Smith IE, Turner NC (2015) Mutation tracking in circulating tumor DNA predicts relapse in early breast cancer. Sci Transl Med 7:302ra133

Goetz MP, Toi M, Campone M, Sohn J, Paluch-Shimon S, Huober J, Park IH, Tredan O, Chen SC, Manso L, Freedman OC, Garnica Jaliffe G, Forrester T, Frenzel M, Barriga S, Smith IC, Bourayou N, Di Leo A (2017) MONARCH 3: Abemaciclib as initial therapy for advanced breast cancer. J Clin Oncol 35:3638–3646

Goetz MP, Hamilton EP, Campone M, Hurvitz SA, Cortes J, Johnston S, Llombart-Cussac A, Kaufman PA, Toi M, Jerusalem G, Graham H, Wang H, Jansen VM, Litchfield LM, Martin M (2023) Landscape of baseline and acquired genomic alterations in circulating tumor DNA with abemaciclib alone or with endocrine therapy in advanced breast cancer. Clin Cancer Res

Gradishar WJ, Moran MS, Abraham J, Abramson V, Aft R, Agnese D, Allison KH, Anderson B, Burstein HJ, Chew H, Dang C, Elias AD, Giordano SH, Goetz MP, Goldstein LJ, Hurvitz SA, Jankowitz RC, Javid SH, Krishnamurthy J, Leitch AM, Lyons J, Mortimer J, Patel SA, Pierce LJ, Rosenberger LH, Rugo HS, Schneider B, Smith ML, Soliman H, Stringer-Reasor EM, Telli ML, Wei M, Wisinski KB, Young JS, Yeung K, Dwyer MA, Kumar R (2023) NCCN guidelines (R) insights: breast cancer, version 4.2023. J Natl Compr Cancer Netw 21:594–608

Hindson CM, Chevillet JR, Briggs HA, Gallichotte EN, Ruf IK, Hindson BJ, Vessella RL, Tewari M (2013) Absolute quantification by droplet digital PCR versus analog real-time PCR. Nat Methods 10:1003–1005

Hortobagyi GN, Stemmer SM, Burris HA, Yap YS, Sonke GS, Paluch-Shimon S, Campone M, Blackwell KL, Andre F, Winer EP, Janni W, Verma S, Conte P, Arteaga CL, Cameron DA, Petrakova K, Hart LL, Villanueva C, Chan A, Jakobsen E, Nusch A, Burdaeva O, Grischke EM, Alba E, Wist E, Marschner N, Favret AM, Yardley D, Bachelot T, Tseng LM, Blau S, Xuan F, Souami F, Miller M, Germa C, Hirawat S, O'Shaughnessy J (2016a) Ribociclib as first-line therapy for HR-positive, advanced breast cancer. N Engl J Med 375:1738–1748

Hortobagyi GN, Chen D, Piccart M, Rugo HS, Burris HA 3rd, Pritchard KI, Campone M, Noguchi S, Perez AT, Deleu I, Shtivelband M, Masuda N, Dakhil S, Anderson I, Robinson DM, He W, Garg A, McDonald ER 3rd, Bitter H, Huang A, Taran T, Bachelot T, Lebrun F, Lebwohl D, Baselga J (2016b) Correlative analysis of genetic alterations and everolimus benefit in hormone receptor-positive, human epidermal growth factor receptor 2-negative advanced breast cancer: results from BOLERO-2. J Clin Oncol 34:419–426

Hu Q, Hong Y, Qi P, Lu G, Mai X, Xu S, He X, Guo Y, Gao L, Jing Z, Wang J, Cai T, Zhang Y (2021) Atlas of breast cancer infiltrated B-lymphocytes revealed by paired single-cell RNA-sequencing and antigen receptor profiling. Nat Commun 12:2186

Jeselsohn R, Yelensky R, Buchwalter G, Frampton G, Meric-Bernstam F, Gonzalez-Angulo AM, Ferrer-Lozano J, Perez-Fidalgo JA, Cristofanilli M, Gomez H, Arteaga CL, Giltnane J, Balko JM, Cronin MT, Jarosz M, Sun J, Hawryluk M, Lipson D, Otto G, Ross JS, Dvir A, Soussan-Gutman L, Wolf I, Rubinek T, Gilmore L, Schnitt S, Come SE, Pusztai L, Stephens P, Brown M, Miller VA (2014) Emergence of constitutively active estrogen receptor-alpha mutations in pretreated advanced estrogen receptor-positive breast cancer. Clin Cancer Res 20:1757–1767

Jeselsohn R, Bergholz JS, Pun M, Cornwell M, Liu W, Nardone A, Xiao T, Li W, Qiu X, Buchwalter G, Feiglin A, Abell-Hart K, Fei T, Rao P, Long H, Kwiatkowski N, Zhang T, Gray N, Melchers D, Houtman R, Liu XS, Cohen O, Wagle N, Winer EP, Zhao J, Brown M (2018) Allele-specific chromatin recruitment and therapeutic vulnerabilities of ESR1 activating mutations. Cancer Cell 33:173–186 e175

Karakas B, Bachman KE, Park BH (2006) Mutation of the PIK3CA oncogene in human cancers. Br J Cancer 94:455–459

Karnik PS, Kulkarni S, Liu XP, Budd GT, Bukowski RM (1994) Estrogen receptor mutations in tamoxifen-resistant breast cancer. Cancer Res 54:349–353

Li S, Shen D, Shao J, Crowder R, Liu W, Prat A, He X, Liu S, Hoog J, Lu C, Ding L, Griffith OL, Miller C, Larson D, Fulton RS, Harrison M, Mooney T, McMichael JF, Luo J, Tao Y, Goncalves R, Schlosberg C, Hiken JF, Saied L, Sanchez C, Giuntoli T, Bumb C, Cooper C, Kitchens RT, Lin A, Phommaly C, Davies SR, Zhang J, Kavuri MS, McEachern D, Dong YY, Ma C, Pluard T, Naughton M, Bose R, Suresh R, McDowell R, Michel L, Aft R, Gillanders W, DeSchryver K, Wilson RK, Wang S, Mills GB, Gonzalez-Angulo A, Edwards JR, Maher C, Perou CM, Mardis ER, Ellis MJ (2013) Endocrine-therapy-resistant ESR1 variants revealed by genomic characterization of breast-cancer-derived xenografts. Cell Rep 4:1116–1130

Liao H, Huang W, Pei W, Li H (2020) Detection of ESR1 mutations based on liquid biopsy in Estrogen receptor-positive metastatic breast cancer: clinical impacts and prospects. Front Oncol 10:587671

Lumachi F, Santeufemia DA, Basso SM (2015) Current medical treatment of estrogen receptor-positive breast cancer. World J Biol Chem 6:231–239

Martin LA, Ribas R, Simigdala N, Schuster E, Pancholi S, Tenev T, Gellert P, Buluwela L, Harrod A, Thornhill A, Nikitorowicz-Buniak J, Bhamra A, Turgeon MO, Poulogiannis G, Gao Q, Martins V, Hills M, Garcia-Murillas I, Fribbens C, Patani N, Li Z, Sikora MJ, Turner N, Zwart W, Oesterreich S, Carroll J, Ali S, Dowsett M (2017) Discovery of naturally occurring ESR1 mutations in breast cancer cell lines modelling endocrine resistance. Nat Commun 8:1865

Merenbakh-Lamin K, Ben-Baruch N, Yeheskel A, Dvir A, Soussan-Gutman L, Jeselsohn R, Yelensky R, Brown M, Miller VA, Sarid D, Rizel S, Klein B, Rubinek T, Wolf I (2013) D538G mutation in estrogen receptor-alpha: a novel mechanism for acquired endocrine resistance in breast cancer. Cancer Res 73:6856–6864

Moynahan ME, Chen D, He W, Sung P, Samoila A, You D, Bhatt T, Patel P, Ringeisen F, Hortobagyi GN, Baselga J, Chandarlapaty S (2017) Correlation between PIK3CA mutations in cell-free DNA and everolimus efficacy in HR+, HER2- advanced breast cancer: results from BOLERO-2. Br J Cancer 116:726–730

O'Leary B, Cutts RJ, Liu Y, Hrebien S, Huang X, Fenwick K, Andre F, Loibl S, Loi S, Garcia-Murillas I, Cristofanilli M, Huang Bartlett C, Turner NC (2018) The genetic landscape and clonal evolution of breast cancer resistance to Palbociclib plus Fulvestrant in the PALOMA-3 trial. Cancer Discov 8:1390–1403

Osborne CK, Schiff R (2011) Mechanisms of endocrine resistance in breast cancer. Annu Rev Med 62:233–247

Overman MJ, Modak J, Kopetz S, Murthy R, Yao JC, Hicks ME, Abbruzzese JL, Tam AL (2013) Use of research biopsies in clinical trials: are risks and benefits adequately discussed? J Clin Oncol 31:17–22

Pan H, Gray R, Braybrooke J, Davies C, Taylor C, McGale P, Peto R, Pritchard KI, Bergh J, Dowsett M, Hayes DF, Ebctcg (2017) 20-year risks of breast-cancer recurrence after stopping endocrine therapy at 5 years. N Engl J Med 377:1836–1846

Phallen J, Sausen M, Adleff V, Leal A, Hruban C, White J, Anagnostou V, Fiksel J, Cristiano S, Papp E, Speir S, Reinert T, Orntoft MW, Woodward BD, Murphy D, Parpart-Li S, Riley D, Nesselbush M, Sengamalay N, Georgiadis A, Li QK, Madsen MR, Mortensen FV, Huiskens J, Punt C, van Grieken N, Fijneman R, Meijer G, Husain H, Scharpf RB, Diaz LA Jr, Jones S, Angiuoli S, Orntoft T, Nielsen HJ, Andersen CL, Velculescu VE (2017) Direct detection of early-stage cancers using circulating tumor DNA. Sci Transl Med 9

Raheem F, Karikalan SA, Batalini F, El Masry A, Mina L (2023) Metastatic ER+ breast cancer: mechanisms of resistance and future therapeutic approaches. Int J Mol Sci 24

Razavi P, Chang MT, Xu G, Bandlamudi C, Ross DS, Vasan N, Cai Y, Bielski CM, Donoghue MTA, Jonsson P, Penson A, Shen R, Pareja F, Kundra R, Middha S, Cheng ML, Zehir A, Kandoth C, Patel R, Huberman K, Smyth LM, Jhaveri K, Modi S, Traina TA, Dang C, Zhang W, Weigelt B, Li BT, Ladanyi M, Hyman DM, Schultz N, Robson ME, Hudis C, Brogi E, Viale A, Norton L, Dickler MN, Berger MF, Iacobuzio-Donahue CA, Chandarlapaty S, Scaltriti M, Reis-Filho JS, Solit DB, Taylor BS, Baselga J (2018) The genomic landscape of endocrine-resistant advanced breast cancers. Cancer Cell 34:427–438 e426

Robertson JFR, Bondarenko IM, Trishkina E, Dvorkin M, Panasci L, Manikhas A, Shparyk Y, Cardona-Huerta S, Cheung KL, Philco-Salas MJ, Ruiz-Borrego M, Shao Z, Noguchi S, Rowbottom J, Stuart M, Grinsted LM, Fazal M, Ellis MJ (2016) Fulvestrant 500 mg versus anastrozole 1 mg for hormone receptor-positive advanced breast cancer (FALCON): an international, randomised, double-blind, phase 3 trial. Lancet 388:2997–3005

Robinson DR, Wu YM, Vats P, Su F, Lonigro RJ, Cao X, Kalyana-Sundaram S, Wang R, Ning Y, Hodges L, Gursky A, Siddiqui J, Tomlins SA, Roychowdhury S, Pienta KJ, Kim SY, Roberts JS, Rae JM, Van Poznak CH, Hayes DF, Chugh R, Kunju LP, Talpaz M, Schott AF, Chinnaiyan AM (2013) Activating ESR1 mutations in hormone-resistant metastatic breast cancer. Nat Genet 45:1446–1451

Roodi N, Bailey LR, Kao WY, Verrier CS, Yee CJ, Dupont WD, Parl FF (1995) Estrogen receptor gene analysis in estrogen receptor-positive and receptor-negative primary breast cancer. J Natl Cancer Inst 87:446–451

Schiavon G, Hrebien S, Garcia-Murillas I, Cutts RJ, Pearson A, Tarazona N, Fenwick K, Kozarewa I, Lopez-Knowles E, Ribas R, Nerurkar A, Osin P, Chandarlapaty S, Martin LA, Dowsett M, Smith IE, Turner NC (2015) Analysis of ESR1 mutation in circulating tumor DNA demonstrates evolution during therapy for metastatic breast cancer. Sci Transl Med 7:313ra182

Siegel RL, Miller KD, Wagle NS, Jemal A (2023) Cancer statistics, 2023. CA Cancer J Clin 73: 17–48

Slamon DJ, Neven P, Chia S, Fasching PA, De Laurentiis M, Im SA, Petrakova K, Bianchi GV, Esteva FJ, Martin M, Nusch A, Sonke GS, De la Cruz-Merino L, Beck JT, Pivot X, Vidam G, Wang Y, Rodriguez Lorenc K, Miller M, Taran T, Jerusalem G (2018) Phase III randomized study of Ribociclib and Fulvestrant in hormone receptor-positive, human epidermal growth factor receptor 2-negative advanced breast cancer: MONALEESA-3. J Clin Oncol 36: 2465–2472

Sledge GW Jr, Toi M, Neven P, Sohn J, Inoue K, Pivot X, Burdaeva O, Okera M, Masuda N, Kaufman PA, Koh H, Grischke EM, Conte P, Lu Y, Barriga S, Hurt K, Frenzel M, Johnston S, Llombart-Cussac A (2020) The effect of Abemaciclib plus Fulvestrant on overall survival in hormone receptor-positive, ERBB2-negative breast cancer that progressed on endocrine therapy-MONARCH 2: a randomized clinical trial. JAMA Oncol 6:116–124

Spoerke JM, Gendreau S, Walter K, Qiu J, Wilson TR, Savage H, Aimi J, Derynck MK, Chen M, Chan IT, Amler LC, Hampton GM, Johnston S, Krop I, Schmid P, Lackner MR (2016) Heterogeneity and clinical significance of ESR1 mutations in ER-positive metastatic breast cancer patients receiving fulvestrant. Nat Commun 7:11579

Takeshita T, Iwase H (2019) dPCR mutational analyses in cell-free DNA: a comparison with tissues. Methods Mol Biol 1909:105–118

Takeshita T, Yamamoto Y, Yamamoto-Ibusuki M, Inao T, Sueta A, Fujiwara S, Omoto Y, Iwase H (2015) Droplet digital polymerase chain reaction assay for screening of ESR1 mutations in 325 breast cancer specimens. Transl Res 166:540–553 e542

Takeshita T, Yamamoto Y, Yamamoto-Ibusuki M, Inao T, Sueta A, Fujiwara S, Omoto Y, Iwase H (2016) Clinical significance of monitoring ESR1 mutations in circulating cell-free DNA in estrogen receptor positive breast cancer patients. Oncotarget 7:32504–32518

Takeshita T, Yamamoto Y, Yamamoto-Ibusuki M, Tomiguchi M, Sueta A, Murakami K, Omoto Y, Iwase H (2017) Analysis of ESR1 and PIK3CA mutations in plasma cell-free DNA from ER-positive breast cancer patients. Oncotarget 8:52142–52155

Takeshita T, Yamamoto Y, Yamamoto-Ibusuki M, Tomiguchi M, Sueta A, Murakami K, Iwase H (2018) Clinical significance of plasma cell-free DNA mutations in PIK3CA, AKT1, and ESR1 gene according to treatment lines in ER-positive breast cancer. Mol Cancer 17:67

Takeshita T, Iwamoto T, Niikura N, Watanabe K, Kikawa Y, Kobayashi K, Iwakuma N, Okamura T, Tada H, Ozaki S, Okuno T, Toh U, Yamamoto Y, Tsuneizumi M, Ishiguro H, Masuda N, Saji S (2024) Identifying prognostic biomarkers for palbociclib add-on therapy in fulvestrant-resistant breast cancer using cell-free DNA sequencing. ESMO Open 9:102385

The Japanese Breast Cancer Society Clinical Practice Guidlines for Breast Cancer (2022). From https://jbcs.xsrv.jp/guideline/2022/

Toy W, Shen Y, Won H, Green B, Sakr RA, Will M, Li Z, Gala K, Fanning S, King TA, Hudis C, Chen D, Taran T, Hortobagyi G, Greene G, Berger M, Baselga J, Chandarlapaty S (2013) ESR1 ligand-binding domain mutations in hormone-resistant breast cancer. Nat Genet 45:1439–1445

Turner NC, Swift C, Kilburn L, Fribbens C, Beaney M, Garcia-Murillas I, Budzar AU, Robertson JFR, Gradishar W, Piccart M, Schiavon G, Bliss JM, Dowsett M, Johnston SRD, Chia SK (2020) ESR1 mutations and overall survival on Fulvestrant versus Exemestane in advanced hormone receptor-positive breast cancer: a combined analysis of the phase III SoFEA and EFECT trials. Clin Cancer Res 26:5172–5177

Untch M, Fasching PA, Haidinger R, Harbeck N, Jackisch C, Luftner D, Muller V, Schumacher-Wulf E, Wurstlein R, Thomssen C (2022) Advanced breast cancer: AGO recommendations 2022 – focus on ABC6 consensus. Geburtshilfe Frauenheilkd 82:1044–1054

Weis KE, Ekena K, Thomas JA, Lazennec G, Katzenellenbogen BS (1996) Constitutively active human estrogen receptors containing amino acid substitutions for tyrosine 537 in the receptor protein. Mol Endocrinol 10:1388–1398

Zhang QX, Borg A, Wolf DM, Oesterreich S, Fuqua SA (1997) An estrogen receptor mutant with strong hormone-independent activity from a metastatic breast cancer. Cancer Res 57:1244–1249

Zhang K, Hong R, Xu F, Xia W, Kaping L, Qin G, Zheng Q, Lu Q, Shi YX, Yuan ZY, Wang S (2018) Clinical value of circulating ESR1 mutations for patients with metastatic breast cancer: a meta-analysis. Cancer Manag Res 10:2573–2580

# Tumor Microenvironment Modulation and Its Impact on Immunotherapy in Triple-Negative Breast Cancer

Naila Malkani, Muhammad Usman Rashid, Ammara Ramzan, and Komal Khalid

**Abstract**

Triple-negative breast cancer (TNBC) is known as a refractory and inoperable breast cancer subtype with the worst prognosis. The poor treatment response due to the lack of specific targetable receptors and its extremely invasive nature makes it a significant point of scientific interest. The tumor microenvironment (TME) in TNBC is a dynamic and diverse landscape composed of various cellular components involving immune cells (macrophages, dendritic cells, T lymphocytes, natural killer cells, and cytokines) and nonimmune cells (endothelial cells, cancer-associated fibroblasts, vascular smooth muscle cells, and extracellular matrix). TME functions paradoxically; being both an immunoreactive and an immunosuppressive factor, it contributes significantly to the immunomodulation and tumor development in TNBC. Novel immunotherapeutic strategies including adoptive cell transfer, cytokines, checkpoint inhibitors, monoclonal antibodies, vaccines, and TME modulation have been developed recently to reshape the TME and treat TNBC. Immunomodulatory medications can adjust the TME by suppressing angiogenesis, promoting infiltration of immune cells, blocking T regulatory cells, and boosting the efficacy of therapies. The intricate interplay between the immunological TME and cancer impacts the efficacy of immunotherapy and other anticancer treatments. This has revolutionized the landscape of cancer treatment for various solid tumor types, particularly TNBC. This chapter explores the implications of TME modulation for the immunotherapy of TNBC. Understanding the dynamics of the TME in TNBC is crucial for developing targeted

N. Malkani (✉) · A. Ramzan · K. Khalid
Department of Zoology, GC University, Lahore, Pakistan
e-mail: nailamalkani@gcu.edu.pk

M. U. Rashid
Department of Basic Sciences Research, Shaukat Khanum Memorial Cancer Hospital and Research Centre (SKMCH & RC), Lahore, Pakistan



Interdisciplinary Cancer Research, https://doi.org/10.1007/16833_2024_280
Published online: 31 May 2024

immunotherapeutic strategies and advancing our knowledge of this aggressive breast cancer subtype.

**Keywords**

Immunomodulation · Immunotherapy · Triple-negative breast cancer · Tumor microenvironment

# 1 Introduction

Breast cancer stands as a complex and diverse adversary in the realm of neoplastic diseases (Testa et al. 2020). It arises from the abnormal growth of cells inside the breast glands or ducts (Ataollahi et al. 2015). Breast tumors, often originating from hyperactive ducts, evolve into either benign or metastatic cancers when persistently stimulated by various carcinogenic factors. Their reach extends beyond the confines of the breast, affecting adjacent organs such as the brain, bones, liver, and lungs (Ibragimova et al. 2023; Lourenço et al. 2022). The risk of acquiring breast cancer is influenced by multiple factors encompassing age at menarche, reproductive health, age at menopause, family history of cancer, lifestyle choices, alcohol consumption, body weight, and smoking habits (Saeed et al. 2014). In breast cancer initiation and progression, the tumor microenvironment (TME) plays a vital role, with macrophages and stromal influences contributing significantly. This intricate process involves various cell types, making it a multistep journey (Sun et al. 2017).

Early detection of breast cancer emerges as a beacon of hope, promising a more favorable prognosis and an increased chance of survival (Ginsburg et al. 2020). In the exploration of potential treatments, understanding the modulation of the TME becomes paramount, especially in the context of triple-negative breast cancer (TNBC) where conventional therapeutic options may fall short. This chapter delves into the implications of TME modulation for the immunotherapy of TNBC, aiming to shed light on a path forward in the battle against this formidable disease.

## 1.1 Epidemiology of Breast Cancer

Breast cancer is the most commonly diagnosed malignancy among women worldwide (Bilani et al. 2020). This pervasive disease is on the rise, with increasing incidence, morbidity, and mortality rates not just in high-income countries but also in middle- and low-income countries (Shoukat and Shah 2023). The World Health Organization (WHO) reports that almost 2.5 million women are diagnosed with breast cancer each year. In Pakistan, it is estimated that one in nine women has a lifetime risk of developing breast cancer, signaling an elevated disease risk (Ghoncheh et al. 2015).

While breast cancer can also affect men, it remains exceedingly rare, with a 1 in 1000 risk of development (Miller et al. 2016). GLOBOCAN 2020 ranks breast cancer as the most frequent cancer in women globally, leading in terms of incidence, prevalence, and mortality. The statistics paint a concerning picture, with an estimated 1,948,321 (27.5%) new cases, 5,135,125 (17.2%) deaths, and 6,608,382 (33.1%) prevalent cases reported in females aged 0–74 worldwide (Sung et al. 2021). Łukasiewicz et al. (2021) reveal an alarming trend, demonstrating a twofold rise in breast cancer incidence and mortality rates throughout the past three decades. Forecasts for 2030 are ominous, globally predicting ~2.7 million new cases and 0.87 million fatalities each year (Łukasiewicz et al. 2021).

Despite higher incidence rates in developed regions, a staggering 63% of all deaths in 2020 occurred in Asia and Africa (Ferlay et al. 2020). Among these, Pakistan bears a significant burden, reporting one of the highest breast cancer incidence rates in Asia. In addition to being the most frequent cancer, it also accounts for the majority of cancer-related fatalities in Pakistan (Khan et al. 2021). Among 45–55 years old, this disease has become the primary cause of mortality (Ataollahi et al. 2015). GLOBOCAN 2020 statistics for Pakistan highlight 23,406 (28.7%) new cases, 11,745 (23.8%) deaths, and 54,374 (33%) prevalent cases recorded among females (Sung et al. 2021).

Worldwide, the survival rate for breast cancer varies greatly, ranging from 40% in low-income countries to 80% in high-income countries. This discrepancy is primarily attributed to the differences in awareness, early detection, and treatment available across different economic strata (Qasim et al. 2020). As breast cancer continues its alarming rise in prevalence and mortality, it underscores the urgency of studying the mechanisms driving breast tumorigenesis and identifying novel therapeutic targets.

## 2 Types of Breast Cancer

Breast cancer manifests in various forms, reflecting its ability to emerge in various regions of the breast, including the lobules, ducts, and tissues nestled between them. The specific type of breast cancer is intricately linked to the nature of the affected cells (Feng et al. 2018). It can be classified into different types and subtypes based on the histological and molecular analysis (Makki 2015) as represented in Fig. 1. This classification is not a mere academic exercise; it serves as a crucial foundation for tailoring effective treatment plans. Each type carries its distinct characteristics, responding uniquely to therapeutic interventions. A comprehensive grasp of these variations allows healthcare professionals to navigate the intricate landscape of breast cancer, ensuring personalized and effective care for each patient. The following sections delve into the diverse types and subtypes, shedding light on their characteristics and implications for treatment strategies.

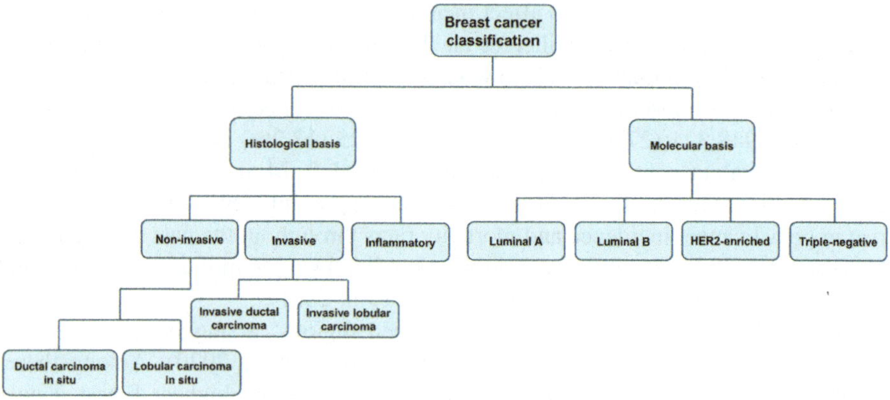


**Fig. 1** Breast cancer classification based on histological and molecular analysis. The diagram illustrates the major subtypes, including ductal and lobular carcinomas, as well as less common histological variants. Molecular classifications, such as hormone receptor status (estrogen receptor (ER), progesterone receptor (PR)), human epidermal growth factor receptor 2 (HER2) status, and gene expression profiling (e.g., luminal, HER2-enriched, and basal-like), are depicted to highlight the heterogeneity within breast cancer. Understanding these classifications is critical for tailoring therapeutic strategies and predicting patient outcomes

## 2.1 Classification Based on the Histological Analysis

In histopathological practice, the classification of breast cancer involves scrutinizing various factors such as cell number, secretion type, location, cell characteristics, architectural features, and immunohistochemical profile. This approach helps distinguish between lobular and ductal tumors, irrespective of their specific location in mammary tissue (Makki 2015; Nounou et al. 2015). Based on histopathological features, breast cancer can be categorized into two primary types including noninvasive (or in situ) and invasive breast cancer.

### 2.1.1 Noninvasive Breast Cancer

#### Ductal Carcinoma In Situ (DCIS)

DCIS represents a pre-invasive/noninvasive condition confined to the normal ducts in breast tissue. It progresses through ducts and lobular acini, forming lesions known as lobular cancer. Early detection and treatment are imperative for patient survival (Feng et al. 2018).

#### Lobular Carcinoma In Situ (LCIS)

LCIS comprises small, uniform neoplastic cells originating from the terminal duct lobular unit and proliferates interlobularly. It serves as a common risk factor and nonobligatory precursor for breast cancer. It predominantly occurs in premenopausal women and tends to be multicentric in 60–80% of cases and bilateral in 40–20% of cases. Women with LCIS face a 6.9–12 times higher relative risk of developing invasive carcinoma (Makki 2015; Wen and Brogi 2018).

### 2.1.2 Invasive/Infiltrating Breast Cancer

Invasive/infiltrating breast cancer spreads to the surrounding stromal tissue invading beyond the typical breast ducts and lobules. It typically affects two thirds of women, aged 55 years or above. It can metastasize to other body parts such as bones, lymph nodes, and lungs. It can be further categorized into invasive ductal carcinoma (IDC) and invasive lobular carcinoma (ILC) (Feng et al. 2018).

#### Invasive Ductal Carcinoma (IDC)

IDC is the most widespread breast cancer type and accounts for 50–80% of all diagnosed cases (Henry and Cannon-Albright 2019). These carcinomas constitute a heterogeneous group, representing malignant ductal proliferation and stromal invasion (Makki 2015). IDC includes various histological subtypes, with IDC nonspecific type (NST) being the most common type, constituting almost 40–75% of all invasive breast cancers (Masood 2016).

#### Invasive Lobular Carcinoma (ILC)

ILC is the second most common breast cancer type and comprises almost 15% of all cases (do Nascimento and Otoni 2020). Though it can affect women of all ages, it is more frequent in older postmenopausal women (Akram et al. 2017). Unlike ductal carcinomas, lobular carcinomas develop as single cells and organize into single sheets, showing distinctive pathological, genetic, and molecular abnormalities (Feng et al. 2018).

#### Inflammatory Breast Cancer (IBC)

IBC, accounting for 1–5% of all invasive/infiltrating breast cancer cases, is characterized by inflammation-like swelling, red/purple skin color, and thickening or pitting of the breast tissue. Symptoms occur in cancer cells obstructing lymph vessels, making them undetectable on mammograms. More common in younger, overweight, and African American women, IBC tends to be more malignant and spreads faster than other frequent types of breast cancers (Lim et al. 2018).

## 2.2 Classification Based on Molecular Analysis

Based on extensive gene expression profile analyses, four therapeutically relevant molecular subgroups were identified including luminal A, luminal B, enriched HER2 (HER2+), and triple-negative breast cancer (TNBC) (Harbeck et al. 2019).

### 2.2.1 Luminal A

Luminal A, representing 50% of IBC cases (Makki 2015), is ER($\geq$1%)/PR($\geq$20%) positive, HER2($\leq$10%) negative and has a low Ki-67 (<14%) level. It is a slow-growing, low-grade carcinoma and often has a good prognosis, with less common lymph node involvement. Hormonal therapies, such as tamoxifen (estrogen receptor modulator) and anastrozole (aromatase inhibitors), are frequently used for treatment (do Nascimento and Otoni 2020).

### 2.2.2 Luminal B

Comprising 20–30% of invasive breast cancer cases (Makki 2015), luminal B is categorized into **(HER2−)** involving ER (≥1%) positive, PR (<20%)/HER2 (≤10%) negative or high Ki-67 (≥20%) level, and **(HER2+)** including ER (≥1%)/ HER2 (>10%) positive or any level of PR. It is considered the most malignant type of ER+ breast cancer. It is usually unresponsive to hormonal treatment, requiring additional therapies like specific target therapy (when HER2+) and chemotherapy (when HER2+/−) (Feng et al. 2018; Fragomeni et al. 2018).

### 2.2.3 HER2-Enriched Breast Cancer

Accounting for 15–20% of invasive breast cancer cases (Cho 2016), HER2-enriched breast cancer indicates HER2 (>10%) positivity, while ER (<1%) and PR (<20%) are negative. Despite luminal tumors, HER2-enriched tumors may grow fast and have a poor prognosis. Drugs that suppress HER2 activity, particularly Trastuzumab, a human monoclonal antibody, and Lapatinib, a tyrosine kinase inhibitor, have a favorable effect on these tumors (Tsang and Gary 2020; Llombart-Cussac et al. 2017).

### 2.2.4 Triple-Negative/Basal-Like Breast Cancer (TNBC)

Constituting for almost 20% of all breast cancer cases, TNBC is characterized as ER (<1%), PR (<20%), and HER2 (≤10%) negative. It is a high-grade breast cancer that frequently behaves more aggressively than any other type of breast cancer (Feng et al. 2018). These tumors are quite prolific (Tsang and Gary 2020).

## 3 TNBC

TNBC emerges as the most malignant, aggressive, and heterogeneous disease closely tied to immunomodulation. It is known for impaired expression of hormonal receptors ER/PR and HER2. It exhibits a higher proliferation rate and a greater chance of metastasis to the brain, liver, and lungs. With a higher recurrence rate and an overall poor prognosis compared to luminal tumors, TNBC is associated with higher-grade tumors and advanced disease stages. Its poor responsiveness to treatment and extremely invasive nature make it a focal point of scientific interest. Gene expression and ontological analysis reveal six TNBC subtypes: mesenchymal (M), basal-like 1 (BL-1), luminal androgen receptor (LAR), basal-like 2 (BL-2), mesenchymal stem-like (MSL), and immunomodulatory (IM) (Bou Zerdan et al. 2022).

### 3.1 Epidemiology

The disease burden of TNBC is on a concerning rise among women globally, marking it as a significant health challenge. This particular subtype of breast cancer exhibits a higher prevalence in premenopausal women, particularly those of American/Hispanic and African origin (Thakur et al. 2018). Notably, the incidence of TNBC is lower in women under the age of 40 (12.2%) than those between the ages of 40 and 59 (50.9%). In 2018, an estimated 2,088,849 TNBC cases were reported, indicating its aggressive prevalence among women (Almansour 2022).

The frequency of TNBC varies across different populations, with Indian women having the highest incidence (27.9%), followed by Indonesian (25%), Algerian (20.80%), Pakistani (18%), American (13.1%), and Chinese women (12.7%). The lowest incidence of TNBC was observed in the Netherlands (6.8%) (Thakur et al. 2018). According to Bauer et al. (2007), a study in California between 1999 and 2003 noted 12.4% of cases to be TNBC, with an average age of 54 years among these patients.

TNBC patients have lower survival rates relative to other subtypes of breast cancer, with a 40% death rate within the first 5 years following diagnosis. After recurrence, this death rate increases to 75% within 3 months. TNBC exhibits a highly invasive nature, with 46% of individuals developing distant metastases. The median survival period is a mere 13.3 months following metastasis, and the risk of recurrence is less than 25% after surgery. The brain and visceral organs are frequently affected by metastasis, commonly occurring in the third year post-diagnosis (Yin et al. 2020; Zhang et al. 2015).

In comparison to non-TNBC patients, TNBC patients experience a shorter average relapse time, ranging from 19 to 40 months, whereas non-TNBC patients have a relapse time of 35–67 months. Despite adjuvant treatment, the 5-year survival rate is around 30% for metastatic TNBC. Furthermore, the 5-year survival rate is 11% for tumors that have migrated to distant organs and 65% for regional tumors (Kohler et al. 2015).

The challenging characteristics of TNBC underscore the critical need for research and targeted interventions to improve outcomes for individuals affected by this aggressive breast cancer subtype.

### 3.2 Risk Factors

The risk factors associated with TNBC encompass a range of influences, including age, sex, mutations, race, physical activity, smoking, diet, genetic history, breast tissue density, history of radiotherapy, drug use, and an increased body mass index (Almansour 2022).

### 3.3 Genetic Factors

Genetic factors are pivotal contributors to TNBC. About 70% of breast tumors in women with hereditary *BRCA1* mutations lack the ER, PR, and HER2 expression. Germline mutations in genes such as *BRCA1* and *BRCA2* are strongly linked to TNBC, with this phenotype present in 60–80% of breast tumors carrying a *BRCA1* mutation. Studies report that approximately 29% of TNBC patients (Ashkenazi Jewish ancestry) harbor *BRCA1* mutations, while *BRCA2* mutations are less commonly associated, present in an estimated 4–9% of TNBC individuals (Afghahi et al. 2016).

Research by Long et al. 2016, indicated that 16–23% of breast tumors in *BRCA2* mutation carriers exhibit a TNBC phenotype. In Pakistan, Rashid et al. 2016, reported that 37% of TNBC patients had *BRCA1* mutations, and only 1% had *BRCA2* mutations in Pakistan. The profile of *BRCA1*-associated tumors aligns closely with TNBC, while *BRCA2*-related tumors more specifically resemble the luminal B subtype (Incorvaia et al., 2020).

Beyond *BRCA1* and *BRCA2*, other DNA repair pathway genes such as *RAD51D, BARD1, PALB2, BRIP1,* and *RAD51C* play roles in TNBC development, contributing to a < 20% lifetime risk for all breast cancers among Caucasians. Corso et al. (2018) and Shahbandi et al. (2020) further identified mutations in genes such as *STK11, XRCC2, PTEN, TP53,* and *CDH1,* linking them to overall TNBC occurrence and breast cancer. Understanding these genetic factors is crucial for advancing our knowledge of TNBC and developing targeted interventions.

## 4 Tumor Microenvironment (TME) In TNBC

The TME stands as a complex, dominant, and heterogeneous component within solid tumors. It houses tumor cells that employ circulatory and lymphatic systems to interact with neighboring cells, exerting influence on cancer occurrence, growth, and progression (Arneth 2019). This intricate environment comprises a heterogeneous population, including immune cells (macrophages, T lymphocytes, natural killer cells, cytokines, and dendritic cells) and nonimmune cells (endothelial cells, cancer-associated fibroblasts, vascular smooth muscle cells, and extracellular matrix), as depicted in Fig. 2 (Fan and He 2022).

Within the TME, these factors interact through interconnected signaling pathways, creating a highly intricate network. Any modification in this network can impact the metabolism and behavior of tumor cells, thereby influencing tumor development, progression, and clinical outcomes (Huang et al. 2017). The TME functions paradoxically, serving as both an immunoreactivity and immunosuppressive factor, contributing significantly to the immunomodulation and tumor development of TNBC (Zheng et al. 2021). Understanding the dynamics of the TME in TNBC is crucial for developing targeted therapeutic strategies and advancing our knowledge of this aggressive breast cancer subtype.

**Fig. 2** Key components of the tumor microenvironment (TME) in TNBC. The TME encompasses diverse cellular and noncellular elements, including immune cells, fibroblasts, blood vessels, extracellular matrix (ECM), and signaling molecules. Understanding the dynamic interactions within the TME is crucial for unravelling the complexities of TNBC progression and devising targeted therapeutic strategies

## 4.1 Components of the TME in TNBC

The TME in TNBC is a dynamic and diverse landscape composed of various cellular components. Among these, tumor-infiltrating lymphocytes (TILs) perform a crucial role in influencing the immune response and tumor behavior.

### 4.1.1 Tumor-Infiltrating Lymphocytes (TILs)

TILs are a primary and diverse component within TME, mainly comprising lymphocytes. In TNBC, they exhibit intrinsic immunogenicity (Valenza et al. 2023). van den Ende et al. (2023) reported that within the TME, TNBC displays a higher incidence of TIL infiltration, which is linked to reduced Ki-67 levels, suggesting less malignant tumor cell growth and a favorable response to therapy.

The TIL population encompasses various subtypes, including:

CD38+/CD138+ plasma cells: Specific types of B cells that produce antibodies (immunoglobulins) and enhance the adaptive immune response.

CD3+ T cells: Mature T cells co-differentiated with antigens expressed on their surface, playing a critical role in the adaptive immune response.

CD20+ B cells: Cells that contribute to the immune response by producing antibodies and interacting with other immune cells (Valenza et al. 2023).

The CD3 + T cell population is further categorized into specific subtypes, each with its distinct functions:

**Foxp3+ Regulatory T Lymphocytes (Foxp3+ Tregs)** T cells play a fundamental role in sustaining immunological homeostasis and inhibiting immune responses in normal conditions. In the context of TNBC, their increased presence contributes to an immunosuppressive milieu, limiting the activation of CD8+ and CD4+ T cells. In healthy peripheral blood, they constitute around 10% of all CD4+ T lymphocytes. However, in tumor lesions, this percentage increases to 30% or 50%, obstructing the activation of CD4+ and CD8+ T cells (Stovgaard et al. 2018; Fan and He 2022).

These Tregs play a fundamental role in angiogenesis and immune suppression. Tregs express Foxp3 (immunosuppressive marker) on their surfaces to determine their lineage and modulate immunosuppressing signals (Lei et al. 2020). Studies, such as the one by Zahran et al. (2021), have associated Foxp3+ Tregs with a poor prognosis in TNBC patients.

**CD8+ Cytotoxic T Lymphocytes (CD8+ TILs)** Essential for promoting an anticancer immune response and contributing to the destruction of tumor cells. Activated by major histocompatibility complex class I (MHC-I) molecules on tumor cells, they produce tumor necrosis factor α (TNFα) and interferon-γ (IFN-γ), contributing to the destruction of tumor cells. The overall number of CD8+ cells shows an inverse relationship with ER/PR expression, highlighting their significance in the context of hormone receptor status. The density of CD8+ TILs in TNBC is associated with enhanced survival and a better prognosis (Raskov et al. 2021).

**CD4+ Cytotoxic T Lymphocytes (CD4+ TILs)** Further categorized into CD4+ regulatory and CD4+ helper T lymphocytes. CD4+ helper T lymphocytes, though infrequent in breast tumors, regulate various immune responses and assist CD8+ T cells, macrophages, and B cells. In TNBC, their role is significant in coordinating the immune response (Stovgaard et al. 2018).

The intricate interplay of these TIL subtypes contributes to the immunomodulation of breast cancer, particularly in TNBC, impacting various pathways within the TME. Understanding these dynamics is essential for developing targeted immunotherapies and improving treatment outcomes in TNBC.

### 4.1.2 Tumor-Associated Macrophages (TAMs)

TAMs play a fundamental role in the TME. Derived from blood monocytes, they differentiate into macrophages, representing a significant component of the TME. Tumor cells release various cytokines and chemokines, such as CCL2, to recruit macrophages into solid tumor tissue, shaping the TME.

TAMs make up a substantial proportion, ranging from 50 to 80%, of mesenchymal cells in the TME and are actively involved in tumor initiation and development (Lu and Ma 2020). They are categorized into subtypes:

M1-type macrophages induce an antitumor (Th-1) immune response, promoting tumor destruction by secreting proinflammatory cytokines.

M2-type macrophages release TGFβ and IL-10 to prevent Th-1 immune responses, also contributing to angiogenesis, tumor development, invasion, migration, and metastasis (Santoni et al. 2018; Deepak et al. 2020).

TAMs are associated with cancer occurrence, directly or indirectly. They secrete various factors, including BFGF-2, PDGF, EGF, TGFβ, vascular endothelial growth factor (VEGF), IL-8, MMPs, COX-2, IL-13, TGFβ, CTL, and NK, influencing tumor proliferation and differentiation (Birbrair et al. 2014; Santoni et al. 2018).

In TNBC, TAMs have been implicated in regulating programmed death ligands (PD-1/PD-L1) expression in the TME, contributing to the poor prognosis of TNBC patients. They actively participate in the emergence and growth of tumors by influencing the expression of PD-1/PD-L1 directly or indirectly in the tumor environment (Hollmén et al. 2015; Santoni et al. 2018).

### 4.1.3 Natural Killer (NK) Cells

NK cells are innate immune cytotoxic lymphocytes that play a fundamental role in enhancing antitumor immunity and mediating immunosurveillance within the TNBC microenvironment. In humans, NK cells are identified as CD3-CD56+ cells and account for 5–15% of circulatory lymphocytes. Within the TME, NK cells perform functions similar to CD8+ T cells (Lei et al. 2020).

NK cells can also be recruited to cancerous or inflamed tissues by chemokines released by dendritic cells (DCs). Once in the TME, NK cells release granzymes and perforin, contributing to apoptosis or tumor cell death. Additionally, they secrete diverse proinflammatory cytokines and chemokines, involving CCL5, TNF-α, IL-6, IFN-γ, and GM-CSF, to provoke an antitumor effect (Guillerey et al. 2016). These cells efficiently detect and eliminate tumor cells through the downregulation mechanism of MHC-I. This downregulation process is utilized by cancer cells to evade T cell identification (Cornel et al. 2020). Studies, such as the one by Thacker et al. (2023), have indicated that in TNBC mice, the depletion of NK cells increased the anti-PD-L1 chemotherapy or antibody response. This emphasizes the critical role of NK cells in the immune response against TNBC, particularly in the context of immunotherapy targeting PD-L1 (Thacker et al. 2023).

### 4.1.4 Dendritic Cells (DCs)

DCs play a central role in the TME and primarily function as antigen-presenting cells (APCs). These cells display antigens and transmit co-stimulatory signals to activate T cells and communicate with B and NK cells. As APCs, dendritic cells capture, process, and present antigens to T cells, in order to initiate an adaptive immune response. This interplay is crucial for the activation and modulation of T cell responses against tumor cells. Furthermore, DCs play a role in communication with B cells and NK cells, further inducing the overall antitumor immune effect within the TME (Kim et al. 2021).

The intricate network involving dendritic cells and other immune cells underscores the complexity of the TME and its influence on the modulation of immune responses against tumors, including those characteristic of TNBC.

### 4.1.5 Tumor-Associated Neutrophils (TANs)

TANs constitute almost 60% of all circulating white blood cells and serve as the initial line of defense at inflammation or infection sites. In addition to releasing a range of chemokines and peroxidases, such as myeloperoxidase (MPO), activated neutrophils interact with mesenchymal cells, macrophages, lymphocytes, DCs, and NKs into the inflammatory areas (Kim and Bae 2016; Huang et al. 2020). In TNBC, TANs induce migration, proliferation, and metastasis of tumor (Nagaraj et al. 2010).

Moreover, TANs release tumor-promoting factors in response to stimulation by TNBC cells which secrete granulocyte-macrophage-colony-stimulating factor (GM-CSF). This includes factors that improve tumor cell invasion and induce angiogenesis (Rymaszewski et al. 2014). The interplay between TANs and TME contributes to the aggressive behavior of TNBC.

### 4.1.6 Cancer-Associated Fibroblasts (CAFs)

CAFs are substantial constituent of the tumor stroma and play an essential role in promoting communication between the TME and cancer cells. Tissue-resident fibroblasts are frequently the source of these cells, but these cells have diverse origins, involving pericytes, adipocytes, stellate cells, endothelial cells, and bone marrow–derived mesenchymal stem cells (Anderson and Simon 2020).

CAFs contribute to tumor cell proliferation by maintaining continuous propagation and growth at metastatic sites. Furthermore, they release a variety of growth factors, such as fibroblast-secreted protein-1 (FSP1) and cytokines that induce metastatic processes in breast and colon cancer (Shiga et al. 2015). The role of CAFs in TNBC involves promoting the development of tumors by activating transforming growth factor-β (TGFβ) (Takai et al. 2016). This underlines the role of CAFs in the development of TNBC and the dynamic interplay within the TME.

### 4.1.7 Cytokines

A diverse array of cytokines and chemokines, involving TGF-, TNF-, INF-, IL-2, IL-8, IL-6, IL-1, IL-18, IL-15, IL-6, IL-8, CXCL10, CCL5, and CCL2, are found in TNBC, thereby influencing tumor metastasis, drug resistance, immune suppression, and overall poor prognosis (Zhang et al. 2017; Ayers et al. 2017; Santoni et al. 2018).

Various cytokines from the TGF superfamily promote the development of Tregs and prevent cytotoxic T lymphocytes (CTLs) from acting as effectors in the TME, contributing to immune evasion (Santoni et al. 2018).

Inflammatory cytokines like IL-6 interfere with the expression of cell surface antigens and adhesion molecules to regulate a range of cellular functions, such as immune response, differentiation, migration and proliferation, differentiation, and immune response. Elevated levels of IL-6 in TNBC cells stimulate the JAK2-STAT3 signaling pathway, as well as the expression of CCL5 and VEGF, promoting TNBC growth and metastasis (Wang et al. 2016). Additionally, IL-8, known for its oncogeneic and angiogenic properties, is highly upregulated in TNBC (Long et al. 2016; Jin et al. 2017).

### 4.1.8 Chemokines

Chemokines, low molecular weight proteins such as CXCL2, CXCL13, CCL14, CCL21, CXCL1, CCL28, CXCL11, CXCL9, CXCL12, and CXCL10, play a key role in promoting chemotaxis. In TNBC, chemokines and their receptors exert control over cell senescence, proliferation, and survival. Specific chemokines like CXCL12, CXCL3, and CXCL2 mediate several signaling pathways such as mTOR and deregulated MAPK and AKT (Mehraj et al. 2022), impacting the progression of various malignancies. Mesenchymal stem cells contribute to increased production of CXCL8, CCL2, and CCL5, promoting cell migration, proliferation, invasion, and metastasis in heterogeneous TNBC by upregulating cyclin D1 and c-Myc expression (Mi et al. 2011).

### 4.1.9 B cells

B cells are the specialized type of immune cells that are frequently present in invasive tumor margins, draining lymph nodes and lymphoid structures. They are also responsible for presenting antigens, producing antibody and secreting cytokines (Arneth 2019). In TNBC, B cells are related to positive prognosis. These cells release cytokines that coordinate with cytotoxic T lymphocytes (CTLs) and function as APCs. However, some studies suggest that B cells may also have a pro-tumorigenic role by attracting myeloid-derived suppressor cells (MDSCs), activating angiogenic factors, promoting vessel formation, and inhibiting CTL activity (Lei et al. 2020).

### 4.1.10 Myeloid-Derived Suppressor Cells (MDSCs)

MDSCs constitute a diverse subpopulation of immature myeloid cells (IMCs) within the TME and are associated with pro-tumorigenic immune responses. Tumor-derived signals stimulate the growth, recruitment, and activation of these cells, leading to the blockage of both adaptive and innate antitumor immune responses (Mehdizadeh et al. 2023). MDSCs induce tumor development through various immunosuppressive mechanisms, involving the secretion of cytokines (such as G-CSF, IL-6, and TGFβ1), increased ROS production, and metabolite depletion. In the TME, MDSCs hinder the activity of NKs and cytotoxic T cells by recruiting Treg. Furthermore, MDSCs contribute to angiogenesis by releasing vascular endothelial growth factor (VEGF), matrix metalloproteinase 9 (MMP9), and prokineticin 2. Their presence may also induce cancer cell migration toward endothelial cells, enhancing TNBC metastasis (Liu et al. 2023).

## 4.2 Modulation of TME

The TME presents both challenges and opportunities for cancer therapy, functioning as a double-edged sword (Han et al. 2021). Modulating the TME involves strategies aimed at inhibiting angiogenesis, inducing immune cell infiltration (specifically recruiting CD8+, CD4 + T cells, and NK cells), restricting tumor metastasis, and enhancing the effectiveness of immunotherapies (Han et al. 2021). Researchers have extensively explored various aspects of the TME to understand how its manipulation can reduce cancer cell proliferation and migration.

Recent years have witnessed innovative therapeutic approaches to reshape the TME and treat TNBC. These approaches include the use of molecules that can impede signaling pathways, inhibitors of poly(ADP-ribose) polymerase (PARP), antibodies targeting growth receptors, and agents that reactivate the immune system. Immunotherapeutic strategies involve blocking the actions of inhibitory receptors like cytotoxic T lymphocyte antigen 4 (CTLA-4), programmed death protein 1 (PD-1), and programmed death ligand 1 (PD-L1). These interventions aim to create a more favorable TME for effective cancer treatment (Galvan Morales et al. 2020).

Immune crosstalk in TNBC is an intricate interplay of various components within the TME, and understanding these interactions is essential for creating personalized and effective therapeutic interventions (Retecki et al. 2021). Figure 3 likely illustrates the intricate network of immune responses and cellular interactions within the TNBC microenvironment, highlighting the potential points of modulation for therapeutic benefit.

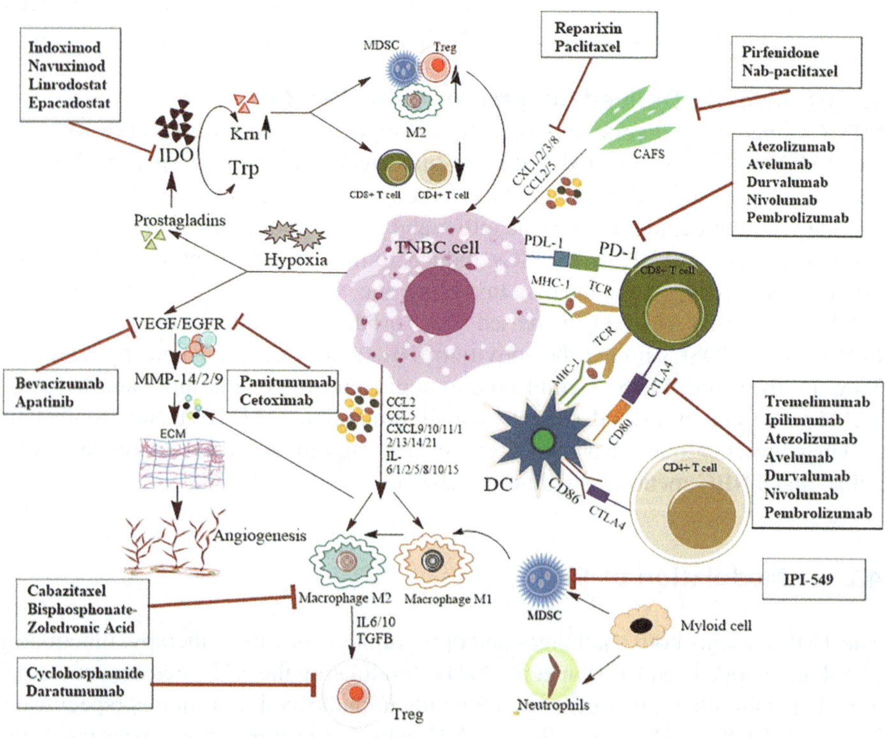


**Fig. 3** The intricate network of immune crosstalk. The diagram illustrates the immunomodulatory potential of various drugs within the TME of TNBC

PD-1 is a crucial immunological checkpoint inhibitor and is expressed on NK, B, T, and myeloid cells. PD-L1 is presented by TNBC cells and inhibits T cell responses by binding to PD-1 on activated T cells. FDA-approved PD-1/PD-L1 inhibitors include atezolizumab, durvalumab, pembrolizumab, cemiplimab, nivolumab, and avelumab (Mansour et al. 2017).

Another immune checkpoint inhibitor is CTLA-4. It is an inhibitory receptor of T cells and is mainly expressed on CD4+ and CD8+ activated T cells. CTLA-4 inhibits the T cell immune response by interacting with both CD80 (B7-1) and CD86 (B7-2). High expression is observed in TNBC. Clinical trials are in progress to determine the efficacy of the anti-CTLA-4 antibodies (ipilimumab and tremelimumab) in TNBC (Yi et al. 2021).

Tumor growth requires new blood vessels to alleviate oxygen shortage. Therefore, targeting angiogenesis seemed to be a promising strategy. Anti-angiogenic therapy has become a prominent strategy to target VEGF-vascular endothelial growth factor (VEGFR) signaling axis, including bevacizumab (a VEGF-neutralizing antibody), aflibercept (acts as a decoy receptor for VEGF), sorafenib (a tyrosine kinase inhibitor), and ramucirumab (antibody blocking VEGF binding to its receptor) (Anderson and Simon 2020).

Recent studies show that cetuximab (an anti-epidermal growth factor receptor (EGFR) antibody) prevents EGFR activation, blocking multiple signaling pathways (Raf/MEK/ERK, JAK/STAT, and PI3K/Akt) to restrict TNBC aggressiveness. It improves the prognosis of individuals with TNBC (Costa et al. 2017).

Several chemokine receptor inhibitors are being investigated to treat various primary tumors. Reparixin, paclitaxel, and balaxifortide drugs are inhibitors of chemokines (CXCL8 and CCL2) and their receptors (CXCR1/2/4) (Ginestier et al. 2010; Schott et al. 2017).

MDSCs' immunosuppressive function can be blocked by directly targeting phosphatidylinositol 3-kinase (PI3K) gamma and delta which in turn inhibit NF-κB and activate CCAAT/enhancer-binding protein beta (CEBPbeta). The IPI-549 drug is a PI3K gamma inhibitor targeting MDSC cells to inhibit immuno-suppressive function in TNBC patients (Liu et al. 2023). Another drug atovaquone also inhibits or lowers the level of immunosuppressive cytokines (IL-10 and TGFβ), Tregs and MDSC, to suppress TNBC growth (Gupta et al. 2021).

Treg modulation is also noted recently. Daratumumab reduces Treg population, leading to an increase in cell counts of CD8+ T cell and T helper cells (Krejcik et al. 2016). Annexin 1 expression blocks Boc1 in TNBC, reducing the Treg population (Bai et al. 2020).

These approaches collectively aim to modulate the tumor microenvironment, inhibit angiogenesis, and promote antitumor immune responses in TNBC. The combination of these strategies reflects the multifaceted nature of cancer treatment, emphasizing the importance of targeting various components for improved therapeutic outcomes. Further therapeutic strategies for targeting specific TME components in TNBC are under investigation.

**Cancer-Associated Fibroblasts (CAFs) Targeting**

Several therapeutic strategies have been explored to target cancer-associated fibroblasts (CAFs) in the context of TNBC, aiming to disrupt the tumor-stromal interactions and inhibit the tumor-promoting functions of CAFs:

*Pirfenidone*: Acts as an anti-fibrotic agent and inhibitor of TGFβ, directly targeting CAFs to decrease TNBC growth.

*Nab-paclitaxel*: Suppresses the tumor-promoting properties of CAFs and targets the cancer-stromal connection, resulting in a reduction of TNBC growth (Takai et al. 2016).

*Tranilast*: Acts as a TGFβ inhibitor and is used as an antihistamine medication. In TNBC mouse models, it targets BCAFs, enhances T cell perfusion, and reduces extracellular matrix (ECM) components.

*Combination therapy* of tranilast with liposomal doxorubicin increases blocking antibodies of immune checkpoints and stimulates the expression of immuno-stimulatory macrophage M1, leading to increased treatment efficacy in TNBC (Mollah and Varamini 2021).

*Cisplatin*: Involved in the inactivation of BCAFs. In a clinical trial, nivolumab in combination with short-term cisplatin therapy was used to treat metastatic TNBC (Voorwerk et al. 2019).

*Poly(ADP-ribose) polymerase (PARP) Inhibitors*: Inhibit PARP, a protein involved in programmed cell death and DNA repair. PARP inhibitors have recently shown sustained antitumor responses, particularly in individuals with advanced germline *BRCA1/2* mutations with breast cancer. Ongoing phase I/II trials are examining the safety and effectiveness of niraparib with pembrolizumab, as well as olaparib with durvalumab (anti-PD-L1 antibody) for advanced TNBC. Rucaparib, another PARP1 inhibitor, is used to treat TNBC (Fong et al. 2009; Galvan Morales et al. 2020). These approaches underscore the diverse strategies employed to target specific components of the TME, particularly CAFs, and enhance therapeutic outcomes for individuals with TNBC. The ongoing exploration of combination therapies reflects the complexity of TNBC and the need for multifaceted treatment strategies. Immunotherapeutic modulation in TNBC is further detailed in Table 1.

## 5 Immunotherapy Implications

The emergence of immunotherapy has marked a paradigm shift in the approaches to treat cancer, moving away from traditional cytotoxic drugs toward approaches that harness the body's immune system against tumors (Salemme et al. 2021). The immune system performs a significant role in preventing tumor progression, and immunotherapy has made significant strides in improving outcomes for different solid tumor types (Sukumar et al. 2021).

During the initiation and progression of cancer, immune cells have a dual role. They not only work to impede tumor development by eradicating immunogenic

**Table 1** Immunotherapeutic modulation of triple-negative breast cancer (TNBC)

| Immunotherapeutic approach | Drug/molecule | TME modulation and effect on TNBC | Clinical trials | Study (population) | Current status (2023) |
|---|---|---|---|---|---|
| **Chemokines CXCL8-CXCR1/2 CCL-2 CXCR-4 antagonist)** **MOA:** Stimulation of the immune system against cancer through blocking TNBC intracellular signal transduction pathways, inhibits cancer cell homing, recruitment, metastasis, and reduces drug resistance https://doi.org/10.1016/j.semcancer.2022.03.008 https://doi.org/10.1158/1078-0432.CCR-16-2748 | Reparixin/paclitaxel | Reduction in tumor cells | Phase I (NCT02370238) | Belgium, Czechia, France, Italy, Poland, Spain, and the USA | Completed |
| | Balaxifortide | Reduction in metastasis of cancer cells | – | – | No results found for TNBC |
| **Cytokine-based therapies (IL-12)** **MOA:** Administration of cytokines to modulate immune responses | IL-12 monotherapy in treatment refractory | Tumor regression | Phase I | Belgium, Canada, France, Italy, Poland, the United Kingdom, and the USA | Completed |
| | Intratumoral tavokino gene teleplasmid, a plasmid encoding IL-12 | Enhanced tumor immunogenicity | Phase II | | |
| | NKTR-214 bempegaldesleukin, an engineered cytokine | High CD8+ T and NK cells level in TME | Phase III | | |

(continued)

**Table 1** (continued)

| Immunotherapeutic approach | Drug/molecule | TME modulation and effect on TNBC | Clinical trials | Study (population) | Current status (2023) |
|---|---|---|---|---|---|
| against cancer https://doi.org/10.1038/s41523-022-00486-y | Combining NKTR214 + nivolumab + ipilimumab | No toxicities were observed | Phase IV (NCT02983045) | | |
| **Exosomes** **MOA:** Act as carriers to transport various substances and activate immune response against cancer https://doi.org/10.3389/fmolb.2022.1022725 | Disintegrin and metalloproteinase 15 | Improved anticancer effects, without adverse effects | – | The USA | No results found for TNBC |
| | Folate (FA) plus erastin | Enhances ferroptosis and inhibits the proliferation and migration of TNBC cells | | | |
| | Mesothelin (MSLN) | Inhibits the growth of MSLN-positive TNBC cells without obvious side effects in vivo | | | |
| **Immune checkpoint inhibitors** **PD-1/PD-L1 inhibitors** **MOA:** Blockade of immune checkpoints to enhance T cell activity and antitumor response https://doi.org/10.1016/j.drup.2017.07.002 https://doi.org/10.3389/ | Atezolizumab | Binds preferentially to PD-L1 and disrupts its association with B7.1 and PD-1 receptors, making T cells more susceptible to the tumor | Phase III (NCT02425891) | Argentina, Australia, Austria, Bosnia, Brazil, Canada, Chile, Colombia, Costa Rica, Czechia, Estonia, Finland, France, Germany, Guatemala, Herzegovina, Hong Kong, Italy, and the USA | Completed |
| | | | Phase III (NCT03125902) | Argentina, Australia, China, Greece, India, | Completed |

| | | | | | |
|---|---|---|---|---|---|
| fonc.2022.919072<br>https://doi.org/10.1038/s41523-022-00486-y<br>https://doi.org/10.1186/s12967-018-1514-7<br>https://doi.org/10.3390/ijms232314937<br>https://doi.org/10.1016/j.critrevonc.2018.01.005<br>https://doi.org/10.1155/2020/8209173 | | | | Israel, Italy, Japan, Morocco, Romania, Russia, Saudi Arabia, Slovakia, Spain, Turkey, and the USA | |
| | | | Phase 1b/II (NCT02708680) | Georgia and the USA | Completed |
| | | | Phase II (NCT03164993) | Denmark and Norway | Active, not recruiting (phase II) |
| | | | Phase IIb (NCT01898117) | Netherlands | Active not recruiting |
| | | | Phase II (NCT02530489) | The USA | Active, Not recruiting |
| | Avelumab | Exhibits inhibitory effect in TNBC cells by targeting the immunoregulatory PD-1 receptors on T cell surface and upregulation of immune relevant genes implicated in the T cell cytotoxicity and PD-1/PD-L1 pathways | Phase 1b/II (NCT02554812) | Australia, France, Japan, Poland, Taiwan, the United Kingdom, and the USA | Terminated |
| | Durvalumab | Significantly improves survival by inhibiting cancer | NCT02628132 | Saudi Arabia | Completed |
| | | | Phase I (NCT03199040) | The USA | Active, not recruiting |

(continued)

**Table 1** (continued)

| Immunotherapeutic approach | Drug/molecule | TME modulation and effect on TNBC | Clinical trials | Study (population) | Current status (2023) |
|---|---|---|---|---|---|
| | Nivolumab | Significant clinical efficacy against several types of malignancies | Phase II (NCT03316586) | The USA | Completed |
| | | | Phase II (NCT02499367) | Netherlands | Active, not recruiting |
| | | | Phase II (NCT03818685) | France | Active, not recruiting |
| | Pembrolizumab | Toxicity and promising antitumor activity related to positive correlation with tumor PD-L1 expression and stromal tumor-infiltrating lymphocyte levels; common adverse events: neutropenia | Phase 1b (NCT02622074) | | Completed |
| | | | NCT03036488 (Phase III) | Australia, Brazil, Canada, Colombia, France, Germany, Ireland, Israel, Japan, Korea, Portugal, Singapore, Sweden, Taiwan, Turkey, the United Kingdom, and the USA | Active, not recruiting |
| | | | Phase II (NCT02734290) | The USA | Active, not recruiting |
| | | | Phase II (NCT03145961) | The USA | Active, not recruiting |
| | Atezolizumab + paclitaxelb | Higher overall survival in the patients treated with atezolizumab and nab-paclitaxel<br>Adverse events include neutropenia, peripheral | Phase II (NCT02530489) (Phase II) | The USA | Active, not recruiting |

| | | | | | |
|---|---|---|---|---|---|
| | | neuropathy, decreased neutrophil count, and fatigue | | | |
| | Pembrolizumab combined with eribulin | Well-tolerated combination with promising antitumor activity, higher objective response rate in patients with PD-L1-positive tumors<br>Adverse events include fatigue, nausea, peripheral sensory neuropathy, alopecia, and constipation | Phase II (NCT05402722) | China | Recruiting |
| **Antiangiogenic therapy**<br>**MOA:**<br>It has become a prominent strategy to target VEGF-VEGFR signaling axis by neutralizing VEGF, blocking its receptor which result as inhibition of angiogenesis and immune stimulation https://doi.org/10.1186/s12967-018-1514-7 | Bevacizumab | Blocks the growth of new blood vessels that inhibit TNBC cells | Phase II (NCT02456857) | The USA | Completed |
| | | | Phase II (NCT00733408) | The USA | Completed |
| | Apatinib | Promotes apoptosis and suppresses migration and invasion of TNBC cells | Phase II (NCT05447702) | China | Recruiting |
| | Aflibercept/ramucirumab | | – | – | No results for TNBC |

(continued)

**Table 1** (continued)

| Immunotherapeutic approach | Drug/molecule | TME modulation and effect on TNBC | Clinical trials | Study (population) | Current status (2023) |
|---|---|---|---|---|---|
| **Anti-EGFR therapy**<br>**MOA:**<br>It works as anticancer drug by preventing EGFR activation, blocking multiple signaling transduction pathways, and limiting tumor aggressiveness https://doi.org/10.1186/s12967-018-1514-7 | Panitumumab | Helps the body's immune system attack the cancer and may interfere with the ability of tumor cells to grow and spread | Phase II (NCT02876107) | The USA | Active, not recruiting |
| | | | Phase II (NCT02593175) | | |
| | Cetuximab | Reduces the CSC population in TNBC tumors. Prevents the EGFR activation and blocks its interaction to ligand. Inhibits multiple signalling pathways, including Raf/MEK/ERK, JAK/STAT and PI3K/Akt, which appears to be adequate to restrict TNBC aggressive behavior | Phase II (NCT00463788) | Australia, Austria, Belgium, Germany, Ireland, Israel, Portugal, Spain, and the United Kingdom | Completed |
| | | | Phase II (NCT01097642) | The USA | Completed |
| | | | Phase II (NCT00633464) | Austria, Czech Republic, France, Greece, Italy, and Poland | Completed |
| **CTLA-4**<br>**MOA:**<br>Exhibiting anticancer activity by suppressing the expression of CTLA-4 on activated T cells. This increases T cell activation and blocks TNBC | Tremelimumab | Might achieve stronger immune responses for the treatment of TNBC patients in clinical trials | Phase II (NCT02527434) | Belgium, Korea, the Netherlands, Poland, and the United States | Completed |
| | Ipilimumab | Shows potentially curative and durable efficacy in tumor regression | Phase II (NCT03818685) | France | Active, not recruiting |

| | | | | | |
|---|---|---|---|---|---|
| progression https://doi.org/10.1016/j.critrevonc.2018.01.005 https://doi.org/10.1186/s12967-018-1514-7 | | | | | |
| **PARP MOA:** Inhibit PARP, a protein involved in programmed cell death and DNA repair resulting in a reduction of TNBC growth https://doi.org/10.1186/s12967-018-1514-7 https://doi.org/10.1155/2020/8209173 | Olaparib | Olaparib treatment increased sub-G1 cell death and G2/M arrest in TNBC cells | Phase II (NCT05128734) | Canada | Not yet recruiting |
| | Veliparib | Significantly improved progression-free survival in patients with BRCA-like metastatic TNBC | NCT02985658 | The USA | No longer available |
| | Talazoparib | Significantly affects cell proliferation and migration, induction of apoptosis and necrosis in TNBC | Phase II (NCT02401347) | The USA | Completed |
| | | | Phase I (NCT05035745) | Singapore | Recruiting |
| | | | Phase II (NCT04849364) | The USA | Recruiting |
| | | | Phase 1b/2 (NCT04039230) | The USA | Recruiting |
| | | | Phase II (NCT05288127) | Malaysia | Recruiting |
| | | | Phase II (NCT03911973) | The USA | Active, not recruiting |
| | | | Phase II (NCT04690855) | The USA | Active, not recruiting |

(continued)

**Table 1** (continued)

| Immunotherapeutic approach | Drug/molecule | TME modulation and effect on TNBC | Clinical trials | Study (population) | Current status (2023) |
|---|---|---|---|---|---|
| | | | Phase 1b/2 (NCT03875313) | The USA | Terminated |
| **IDO inhibitors** **MOA:** By inhibiting the Kyn pathway, IDO1 inhibitors reduce the immunosuppressive effects, which may help overcome associated immune resistance and enhance antitumor activity http://creativecommons.org/licenses/bync-nd/4.0 https://doi.org/10.1016/j.critrevonc.2018.01.005 | Indoximod<br>Navoximod<br>Linrodostat<br>Epacadostat | Exhibited a therapeutic potential in TNBC cells, and pro-inflammatory cytokines could affect the effectiveness of indoximod | – | – | No results for TNBC |
| **Epigenetics inhibitors** **MOA:** They have the ability to reverse immune suppression through several mechanisms such as the enhancement of tumor-associated antigen | Birabresib (BETi) | Potent BETi antitumor efficacy. | (Phase Ib) NCT02259114 (Phase IB) | – | Completed |
| | Entinostat (HDACi) + azacitidine (DNMTi) | Synergistic antitumor activity | Phase II (NCT01349959) | The USA | Active, not recruiting |
| | Chidamide (HDACi | Suppresses the growth of TNBC cells. | NCT04582955 | China | Unknown status |
| | | | NCT03361800 | The USA | Terminated |

| | | | | | |
|---|---|---|---|---|---|
| expression and antigen-processing components, and modification of chemokine profiles that elicit a host immunological response https://doi.org/10.3390/cancers13164139 https://doi.org/10.1016/j.critrevonc.2018.01.005 | Entinostat (HDACi) | It has also been shown to reduce the percentage of tumor-initiating cells in TNBC cells | | | |
| **Adoptive cell therapies** **MOA:** Infusion of engineered or activated immune cells to target cancer cells https://doi.org/10.1155/2020/8209173 https://doi.org/10.1186/s12967-018-1514-7 | Anti-MUC1 CAR T cells | Significantly reduced *TNBC* tumor growth in a xenograft model. The productive engagement of TCR/MHC/antigen complexes on the target cell surface triggers the CTL's effector functions and induces the destruction of the target cell through the release of inflammatory cytokines, including tumor necrosis factor-α (TNFα), IFN-γ, FAS ligand (FASL), TNF-related apoptosis inducing ligand | Phase I/II (NCT02587689) | China | Unknown status |
| | NKR-2 CAR T cells | | NCT03018405 (Phase I) | The USA Belgium | Unknown status |
| | ROR1 + CAR T cells | | Phase I (NCT02706392) | The USA | Terminated |

(continued)

**Table 1** (continued)

| Immunotherapeutic approach | Drug/molecule | TME modulation and effect on TNBC | Clinical trials | Study (population) | Current status (2023) |
|---|---|---|---|---|---|
| **Combination therapies** | PD-1/PD-L1 inhibitors + chemotherapy | MOA: simultaneous use of multiple immunotherapeutic agents or combining them with standard treatments | | | |
| **PDL1 + CTL-4** https://doi.org/10.1186/s12967-018-1514-7 https://doi.org/10.1016/j.critrevonc.2018.01.005 | Tremelimumab + durvalumab | Ability to increase the efficacy of immunotherapy and to slow down primary tumor outgrowth and metastasis | Phase Ib (NCT02658214) | Japan and Korea | Completed |
| **PD-1 + PARP** https://doi.org/10.1186/s12967-018-1514-7 https://doi.org/10.1016/j.critrevonc.2018.01.005https://doi.org/10.3390/ijms232314937 | Pembrolizumab + niroparib | Exerts a supplementary and increases antitumor effect in combination with PARP inhibitors | Phase I/II (NCT02657889) | The USA | Completed |
| **PDL1+ PARP** https://doi.org/10.1186/s12967-018-1514-7 https://doi.org/10.3390/ijms232314937 https://doi.org/10.1016/j.critrevonc.2018.01.005 | Durvalumab + olaparib | PARP inhibitors can upregulate PD-L1 expression and enhance cancer-associated immunosuppression | Phase II (NCT03167619) | The USA | Completed |

| | | | | | |
|---|---|---|---|---|---|
| **HDACi + anti PD-1** https://doi.org/10.3390/cancers13164139 | Atezolizumab + entinostat | Not prolonged progression-free survival, greater toxicity | Phase II (NCT02708680) | Georgia and the USA | Completed |
| | Panobinostat + PRDOO1 | It induces the apoptosis in cancer cells | Phase Ib (NCT02890069) | The Netherlands, Taiwan, UK, and the USA | Completed |
| **STING agonist + anti PD-1** https://doi.org/10.1186/s12967-018-1514-7 | MK-1454 (i.t.) +/– pembrolizumab (i.v.) | Exhibited anticancer activity. | – | – | No results for TNBC |
| | ADU-S100 (i.t.) + PDR001(i.v.) (spartalizumab) | This combination displays the antitumor activity in PD-1–naive TNBC and PD-1-relapsed/refractory melanoma | – | – | No results for TNBC |
| **Anti-MDSCs** **MOA:** By limiting the MDSC population, reducing immunosuppressive cytokines and chemokines, these drugs could elicit tumor suppression https://doi.org/10.3389/fimmu.2023.1199273 https://doi.org/10.3390/ijms22105150 | IPI-549 | Target MDSC cells or PI3K gamma inhibitor in TNBC patients to inhibit MDSCs immunosuppressive function | Phase 1/1b (NCT03719326) | Australia and the USA | Completed |
| | Atovaquone | Inhibits or reduces the level of MDSCs, Tregs, and immunosuppressive cytokines (IL-10 and TGFβ) to suppress TNBC growth | – | – | No results for TNBC |
| **Anti-Tregs** **MOA:** It targets | Cyclophosphamide | The suppression of CCR2-positive Tregs by cyclophosphamide | Phase III (NCT04335669) | Denmark and Sweden | Recruiting |
| | | | | The USA | Withdrawn |

(continued)

**Table 1** (continued)

| Immunotherapeutic approach | Drug/molecule | TME modulation and effect on TNBC | Clinical trials | Study (population) | Current status (2023) |
|---|---|---|---|---|---|
| immunosuppressive Treg cells and increases the expression of CD4+ and CD8+ T effector cells to inhibit tumor | | results in a reduction in immunosuppressive effects and an increase in cytotoxic CD8+ T cell activation | Phase II (NCT03554109) | | |
| https://doi.org/10.1002/cncr.34084 https://doi.org/10.1182/blood-2015-12-687749 | Daratumumab | Treg population can be reduced by treating the patients with daratumumab, which leads to increase in the number of CD8+ T cell and T helper cells | Phase I/II (NCT03098550) | Canada, France, Germany, Italy, Switzerland, and the USA | Completed |
| **Anti-CAFs** **MOA:** Acts as an anti-fibrotic agent and TGFβ inhibitors which enhances T cell perfusion which stimulate immune response https://doi.org/10.18632/oncotarget.12658 https://doi.org/10.3390/biomedicines9121921 | Pirfenidone | It suppresses the tumor by acting as an anti-fibrotic and TGFβ antagonist | – | – | No results for TNBC |
| | | | NCT03907800 | China | Recruiting |
| | | | Phase III (NCT04085276) | China | Active, not recruiting |
| | | | Phase III (NCT04251533) | Colombia, Croatia, France, Germany, Hungary, India, Italy, Malaysia, Peru, Poland, Slovakia, Slovenia, Spain, Switzerland, Taiwan, Turkey, the United Kingdom, and the USA | Active, not recruiting |

| | | | | | |
|---|---|---|---|---|---|
| | | | NCT02530489 | The USA | Active, not recruiting |
| | | | NCT04159142 | China | Unknown status |
| | Cisplatin | It is also involved in the inactivation of BCAFs | Phase II (NCT01091454) | The USA | Completed |
| | | | Phase II (NCT01982448) | The USA | Completed |
| | | | Phase I (NCT00998036) | The USA | Completed |
| | | | Phase II (NCT03012477) | The USA | Complete |
| | | | Phase III (NCT04595565) | Autria, Germany, France, Belgium, Italy, and Ireland | Recruiting |
| | | | Phase III (NCT02445391) | South Africa and the USA | Active, not recruiting |
| | | | Phase Ib/II (NCT01918306) | The USA | Terminated |
| **Anti-TGF β** **MOA:** It works by downregulation or inhibition of TGFβ and related signaling pathways which result as anticancer activity https://doi.org/10.3390/biomedicines9121921 | Tranilast | It is a TGFβ inhibitor and is used as an anti-histamine medication. It enhanced T cell perfusion and infiltration by reducing ECM components | – | – | No results for TNBC |

(continued)

**Table 1** (continued)

| Immunotherapeutic approach | Drug/molecule | TME modulation and effect on TNBC | Clinical trials | Study (population) | Current status (2023) |
|---|---|---|---|---|---|
| **Anti-TAMs**<br>**MOA:**<br>TAM-targeting therapies are important for antitumor strategies because they encourage TAM phagocytosis, blocking TAM recruitment, reprogram TAMs and suppressing immunosuppressive TME<br>https://doi.org/10.1038/s41416-020-0909-4<br>https://doi.org/10.3389/fonc.2022.772615 | Cabacitaxel | It has the ability to penetrate the blood–brain barrier. It differentiates M2 macrophages into M1 macrophages and induces the activation of NF-kB signaling pathway. It is demonstrated that CD47 blockade and cabazitaxel combined treatment suppress the tumor metastasis and promote programmed cell removal (PrCR) of TNBC cells in preclinical models | Phase II (NCT01779479) | Germany | Completed |
| | Bisphosphonate-zoledronic acid | It can induce macrophage apoptosis and eradicate TAMs, block famesyl diphosphate (FPP) synthase activity, and promote apoptosis of macrophages | NCT04045522 | China | Unknown status |

| | | | | | |
|---|---|---|---|---|---|
| **Cancer vaccines** MOA: Stimulation of the immune system to recognize and target tumor-specific antigens https://doi.org/10.3390/vaccines11010146 https://doi.org/10.1155/2020/8209173 https://doi.org/10.1186/s12967-018-1514-7 | DC vaccine | Whole apoptotic breast carcinoma merged with DCs. Acts as an antitumor by displaying T cell responses | Phase I/II (NCT02018458) | The USA | Completed |
| | Neoantigen vaccine | Triggers an immune reaction against an antigen to a tumor, which can result in the development and proliferation of T lymphocytes that can identify tumor antigens and suppress tumor growth | Phase I (NCT03199040) | The USA | Active, not recruiting |
| | PVX-410 | Induces an immune response against the targeted tumor cell, which may act to help stimulate an immune response to the targeted tumor cell | Phase I (NCT02348320) | The USA | Completed |
| | | | Phase I (NCT02826434) | The USA | Active, not recruiting |
| | | | Phase I (NCT03362060) | The USA | Active, not recruiting |
| | | | Phase I (NCT03122106) | The USA | Terminated |

neoplastic cells but also contribute to remodeling the immunogenicity of tumors. The complex interplay between immunological TME and cancer impacts the efficacy of immunotherapy and other anticancer treatments (Salemme et al. 2021). This has particularly revolutionized the landscape of cancer treatment for various solid tumor types, particularly breast cancer. In the context of TNBC, which is known for its worst prognosis and resistance to standard therapies, immunotherapy represents a promising avenue for intervention.

Understanding the immunological characteristics and profiles of TNBC and developing innovative immune-modulatory approaches for its treatment are crucial areas of breast cancer research. TNBC, known for its worst prognosis, poor responsiveness, and resistance to standard therapies, presents a challenge that immunotherapy aims to address (Abdou et al. 2022). Despite significant progress, clinical effectiveness is limited to a certain number of patients. Secondary resistance eventually occurs (Salemme et al. 2021) even in those who initially respond.

Immunotherapy encompasses various approaches (Fig. 4), including **adoptive cell transfer** (enhances the patient's immune system's ability cells to recognize and destroy cancer cells), **cytokines** (use of immune-stimulating proteins to boost the body's response against cancer), **checkpoint inhibitors** (blockade of immune

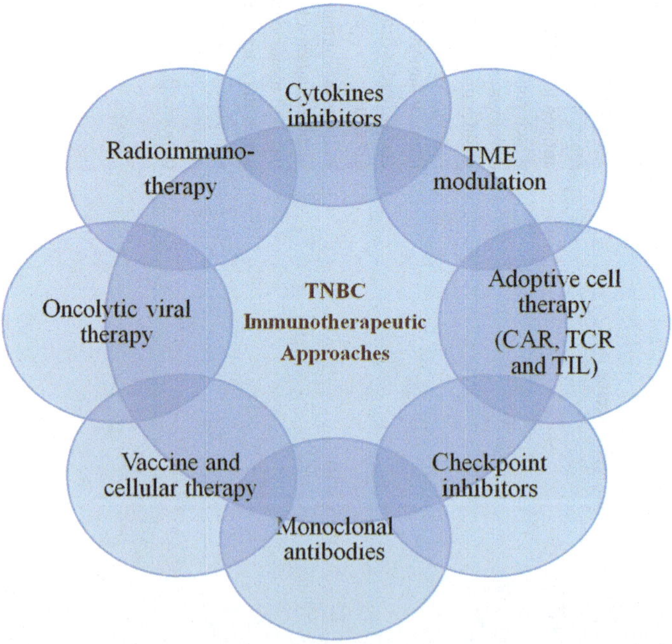


**Fig. 4** Immunotherapeutic approaches in TNBC. The illustration encompasses immune checkpoint inhibitors, adoptive cell therapies, cancer vaccines, and cytokine-based therapies, showcasing the multi-faceted approaches aimed at harnessing the immune system to combat TNBC. Each strategy targets specific aspects of the immune response, underscoring the evolving landscape of immunotherapy in the quest for effective and personalized treatment modalities for TNBC patients

checkpoints to unleash the immune system against cancer cells), **monoclonal antibodies** (engineered antibodies designed to target specific proteins on cancer cells), **TME modulation** (strategies to modify the TME to make it less favorable for cancer growth), and **vaccines** (specialized approaches targeting specific cancer types and vaccine development) (Guanghui et al. 2019; Jia et al. 2017). Clinical effectiveness is limited, and overcoming resistance mechanisms is a challenge. The goal of ongoing research is to refine and broaden the applications of immunotherapy.

The evolving landscape of immunotherapy holds great promise for cancer treatment, particularly in challenging cancers such as TNBC. The ongoing exploration of various immunotherapeutic approaches and the optimization of their synergies are critical for overcoming challenges and improving patient outcomes. Continued research and innovation in immunotherapy contribute to the development of more potent and targeted treatments for TNBC and other cancers.

## 5.1 Immune Checkpoint Blockade

The blockade of immune checkpoints through immunomodulating antibodies has emerged as a novel and promising approach in cancer immunotherapy, particularly for breast cancer (Mansour et al. 2017). In the context of TNBC, several immune checkpoint inhibitors have shown potential for providing durable tumor shrinkage and inducing tumor immunity (Guanghui et al. 2019).

### 5.1.1 Atezolizumab

A humanized IgG1 monoclonal antibody (moAb) selectively binding to PD-L1, disrupting its interaction with B7.1 and PD-1 receptor, enhancing T cells effectiveness against TNBC. In clinical trials, atezolizumab demonstrated improved outcomes, particularly in metastatic TNBC patients, especially those with higher concentrations of PD-L1-positive tumor-infiltrating immune cells (Emens et al. 2019; Heimes and Schmidt 2019). Combining atezolizumab with nab-paclitaxel showed enhanced progression-free survival (PFS) upto 7.2 months in metastatic TNBC (Schmid et al. 2018).

### 5.1.2 Pembrolizumab

A humanized, IgG4 moAb, selectively binding to the PD-1 receptor, blocking its interaction with PD-L1 and PD-L2 ligands, and promoting antitumor immune response by neutralizing T cells. Clinical trials have demonstrated effectiveness in treating early stages of metastatic TNBC. Combining pembrolizumab with chemotherapy showed a clinically meaningful anticancer impact (Nanda et al. 2016).

### 5.1.3 Durvalumab

A human immunoglobulin moAb, inhibiting the interaction of PD-L1 with its ligands, suppressing TNBC. Combining durvalumab with neoadjuvant chemotherapy has shown increased survival outcomes in TNBC patients (Foldi et al. 2021).

### 5.1.4 Nivolumab

A human moAb, targeting the PD-1 receptor on T cell surface, demonstrating high clinical effectiveness in TNBC patients. Combining nivolumab with chemotherapeutic agents like doxorubicin and cisplatin upregulates immune-relevant genes involved in T cell cytotoxicity and PD-1/PD-L1 pathways (Wesolowski et al. 2022).

### 5.1.5 Avelumab

A human IgG1 moAb, targeting PD-L1 and inhibiting the PD-1/PD-L1 pathway (Heery et al. 2017). It induces antibody-dependent cell-mediated cytotoxicity (ADCC), leading to TNBC cell death by enhancing NK-mediated cytotoxicity (Juliá et al. 2018).

### 5.1.6 Tremelimumab

An immune checkpoint inhibitor of CTLA-4 pathway, exhibiting anticancer activity by suppressing the CTLA-4 expression on activated T cells. This increases T cell activation and blocks TNBC progression (Vonderheide et al. 2010).

### 5.1.7 Ipilimumab

A CTLA-4 pathway antagonist, a human IgG1 moAb binding to the CTLA-4 receptor on activated T cells. FDA approved this drug for several malignancies, including TNBC. Combining ipilimumab with cryoablation has shown positive immunological alterations, including T cell activation and an increase in the tumor CD8+ cells ratio (Wesolowski et al. 2022). These immune checkpoint inhibitors represent a diverse array of strategies, each aimed at modulating specific components of the immune response to target and combat TNBC. Clinical trials still continue to examine their efficacy and potential in improving outcomes for TNBC patients.

## 5.2 Treg Blockade

### 5.2.1 Cyclophosphamide

Commonly used to treat autoimmune disorders and various cancers and has shown dual effects based on dosage. High doses lead to immunosuppression, while moderate doses enhance the immunological response. It directly targets immunosuppressive Treg cells and increases the expression of CD4+ and CD8+ T effector cells to inhibit TNBC (Malla et al. 2022). It demonstrated the inhibition of TNBC growth in preclinical models (Pernas et al. 2018).

### 5.2.2 Daratumumab

A novel and high-affinity human moAb targeting the CD38 epitope and used against TNBC. Treatment with daratumumab results in a reduction in the immune-suppressive Treg population, leading to an increase in T helper and CD8-positive T cells concurrently (Krejcik et al. 2016). In preclinical studies, it showed favorable results against TNBC.

## 5.3 Chemokines Blockade

Preclinical research and clinical trials are exploring chemokine receptor inhibitors as potential treatments for various primary tumors; including TNBC (Saxena and Singh 2021). Combining chemokine inhibitors with immune checkpoint inhibitors (ICIs) or chemotherapy has shown promising reductions in metastatic burden.

### 5.3.1 Reparixin

A noncompetitive inhibitor of CXCL8 chemokine receptors CXCR1/2. Interaction with CXCL8 receptors blocks intracellular signal transduction pathways. In a phase I clinical trial, the combined use of reparixin with paclitaxel showed effectiveness and good tolerance in metastatic breast cancer or TNBC patients (Schott et al. 2017).

### 5.3.2 MSX-122

A CXCR4 partial agonist with suppressing CXCR4 functions by competitively interacting with the receptor. It inhibits cAMP signaling, preventing cell homing and recruitment without activating stem cells. Additionally, it exhibits anti-inflammatory and anticancer properties and can orally inhibit breast cancer lung metastasis (Zhao et al. 2022).

### 5.3.3 Balixafortide

A potential orally bioavailable agonist of the CXCR4 chemokine receptor. It blocks the interaction between CXCL12 and stromal-derived factor 1 (SDF1) with the CXCR4 receptor. It also inhibits tumor cell angiogenesis, migration, proliferation, and metastasis and reduces drug resistance.

## 5.4 Cytokines Blockade

Cytokines are essential targets for suppressing tumor responses (Berraondo et al. 2019). While IL-2 and IFN-α are the only two FDA-approved cytokines for cancer treatment (Floros & Tarhini 2015); their high toxicity profile limits their use (Diab et al. 2020). Interleukin-12 (IL-12) indicated a definite antitumor response as a monotherapy to inhibit TNBC growth, and its efficacy is further increased when combined with anti-PD-1 (Abdou et al. 2022).

### 5.4.1 Bazedoxifene

An FDA-approved drug that targets the inflammatory cytokine IL-6, blocking TNBC growth. Combining bazedoxifene with reparixin or SCH527123 (SCH) decreased the proliferative capacity of TNBC cells (Fu and Lin 2018).

### 5.4.2 Tocilizumab

A critical agonist of the inflammatory cytokine IL-8. Inhibition of IL-8 leads to the downregulation of NF-κB, IL-6 signal transducer, and STAT3 activation. It targets angiogenesis and may be beneficial as a treatment for TNBC patients (Alraouji and

Aboussekhra 2021). Combining tocilizumab with maraviroc, a CCR5 agonist, markedly decreased the invasive and migratory properties of TNBC cells and metastasis of breast cancer in a xenograft mouse model of TNBC (Manore et al. 2022).

## 5.5 TAM Blockade

### 5.5.1 Cabazitaxel (CTX)

A natural taxoid that can penetrate the blood–brain barrier. It differentiates M2 macrophages into M1 macrophages and induces the activation of the NF-κB signaling pathway. CD47 blockade combined with cabazitaxel treatment has been shown to suppress tumor metastasis and promote programmed cell removal (PrCR) of TNBC cells in preclinical models (Cao et al. 2021).

### 5.5.2 Bisphosphonate

A group of medicines that induce macrophage apoptosis and eradicate TAMs. Macrophages can readily engulf bisphosphonates via endocytosis. Once inside the macrophages, bisphosphonates block famesyl diphosphate (FPP) synthase activity and promote apoptosis. Numerous clinical trials have demonstrated substantial benefits of bisphosphonate therapy for postmenopausal women with breast cancer or TNBC, indicating its potential as a TAM blockade strategy (Qiu et al. 2022).

**Therapeutic Vaccines**

Therapeutic vaccines aim to improve tumor-specific immunogenic responses via active immunization, targeting recognized antigens (Ags) present in breast tumors. These vaccines can impact tumor progression by modulating the TME through chemokines and by eliciting responses from NK cells and cytotoxic CD8+ T cells (CTL) (Abdou et al. 2022). They have the potential to be beneficial in both therapeutic and preventive settings by inducing T cell responses to tumor-derived antigens and activating memory cells to block cancer recurrence.

When selecting antigens, it is crucial to consider three key criteria: specificity to cancer cells to avoid off-target effects, high immunogenicity to elicit a strong immune response, and essentiality to tumor cell survival (Hosseini et al. 2023). Therapeutic vaccines exist in different modalities, including dendritic cells (DCs), carbohydrates, peptides, recombinant DNA or RNA, and whole cells (Burke et al. 2019). DC vaccines leverage the T cell stimulatory and antigen-presenting properties, along with modulating chemokines and cytokines to control the migration of lymphocytes and inflammation, contributing to the long-lasting antitumor effects of the vaccine (Perez and De Palma 2019). To optimize the results of vaccination therapy, it is crucial to refine the vaccine development process, including selecting appropriate vaccines and delivery systems and exploring various combination therapy approaches. While therapeutic vaccines have shown promising results in preclinical research for TNBC, their clinical efficacy, safety, and translation are yet to be fully understood (Hosseini et al. 2023). Ongoing research is focused on refining the vaccine development process to improve outcomes in breast cancer patients. Therapeutic vaccines represent a promising avenue for cancer treatment, and continued research is essential to unlock their full potential in clinical settings.

## 6 Conclusion and Challenges in TNBC Treatment

TNBC remains the refractory and inoperable breast cancer subtype with the worst prognosis.

(i) *Refractory Nature and Poor Prognosis.*
TNBC is known for its refractory and inoperable nature, posing a significant challenge in treatment. Despite advances, it has the worst prognosis among all the subtypes of breast cancer.

(ii) *Lack of Targetable Receptors and Heterogeneity.*
TNBC lacks specific targetable receptors, limiting the effectiveness of targeted therapies.
High heterogeneity within TNBC subtypes results in varied clinical responses to treatments.

(iii) *Distinct Clinical Responses.*
Patients with different subtypes of TNBC exhibit distinct clinical responses to treatment, complicating personalized therapy.

(iv) *Conventional Treatment Modalities.*
Conventional approaches like chemotherapy, radiation, and surgery remain the mainstay of treatment for TNBC patients.
Despite aggressive care, a significant percentage of patients experience recurrence and reduced survival rates.

(v) *Multidrug Resistance (MDR).*
Molecular mechanism–based multidrug resistance to various anticancer drugs has been reported in TNBC.
MDR significantly hampers the effectiveness of chemotherapy in treating TNBC.

(vi) *Limited Treatment Options.*
TNBC patients face limited treatment options owing to the dearth of specific molecular targets.
The absence of targetable receptors makes it challenging to develop tailored therapies.

(vii) *Immunotherapy Challenges.*
While cancer immunotherapy holds promise, challenges include understanding the intricate activation mechanisms of NK cells and T cells.
Reprogramming cytotoxic T cells and managing cytokines are crucial for the success of immunotherapeutic interventions.

(viii) *Recurrence and Mortality.*
More than 50% of patients of TNBC (stage I–III) develop recurrence, and over 37% pass away within 5 years, highlighting the need for more effective treatments.

(ix) *Research and Innovation.*
Comprehensive molecular insights are essential for developing targeted therapies and understanding the underlying mechanisms of TNBC.

Continued research and innovative approaches are necessary to overcome the challenges associated with TNBC treatment and improve patient outcomes.

## References

Abdou Y, Goudarzi A, Yu JX, Upadhaya S, Vincent B, Carey LA (2022) Immunotherapy in triple negative breast cancer: beyond checkpoint inhibitors. NPJ Breast Cancer 8(1):121

Afghahi A, Telli ML, Kurian AW (2016) Genetics of triple-negative breast cancer: Implications for patient care. Curr Probl Cancer 40(2–4):130–140

Akram M, Iqbal M, Daniyal M, Khan AU (2017) Awareness and current knowledge of breast cancer. Biol Res 50(1):33

Almansour NM (2022) Triple-negative breast cancer: a brief review about epidemiology, risk factors, signaling pathways, treatment and role of artificial intelligence. Front Mol Biosci 9: 836417

Alraouji NN, Aboussekhra A (2021) Tocilizumab inhibits IL-8 and the proangiogenic potential of triple negative breast cancer cells. Mol Carcinog 60(1):51–59

Anderson NM, Simon MC (2020) The tumor microenvironment. Curr Biol 30(16):R921–R925

Arneth B (2019) Tumor microenvironment. Medicina 56(1):15

Ataollahi M, Sharifi J, Paknahad M, Paknahad A (2015) Breast cancer and associated factors: a review. J Med Life 8(Spec Iss 4):6

Ayers M, Lunceford J, Nebozhyn M, Murphy E, Loboda A, Kaufman DR, McClanahan TK (2017) IFN-γ–related mRNA profile predicts clinical response to PD-1 blockade. J Clin Invest 127(8): 2930–2940

Bai F, Zhang P, Fu Y, Chen H, Zhang M, Huang Q, Li D, Li B, Wu K (2020) Targeting ANXA1 abrogates Treg-mediated immune suppression in triple-negative breast cancer. J Immunother Cancer 8(1):e000169

Bauer KR, Brown M, Cress RD, Parise CA, Caggiano V (2007) Descriptive analysis of estrogen receptor (ER)-negative, progesterone receptor (PR)-negative, and HER2-negative invasive breast cancer, the socalled triple-negative phenotype: a population-based study from the California cancer registry. Cancer 109(9):1721–1728

Berraondo P, Sanmamed MF, Ochoa MC, Etxeberria I, Aznar MA, Pérez-Gracia JL, Rodríguez-Ruiz ME, Ponz-Sarvise M, Castañón E, Melero I (2019) Cytokines in clinical cancer immunotherapy. Br J Cancer 120(1):6–15

Bilani N, Zabor EC, Elson L, Elimimian EB, Nahleh Z (2020) Breast cancer in the United States: a cross-sectional overview. J Cancer Epidemiol 2020:6387378

Birbrair A, Zhang T, Wang ZM, Messi ML, Olson JD, Mintz A, Delbono O (2014) Type-2 pericytes participate in normal and tumoral angiogenesis. Am J Phys Cell Phys 307(1):25–38

Bou Zerdan M, Ghorayeb T, Saliba F, Allam S, Bou Zerdan M, Yaghi M, Nahleh Z (2022) Triple negative breast cancer: updates on classification and treatment in 2021. Cancers 14(5):1253

Burke EE, Kodumudi K, Ramamoorthi G, Czerniecki BJ (2019) Vaccine therapies for breast cancer. Surg Oncol Clin 28(3):353–367

Cao X, Li B, Chen J, Dang J, Chen S, Gunes EG, Xu B, Tian L, Muend S, Raoof M (2021) Effect of cabazitaxel on macrophages improves CD47-targeted immunotherapy for triple-negative breast cancer. J Immunother Cancer 9(3):e002022

Cho N (2016) Molecular subtypes and imaging phenotypes of breast cancer. Ultrasonography 35(4):281

Cornel AM, Mimpen IL, Nierkens S (2020) MHC class I downregulation in cancer: underlying mechanisms and potential targets for cancer immunotherapy. Cancers 12(7):1760

Corso G, Veronesi P, Sacchini V, Galimberti V (2018) Prognosis and outcome in *CDH1*-mutant lobular breast cancer. Eur J Cancer Prev 27(3):237–238

Costa R, Shah AN, Santa-Maria CA, Cruz MR, Mahalingam D, Carneiro BA, Chae YK, Cristofanilli M, Gradishar WJ, Giles FJ (2017) Targeting epidermal growth factor receptor in triple negative breast cancer: new discoveries and practical insights for drug development. Cancer Treat Rev 53:111–119

Deepak KGK, Vempati R, Nagaraju GP, Dasari VR, Nagini S, Rao DN, Malla RR (2020) Tumor microenvironment: Challenges and opportunities in targeting metastasis of triple negative breast cancer. Pharmacol Res 153:104683

Diab A, Tannir NM, Bentebibel SE, Hwu P, Papadimitrakopoulou V, Haymaker C, Cho DC (2020) Bempegaldesleukin (NKTR-214) plus nivolumab in patients with advanced solid tumors: phase I dose-escalation study of safety, efficacy, and immune activation (PIVOT-02). Cancer Discov 10(8):1158–1173

do Nascimento RG, Otoni KM (2020) Histological and molecular classification of breast cancer: what do we know? Mastology 30:1–8

Emens LA, Cruz C, Eder JP, Braiteh F, Chung C, Tolaney SM, Kuter I, Nanda R, Cassier PA, Delord J-P (2019) Long-term clinical outcomes and biomarker analyses of atezolizumab therapy for patients with metastatic triple-negative breast cancer: a phase 1 study. JAMA Oncol 5(1): 74–82

Fan Y, He S (2022) The characteristics of tumor microenvironment in triple negative breast cancer. Cancer Manag Res 14:1–17

Feng Y, Spezia M, Huang S, Yuan C, Zeng Z, Zhang L, Ji X, Liu W, Huang B, Luo W (2018) Breast cancer development and progression: Risk factors, cancer stem cells, signaling pathways, genomics, and molecular pathogenesis. Genes Dis 5(2):77–106

Ferlay J, Ervik M, Lam F, Colombet M, Mery L, Piñeros M, Znaor A, Soerjomataram I, Bray F (2020) Global cancer observatory: cancer today. International Agency for Research on Cancer, Lyon

Floros T, Tarhini AA (2015, August) Anticancer cytokines: biology and clinical effects of interferon-α2, interleukin (IL)-2, IL-15, IL-21, and IL-12. Semin Oncol 42(4):539–548

Foldi J, Silber A, Reisenbichler E, Singh K, Fischbach N, Persico J, Adelson K, Katoch A, Horowitz N, Lannin D (2021) Neoadjuvant durvalumab plus weekly nab-paclitaxel and dose-dense doxorubicin/cyclophosphamide in triple-negative breast cancer. NPJ Breast Cancer 7(1):9

Fong PC, Boss DS, Yap TA, Tutt A, Wu P, Mergui-Roelvink M, Mortimer P, Swaisland H, Lau A, O'Connor MJ (2009) Inhibition of poly (ADP-ribose) polymerase in tumors from BRCA mutation carriers. N Engl J Med 361(2):123–134

Fragomeni SM, Sciallis A, Jeruss JS (2018) Molecular subtypes and local-regional control of breast cancer. Surg Oncol Clin 27(1):95–120

Fu S, Lin J (2018) Blocking interleukin-6 and interleukin-8 signaling inhibits cell viability, colony-forming activity, and cell migration in human triple-negative breast cancer and pancreatic cancer cells. Anticancer Res 38(11):6271–6279

Galvan Morales MA, Barrera Rodriguez R, Santiago Cruz JR, Teran LM (2020) Overview of new treatments with immunotherapy for breast cancer and a proposal of a combination therapy. Molecules 25(23):5686

Ghoncheh M, Mohammadian-Hafshejani A, Salehiniya H (2015) Incidence and mortality of breast cancer and their relationship to development in Asia. Asian Pac J Cancer Prev 16(14): 6081–6087

Ginestier C, Liu S, Diebel ME, Korkaya H, Luo M, Brown M, Wicha MS (2010) CXCR1 blockade selectively targets human breast cancer stem cells in vitro and in xenografts. J Clin Invest 120(2):485–497
Ginsburg O, Yip CH, Brooks A, Cabanes A, Caleffi M, Dunstan Yataco JA, Anderson BO (2020) Breast cancer early detection: a phased approach to implementation. Cancer 126:2379–2393
Guanghui R, Xiaoyan H, Shuyi Y, Jun C, Guobin Q (2019) An efficient or methodical review of immunotherapy against breast cancer. J Biochem Mol Toxicol 33(8):e22339
Guillerey C, Huntington ND, Smyth MJ (2016) Targeting natural killer cells in cancer immunotherapy. Nat Immunol 17(9):1025–1036
Gupta N, Gaikwad S, Kaushik I, Wright SE, Markiewski MM, Srivastava SK (2021) Atovaquone suppresses triple-negative breast tumor growth by reducing immune-suppressive cells. Int J Mol Sci 22(10):5150
Han S, Wang W, Wang S, Yang T, Zhang G, Wang D, Ju R, Lu Y, Wang H, Wang L (2021) Tumor microenvironment remodeling and tumor therapy based on M2-like tumor associated macrophage-targeting nano-complexes. Theranostics 11(6):2892
Harbeck N, Penault-Llorca F, Cortes J, Gnant M, Houssami N, Poortmans P, Cardoso F (2019) Breast cancer. Nat Rev Dis Primers 5:66
Heery CR, O'Sullivan-Coyne G, Madan RA, Cordes L, Rajan A, Rauckhorst M, Lamping E, Oyelakin I, Marté JL, Lepone LM (2017) Avelumab for metastatic or locally advanced previously treated solid tumours (JAVELIN Solid Tumor): a phase 1a, multicohort, dose-escalation trial. Lancet Oncol 18(5):587–598
Heimes A-S, Schmidt M (2019) Atezolizumab for the treatment of triple-negative breast cancer. Expert Opin Investig Drugs 28(1):1–5
Henry NL, Cannon-Albright LA (2019) Breast cancer histologic subtypes show excess familial clustering. Cancer 125(18):3131–3138
Hollmén M, Roudnicky F, Karaman S, Detmar M (2015) Characterization of macrophage-cancer cell crosstalk in estrogen receptor positive and triple-negative breast cancer. Sci Rep 5(1):9188
Hosseini M, Seyedpour S, Khodaei B, Loghman A-H, Seyedpour N, Yazdi M-H, Rezaei N (2023) Cancer vaccines for triple-negative breast cancer: a systematic review. Vaccine 11(1):146
Huang A, Cao S, Tang L (2017) The tumor microenvironment and inflammatory breast cancer. J Cancer 8(10):1884
Huang H, Zhang H, Onuma AE, Tsung A (2020) Neutrophil elastase and neutrophil extracellular traps in the tumor microenvironment. In: Tumor microenvironment: state of the science. Springer, Cham, pp 13–23
Ibragimova MK, Tsyganov MM, Kravtsova EA, Tsydenova IA, Litviakov NV (2023) Organ-specificity of breast cancer metastasis. Int J Mol Sci 24(21):15625
Incorvaia L, Fanale D, Bono M, Calò V, Fiorino A, Brando C, Bazan V (2020) *BRCA1/2* pathogenic variants in triple-negative *versus* luminal-like breast cancers: genotype–phenotype correlation in a cohort of 531 patients. Ther Adv Med Oncol 12:1758835920975326
Jia H, Truica CI, Wang B, Wang Y, Ren X, Harvey HA, Song J, Yang J-M (2017) Immunotherapy for triple-negative breast cancer: Existing challenges and exciting prospects. Drug Resist Updat 32:1–15
Jin K, Pandey NB, Popel AS (2017) Crosstalk between stromal components and tumor cells of TNBC via secreted factors enhances tumor growth and metastasis. Oncotarget 8(36):60210
Juliá EP, Amante A, Pampena MB, Mordoh J, Levy EM (2018) Avelumab, an IgG1 anti-PD-L1 immune checkpoint inhibitor, triggers NK cell-mediated cytotoxicity and cytokine production against triple negative breast cancer cells. Front Immunol 9:2140
Khan NH, Duan S-F, Wu D-D, Ji X-Y (2021) Better reporting and awareness campaigns needed for breast cancer in Pakistani women. Cancer Manag Res 13:2125–2129
Kim J, Bae JS (2016) Tumor-associated macrophages and neutrophils in tumor microenvironment. Mediat Inflamm 2016:6058147
Kim CW, Kim K-D, Lee HK (2021) The role of dendritic cells in tumor microenvironments and their uses as therapeutic targets. BMB Rep 54(1):31

Kohler BA, Sherman RL, Howlader N, Jemal A, Ryerson AB, Henry KA, Boscoe FP, Cronin KA, Lake A, Noone A-M (2015) Annual report to the nation on the status of cancer, 1975-2011, featuring incidence of breast cancer subtypes by race/ethnicity, poverty, and state. J Natl Cancer Inst 107(6):djv048

Krejcik J, Casneuf T, Nijhof IS, Verbist B, Bald J, Plesner T, Syed K, Liu K, van de Donk NW, Weiss BM (2016) Daratumumab depletes CD38+ immune regulatory cells, promotes T-cell expansion, and skews T-cell repertoire in multiple myeloma. Blood J Am Soc Hematol 128(3):384–394

Lei X, Lei Y, Li J-K, Du W-X, Li R-G, Yang J, Li J, Li F, Tan H-B (2020) Immune cells within the tumor microenvironment: biological functions and roles in cancer immunotherapy. Cancer Lett 470:126–133

Lim B, Woodward WA, Wang X, Reuben JM, Ueno NT (2018) Inflammatory breast cancer biology: the tumour microenvironment is key. Nat Rev Cancer 18(8):485–499

Liu H, Wang Z, Zhou Y, Yang Y (2023) MDSCs in breast cancer: an important enabler of tumor progression and an emerging therapeutic target. Front Immunol 14:1199273

Llombart-Cussac A, Cortés J, Paré L, Galván P, Bermejo B, Martínez N, Prat A (2017) HER2-enriched subtype as a predictor of pathological complete response following trastuzumab and lapatinib without chemotherapy in early-stage HER2-positive breast cancer (PAMELA): an open-label, single-group, multicentre, phase 2 trial. Lancet Oncol 18(4):545–554

Long X, Ye Y, Zhang L, Liu P, Yu W, Wei F, Yu J (2016) IL-8, a novel messenger to cross-link inflammation and tumor EMT via autocrine and paracrine pathways. Int J Oncol 48(1):5–12

Lourenço C, Conceição F, Jerónimo C, Lamghari M, Sousa DM (2022) Stress in metastatic breast cancer: to the bone and beyond. Cancers 14(8):1881

Lu J, Ma L (2020) The role of tumor-associated macrophages in the development, metastasis and treatment of breast cancer. Pathol Res Pract 216(9):153085. https://doi.org/10.1016/j.prp.2020.153085

Łukasiewicz S, Czeczelewski M, Forma A, Baj J, Sitarz R, Stanisławek A (2021) Breast cancer—epidemiology, risk factors, classification, prognostic markers, and current treatment strategies—an updated review. Cancers 13(17):4287

Makki J (2015) Diversity of breast carcinoma: histological subtypes and clinical relevance. Clin Med Insights: Pathol 8. https://doi.org/10.4137/CPath.S31563

Malla RR, Vasudevaraju P, Vempati RK, Rakshmitha M, Merchant N, Nagaraju GP (2022) Regulatory T cells: their role in triple-negative breast cancer progression and metastasis. Cancer 128(6):1171–1183

Manore SG, Doheny DL, Wong GL, Lo H-W (2022) IL-6/JAK/STAT3 signaling in breast cancer metastasis: biology and treatment. Front Oncol 12:866014

Mansour M, Teo ZL, Luen SJ, Loi S (2017) Advancing immunotherapy in metastatic breast cancer. Curr Treat Options in Oncol 18:1–15

Masood S (2016) Breast cancer subtypes: morphologic and biologic characterization. Womens Health 12(1):103–119

Mehdizadeh R, Shariatpanahi SP, Goliaei B, Rüegg C (2023) Targeting myeloid-derived suppressor cells in combination with tumor cell vaccination predicts anti-tumor immunity and breast cancer dormancy: an in silico experiment. Sci Rep 13(1):5875

Mehraj U, Mushtaq U, Mir MA, Saleem A, Macha MA, Lone MN, Wani NA (2022) Chemokines in triple-negative breast cancer heterogeneity: New challenges for clinical implications. Semin Cancer Biol 86:769–783

Mi Z, Bhattacharya SD, Kim VM, Guo H, Talbot LJ, Kuo PC (2011) Osteopontin promotes CCL5-mesenchymal stromal cell-mediated breast cancer metastasis. Carcinogenesis 32(4):477–487

Miller KD, Siegel RL, Lin CC, Mariotto AB, Kramer JL, Rowland JH, Stein KD, Alteri R, Jemal A (2016) Cancer treatment and survivorship statistics, 2016. CA Cancer J Clin 66(4):271–289

Mollah F, Varamini P (2021) Overcoming therapy resistance and relapse in TNBC: Emerging technologies to target breast cancer-associated fibroblasts. Biomedicines 9(12):1921

Nagaraj S, Schrum AG, Cho H-I, Celis E, Gabrilovich DI (2010) Mechanism of T cell tolerance induced by myeloid-derived suppressor cells. J Immunol 184(6):3106–3116
Nanda R, Chow LQ, Dees EC, Berger R, Gupta S, Geva R, Pusztai L, Pathiraja K, Aktan G, Cheng JD (2016) Pembrolizumab in patients with advanced triple-negative breast cancer: phase Ib KEYNOTE-012 study. J Clin Oncol 34(21):2460
Nounou MI, ElAmrawy F, Ahmed N, Abdelraouf K, Goda S, Syed-Sha-Qhattal H (2015) Breast cancer: conventional diagnosis and treatment modalities and recent patents and technologies. Breast Cancer 9. https://doi.org/10.4137/BCBCR.S29420
Perez CR, De Palma M (2019) Engineering dendritic cell vaccines to improve cancer immunotherapy. Nat Commun 10(1):5408
Pernas S, Martin M, Kaufman PA, Gil-Martin M, Pardo PG, Lopez-Tarruella S, Manso L, Ciruelos E, Perez-Fidalgo JA, Hernando C (2018) Balixafortide plus eribulin in HER2-negative metastatic breast cancer: a phase 1, single-arm, dose-escalation trial. Lancet Oncol 19(6): 812–824
Qasim S, Tayyab H, Zulqadar K, Masood S, Qasim TB, Zubair Z (2020) Breast cancer knowledge and perceived barriers to help seeking among pre-clinical and clinical female medical students of King Edward Medical University, Lahore: a cross-sectional study. BMC Med Educ 20:1–9
Qiu X, Zhao T, Luo R, Qiu R, Li Z (2022) Tumor-associated macrophages: key players in triple-negative breast cancer. Front Oncol 12:772615
Rashid MU, Muhammad N, Bajwa S, Faisal S, Tahseen M, Bermejo JL, Hamann U (2016) High prevalence and predominance of BRCA1 germline mutations in Pakistani triple-negative breast cancer patients. BMC Cancer 16(1):1–10
Raskov H, Orhan A, Christensen JP, Gögenur I (2021) Cytotoxic CD8+ T cells in cancer and cancer immunotherapy. Br J Cancer 124(2):359–367
Retecki K, Seweryn M, Graczyk-Jarzynka A, Bajor M (2021) The immune landscape of breast cancer: strategies for overcoming immunotherapy resistance. Cancers 13(23):6012
Rymaszewski AL, Tate E, Yimbesalu JP, Gelman AE, Jarzembowski JA, Zhang H, Vikis HG (2014) The role of neutrophil myeloperoxidase in models of lung tumor development. Cancers 6(2):1111–1127
Saeed RS, Bakir YY, Ali LM (2014) Are women in Kuwait aware of breast cancer and its diagnostic procedures? Asian Pac J Cancer Prev 15(15):6307–6313
Salemme V, Centonze G, Cavallo F, Defilippi P, Conti L (2021) The crosstalk between tumor cells and the immune microenvironment in breast cancer: implications for immunotherapy. Front Oncol 11:289
Santoni M, Romagnoli E, Saladino T, Foghini L, Guarino S, Capponi M, Giannini M, Cognigni PD, Ferrara G, Battelli N (2018) Triple negative breast cancer: key role of tumor-associated macrophages in regulating the activity of anti-PD-1/PD-L1 agents. Biochim Biophys Acta Rev Cancer 1869(1):78–84
Saxena S, Singh RK (2021) Chemokines orchestrate tumor cells and the microenvironment to achieve metastatic heterogeneity. Cancer Metastasis Rev 40(2):447–476
Schmid P, Adams S, Rugo HS, Schneeweiss A, Barrios CH, Iwata H, Diéras V, Hegg R, Im S-A, Shaw Wright G (2018) Atezolizumab and nab-paclitaxel in advanced triple-negative breast cancer. N Engl J Med 379(22):2108–2121
Schott AF, Goldstein LJ, Cristofanilli M, Ruffini PA, McCanna S, Reuben JM, Perez RP, Kato G, Wicha M (2017) Phase Ib pilot study to evaluate reparixin in combination with weekly paclitaxel in patients with HER-2-negative metastatic breast cancer. Clin Cancer Res 23(18): 5358–5365
Shahbandi A, Nguyen HD, Jackson JG (2020) TP53 mutations and outcomes in breast cancer: reading beyond the headlines. Trends Cancer 6(2):98–110
Shiga K, Hara M, Nagasaki T, Sato T, Takahashi H, Takeyama H (2015) Cancer-associated fibroblasts: their characteristics and their roles in tumor growth. Cancers 7(4):2443–2458
Shoukat Z, Shah AJ (2023) Breast cancer awareness and associated factors among women in Pakistan: A cross-sectional descriptive study. Asian Pac J Cancer Prev 24(5):1561

Stovgaard ES, Nielsen D, Hogdall E, Balslev E (2018) Triple negative breast cancer–prognostic role of immune-related factors: a systematic review. Acta Oncol 57(1):74–82
Sukumar J, Gast K, Quiroga D, Lustberg M, Williams N (2021) Triple-negative breast cancer: promising prognostic biomarkers currently in development. Expert Rev Anticancer Ther 21(2): 135–148
Sun Y-S, Zhao Z, Yang Z-N, Xu F, Lu H-J, Zhu Z-Y, Shi W, Jiang J, Yao P-P, Zhu H-P (2017) Risk factors and preventions of breast cancer. Int J Biol Sci 13(11):1387
Sung H, Ferlay J, Siegel RL, Laversanne M, Soerjomataram I, Jemal A, Bray F (2021) Global cancer statistics 2020: GLOBOCAN estimates of incidence and mortality worldwide for 36 cancers in 185 countries. CA Cancer J Clin 71(3):209–249
Takai K, Le A, Weaver VM, Werb Z (2016) Targeting the cancer-associated fibroblasts as a treatment in triple-negative breast cancer. Oncotarget 7(50):82889
Testa U, Castelli G, Pelosi E (2020) Breast cancer: a molecularly heterogenous disease needing subtype-specific treatments. Med Sci 8(1):18
Thacker G, Henry S, Nandi A, Debnath R, Singh S, Nayak A, Susnik B, Boone MM, Zhang Q, Kesmodel SB (2023) Immature natural killer cells promote progression of triple-negative breast cancer. Sci Transl Med 15(686):eabl4414
Thakur KK, Bordoloi D, Kunnumakkara AB (2018) Alarming burden of triple-negative breast cancer in India. Clin Breast Cancer 18(3):e393–e399
Tsang JY, Gary MT (2020) Molecular classification of breast cancer. Adv Anat Pathol 27(1):27–35
Valenza C, Taurelli Salimbeni B, Santoro C, Trapani D, Antonarelli G, Curigliano G (2023) Tumor infiltrating lymphocytes across breast cancer subtypes: current issues for biomarker assessment. Cancers 15(3):767
van den Ende NS, Nguyen AH, Jager A, Kok M, Debets R, van Deurzen CH (2023) Triple-negative breast cancer and predictive markers of response to neoadjuvant chemotherapy: a systematic review. Int J Mol Sci 24(3):2969
Vonderheide RH, LoRusso PM, Khalil M, Gartner EM, Khaira D, Soulieres D, Dorazio P, Trosko JA, Rüter J, Mariani GL (2010) Tremelimumab in combination with exemestane in patients with advanced breast cancer and treatment-associated modulation of inducible costimulator expression on patient T cells. Clin Cancer Res 16(13):3485–3494
Voorwerk L, Slagter M, Horlings HM, Sikorska K, van de Vijver KK, de Maaker M, Kok M (2019) Immune induction strategies in metastatic triple-negative breast cancer to enhance the sensitivity to PD-1 blockade: the TONIC trial. Nat Med 25(6):920–928
Wang J, Chen H, Chen X, Lin H (2016) Expression of tumor-related macrophages and cytokines after surgery of triple-negative breast cancer patients and its implications. Med Sci Monit 22:115
Wen HY, Brogi E (2018) Lobular carcinoma in situ. Surg Pathol Clin 11(1):123–145
Wesolowski J, Tankiewicz-Kwedlo A, Pawlak D (2022) Modern immunotherapy in the treatment of triple-negative breast cancer. Cancers 14(16):3860
Yi H, Li Y, Tan Y, Fu S, Tang F, Deng X (2021) Immune checkpoint inhibition for triple-negative breast cancer: current landscape and future perspectives. Front Oncol 11:648139
Yin L, Duan J-J, Bian X-W, Yu S-C (2020) Triple-negative breast cancer molecular subtyping and treatment progress. Breast Cancer Res 22:1–13
Zahran AM, El-Badawy O, Kamel LM, Rayan A, Rezk K, Abdel-Rahim MH (2021) Accumulation of regulatory T cells in triple negative breast cancer can boost immune disruption. Cancer Manag Res 13:6019–6029
Zhang L, Fang C, Xu X, Li A, Cai Q, Long X (2015) Androgen receptor, EGFR, and BRCA1 as biomarkers in triple-negative breast cancer: a meta-analysis. Biomed Res Int 2015:357485
Zhang X, Zeng Y, Qu Q, Zhu J, Liu Z, Ning W, Huang JA (2017) PD-L1 induced by IFN-γ from tumor-associated macrophages via the JAK/STAT3 and PI3K/AKT signaling pathways promoted progression of lung cancer. Int J Clin Oncol 22:1026–1033
Zhao R, Liu J, Li Z, Zhang W, Wang F, Zhang B (2022) Recent advances in CXCL12/CXCR4 antagonists and nano-based drug delivery systems for cancer therapy. Pharmaceutics 14(8):1541
Zheng H, Siddharth S, Parida S, Wu X, Sharma D (2021) Tumor microenvironment: key players in triple negative breast cancer immunomodulation. Cancers 13(13):3357

# Claudin 1: An Emerging Target for Triple-Negative Breast Cancer

Isabelle Grillier-Vuissoz, Marine Geoffroy, and Sandra Kuntz

**Abstract**

Claudin 1 is a transmembrane protein known as a key component of the tight junctions playing a pivotal role in cell-cell interaction and homeostasis of epithelial cells. In breast cancer, the expression of claudin 1 is absent or strongly diminished in 76% of triple-negative breast cancer (TNBC). These tumors constitute the most aggressive subtype of breast cancer. They are a heterogeneous group of breast tumors that do not express estrogen receptor (ER) α, progesterone receptor (PR), and human epidermal growth factor receptor-2 (HER2). Therefore, TNBC patients are not eligible for targeted therapies that have been developed for other breast cancer subtypes. TNBCs have a higher tendency to form metastasis and they rapidly acquire chemo-resistance. Therefore, TNBCs represent a major challenge in cancerology, and the main goal of cancer research is to characterize potential specific targets to treat these tumors lacking efficient therapies. Clinical studies reported that a loss of claudin 1 expression correlates with increased aggressiveness and metastasis potential associated with recurrence of disease in invasive breast carcinoma patients. Although the role of claudin 1 in breast cancer is not clearly defined, several in vitro studies suggested that claudin 1 functions as a tumor suppressor. The reexpression of claudin 1 in some TNBC cell line was shown to be sufficient to induce apoptosis, suggesting that claudin 1 expression could constitute a new therapeutic option in "claudin 1–low" TNBC. This review summarizes the role of claudin 1 in TNBC cancer and its impact on developing novel strategies to treat breast cancer.

I. Grillier-Vuissoz (✉) · S. Kuntz
CRAN, UMR 7039, Université de Lorraine, Vandœuvre-lès-Nancy, France
e-mail: isabelle.grillier-vuissoz@univ-lorraine.fr

M. Geoffroy
CRAN, UMR 7039, Université de Lorraine, Vandœuvre-lès-Nancy, France

SQY Therapeutics, UVSQ, Montigny le Bretonneux, France



Interdisciplinary Cancer Research, https://doi.org/10.1007/16833_2024_250
Published online: 29 May 2024

**Keywords**

Apoptosis · Biomarker · Chemotherapy · CLDN1 · Migration · Therapeutic target · Tight junction · Triple-negative breast cancer

# 1 Introduction

## 1.1 Claudin 1: A Tight Junction Protein

In animal tissues, neighboring cells are attached to each other via intercellular junctions which ensure the coherence of the whole tissue. Among intercellular junctions, tight junctions (TJs) are located in the most apical region of cells and form a physiological barrier between the apical and basal compartments. TJs also provide two barrier functions: a paracellular barrier called "gate function," regulating the flow of small ions and molecules between intercellular space of two adjacent cells, and a diffusion barrier called "fence function," forming a boundary between apical and basolateral membranes. The latter allows the maintenance of cell polarity, critical to maintain proper cellular communication. In addition, TJs can modulate cellular proliferation and differentiation (Zihni et al. 2016).

Many proteins contribute to form the TJs. They are divided two types: transmembrane proteins and cytoplasmic proteins. The transmembrane proteins include three main types: the claudins, the tight junction–associated MARVEL proteins (TAMPs) and the the junctional adhesion molecules (JAMs). The cytoplasmic proteins cover the adaptor and the signaling proteins (Fig. 1). The claudin members of the tetraspan transmembrane family form the backbone of the TJs. TAMP family characterized by a MARVEL domain (MAL: myelin and lymphocyte protein and related protein for vesicle trafficking and membrane link) includes occluding, the well-known TAMP. The JAM family belongs to immunoglobulin superfamily. Transmembrane proteins bind to proteins underneath that belong to the cytoplasmic plaque, through various domains. The claudins and the JAMs contain one or several binding sites for PDZ domain (PSD-95, Dlg, ZO-1) while occludin binds to protein of the cytoplasmic plaque through its guanylate kinase (GUK) domain. Adaptor and signaling protein form the cytoplasmic plaque. The adaptor proteins bind to the transmembrane proteins, mostly through PDZ domains. The proteins containing a PDZ domain include mainly the membrane-associated guanylate kinase (MAGUK), protein represented by zonula occludens 1, 2, 3 (ZO) and the protein associated with lin-1 (PALS), the MAGUK inverted protein (MAGI), and the multi-PDZ domain protein (MUPP1). MUPP1 family includes the proteins pals 1–associated tight junction protein (PATJ). Few adaptor proteins do not contain PDZ domains, mainly the cingulin and the symplekine. The adaptor proteins play a crucial structural role as they link the transmembrane proteins to the cytoskeleton, contributing in the maintenance of the architecture of the cells. Signaling protein associated to the TJs includes G proteins (Rab13, Cdc42), kinases (PKC), phosphatases (PP2A), and transcription factors, the most known is ZONAB (Geoffroy 2018; Lemesle 2022).

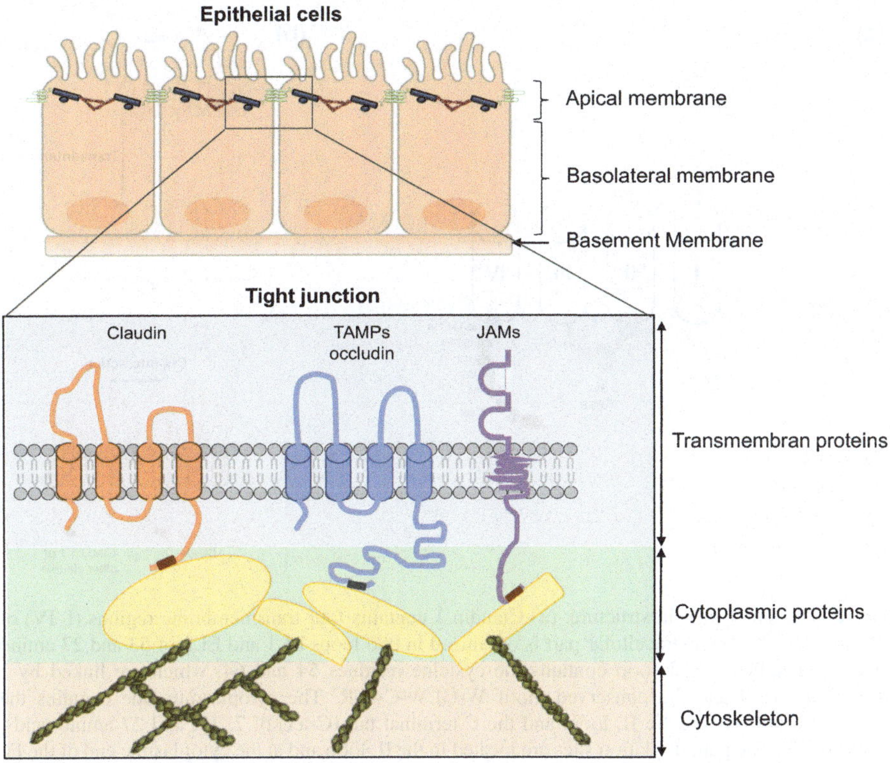


**Fig. 1** Organization of tight junctions in epithelial cells. Tight junctions are located in the most apical region of cells. TJs are composed of transmembrane proteins which are divided into three families: claudin, TAMPs with occluding, and JAMs. Both adaptor and signaling proteins form the cytoplasmic plaque (in yellow). Claudin and JAMs contain a PDZ-binding domain (rectangle in red) that allows them to interact with adaptors and proteins containing a PDZ domain, while occludin binds to proteins of the cytoplasmic plaque through its GUK domain (rectangle in blue). The adaptor proteins link the transmembrane proteins to the cytoskeleton of the cell (in green)

## 1.2 Structure of Claudin 1

The claudin family displays at least 27 members in mammals, and 26 in humans as claudin 13 is not expressed in humans (Mineta et al. 2011). Their molecular weight ranges between 21 and 34 kDa, and claudins are encoded by at least 17 different genes located on 13 different chromosomes. In humans, the gene encoding the claudin 1 protein is located on chromosome 3 (Liu et al. 2016).

Claudin 1 was first identified in 1998 in a fraction of purified chicken liver junctions (Furuse et al. 1998). It plays a key role in the formation of TJs. Indeed, the overexpression of claudin 1 or claudin 2 in mouse fibroblasts lacking TJs is sufficient to reconstitute a network of strands similar to that of TJs (Furuse et al. 1998). Claudin 1 is composed of 211 amino acid with a molecular weight of

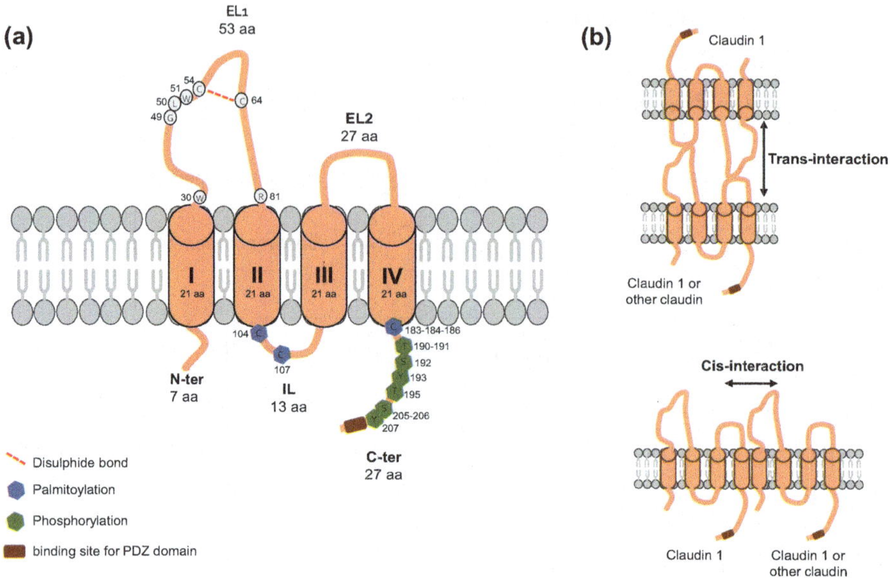


**Fig. 2** Claudin 1 protein structure. (**a**) Claudin 1 contains four transmembrane regions (I–IV) of 21 amino acids. The extracellular part is organized in two loops EL1 and EL2 of 53 and 27 amino acids, respectively. EL1 loop contains the cysteine residues 54 and 64, which are linked by a disulphide bond and the conserved motif W-GLW-C-C-R. The cytoplasmic side contains the N-terminal tail (N-ter), the IL loop, and the C-terminal tail (C-ter) of 7, 13, and 27 amino acids, respectively. The palmitoylation sites are located in the IL loop and at the cytoplasmic end of the IV transmembrane domain. The C-terminal tail of claudin 1 hold several phosphorylation sites and a motif for binding site for PDZ domain. (**b**) Claudin 1 interacts with other claudins in trans by "head-by-head" dimerization of adjacent claudins localized in two different cells or in cis between two claudins inside the same cell membrane

22.744 kDa. The protein contains four transmembrane regions of 21 amino acids (Fig. 2a). The extracellular part is organized in two extracellular loop (EL), EL1 and EL2, of 53 and 27 amino acids, respectively. In EL1, cysteine residues 54 and 64 are linked by a disulfide bond and contain the conserved claudin family motif W-GLW-C-C-R (Fig. 2a). On the cytoplasmic side, the protein starts with a short N-terminal tail of 7 residues and ends with a long C-terminal tail of 27 residues. The intracytoplasmic loop (IL) connecting the second and third transmembrane region is 13 amino acids. The C-terminal tail carries the PDZ domain binding site containing the tyrosine (Y)-valine (V) motif, which is conserved in most claudins (Lal-Nag and Morin 2009; Cukierman et al. 2009). The three-dimensional modeling of claudin 1 shows that the transmembrane domains are organized in α-helix while the two extracellular loops are arranged in β-sheet. EL1 is composed of four antiparallel β-sheets and the EL2 loop of a single β-sheet (Suzuki et al. 2014; Suzuki et al. 2017).

The claudin 1 protein is also the target of posttranslational modifications at its intracytoplasmic C-terminal tail such as palmitoylation and phosphorylation (Fig. 2a). Posttranslational modifications of claudin 1 play an important role in its membrane localization. Thus, the deletion of the C-terminal tail of claudin 1 prevents its localization at TJs (Rüffer and Gerke 2004). Phosphorylation of the claudin 1 C-terminal tail regulates its insertion into the plasma membrane (Van Itallie and Anderson 2006; Günzel and Yu 2013). It has been demonstrated that MAPK, PKA, PKC, and PKG can phosphorylate the C-terminal tail of claudin 1 at several sites (Fig. 2a). For example, in rat lung endothelial cells, MAPK kinase induces phosphorylation of claudin 1 at residue T203 and promotes its insertion into the plasma membrane (Fujibe et al. 2004). Thus, dephosphorylation of claudin 1 by PP2A induces its delocalization from TJs and its accumulation in the cytoplasm (Nunbhakdi-Craig et al. 2002; Nita-Lazar et al. 2010). Several sites of palmitoylation are described and consist of a covalent bond between the cysteines and palmitic acid. The palmitoylation of claudins is also necessary for their localization at TJs. Indeed, membrane-located claudin 1 was also found with palmitoylation-like modifications in MDCK cells (Lynch et al. 2007).

Claudin 1 interacts with other claudins through EL1 and EL2 loops and are responsible for epithelial cells adhesion to each other (Fig. 2b) (Furuse et al. 1998; Balda and Matter 2000). Predominantly, these interactions are homophilic (claudin 1-claudin 1); however, claudin 1 can interact with other claudins in a heterophilic configuration. These interactions could be in trans by "head-by-head" dimerization of adjacent claudins localized in two different cells or in cis between two claudins inside the same cell membrane. The EL1 loop of claudins carries charged aromatic amino acids as well as an ion binding site. This characteristic is crucial because it determines paracellular ion selectivity (Colegio et al. 2003; González-Mariscal et al. 2008; Günzel and Yu 2013). Claudin 1 plays a key role in regulating paracellular permeability since its inhibition in human keratinocytes increases paracellular permeability. In the same way, its overexpression in MDCK cells reduces paracellular permeability (Shi et al. 2018).

## 1.3 Claudin 1 Expression in Breast Tissue

Claudin 1 is ubiquitously expressed in most tissues of the body like intestine, spleen, brain, liver, kidney, testicle, respiratory tract, lung, endothelia, bladder, epidermis, eyes, salivary glands, breast glands, exocrine pancreas, ovaries and prostate (Günzel and Yu 2013; Bhat et al. 2023). During normal mammary gland development, claudin 1 plays a key physiological role with four other claudins 3, 4, 7, and 8. The expression of these different claudins vary according to the hormonal status. In nonpregnant and non-lactating mice, claudin 1 is only detected at TJs. During gestation and lactation, claudin 3 and then claudin 8 replaces claudin 1 at the TJs (Baumgartner et al. 2017). One might raise several questions: what is the impact of mammary tumorigenesis on claudin 1 distribution? Indeed, it has been established by a clinical study that claudin 1 expression is generally decreased in breast cancer (Tokés et al. 2005). Are the different subtypes of breast affected? Currently, breast cancer is classified into three groups of tumors: luminal tumors that express estrogen receptor alpha (ERα) and progesterone receptor (PR), human epidermal growth receptor 2 (HER2)-"enriched" tumors that overexpress human epidermal growth receptor 2, and triple-negative breast cancer (TNBC) that do not express ERα, PR, and HER2. Lu et al. examined 222 invasive breast carcinomas and showed that claudin 1 is underexpressed in 95% of luminal tumors, 86% of HER2-enriched tumors and 78% of TNBC. These tumors are called "claudin 1–low" tumors. In 16% of cases, claudin 1 is overexpressed and are called "claudin 1–high" tumors that cover 5% of luminal, 14% of HER2-enriched, and 22% of triple-negative tumors (Lu et al. 2013).

## 2 Claudin 1 Role in Triple-Negative Breast Cancer

### 2.1 Lack of Claudin 1 Expression in TNBC Correlates with Poor Prognostic

TNBC accounts for 15–20% of breast cancer and constitutes the most challenging type of breast cancer. The prognosis of TNBC is poor with a higher risk of recurrence, and a higher occurrence of metastasis compared to other subtypes of breast cancer (Szymiczek et al. 2021). Frequently, at the time of the diagnosis, metastatic cells already reach the lymph nodes (Tan et al. 2008; Foulkes et al. 2010). TNBC is a heterogenous group of tumors characterized by the absence of ERα, PR, and HER 2; three receptors are currently used for the diagnosis and treatment of breast cancer (Derakhshan and Reis-Filho 2022). Therefore, TNBC patients are not eligible for targeted therapies that have been developed for the other breast cancer subtypes. Recently, PARP inhibitors have been approved for BRCA-mutant TNBC (Robson et al. 2017), and immune modulators are raising hope (Schmid et al. 2018); however, targeted treatments are still limited (Bhat et al. 2023). Currently, TNBC treatment relies on the combination of surgery, conventional chemotherapy, and radiotherapy (Al-Mahmood et al. 2018). A major pitfall of TNBC treatment is the rapid acquisition of chemoresistance. About 60–70% of TNBC evolves into chemotherapy resistance, leading to frequent relapses (Liedtke et al. 2008; Foulkes et al. 2010; von Minckwitz et al. 2012).

In order to identify molecular signatures of TNBC, several genomic and transcriptomic studies classified them into various subgroups (Lehmann et al. 2011; Burstein et al. 2015; Ahn et al. 2016). According to their classifications, TNBCs are often divided into six groups: basal-like 1 (BL1), basal-like 2 (BL2), mesenchymal (M), mesenchymal stem-like (MSL), immunomodulatory (IM), and luminal androgen receptor (LAR). Basically, the BL1 subtype is characterized by a strong expression of the cell division and the DNA damage rescue pathways. The BL2 subtype exhibits a high expression of growth factor signaling pathways and it often expresses myoepithelial markers. The IM subtype exhibits gene expressions associated with immune escapement. The M subtype often presents a sarcoma-like cell morphology. The MSL subtype displays a strong epithelial-to-mesenchymal transition (EMT) and a high level of stemness-related genes. Finally, the LAR subtype expressed androgen receptor and often the prolactin receptor. The aim of these characterizations was to identify potential targets in order to develop personalized therapies to treat TNBC. Based on this approach, several publications including our work highlighted claudin 1 as potential biomarker and/or therapeutic target for TNBC.

Various clinical studies established that most of the TNBC presented a low claudin 1 expression, correlated with a poor prognosis, a high risk of recurrence, invasiveness, and malignant potential (Morohashi et al. 2007; Gerhard et al. 2012; Lu et al. 2013; Ma et al. 2014). Thus, Gerhard et al. showed that claudin 1 is repressed in 87.4% of 103 TNBC tumors (Gerhard et al. 2012). Likewise, two independent immunohistochemical analyses based on 226 and 173 TNBC patients, respectively, revealed that 77% of TNBC presented a low level of claudin 1 (Lu et al. 2013; Ma et al. 2014). Interestingly, a positive correlation between the lack of claudin 1 expression and lymph node metastasis was noted. The "claudin 1–low" TNBC patients harbor a worse relapse-free survival (RFS) and overall survival (OS) compared to "claudin 1–high" patients. Indeed, the RFS of claudin 1–negative patients represents 40% against 90% for claudin 1–positive patients 80 months post-diagnosis (Ma et al. 2014). In addition, our bioinformatic analysis on clinical databases confirmed a strong correlation between claudin 1 expression level and OS (Lemesle et al. 2022). This study highlights that only 50% of TNBC "claudin 1–low" patients survived, while over 80% of TNBC "claudin 1–high" patients are still alive after 80 months after diagnosis (Fig. 3a). Finally, all clinical data highlight that lack of claudin 1 expression in TNBC patients correlates with worse prognosis compared to TNBC "claudin 1–high" patients. The role of claudin 1 in TNBC is not yet clearly defined, but several studies suggest that claudin 1 could act as a tumor suppressor in breast cancer lacking claudin 1 expression (Zhou et al. 2015b).

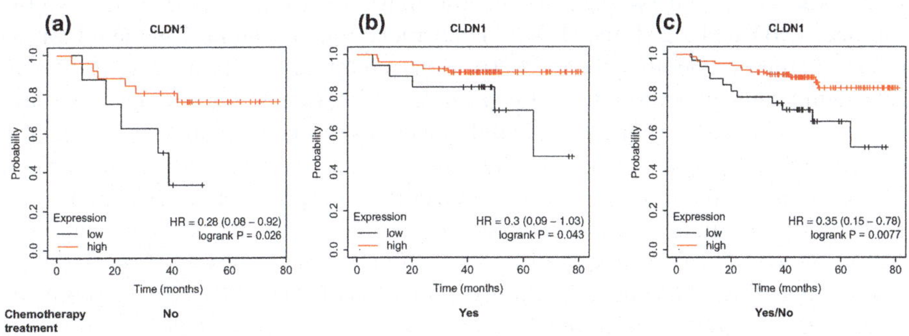


**Fig. 3** The overall survival is lower in "claudin 1–low" TNBC patients. Kaplan-Meier overall survival curve is based on claudin 1 expression and presence and/or absence of chemotherapy treatment. The number of surviving patients is indicated for claudin 1–low (black) and high (red) status at various times (0, 20, 40, 60, and 80 months). Logrank p-values <0.05 were considered statistically significant (Lemesle et al. 2022)

## 2.2 Claudin 1 Plays a Role in the Inhibition of Migration in "Claudin 1–Low" TNBC

In the first five years post-diagnosis, the risk of recurrence for TNBC patients is higher compared to non-TNBC patients and is caused by the excess of metastasis spread (Dent et al. 2009). To acquire an invasive phenotype for metastatic progression, the tumor cells change their behavior and undergo the epithelial-to-mesenchymal transition (EMT). During the epithelial-to-mesenchymal transition, cancer cells lose their epithelial phenotype, and the gain of mesenchymal features allows them to migrate and promote the intravasation of environment tissues. A key step to this process is the complete disruption of cell-cell adhesion junctions and apical-basal polarity, leading to a total EMT associated with an individual migration (Fig. 4). However, recent studies indicate the existence of an EMT hybrid state defined as an intermediate state (Pastushenko and Blanpain 2019). This partial EMT is correlated by a co-expression of epithelial and mesenchymal markers and is associated with increased migration, invasion, and a collective migration of the tumor cells localized at the leading edge of the tumor.

As claudin 1 is the backbone of the TJs, several works highlighted that the absence or the underexpression of claudin 1 in tumors contributes to the EMT and promotes invasiveness and metastasis (Myal et al. 2010; Li 2021). Numerous studies show a correlation between loss of claudin 1 and metastasis processes in breast cancer. A comparative study of 83 breast cancer cases demonstrated a relationship between the decrease of claudin 1 expression and the lymph node metastasis-positive patients (Morohashi et al. 2007). The same conclusion was drawn from 173 TNBC patients (Ma et al., 2014). Moreover, several studies demonstrated an inverse correlation between claudin 1 expression and EMT/migration/invasion processes in TNBC cell lines (Li et al. 2017; Mattern et al. 2019; Chiang et al. 2019). In accordance with these data, we demonstrated in two "claudin 1–low" TNBC cell lines (MDA-MB-231 and Hs578T) the crucial role of claudin 1 in the inhibition of cell migration (Geoffroy et al. 2020). Both inhibition of cell migration and cell-cell aggregation were correlated to the level of claudin 1. Interestingly, we also observed an increase in adherens and tight junctions formation. Furthermore, we also reported the disappearance of actin stress fibers, consistent with reduced motility of TNBC cells. All these data suggest that claudin 1 reexpression in "claudin 1–low" TNBC may constitute a therapeutic strategy to reduce the invasive phenotype of TNBC tumors. Nevertheless, the role of claudin 1 remains unsolved. Indeed, few studies have showed that the migration and invasion of MDA-MB-231 cells could be correlated to an increase of claudin 1 expression. A transcriptional analysis indicated that the overexpression of claudin 1 leads to an alteration of some genes involved in EMT, suggesting a pro-migratory effect of claudin 1 (Majer et al. 2016). Another study showed that stanniocalicin 2 inhibits metastasis formation in vitro and in vivo through the inhibition of PKC/claudin 1 signaling pathway (Hou et al. 2015). Likewise, the coagulation factor FVIIa enhances the migratory and invasive potential of MDA-MB-231 cells by nuclear accumulation of β-catenin that, in turn, promotes the expression of cyclin D1, c-Myc, COX-2, MMP-7, MMP-14,

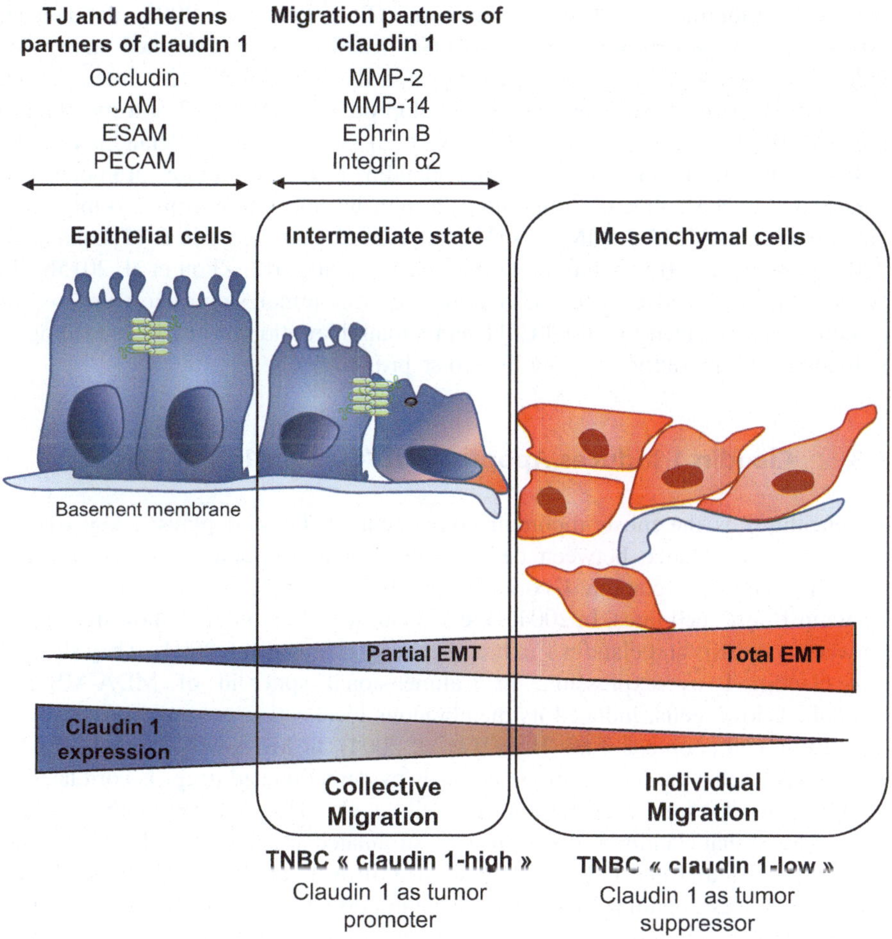


**Fig. 4** Claudin 1 in epithelial mesenchymal transition (EMT). EMT is a critical process in cancer cells characterized by the disruption of epithelial cell-cell junctions and apical-basal polarity loss leading to the conversion into invasive mesenchymal cells. Partial EMT is an intermediate state expressing different levels of epithelial and mesenchymal markers. Claudin 1 exhibits the opposite role in cell migration. Claudin 1 acts as a tumor promoter in "claudin 1–high" TNBC and contributes to the collective migration through its interaction with pro-migration partners such as MMP-2, MMP-14, Ephrin B, and Integrin α2. In "claudin 1–low" TNBC, the loss of claudin 1 is associated with a total EMT and a total disruption of junctions leading to individual cell migration

and claudin 1 (Roy et al. 2017). Such contradictory results could be explained by an interaction of claudin 1 with various partners in the cell. In non-tumorigenic mammary epithelial cells, claudin 1 plays a physiological role in the maintenance of the TJs and polarity. Classical partners of claudin 1 at the membrane are the TAMPs such as occludin, JAMs, endothelial cell selective adhesion molecule (ESAM), a related JAM molecule, or platelet endothelial cell adhesion molecule (PECAM) (Fig. 4). During the migratory process, claudin 1 could interact with pro-migratory partner contributing to collective migration (Giampieri et al. 2009; Zhou et al. 2015b). So, in epidermoid cancer cells OSC-4 and in the transformed embryonic kidney cells HEK293 T, it has been reported that claudin 1 could interact with pro-invasive partners such as matrix metalloproteinase (MMP)-2 and membrane-type (MMP)-1 (Miyamori et al. 2001; Oku et al. 2006). Also, in "claudin 1–high" TNBC patients, it was proposed that claudin 1 acts as tumor promoter and contributes to the collective migration through its interaction with pro-migration partners such as MMP-2, MMP-14 (Miyamori et al. 2001), ephrin B (Tanaka et al. 2005; Zhou et al. 2015b), and integrin α2 (Ding et al. 2012; Zhou et al. 2015b). In "claudin 1–low" TNBC patients, claudin 1 acts as tumor suppressor: the loss of claudin 1 is associated to a total EMT and a total disruption of junctions leading to individual cell migration, causing the worse prognostic.

## 2.3 Claudin 1 Induces Apoptosis in "Claudin 1–Low" TNBC

As claudin 1 is the most important component of TJs and plays a key role in regulating the balance between cell proliferation and apoptosis, many research teams investigate if claudin 1 could be involved in apoptotic function in TNBC "claudin 1–low" cell lines. In 2004, Hoevel et al. were the first to demonstrate a link between apoptosis and claudin 1 in breast cancer (Hoevel et al. 2004). They showed that claudin 1 overexpression, in tridimensional spheroid of MDA-MB-361 "claudin 1–low" cells, induced its membranous localization in association with an increase in cell apoptosis rate (Hoevel et al. 2004). In MDA-MB-453 "claudin 1–low" TNBC cells, the induction of apoptosis by demethylated drugs is correlated to membranous claudin 1 reexpression (Di Cello et al. 2013). More recently, our data also revealed that claudin 1 overexpression stimulated apoptosis in "claudin 1–low" MDA-MB-231 and Hs578T TNBC cells (Geoffroy et al. 2017). Likewise, in the migration process, claudin 1 partners play an important role in this function. A study suggests that the recruitment of the death receptor Fas by claudin 1 at the membrane is important for enhancing cisplatin-induced apoptosis by the extrinsic pathway in epithelial cancer cells KLE and HepG2 (Fortier et al. 2013). In the same way, the treatment of normal mammary epithelial cell line EpH4 with a claudin-disrupting mimetic induces apoptosis and mislocalization of claudin 4 and occludin in the death-inducing signaling complex (DISC) (Beeman et al. 2012). Using data mining, two other potential partners of claudin 1 have been revealed and could be involved in the apoptotic process. The first one is the tetraspan plasma membrane protein PERP (p53 apoptosis effector related to PMP22), a key component of the desmosome. Indeed, the Gene MANIA database suggests that PERP and claudin 1 are co-expressed and could interact together in tumoral cells (Lemesle 2022). This potential interaction is highly interesting as PERP has been described as an apoptosis-specific effector of p53. Moreover, its expression is specifically upregulated under conditions of p53-dependent apoptosis (Roberts and Paraoan 2020). PERP knockdown in mice reveals its key role in p53-dependent apoptosis and its role as a tumor suppressor in ultraviolet B (UVB)-induced squamous cell skin carcinoma model (Beaudry et al. 2010). The second potential partner is Ras homolog family member B (RhoB), a member of Rho GTPase family. The Biogrid database reveals that RhoB could interact with ZO-1, ZO-2, and PATJ which are linked with claudin 1 through a PDZ-binding motif (Lemesle 2022). Moreover, by a proximity-dependent biotinylation approach, an interaction has been described between RhoB and claudin 1 (Go et al. 2021). In general, Rho GTPases contribute to cell polarity and cell migration process through their ability to regulate actin cytoskeleton dynamics (Ebnet and Gerke 2022). RhoB exhibits tumor-suppressive potential depending on the context, due to its function in the cellular response to DNA damage and the associated induction of apoptosis (Huang and Prendergast 2006). Indeed, the stable inhibition of RhoB reduced apoptotic response and protected cells from ionizing radiation-induced cell death (Srougi and Burridge 2011).

## 3 Claudin 1 as Therapeutic Target

### 3.1 Claudin 1 Role in Chemosensitivity

Little is known about the role of claudin 1 on the chemosensitivity in breast cancer. Remarkably, in the hormone dependent cell line MCF-7, claudin 1 overexpression increased the sensitivity to cisplatin, etoposide, and tamoxifen (Zhou et al. 2015a). Moreover, a clinical study reported that anthracycline and taxane neoadjuvant chemotherapy induced claudin 1 expression and diminished Ki-67 in invasive breast tumors (Skálová et al. 2019). Recently, by analyzing the Breast CancerGene-Expression database, we showed that the TNBC group is composed of two distinct subpopulations. The first one is the "claudin 1–low" tumors expressing low levels of claudin 1 mRNA and the second one is the "claudin 1–high" tumors expressing high levels of claudin 1 mRNA. More importantly, we revealed a strong correlation between the level of claudin 1 expression and TNBC patients' OS, in the presence of chemotherapy (Fig. 3b,c) (Lemesle et al. 2022). The OS of "claudin 1–low" patients is always lower than "claudin 1–high" patients. Of note, the OS dropped less rapidly when they received chemotherapy, suggesting a partial response to chemotherapy. In addition, our team investigated the sensitivity of TNBC cells regarding their claudin 1 status. The effects of three agents were evaluated: 5-fluorouracil (5-FU), paclitaxel (PTX), and doxorubicin (DOX) on two "claudin 1–low" cells and one "claudin 1–high" cell. These compounds belong to the three main classes of chemotherapeutic agents commonly used for TNBC patients. We demonstrated that all chemotherapeutic agents were more effective in decreasing cell viability and inducing apoptosis in TNBC "claudin 1–high" HCC1806 cells compared to "claudin 1–low" MDA-MB-231 and Hs578T cells. Then, we established that claudin 1 expression increased the sensitivity of TNBC cell lines to these agents, using RNA interference and stable overexpressing models (Lemesle et al. 2022). Altogether, these data are in agreement with the key role of claudin 1 in the chemosensitivity of TNBC cells. Thus, one might suggest that reexpression of claudin 1 in TNBC cells could be a promising perspective to treat TNBC patients.

## 3.2 Future Therapeutic Directions

New horizons to improve the clinical outcome can be developed by expressing endogenous claudin 1 in the TNBC cells. The first direction is to target epigenetic alterations which lead to the underexpression of claudin 1 in TNBC. As a matter of fact, epigenetic regulation of claudin 1 expression has been reported in colorectal cancer, podocyte, and breast cancer (Di Cello et al. 2013; Hasegawa et al. 2013; Hahn-Strömberg et al. 2017). Indeed, hypermethylation of CpG islands in the claudin 1 promoter region reduces its expression in breast cancer (Di Cello et al. 2013). Using the Cancer Genome Atlas (TCGA) database, they showed that claudin 1 gene is part of a highly methylated gene cluster in breast tumors, and they found a high number of CpG islands on the 500 pair of base portions surrounding the ATG initiation codon. They, also highlighted a negative correlation between the expression of claudin 1 and the methylation of its promoter at the level of these CpG islands. In this context, the use of demethylating agents, 5-azacytidine or decitabine, induced claudin 1 reexpression in several breast cancer cell lines ER+ or ER−, and led to an increase of apoptotic signal (Di Cello et al. 2013). This prompted us to the second direction that would be to induce endogenous claudin 1, and thereby stimulated apoptotic program. Thus, our team investigated if 5-FU, DOX, and PTX could induce endogenous claudin 1 expression in MDA-MB-231 and Hs578T cells as 5-azacytidine did (Fig. 5a). We observed that DOX and PTX induced endogenous claudin 1 expression only in MDA-MB-231 cell line. Of note, 5-FU could not induce endogenous expression in both cell lines. Interestingly, this endogenous claudin 1 expression was correlated to a pro-apoptotic response in both TNBC "claudin 1–low" cells. Remarkably, the silencing of this endogenous expression of claudin 1 by RNA interference approach strongly diminishes the pro-apoptotic effect of DOX and PTX and strengthens the importance of endogenous claudin 1 in the proapoptotic response (Lemesle et al. 2022). Regarding these previous data, we investigated if endogenous claudin 1 expression stimulated by 5-azacytidine could sensitize TNBC "claudin 1–low" cells MDA-MB-231 and Hs578T to 5-FU (Fig. 5b). Remarkably, we observed that MDA-MB-231 cells exhibit a strong apoptotic signal in correlation with claudin 1 expression. Nevertheless, the endogenous expression of the claudin 1 in Hs578T could not efficiently stimulate apoptosis. This discrepancy could be explained by a weaker 5-FU uptake in Hs578T cells, as it is established that the transporter-mediated uptake of 5-FU is negligible in this cell line compared to other TNBC cell lines including the MDA-MB-231 cells (Noguchi et al. 2021). Altogether, these data suggested that low claudin 1 expression could predict a poor response in TNBC and these patients could be oriented toward chemotherapeutics treatments which stimulated claudin 1 expression with compounds such as taxanes and anthracycline. A third direction consists of identifying new uses for existing drugs known as drug repositioning in order to induce claudin 1 expression. Thus, we investigated the anticancer effects of troglitazone (TGZ) derivatives, a class of compound that greatly enhanced the treatment of diabetes in early 2000. Unfortunately, TGZ exhibited severe hepatotoxicity side effects and was withdrawn from the market.

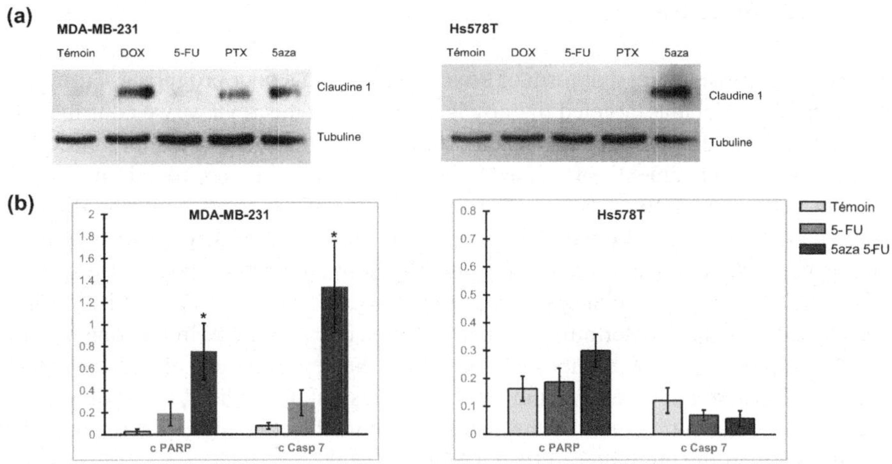


**Fig. 5** Claudin 1 as a therapeutic target for TNBC. (**a**) MDA-MB-231 and Hs578T TNBC "claudin 1–low" cells were treated with 86,2 nM of Doxorubicin (DOX), 50 μM of 5-fluorouracil (5-FU), 5 nM of paclitaxel (PTX), and 20 μM of 5-azacytidine (5-aza) for 72 h and subjected to western blot analysis with claudin 1 antibody. α-tubulin was used as loading control. (**b**) MDA-MB-231 and Hs578T cells were pretreated 6 h with 5 μM of 5-azacytidine (5-aza), followed by 72 h with 50 μM of 5-fluorouracil (5-FU). Apoptotic markers cleaved PARP (c PARP), and cleaved caspase 7 (c Casp7) were analyzed by western blotting. α-tubulin was used as loading control. Relative protein level corresponds to cleaved PARP and caspase 7 band intensity value adjusted to α-tubulin. The values represent the means ± SEM of four to five different experiments. Multiple comparisons were performed with ANOVA test followed by Bonferroni's post hoc test. A $p$ value of less than 0.05 was considered statistically significant with $*p < 0.05$

We synthesized Δ2-TGZ, a TGZ derivative, with a double bond adjoining the terminal thiazolidinedione-2,4-dione ring exhibiting lower hepatotoxicity (Grillier-Vuissoz et al. 2012). The anticancer effects of this compound were more effective on TNBC than on hormone-dependent cell lines (Lecomte et al. 2008; Colin et al. 2010; Salamone et al. 2012). In the search of the underlying mechanism of action, we discovered that claudin 1 was a target of Δ2-TGZ. Using a microarray approach on (ER+) breast cancer cells line MCF-7, we showed that among 33,297 genes, claudin 1 was the most upregulated after 12 h of treatment. Our teams showed that Δ2-TGZ induced an upregulation of claudin 1 in MDA-MB-231 and Hs578T cells. Remarkably, by a silencing approach, we demonstrated again the key role of claudin 1 in the proapoptotic effect of this compound. Of note, no upregulation was found in normal cell line MCF-10 A (Geoffroy et al. 2017).

Finally, all these data strongly strengthen the hypothesis that claudin 1 could be a promising therapeutic target for the treatment of TNBC subpopulation. Several strategies to induce claudin 1 endogenous expression could benefit the patient and need to be further investigated.

# 4 Conclusion

TNBC is the most aggressive form of breast cancer with a poor prognosis. Currently, no available molecular targeted therapy is available. In this review, we contribute to better understanding of the anticancer role of claudin 1. In the short term, the therapeutic management of TNBC patients could be improved by using claudin 1 as a predictive marker. The "claudin 1–low" patients could be directed to a chemotherapy combination of taxane and anthracycline known to induce an endogenous claudin 1 expression and thereby stimulate a pro-apoptotic response. Also, the screening for claudin 1 expression in the tumor or in blood exosome vesicles could help to determine the chemotherapeutic treatment. In the longer term, the development of new compounds based on the repositioning of old molecules such as TGZ derivatives, able to induce the expression of claudin 1, could play a key role in the therapeutic panel of the patients. Altogether, we highlight that claudin 1 could be a promising target to treat TNBC patients.

**Acknowledgments and Funding Information** We thank Alexandra Kleinclauss for technical support and Marine Lemesle for her Ph.D. work. Part of the results cited in this review was performed at CRAN, UMR 7039, CNRS, Université de Lorraine, and L2CM, CNRS, Université de Lorraine labs. These studies were supported by grants from the "Université de Lorraine," the "Conseil Régional du Grand Est," and the "Ligue Contre le Cancer."

**Compliance with Ethical Standards** The author(s) confirm that this chapter has no conflict of interest.

## References

Ahn SG, Kim SJ, Kim C, Jeong J (2016) Molecular classification of triple-negative breast cancer. J Breast Cancer 19:223–230. https://doi.org/10.4048/jbc.2016.19.3.223

Al-Mahmood S, Sapiezynski J, Garbuzenko OB, Minko T (2018) Metastatic and triple-negative breast cancer: challenges and treatment options. Drug Deliv Transl Res 8:1483–1507. https://doi.org/10.1007/s13346-018-0551-3

Balda MS, Matter K (2000) Transmembrane proteins of tight junctions. Semin Cell Dev Biol 11: 281–289. https://doi.org/10.1006/scdb.2000.0177

Baumgartner HK, Rudolph MC, Ramanathan P, Burns V, Webb P, Bitler BG, Stein T, Kobayashi K, Neville MC (2017) Developmental expression of Claudins in the mammary gland. J Mammary Gland Biol Neoplasia 22:141–157. https://doi.org/10.1007/s10911-017-9379-6

Beaudry VG, Jiang D, Dusek RL, Park EJ, Knezevich S, Ridd K, Vogel H, Bastian BC, Attardi LD (2010) Loss of the p53/p63 regulated desmosomal protein Perp promotes tumorigenesis. PLoS Genet 6:e1001168. https://doi.org/10.1371/journal.pgen.1001168

Beeman N, Webb PG, Baumgartner HK (2012) Occludin is required for apoptosis when claudin-claudin interactions are disrupted. Cell Death Dis 3:e273. https://doi.org/10.1038/cddis.2012.14

Bhat Y, Thrishna MR, Banerjee S (2023) Molecular targets and therapeutic strategies for triple-negative breast cancer. Mol Biol Rep. https://doi.org/10.1007/s11033-023-08868-6

Burstein MD, Tsimelzon A, Poage GM, Covington KR, Contreras A, Fuqua SAW, Savage MI, Osborne CK, Hilsenbeck SG, Chang JC, Mills GB, Lau CC, Brown PH (2015) Comprehensive genomic analysis identifies novel subtypes and targets of triple-negative breast cancer. Clin Cancer Res 21:1688–1698. https://doi.org/10.1158/1078-0432.CCR-14-0432

Chiang S-K, Chang W-C, Chen S-E, Chang L-C (2019) DOCK1 regulates growth and motility through the RRP1B-Claudin-1 pathway in Claudin-low breast cancer cells. Cancers (Basel) 11: 1762. https://doi.org/10.3390/cancers11111762

Colegio OR, Van Itallie C, Rahner C, Anderson JM (2003) Claudin extracellular domains determine paracellular charge selectivity and resistance but not tight junction fibril architecture. Am J Physiol Cell Physiol 284:C1346–C1354. https://doi.org/10.1152/ajpcell.00547.2002

Colin C, Salamone S, Grillier-Vuissoz I, Boisbrun M, Kuntz S, Lecomte J, Chapleur Y, Flament S (2010) New troglitazone derivatives devoid of PPARγ agonist activity display an increased antiproliferative effect in both hormone-dependent and hormone-independent breast cancer cell lines. Breast Cancer Res Treat 124:101–110. https://doi.org/10.1007/s10549-009-0700-y

Cukierman L, Meertens L, Bertaux C, Kajumo F, Dragic T (2009) Residues in a highly conserved Claudin-1 motif are required for hepatitis C virus entry and mediate the formation of cell-cell contacts. J Virol 83:5477–5484. https://doi.org/10.1128/JVI.02262-08

Dent R, Hanna WM, Trudeau M, Rawlinson E, Sun P, Narod SA (2009) Pattern of metastatic spread in triple-negative breast cancer. Breast Cancer Res Treat 115:423–428. https://doi.org/10.1007/s10549-008-0086-2

Derakhshan F, Reis-Filho JS (2022) Pathogenesis of triple-negative breast cancer. Annu Rev Pathol 17:181–204. https://doi.org/10.1146/annurev-pathol-042420-093238

Di Cello F, Cope L, Li H, Jeschke J, Wang W, Baylin SB, Zahnow CA (2013) Methylation of the claudin 1 promoter is associated with loss of expression in estrogen receptor positive breast cancer. PLoS One 8:e68630. https://doi.org/10.1371/journal.pone.0068630

Ding L, Lu Z, Foreman O, Tatum R, Lu Q, Renegar R, Cao J, Chen Y-H (2012) Inflammation and disruption of the mucosal architecture in Claudin-7-deficient mice. Gastroenterology 142:305–315. https://doi.org/10.1053/j.gastro.2011.10.025

Ebnet K, Gerke V (2022) Rho and Rab family small GTPases in the regulation of membrane polarity in epithelial cells. Front Cell Dev Biol 10:948013. https://doi.org/10.3389/fcell.2022.948013

Fortier A-M, Asselin E, Cadrin M (2013) Keratin 8 and 18 loss in epithelial cancer cells increases collective cell migration and cisplatin sensitivity through Claudin1 up-regulation. J Biol Chem 288:11555–11571. https://doi.org/10.1074/jbc.M112.428920

Foulkes WD, Smith IE, Reis-Filho JS (2010) Triple-negative breast cancer. N Engl J Med 363: 1938–1948. https://doi.org/10.1056/NEJMra1001389

Fujibe M, Chiba H, Kojima T, Soma T, Wada T, Yamashita T, Sawada N (2004) Thr203 of claudin-1, a putative phosphorylation site for MAP kinase, is required to promote the barrier function of tight junctions. Exp Cell Res 295:36–47. https://doi.org/10.1016/j.yexcr.2003.12.014

Furuse M, Sasaki H, Fujimoto K, Tsukita S (1998) A single gene product, Claudin-1 or −2, reconstitutes tight junction strands and recruits Occludin in fibroblasts. J Cell Biol 143:391–401. https://doi.org/10.1083/jcb.143.2.391

Geoffroy M (2018) Role of claudin 1 in TNBC and in involvement in anticancer effects of TGZ derivatives

Geoffroy M, Kleinclauss A, Grandemange S, Hupont S, Boisbrun M, Flament S, Grillier-Vuissoz I, Kuntz S (2017) Pro-apoptotic effect of Δ2-TGZ in "claudin-1-low" triple-negative breast cancer cells: involvement of claudin-1. Breast Cancer Res Treat 165:517–527. https://doi.org/10.1007/s10549-017-4378-2

Geoffroy M, Kleinclauss A, Kuntz S, Grillier-Vuissoz I (2020) Claudin 1 inhibits cell migration and increases intercellular adhesion in triple-negative breast cancer cell line. Mol Biol Rep 47:7643–7653. https://doi.org/10.1007/s11033-020-05835-3

Gerhard R, Ricardo S, Albergaria A, Gomes M, Silva AR, Logullo ÂF, Cameselle-Teijeiro JF, Paredes J, Schmitt F (2012) Immunohistochemical features of claudin-low intrinsic subtype in metaplastic breast carcinomas. Breast 21:354–360. https://doi.org/10.1016/j.breast.2012.03.001

Giampieri S, Manning C, Hooper S, Jones L, Hill CS, Sahai E (2009) Localized and reversible TGFbeta signalling switches breast cancer cells from cohesive to single cell motility. Nat Cell Biol 11:1287–1296. https://doi.org/10.1038/ncb1973

Go CD, Knight JDR, Rajasekharan A, Rathod B, Hesketh GG, Abe KT, Youn J-Y, Samavarchi-Tehrani P, Zhang H, Zhu LY, Popiel E, Lambert J-P, Coyaud É, Cheung SWT, Rajendran D, Wong CJ, Antonicka H, Pelletier L, Palazzo AF, Shoubridge EA, Raught B, Gingras A-C (2021) A proximity-dependent biotinylation map of a human cell. Nature 595:120–124. https://doi.org/10.1038/s41586-021-03592-2

González-Mariscal L, Tapia R, Chamorro D (2008) Crosstalk of tight junction components with signaling pathways. Biochim Biophys Acta 1778:729–756. https://doi.org/10.1016/j.bbamem.2007.08.018

Grillier-Vuissoz I, Mazerbourg S, Boisbrun M, Kuntz S, Chapleur Y, Flament S (2012) PPAR-γ-independent activity of thiazolidinediones: a promising mechanism of action for new anticancer drugs? J Carcinogenesis Mutagenesis. https://doi.org/10.4172/2157-2518.S8-002

Günzel D, Yu ASL (2013) Claudins and the modulation of tight junction permeability. Physiol Rev 93:525–569. https://doi.org/10.1152/physrev.00019.2012

Hahn-Strömberg V, Askari S, Ahmad A, Befekadu R, Nilsson TK (2017) Expression of claudin 1, claudin 4, and claudin 7 in colorectal cancer and its relation with CLDN DNA methylation patterns. Tumour Biol 39:1010428317697569. https://doi.org/10.1177/1010428317697569

Hasegawa K, Wakino S, Simic P, Sakamaki Y, Minakuchi H, Fujimura K, Hosoya K, Komatsu M, Kaneko Y, Kanda T, Kubota E, Tokuyama H, Hayashi K, Guarente L, Itoh H (2013) Renal tubular Sirt1 attenuates diabetic albuminuria by epigenetically suppressing Claudin-1 overexpression in podocytes. Nat Med 19:1496–1504. https://doi.org/10.1038/nm.3363

Hoevel T, Macek R, Swisshelm K, Kubbies M (2004) Reexpression of the TJ protein CLDN1 induces apoptosis in breast tumor spheroids. Int J Cancer 108:374–383. https://doi.org/10.1002/ijc.11571

Hou J, Wang Z, Xu H, Yang L, Yu X, Yang Z, Deng Y, Meng J, Feng Y, Guo X, Yang G (2015) Stanniocalicin 2 suppresses breast cancer cell migration and invasion via the PKC/claudin-1-mediated signaling. PLoS One 10:e0122179. https://doi.org/10.1371/journal.pone.0122179

Huang M, Prendergast GC (2006) RhoB in cancer suppression. Histol Histopathol 21:213–218. https://doi.org/10.14670/HH-21.213

Lal-Nag M, Morin PJ (2009) The claudins. Genome Biol 10:235. https://doi.org/10.1186/gb-2009-10-8-235

Lecomte J, Flament S, Salamone S, Boisbrun M, Mazerbourg S, Chapleur Y, Grillier-Vuissoz I (2008) Disruption of ERα signalling pathway by PPARγ agonists: evidences of PPAR-γ-independent events in two hormone-dependent breast cancer cell lines. Breast Cancer Res Treat 112:437–451. https://doi.org/10.1007/s10549-007-9886-z

Lehmann BD, Bauer JA, Chen X, Sanders ME, Chakravarthy AB, Shyr Y, Pietenpol JA (2011) Identification of human triple-negative breast cancer subtypes and preclinical models for selection of targeted therapies. J Clin Invest 121:2750–2767. https://doi.org/10.1172/JCI45014

Lemesle M (2022) Claudin 1: a way to sensitize triple-negative breast cancer cells to chemotherapy

Lemesle M, Geoffroy M, Alpy F, Tomasetto C-L, Kuntz S, Grillier-Vuissoz I (2022) CLDN1 sensitizes triple-negative breast cancer cells to chemotherapy. Cancers 14:5026. https://doi.org/10.3390/cancers14205026

Li J (2021) Context-dependent roles of Claudins in tumorigenesis. Front Oncol 11:676781. https://doi.org/10.3389/fonc.2021.676781

Li X-M, Liu W-L, Chen X, Wang Y-W, Shi D-B, Zhang H, Ma R-R, Liu H-T, Guo X-Y, Hou F, Li M, Gao P (2017) Overexpression of TMPRSS4 promotes tumor proliferation and aggressiveness in breast cancer. Int J Mol Med 39:927–935. https://doi.org/10.3892/ijmm.2017.2893

Liedtke C, Mazouni C, Hess KR, André F, Tordai A, Mejia JA, Symmans WF, Gonzalez-Angulo AM, Hennessy B, Green M, Cristofanilli M, Hortobagyi GN, Pusztai L (2008) Response to neoadjuvant therapy and long-term survival in patients with triple-negative breast cancer. JCO 26:1275–1281. https://doi.org/10.1200/JCO.2007.14.4147

Liu F, Koval M, Ranganathan S, Fanayan S, Hancock WS, Lundberg EK, Beavis RC, Lane L, Duek P, McQuade L, Kelleher NL, Baker MS (2016) Systems proteomics view of the endogenous human Claudin protein family. J Proteome Res 15:339–359. https://doi.org/10.1021/acs.jproteome.5b00769

Lu S, Singh K, Mangray S, Tavares R, Noble L, Resnick MB, Yakirevich E (2013) Claudin expression in high-grade invasive ductal carcinoma of the breast: correlation with the molecular subtype. Mod Pathol 26:485–495. https://doi.org/10.1038/modpathol.2012.187

Lynch RD, Francis SA, McCarthy KM, Casas E, Thiele C, Schneeberger EE (2007) Cholesterol depletion alters detergent-specific solubility profiles of selected tight junction proteins and the phosphorylation of occludin. Exp Cell Res 313:2597–2610. https://doi.org/10.1016/j.yexcr.2007.05.009

Ma F, Ding X, Fan Y, Ying J, Zheng S, Lu N, Xu B (2014) A CLDN1-negative phenotype predicts poor prognosis in triple-negative breast cancer. PLoS One 9:e112765. https://doi.org/10.1371/journal.pone.0112765

Majer A, Blanchard AA, Medina S, Booth SA, Myal Y (2016) Claudin 1 expression levels affect miRNA dynamics in human basal-like breast cancer cells. DNA Cell Biol 35:328–339. https://doi.org/10.1089/dna.2015.3188

Mattern J, Roghi CS, Hurtz M, Knäuper V, Edwards DR, Poghosyan Z (2019) ADAM15 mediates upregulation of Claudin-1 expression in breast cancer cells. Sci Rep 9:12540. https://doi.org/10.1038/s41598-019-49021-3

Mineta K, Yamamoto Y, Yamazaki Y, Tanaka H, Tada Y, Saito K, Tamura A, Igarashi M, Endo T, Takeuchi K, Tsukita S (2011) Predicted expansion of the claudin multigene family. FEBS Lett 585:606–612. https://doi.org/10.1016/j.febslet.2011.01.028

Miyamori H, Takino T, Kobayashi Y, Tokai H, Itoh Y, Seiki M, Sato H (2001) Claudin promotes activation of pro-matrix metalloproteinase-2 mediated by membrane-type matrix metalloproteinases. J Biol Chem 276:28204–28211. https://doi.org/10.1074/jbc.M103083200

Morohashi S, Kusumi T, Sato F, Odagiri H, Chiba H, Yoshihara S, Hakamada K, Sasaki M, Kijima H (2007) Decreased expression of claudin-1 correlates with recurrence status in breast cancer. Int J Mol Med 20:139–143

Myal Y, Leygue E, Blanchard AA (2010) Claudin 1 in breast tumorigenesis: revelation of a possible novel "claudin high" subset of breast cancers. J Biomed Biotechnol 2010:956897. https://doi.org/10.1155/2010/956897

Nita-Lazar M, Rebustini I, Walker J, Kukuruzinska MA (2010) Hypoglycosylated E-cadherin promotes the assembly of tight junctions through the recruitment of PP2A to adherens junctions. Exp Cell Res 316:1871–1884. https://doi.org/10.1016/j.yexcr.2010.02.008

Noguchi S, Takagi A, Tanaka T, Takahashi Y, Pan X, Kibayashi Y, Mizokami R, Nishimura T, Tomi M (2021) Fluorouracil uptake in triple-negative breast cancer cells: negligible contribution of equilibrative nucleoside transporters 1 and 2. Biopharm Drug Dispos 42:85–93. https://doi.org/10.1002/bdd.2261

Nunbhakdi-Craig V, Machleidt T, Ogris E, Bellotto D, White CL, Sontag E (2002) Protein phosphatase 2A associates with and regulates atypical PKC and the epithelial tight junction complex. J Cell Biol 158:967–978. https://doi.org/10.1083/jcb.200206114

Oku N, Sasabe E, Ueta E, Yamamoto T, Osaki T (2006) Tight junction protein claudin-1 enhances the invasive activity of oral squamous cell carcinoma cells by promoting cleavage of laminin-5 gamma2 chain via matrix metalloproteinase (MMP)-2 and membrane-type MMP-1. Cancer Res 66:5251–5257. https://doi.org/10.1158/0008-5472.CAN-05-4478

Pastushenko I, Blanpain C (2019) EMT transition states during tumor progression and metastasis. Trends Cell Biol 29:212–226. https://doi.org/10.1016/j.tcb.2018.12.001

Roberts O, Paraoan L (2020) PERP-ing into diverse mechanisms of cancer pathogenesis: regulation and role of the p53/p63 effector PERP. Biochim Biophys Acta Rev Cancer 1874:188393. https://doi.org/10.1016/j.bbcan.2020.188393

Robson M, Im S-A, Senkus E, Xu B, Domchek SM, Masuda N, Delaloge S, Li W, Tung N, Armstrong A, Wu W, Goessl C, Runswick S, Conte P (2017) Olaparib for metastatic breast cancer in patients with a germline BRCA mutation. N Engl J Med 377:523–533. https://doi.org/10.1056/NEJMoa1706450

Roy A, Ansari SA, Das K, Prasad R, Bhattacharya A, Mallik S, Mukherjee A, Sen P (2017) Coagulation factor VIIa-mediated protease-activated receptor 2 activation leads to β-catenin accumulation via the AKT/GSK3β pathway and contributes to breast cancer progression. J Biol Chem 292:13688–13701. https://doi.org/10.1074/jbc.M116.764670

Rüffer C, Gerke V (2004) The C-terminal cytoplasmic tail of claudins 1 and 5 but not its PDZ-binding motif is required for apical localization at epithelial and endothelial tight junctions. Eur J Cell Biol 83:135–144. https://doi.org/10.1078/0171-9335-00366

Salamone S, Colin C, Grillier-Vuissoz I, Kuntz S, Mazerbourg S, Flament S, Martin H, Richert L, Chapleur Y, Boisbrun M (2012) Synthesis of new troglitazone derivatives: anti-proliferative activity in breast cancer cell lines and preliminary toxicological study. Eur J Med Chem 51:206–215. https://doi.org/10.1016/j.ejmech.2012.02.044

Schmid P, Adams S, Rugo HS, Schneeweiss A, Barrios CH, Iwata H, Diéras V, Hegg R, Im S-A, Shaw Wright G, Henschel V, Molinero L, Chui SY, Funke R, Husain A, Winer EP, Loi S, Emens LA, IMpassion130 Trial Investigators (2018) Atezolizumab and nab-paclitaxel in advanced triple-negative breast cancer. N Engl J Med 379:2108–2121. https://doi.org/10.1056/NEJMoa1809615

Shi J, Barakat M, Chen D, Chen L (2018) Bicellular tight junctions and wound healing. Int J Mol Sci 19:3862. https://doi.org/10.3390/ijms19123862

Skálová H, Hájková N, Majerová B, Bártů M, Povýšil C, Tichá I (2019) Impact of chemotherapy on the expression of claudins and cadherins in invasive breast cancer. Exp Ther Med 18:3014–3024. https://doi.org/10.3892/etm.2019.7930

Srougi MC, Burridge K (2011) The nuclear guanine nucleotide exchange factors Ect2 and Net1 regulate RhoB-mediated cell death after DNA damage. PLoS One 6:e17108. https://doi.org/10.1371/journal.pone.0017108

Suzuki H, Nishizawa T, Tani K, Yamazaki Y, Tamura A, Ishitani R, Dohmae N, Tsukita S, Nureki O, Fujiyoshi Y (2014) Crystal structure of a claudin provides insight into the architecture of tight junctions. Science 344:304–307. https://doi.org/10.1126/science.1248571

Suzuki H, Tani K, Fujiyoshi Y (2017) Crystal structures of claudins: insights into their intermolecular interactions: crystal structures of claudins. Ann N Y Acad Sci 1397:25–34. https://doi.org/10.1111/nyas.13371

Szymiczek A, Lone A, Akbari MR (2021) Molecular intrinsic versus clinical subtyping in breast cancer: a comprehensive review. Clin Genet 99:613–637. https://doi.org/10.1111/cge.13900

Tan DSP, Marchió C, Jones RL, Savage K, Smith IE, Dowsett M, Reis-Filho JS (2008) Triple negative breast cancer: molecular profiling and prognostic impact in adjuvant anthracycline-treated patients. Breast Cancer Res Treat 111:27–44. https://doi.org/10.1007/s10549-007-9756-8

Tanaka M, Kamata R, Sakai R (2005) Phosphorylation of ephrin-B1 via the interaction with claudin following cell-cell contact formation. EMBO J 24:3700–3711. https://doi.org/10.1038/sj.emboj.7600831

Tokés A-M, Kulka J, Paku S, Máthé M, Páska C, Lódi C, Kiss A, Schaff Z (2005) The expression of five different claudins in invasive breast carcinomas: comparison of pT1pN1 and pT1pN0 tumors. Pathol Res Pract 201:537–544. https://doi.org/10.1016/j.prp.2005.05.005

Van Itallie CM, Anderson JM (2006) Claudins and epithelial paracellular transport. Annu Rev Physiol 68:403–429. https://doi.org/10.1146/annurev.physiol.68.040104.131404

von Minckwitz G, Untch M, Blohmer J-U, Costa SD, Eidtmann H, Fasching PA, Gerber B, Eiermann W, Hilfrich J, Huober J, Jackisch C, Kaufmann M, Konecny GE, Denkert C, Nekljudova V, Mehta K, Loibl S (2012) Definition and impact of pathologic complete response on prognosis after Neoadjuvant chemotherapy in various intrinsic breast cancer subtypes. JCO 30:1796–1804. https://doi.org/10.1200/JCO.2011.38.8595

Zhou B, Blanchard A, Wang N, Ma X, Han J, Schroedter I, Leygue E, Myal Y (2015a) Claudin 1 promotes migration and increases sensitivity to tamoxifen and anticancer drugs in luminal-like human breast cancer cells MCF7. Cancer Investig 33:429–439. https://doi.org/10.3109/07357907.2015.1060996

Zhou B, Moodie A, Blanchard A, Leygue E, Myal Y (2015b) Claudin 1 in breast cancer: new insights. J Clin Med 4:1960–1976. https://doi.org/10.3390/jcm4121952

Zihni C, Mills C, Matter K, Balda MS (2016) Tight junctions: from simple barriers to multifunctional molecular gates. Nat Rev Mol Cell Biol 17:564–580. https://doi.org/10.1038/nrm.2016.80

# Histaminergic System as a Promising Target for Breast Cancer Treatment

Melisa B. Nicoud, Mónica A. Táquez Delgado, Ignacio Ospital, Helena A. Sterle, Karina Formoso, and Vanina A. Medina

**Abstract**

Breast cancer is a heterogeneous disease that remains the deadliest cancer for women worldwide, which highlights the need of new effective therapeutic approaches especially in hard-to-treat breast cancer subtypes. From cell lines to animal models and human clinical studies, there is overwhelming evidence supporting the key role of the histaminergic system in breast cancer development and spread and in the response to cancer therapies.

In this chapter, after briefly summarizing the most relevant aspects of histamine metabolism and histamine receptors' characteristics, we will address the latest evidence on the role of the histaminergic system in the complex breast cancer biology, with particular emphasis on the identification of potential molecular targets for breast cancer therapies. Later on, we will discuss the contradictory roles of histamine in regulating the phenotype and functions of different immune cells and cancer cells. This review further highlights the immunomodulatory role

M. B. Nicoud · M. A. Táquez Delgado · I. Ospital · V. A. Medina (✉)
Laboratory of Tumor Biology and Inflammation, Institute for Biomedical Research (BIOMED), School of Medical Sciences, Pontifical Catholic University of Argentina (UCA), and the National Scientific and Technical Research Council (CONICET), Buenos Aires, Argentina
e-mail: vanina_medina@uca.edu.ar

H. A. Sterle
Neuroimmunomodulation and Molecular Oncology Laboratory, Institute for Biomedical Research (BIOMED), School of Medical Sciences, Pontifical Catholic University of Argentina (UCA), and the National Scientific and Technical Research Council (CONICET), Buenos Aires, Argentina

K. Formoso
Laboratory of Tumor Biology and Inflammation, Institute for Biomedical Research (BIOMED), School of Medical Sciences, Pontifical Catholic University of Argentina (UCA), and the National Scientific and Technical Research Council (CONICET), Buenos Aires, Argentina

Institut des Maladies Métaboliques et Cardiovasculaires, Inserm, Université Paul Sabatier, Toulouse, France



Interdisciplinary Cancer Research, https://doi.org/10.1007/16833_2022_104
Published online: 18 January 2023

of histamine and histamine receptors in breast cancer with implications in cancer immunotherapy. In this line, immune cells and also cancer cells can produce and respond to histamine, which denotes the complexity of the histamine/histamine receptor axis modulation of the antitumor immunity.

To date, checkpoint inhibitor immunotherapy has the potential to generate tumor-specific and durable responses, revolutionizing the management of breast cancer. However, improvements in terms of rational immunotherapy combinations should be made to overcome reduced clinical responses, the lack of predictive biomarkers, and counteract acquired therapy resistance. In this scenario, the histaminergic system provides a promising strategy for the potential therapeutic exploitation of new immunomodulatory drug targets.

**Keywords**

Anticancer treatment · Antitumor immunity · Breast cancer · Histamine receptors · Immunomodulatory drugs · Immunotherapy

## 1 Introduction

Breast cancer (BC) is the most prevalent type of cancer in females, accounting for the highest number of deaths. Only in 2020 BC represented 29% of cancer cases and 15% of cancer deaths in females worldwide (Harbeck et al. 2019; Sung et al. 2021). Nevertheless, the pattern of incidence and mortality varies greatly around the world. For instance, developed countries have a greater incidence than middle- and low-income countries, which could be related to a higher prevalence of risk factors, being the increasing aging of the populations the most important one. In lower-income nations, on the other hand, cancer mortality is higher because health care and public cancer centers are scarcer, resulting in a delayed diagnosis and, thus, a worse prognosis (Duggan et al. 2021) (Fig. 1).

BC is a highly heterogeneous neoplasia whose classification has shifted over time and is frequently called into question. However, it can be roughly classified based on its histopathological appearance and the expression of hormone receptors and growth factors (Tsang and Tse 2020). In this line, the presence of estrogen receptor (ER), progesterone receptor (PR), and human epidermal growth factor receptor 2 (HER2) receptors is an important factor to take into consideration. The majority of BC express ER and PR, and when co-expressed, they represent the least aggressive type of BC. HER2 is found in a smaller percentage of cases and determines a more aggressive type of cancer, with a higher mortality rate. Finally, as the name implies, triple negative breast cancer (TNBC) lacks all those receptors. In particular, TNBC has a poor prognosis due to both its aggressiveness and the scarcity of available treatments (Sotiriou et al. 2003; Cleator et al. 2007; Brouckaert et al. 2012). In addition, different molecular subtypes of BC have been identified based on gene expression profiles of human tumors, which include luminal A, luminal B,

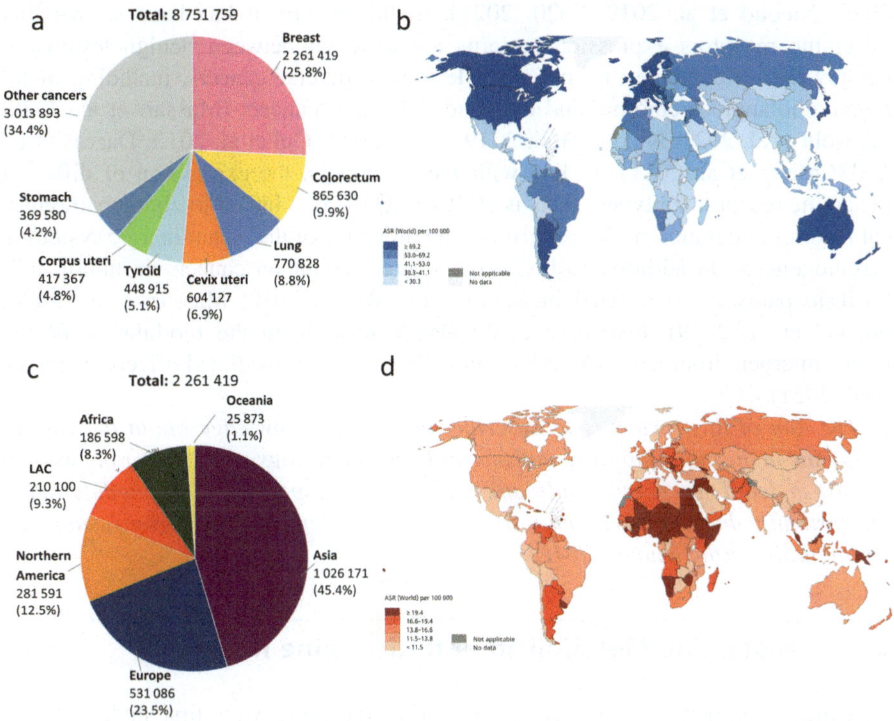


**Fig. 1** Epidemiology of breast cancer. (**a**) Estimated number of new cases for each type of cancer for females in 2020. (**b**) Estimated number of new cases of BC for females by continent. (**c**) Map depicting the estimated age-standardized incidence rate for BC in 2020, LAC, Latin America and the Caribbean. (**d**) Map depicting the estimated age-standardized mortality rate for BC in 2020. All the graphs were obtained from GLOBOCAN 2020 (https://gco.iarc.fr/today)

basal-like, normal breast-like, and HER2 positive subgroups, which imply different biological features, incidence, and prognosis (Prat et al. 2015).

Both BC detection and therapy have been substantially improved in the previous decade, leading to lower mortality rates, particularly in affluent nations. However, mortality remains high, and treatments are far from ideal, carrying great secondary effects due to their high toxicity (Chaudhry et al. 2021). Among the various new therapies being researched, histamine and histamine receptor modulation have emerged as a promising tool to fight against cancer progression. Histamine, 2-(4-imidazolyl)-ethylamine, is one of the most extensively investigated molecules in biomedicine (Tiligada and Ennis 2020; O'Mahony et al. 2011). Our laboratory along with others have shown that the treatment with histamine can alter cell proliferation, apoptosis, differentiation, migration, and immune response resulting in both pro- and antitumorigenic effects depending on the tumor type (Medina and Rivera 2010; Massari et al. 2020; Sarasola et al. 2021; Nguyen and Cho 2021). These contrasting effects can be attributed to the differential expression of the proteins responsible to mediate or block histamine biological effect (Sarasola et al.

2021; Nicoud et al. 2019, 2020, 2021). In this regard, it has been shown that histamine receptors expression patterns are different between benign lesions or healthy tissues compared to malignant lesions in diverse cancers, including melanoma, cholangiocarcinoma, and oral and colorectal cancers (Massari et al. 2020; Sarasola et al. 2021; Nicoud et al. 2019, 2020, 2021; Cui et al. 2013; Darvas et al. 2003; Meng et al. 2011). In line with these findings, the expression of different histamine receptor subtypes, such as H1R and H4R, was linked to clinicopathological features and tumor grade, confirming the influence of the histaminergic system in carcinogenesis. In addition to histamine's direct effect on cancer via tumor cell-intrinsic pathways (reviewed in Medina and Rivera 2010; Massari et al. 2020; Nicoud et al. 2019), histamine could also contribute to the modulation of the tumor microenvironment (TME) by controlling immune-mediated effects (Sarasola et al. 2021).

*The aim of this review is to describe the current knowledge about the role of histamine in the molecular mechanisms and immunobiology of BC. Later on, we will discuss in detail the current landscape and clinical perspectives of the modulation of the histamine-histamine receptor axis on BC development and progression and additionally on the immune TME.*

## 2 Histamine Metabolism and Histamine Receptors

Histamine's properties, and later on those of its receptors, were first studied a little more than a century ago. Windaus and Vogt synthesized histamine in 1907 by removing the carboxyl group from the amino acid histidine. In 1910, Dale and Laidlaw isolated histamine from mold, which was capable of causing vasodilation; smooth muscle contraction in the airways, uterus, and intestine; stimulated heart rate; and contractility. At the same time, they demonstrated that it could elicit physiological responses associated with allergic reactions. Finally, in 1927, Best and colleagues discovered histamine in vertebrate tissue, particularly the lungs (Best 1929; Dale and Laidlaw 1910; Lieberman 2011). Numerous studies have since then revealed that histamine plays a critical role in a wide range of biological processes, both physiological and pathological, including inflammation, allergy, cancer, neurotransmission, and immune system regulatory mechanisms, among others. Along with its broad functions, all histamine receptor subtypes are promising or already well-established drug targets. In this section, we summarize the most relevant aspects of histamine metabolism and histamine receptors' characteristics.

Histamine synthesis is now known to occur in the cytoplasm of several types of cells, including mast cells, basophils, T cells, dendritic cells (DCs), macrophages, epithelial cells, neurons, and, more recently described, cardiomyocytes (Huang et al. 2018; Neumann et al. 2021). This process is carried out by the histidine decarboxylase (HDC) mediated decarboxylation of the amino acid L-histidine (Huang et al. 2018; Darvas and Falus 2004) and occurs in most cell types on a constant basis, but it can also be induced by the cytokines IL-3, IL-18, and IL-33 (Fig. 2).

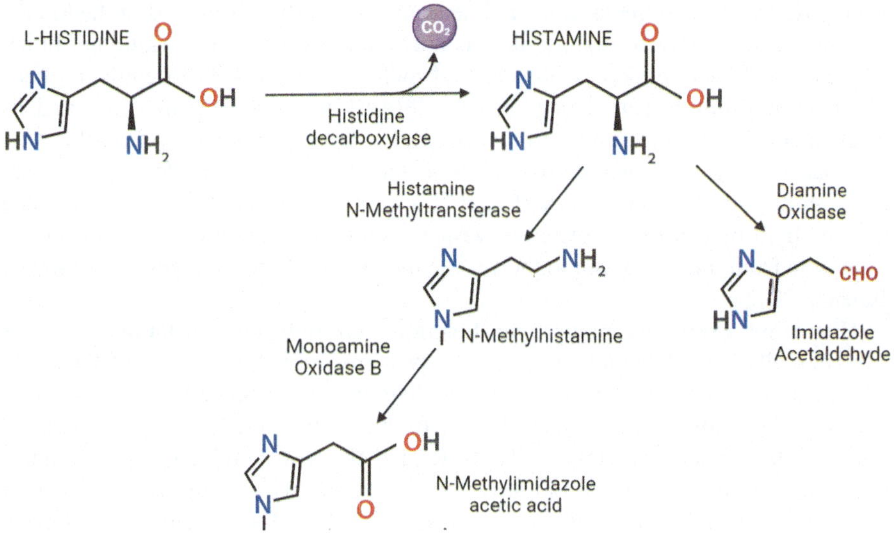


**Fig. 2** Histamine synthesis and degradation pathways. Histamine is synthesized by the histidine decarboxylase (HDC) that decarboxylates the amino acid L-histidine. Histamine is then degraded by two independent pathways. The enzyme histamine N-methyltransferase (HNMT) methylates the imidazole ring producing N-methylhistamine that is later oxidized by the monoamine oxidase B (MAO B) in the mitochondria. The enzyme diamine oxidase (DAO) carries out the oxidative deamination of the first amino group forming imidazole acetaldehyde

Following its synthesis and consequent activity, histamine has to be inactivated, a process that can occur through two different metabolic pathways. The first one is carried out by the enzyme histamine N-methyltransferase (HNMT) and involves the methylation of the imidazole ring, thus producing N-methylhistamine that is further oxidized by the monoamine oxidase B (MAO B) in the mitochondria (Maršavelski and Vianello 2017). The second pathway implies the oxidative deamination of the first amino group forming imidazole acetaldehyde, and it is catalyzed by the enzyme diamine oxidase (DAO). While HNMT shows great specificity toward histamine, the DAO has a broader list of substrates (reviewed in Schwelberger and Sánchez-Jiménez 2013).

Once it is synthesized, histamine is stored in intracellular granules until, upon external stimulation of mast cells, the major producer of histamine, they degranulate, releasing the stored histamine. Histamine is positively charged and thus a cytosolic and transmembrane transport system is required for its uptake, storage, and release. This system comprises organic cation transporters (OCTs), plasma membrane monoamine transporters (PMAT), and vesicular monoamine transporters (VMAT) (Schwelberger and Sánchez-Jiménez 2013). OCTs participate in the transport of endogenous amines or xenobiotic compounds in the liver, kidneys, intestine, and other organs important for the elimination of these products (Schwelberger and Sánchez-Jiménez 2013). There are three subtypes of OCT: OCT1, OCT2, and OCT3. In humans, histamine is transported only by OCT2 and OCT3. OCT2 is mainly detected in epithelial cells and neurons. It is highly expressed in the kidney and in a wide variety of organs such as the small intestine, lung, skin, and brain, whereas OCT3 expression is broader and includes, in addition to epithelial cells and neurons, muscle cells and glial cells. PMAT is a cation transporter, and its mechanism of action is similar to that of OCT3. It facilitates passive translocation of positive organic ions across plasma membranes in an electrogenic and concentration-dependent manner (Engel et al. 2004). Finally, VMAT is integrated into the lipid bilayer of the secretory vesicle membrane and allows the transport of histamine from the cytoplasm to the vesicle membrane (Schwelberger and Sánchez-Jiménez 2013).

Histamine exerts its pleiotropic functions through four histamine receptor subtypes, all of which belong to the large G-protein-coupled receptor (GPCR) family and exist in a constant balance between their inactive and active conformation, displaying constitutive activity. Each receptor has been named in the chronological order of their discovery H1R, H2R, H3R, and the last identified H4R. Counter-intuitively the first two receptors have the lowest affinity among the four receptor subtypes (Alexander et al. 2017; Panula et al. 2015). The different GPCR have several signaling pathways and expression patterns (Table 1).

The histamine H1 receptor (H1R) was the first and the most studied receptor subtype to whom the first antihistamine drugs were targeted. Nevertheless, it was not until 1980 with the development of loratadine, a second-generation antihistamine, that more specific drugs were accomplished that avoided drowsiness as a secondary effect. H1R is expressed ubiquitously including brain, heart, smooth muscle, and liver, among others. Upon histamine activation, H1R stimulates phospholipase C (PLC) through Gαq/11, leading to the formation of inositol 1,4,5-triphosphate (IP3) and 1,2-diacylglycerol (DAG) and the consequent increase in cytosolic $Ca^{2+}$ (Panula et al. 2015; Hill et al. 1997). This activation can lead to different biological effects depending on the tissue and cell type involved. H1R has been described to mediate smooth muscle cells relaxation, vasodilation, and increased vascular permeability. Therefore, most H1R antagonists/inverse agonists, including mepyramine, fexofenadine, loratadine, diphenhydramine, and astemizole, are widely used for the treatment of allergic diseases (Seifert et al. 2013; Strasser et al. 2013).

Histamine H2 receptor (H2R) existence was confirmed in 1972, and its expression has been reported in almost all peripheral tissues and in the central nervous system (CNS). Activation of the H2R leads to cyclic adenosine monophosphate (cAMP) production through the stimulation of Gαs and adenylyl cyclase (AC). H2R further participates in the phosphoinositide second messenger signaling pathway (Panula et al. 2015; Bristow et al. 1982; Dy and Schneider 2004). Many of the H1R-mediated effects can be balanced by the H2R, including the relaxation of smooth muscle cells, causing vasodilation. Among the numerous functions described for this receptor, the main ones are related to gastric acid secretion from parietal cells, and chronotropic and inotropic effects in the heart. Most H2R antagonists/inverse agonists, including cimetidine, famotidine, and nizatidine, are clinically used to inhibit histamine-induced gastric acid secretion (Seifert et al. 2013; Strasser et al. 2013).

**Table 1** Molecular and functional characteristics of histamine receptors

| Histamine receptors | Characteristics | | | |
|---|---|---|---|---|
| | Agonists | Antagonists/ Inverse agonists | Expression | Physiological effects |
| 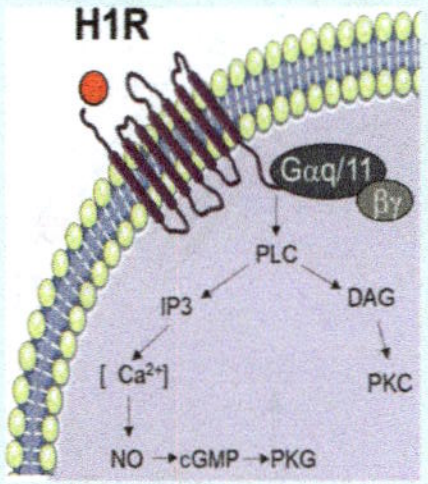 | 2-[(3-trifluoromethyl) phenyl] histamine<br>Histaprodifen<br>Methylhistaprodifen | Mepyramine<br>Terfenadine<br>Cyproheptadine<br>Loratadine<br>Diphenhydramine<br>Triprolidine<br>Astemizole<br>Chlorpheniramine | Brain, smooth muscle, skin, heart, gastrointestinal tract, eosinophils, neutrophils, DCs, monocytes, T and B cells | Relaxation of smooth muscle cells, vasodilation, increase vascular permeability |
| 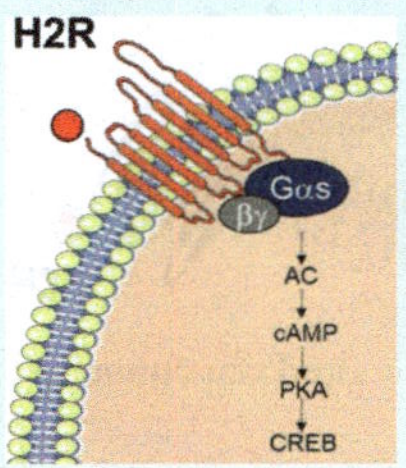 | Amthamine<br>Impromidine | Cimetidine<br>Ranitidine<br>Tiotidine<br>Famotidine | Brain, smooth muscle, skin, heart, gastrointestinal tract, eosinophils, neutrophils, DCs, monocytes, T and B cells | Gastric acid secretion, heart chronotropic and inotropic effects, inhibition of basophils' chemotaxis |

(continued)

**Table 1** (continued)

| Histamine receptors | Characteristics | | | |
|---|---|---|---|---|
| | Agonists | Antagonists/ Inverse agonists | Expression | Physiological effects |
| **H3R** 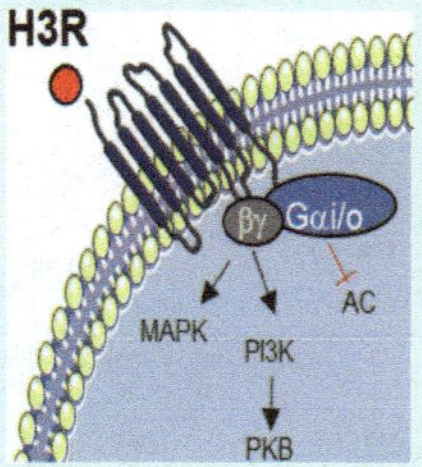 | (R)-α-methylhistamine<br>Imetit<br>Immepip | Clobenpropit<br>JNJ5207852<br>JNJ10181457<br>OUP-186<br>Pitolisant<br>Thioperamide | CNS, brain, eosinophils, DCs, monocytes, gastric mucosa | Regulating the release of histamine from histaminergic neurons and other neurotransmitters |
| **H4R** 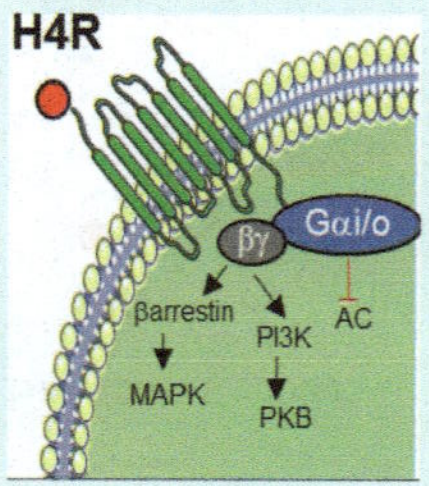 | VUF 8430<br>Clobenpropit<br>JNJ28610244<br>4-methylhistamine<br>ST-1006<br>Imetit | JNJ7777120<br>Thioperamide<br>JNJ10191584<br>VUF 6002<br>A-987306<br>A-940894<br>OUP-16 | Bone marrow, eosinophils, basophils, neutrophils, DCs, monocytes, T and B cells, mast cells, spleen, thymus, lung, small and large intestines, heart | Inflammatory effects, immunomodulatory pathways |

Abbreviations: *PLC* phospholipase C, *IP3* inositol 1,4,5-triphosphate, *DAG* 1,2-diacylglycerol, *PKC* protein kinase C, *NO* nitric oxide, *cGMP* cyclic guanosine monophosphate, *PKG* protein kinase G, *AC* adenylyl cyclase, *cAMP* cyclic adenosine monophosphate, *PKA* cAMP-dependent protein kinase, *CREB* cAMP response element-binding, *MAPK* mitogen-activated protein kinase, *PI3K* phosphoinositide 3-kinase, *PKB* protein kinase C, *DCs* dendritic cells, *CNS* central nervous system

Regarding histamine H3 receptor (H3R), it was first described in 1983. It acts through Gαi/o and, contrary to H1R and H2R, its activation leads to inhibition of cAMP production, accumulation of $Ca^{2+}$, and stimulation of mitogen-activated protein kinase (MAPK) pathway (Panula et al. 2015; Shahid et al. 2009; Dimitriadou et al. 1994). H3R is mainly expressed in the CNS and regulates the release of histamine from histaminergic neurons and other neurotransmitters (Sarasola et al. 2021; Bäumer and Rossbach 2010; Branco et al. 2018; Rouleau et al. 2000). H3R blocking ligands are being investigated for the treatment of several CNS disorders, including Alzheimer's disease, obesity, and schizophrenia (Amini et al. 2015; Dauvilliers et al. 2013; Kantor and Jurkiewicz 2006; Kasteleijn-Nolst Trenité et al. 2013; Szakacs et al. 2017). In fact, pitolisant is a first-in-class Food and Drug Administration (FDA)- and European Medicines Agency (EMA)-approved H3R inverse agonist/antagonist for the treatment of narcolepsy-related daytime sleepiness (Dauvilliers et al. 2013; Guevarra et al. 2020).

Histamine H4 receptor (H4R) is a Gαi/o-coupled receptor that mediates the inhibition of the AC and downstream cAMP responsive elements (CRE) as well as the activation of MAPK and PLC with $Ca^{2+}$ mobilization (Sarasola et al. 2021; Panula et al. 2015; Seifert et al. 2013; Shahid et al. 2009). H4R is predominantly expressed in immune cells and has been found in other tissues such as the lung, small and large intestines, and heart (Sarasola et al. 2021; Nguyen and Cho 2021; Nicoud et al. 2019). Therefore, the H4R is involved in inflammatory and immunomodulatory processes, orchestrating both the innate and adaptive immunity (Wendell et al. 2020). To date there are no ligands approved for the H4R, but several clinical trials are taking place (Kollmeier et al. 2018; Murata et al. 2015; Schaper-Gerhardt et al. 2020; Thurmond et al. 2014, 2016, 2017; Werfel et al. 2019).

Interestingly, all four histamine receptors have been found to be expressed in several human cancers.

## 3 Role of Histamine and Histamine Receptors in Breast Cancer Biology and Therapeutic Approaches

In this section, we address the latest evidence on the role of histaminergic systems in the complex BC biology, with particular emphasis on the identification of potential molecular targets for BC therapies.

Histamine plays a role in various physiological and pathological processes involved in the development and functioning of the mammary gland (Medina and Rivera 2010; Davio et al. 1994; Falus et al. 2010; Medina 2013). Higher histamine concentration was observed in human breast tumors, human BC cell lines, and N-nitroso-N-methylurea (NMU)-induced rat mammary adenocarcinomas compared with their normal counterparts, suggesting a link between endogenous histamine levels and BC progression. HDC, the key enzyme involved in histamine synthesis, showed altered expression and activity in BC, associated with changes in endogenous histamine levels (Medina 2013; Medina et al. 2006; Rivera et al. 2000; Sieja et al. 2005; von Mach-Szczypiński et al. 2009). In this sense, in a Han Chinese population, He et al. analyzed the clinical associations between BC and single nucleotide polymorphisms in *HDC, HNMT,* and *H3R* that have previously been found to result in altered expression and function of these proteins. Even though no relationship between the polymorphisms of *HNMT* or *HRH3* and BC has been found in this study, a significant association between an HDC polymorphism variant (rs7164386) and BC risk was demonstrated based on genotype-allele association analyses. These findings highlight the important role of HDC in the pathogenesis of BC (He et al. 2014). Accordingly, bioinformatic analyses of The Cancer Genome Atlas (TCGA) transcriptomic data show that HDC expression is downregulated in BC compared with normal tissue, and in addition, the most aggressive cancer subtypes including TNBC exhibited the lowest expression levels of the HDC mRNA. Furthermore, higher HDC expression was significantly associated with better disease-free survival in TNBC patients as well as in patients with other BC subtypes (Massari et al. 2020; Nicoud et al. 2019, 2020).

The analysis of the HDC and HNMT mRNA expression levels in TCGA Pan-Cancer Atlas dataset shows that both enzymes display reduced levels of expression in the aggressive basal-like and HER2-positive tumors compared with luminal A, luminal B, and normal breast-like tumors, which show better prognosis (Fig. 3).

Since the beginning of the 1990s, the presence of H1R and H2R in tissues and cell lines of normal and malignant human BC has been described (Medina 2013; Davio

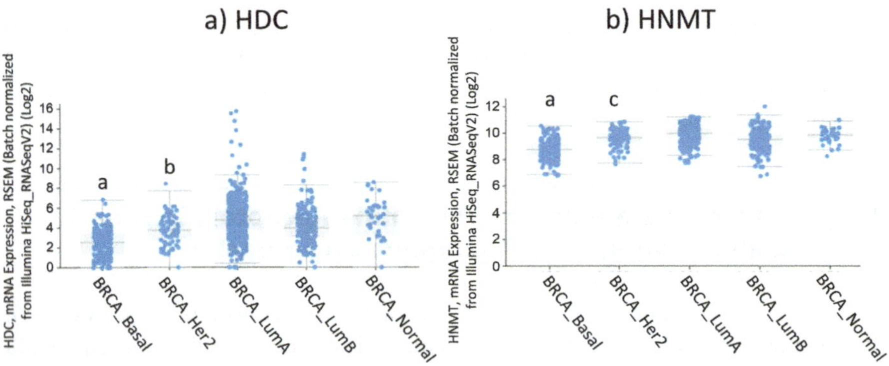


**Fig. 3** Bioinformatic analyses of the expression of the enzymes involved in the synthesis and catabolism of histamine in breast cancer. mRNA expression levels were obtained from breast cancer datasets at the cBioportal (Cancer Cell Line Encyclopedia, Broad 2019 (Cerami et al. 2012; Gao et al. 2013; Ghandi et al. 2019) for Cancer Genomics (TCGA PanCancer Atlas (DeSantis et al. 2019; Hoadley et al. 2018). (**a**) Histidine decarboxylase (HDC) and (**b**) histamine N-methyltransferase (HNMT) mRNA expression in different breast cancer (BRCA) subtypes. Box plots show the expression levels as log-transformed mRNA expression z-scores compared to the expression distribution of all samples (RNA Seq V2 RSEM). BRCA_Basal ($n = 171$), BRCA_Her2 ($n = 78$), BRCA_LumA (Luminal A, $n = 499$), BRCA_LumB (Luminal B, $n = 197$), BRCA_Normal (normal breast-like, $n = 36$). (a) $P < 0.001$ BRCA_Basal vs. All subtypes; (b) $P < 0.001$ BRCA_Her2 vs. BRCA_LumA; (c) $P < 0.005$ BRCA_Her2 vs. BRCA_LumA. Kruskal-Wallis test and Dunn's multiple comparisons test

et al. 1993; Lemos et al. 1995). Subsequently, various investigations were carried out in different animal models to understand the role of both receptors in the proliferation of BC (Medina 2013; Lamas et al. 2017). Different authors have indicated that histamine exerts an inhibitory effect on BC cell proliferation by activating a cAMP-dependent mechanism through the H1R, while the activation of the H2R induces their proliferation through the PCL-dependent pathway (Cricco et al. 1994; Davio et al. 1995; Rivera et al. 1993).

The H1R was the first of the histaminergic family of receptors to be targeted for clinical use. Several antagonists of H1R (originally named antihistamines) are widely used over-the-counter drugs to treat allergic inflammation such as rhinitis, conjunctivitis, urticaria, and even anaphylaxis. They are known to be well-tolerated drugs by a large part of the population because they have shown few associated side effects. In addition, there are several studies that show their effectiveness against various types of cancer.

The survival associated with the use of six common H1R antihistamines (cetirizine, clemastine, desloratadine, ebastine, fexofenadine, and loratadine) was recently reported in a cohort study of Swedish BC patients. The treatments with desloratadine or loratadine were consistently associated with improved survival, regardless of patient age, menopause and hormonal receptor status, or tumor stage (Fritz et al. 2020, 2021). In contrast, in a Danish nationwide cohort study, no association was found between the use of H1R antihistamines and the development of contralateral breast cancer (CBC). Furthermore, the use of cationic amphiphilic antihistamines among patients receiving chemotherapy was not associated with a decrease in CBC risk (Bens et al. 2020). In the same way, a study based on 3,133 BC cases showed no association between the antihistamine use and BC risk (Nadalin et al. 2003).

Interestingly, using different human databases and bioinformatic tools, the H1R expression was reported to be significantly higher in basal-like and HER2 BC subtypes compared with luminal BC. Moreover, this H1R overexpression correlated with shorter overall survival (Fernández-Nogueira et al. 2016). In this line, using another dataset (TCGA Pan-Cancer Atlas dataset), it is confirmed that H1R mRNA expression is higher in HER2-enriched tumors compared with luminal A and B, while its expression in basal-like tumors is lower compared with HER2-enriched and normal-like subtypes but higher than in luminal BC (Fig. 4a). In complete agreement with these data, terfenadine H1R blockade suppressed cell proliferation and activated ERK signaling to initiate apoptotic cell death in vitro. Furthermore, in vivo studies demonstrated that terfenadine administration reduced the tumor growth of basal and trastuzumab-resistant BC cells (Fernández-Nogueira et al. 2018). The combined use of histamine with another H1R antagonist, astemizole, induced cytotoxicity and endoplasmic reticulum stress in MCF-7 BC cells, resulting in the induction of apoptotic or autophagic cell death, reactive oxygen species (ROS) generation, and cAMP release (Jakhar et al. 2016). In addition, the combination of astemizole with calcitriol inhibited the growth of tumors developed with human T-47D BC cells in athymic female mice more effectively than each single treatment (García-Quiroz et al. 2014). Combinations of astemizole and the EGFR inhibitor, gefitinib, resulted in a synergistic antiproliferative effect in BC cells (García-Quiroz et al. 2019).

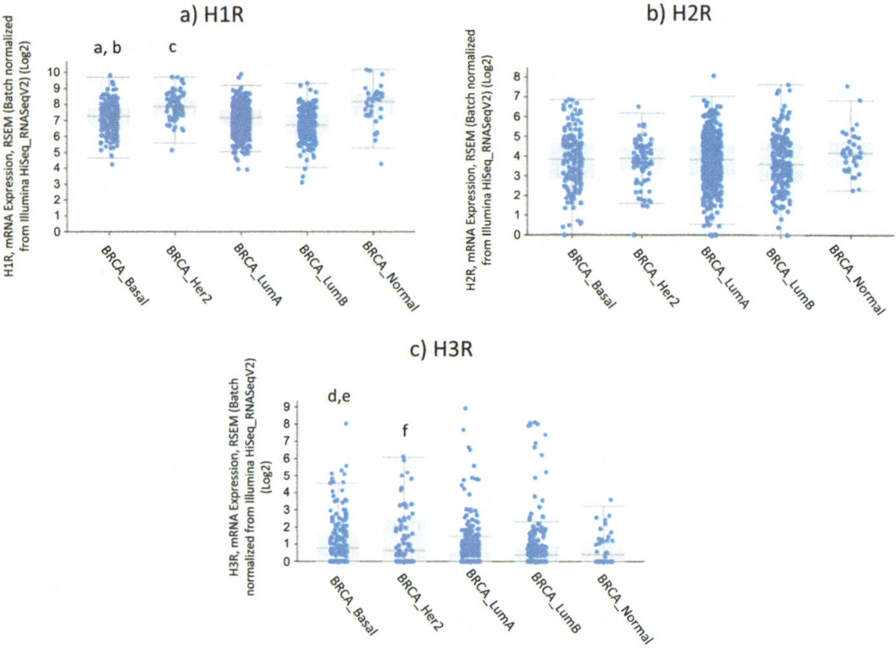


**Fig. 4** Bioinformatic analyses of the expression of histamine receptors in breast cancer. mRNA expression levels were obtained from breast cancer datasets at the cBioportal (Cancer Cell Line Encyclopedia, Broad 2019) (Cerami et al. 2012; Gao et al. 2013; Ghandi et al. 2019), for Cancer Genomics (TCGA PanCancer Atlas) (DeSantis et al. 2019; Hoadley et al. 2018), (**a**) histamine receptor 1 (H1R), (**b**) histamine receptor 2 (H2R), and (**c**) histamine receptor 3 (H3R) mRNA expression in different breast cancer (BRCA) subtypes. Box plots show the expression levels as log-transformed mRNA expression z-scores compared to the expression distribution of all samples (RNA Seq V2 RSEM). BRCA_Basal (n = 171), BRCA_Her2 (n = 78), BRCA_LumA (Luminal A, n = 499), BRCA_LumB (Luminal B, n = 197), BRCA_Normal (normal breast-like, n = 36). (a) $P < 0.0001$ BRCA_Basal vs. BRCA_Her2 and BRCA_LumB; (b) $P = 0.0009$ BRCA_Basal vs. BRCA_Normal; (c) $P < 0.0001$ BRCA_Her2 vs. BRCA_LumA and BRCA_LumB; (d) $P < 0.0001$ BRCA_Basal vs. BRCA_LumA; (e) $P = 0.0019$ BRCA_Basal vs. BRCA_LumB; (f) $P < 0.0001$ BRCA_Her2 vs. BRCA_LumA, Kruskal-Wallis test and Dunn's multiple comparisons

Ebastine, a second-generation antagonist of H1R, impairs cell proliferation, migration, and invasion of BC cells by targeting the promoter of the enhancer of zeste 2 polycomb repressive complex 2 subunit (EZH2) and the treatment with ebastine showed a potent and dose-dependent antitumor effect against TNBC in vivo in murine models of the TNBC patient-derived xenograft BCM3887 and the TNBC cell line SUM159 xenograft (Li et al. 2020). In line with these results, a retrospective study reported that BC patients taking ebastine for allergy treatment improved overall and BC-specific survival rate compared with those not taking the medication (Olsson 2018; Olsson et al. 2015).

The H2R has been shown to play a crucial role in regulating tumor growth and metastasis during early breast tumor development. In NMU-induced experimental mammary carcinomas, treatment with H2R antagonists (ranitidine or cimetidine) decreased tumor incidence (Rivera et al. 2000; Davio et al. 1995; Rivera et al. 1993). Administration of the H2R antagonists, ranitidine and famotidine, showed no significant effects on final tumor weight or growth kinetics over 19–21 days in the 4T1 TNBC model. However, both drugs reduced lung metastases (Vila-Leahey et al. 2016).

A retrospective study was performed to evaluate the relationship between the usage of H2R blockers and the risk of invasive BC. The use of cimetidine and famotidine was not associated with an increased risk of BC, whereas current users of ranitidine had a 2.2-fold increased risk of developing ductal carcinoma (Coogan et al. 2005; Mathes et al. 2008). In addition, the occurrence of the rs2067474 polymorphism of the human *H2R* gene was not correlated with an increased BC risk or with related clinicopathological parameters in the Chinese Han population (Cai et al. 2015). The expression of H2R mRNA shows no significant difference when comparing different BC subtypes (Fig. 4b). Furthermore, some clinical trials carried out with H2R antagonists lead to inconclusive results for BC patients (Bowrey et al. 2000; Parshad et al. 2005). Data presented suggests that H2R antagonists are not therapeutically useful in BC.

Our laboratory first described the expression of H3R not only in human cell lines, but also in benign lesions and breast carcinomas. In BC specimens a high H3R expression was positively correlated with the expression of HDC and proliferating cell nuclear antigen (PCNA), a marker of proliferation. It is important to highlight that histopathologically normal breast tissue surrounding carcinomas revealed a lower or negative expression compared with high expression of H3R in tumor cells of the same patient (Medina et al. 2006, 2008). In relation to these results, the bioinformatic analysis of TCGA Pan-Cancer Atlas dataset showed that H3R expression was higher in the aggressive basal-like and HER2-enriched BC compared with the more favorable clinical outcome luminal A and luminal B BC (Fig. 4c).

In agreement with these results, histamine regulated the proliferation of MDA-MB-231 TNBC cells in a dose-dependent manner, decreasing cell growth at micromolar concentrations while lower nanomolar concentrations increased cell proliferation moderately. The positive effect on cell proliferation was mimicked by H3R agonists (Imetit and R-α methyl histamine) and blocked by the combined treatment with H3R antagonists (Medina et al. 2008). In addition, in vivo administration of the H3R antagonist JNJ10181457 successfully reduced tumor size of MDA-MB-231 xenograft tumors developed in immunodeficient nude mice (Martinel Lamas et al. 2013, 2015a). Consistent with these findings, OUP-186 and JNJ5207852, both H3R antagonists, produced antiproliferative activities on MDA-MB-231 and MCF-7 BC cell lines through the induction of cell apoptosis and cell cycle arrest (Medina et al. 2006, 2008; Tanaka et al. 2016).

Since the discovery of the H4R, numerous studies have focused their efforts on elucidating the role of this receptor in inflammatory and immune-related pathologies, including allergy, atopic dermatitis, and autoimmune diseases. This led to the development of several pharmacological compounds, both specific agonists and antagonists with the ability to modulate H4R activity in these contexts. However, the implications of H4R in the progression of BC have recently begun to be explored. Our laboratory reported for the first time the expression of H4R in human MDA-MB-231 and MCF7 BC cell lines and in BC samples (Medina et al. 2006, 2008). Subsequently, through large-scale genomic analysis of BC patients using TCGA dataset, we found that the *H4R* expression was downregulated in primary tumors compared with normal tissue (Massari et al. 2020). In line with this, and denoting the relevance of the H4R subtype in the progression of BC, it was demonstrated that *H4R* gene polymorphisms (rs623590, rs11662595, and rs1421125) were associated with enhanced malignancy and increased risk of developing the disease in the Han Chinese population (He et al. 2013). Recently, in an exploratory study in 30 samples of TNBC patients, it was shown that H4R immunoreactivity was detected in about 70% of tumors, and its expression was positively correlated with H4R levels in the histologically normal adjacent tissue. Using TCGA data (Pan-Cancer Atlas), it was found that H4R mARN expression was downregulated in basal-like/TNBC compared with luminal A and normal breast-like tumors while a higher H4R expression was associated with improved survival outcomes in these patients. Although further studies in large cohorts are needed to better understand the role of H4R in BC biology, these findings suggest that the H4R could represent a potential prognostic marker in this neoplasia (Speisky et al. 2022).

A large collection of in vitro and in vivo studies has shown that the activation of H4R can modulate processes associated with BC progression and the TME. H4R agonists (histamine, VUF8430, JNJ28610244) reduced cell proliferation and at the same time induced cell cycle arrest, increasing the number of apoptotic and senescent MDA-MB-231 TNBC cells (Martinel Lamas et al. 2013; Medina et al. 2011a). Similarly, JNJ28610244 and VUF8430 inhibited cell growth of the TNBC 4T1 murine cell line (Nicoud et al. 2020). Furthermore, both histamine (5 mg/kg) and JNJ28610244 (10 mg/kg) reduced the growth of tumors developed in nude mice with MDA-MB-231 cells, whereas only histamine increased median survival and tumor apoptosis (Martinel Lamas et al. 2013). On the other hand, in vitro histamine treatment potentiated ionizing radiation-induced effects, increasing DNA damage, cell senescence, and apoptosis. Moreover, it prevented the radiation-induced mesenchymal change and cell migration of TNBC lines by decreasing phospho-Src levels (Medina et al. 2006; Galarza et al. 2016, 2020; Martinel Lamas et al. 2015b). In these cells, histamine pretreatment prevented the radiation-induced mesenchymal changes (e.g., decrease in the epithelial marker E-cadherin, increase in the mesenchymal marker vimentin, and Slug and MMP-2 activity) and reduced the migratory behavior of irradiated cells (Galarza et al. 2016).

Furthermore, conditioned media derived from fibroblasts induced epithelial-mesenchymal transition phenotypic changes in MCF-7 and MDA-MB-231 BC cells, effects that were reversed with histamine treatment (Porretti et al. 2014). In vivo, histamine administration increased the antitumor response to ionizing radiation and doxorubicin chemotherapy in TNBC murine models. Interestingly, histamine protected highly radiosensitive tissues, including the small intestine, salivary glands, and bone marrow from ionizing radiation-induced cytotoxic damage. Even when combined with chemotherapeutic treatment, histamine reduced adverse effects such as hepatic and cardiac toxicity (Porretti et al. 2014; Carabajal et al. 2012; Martinel Lamas et al. 2015c, d; Medina et al. 2007, 2010, 2011b).

## 4 Role of Histamine in Breast Cancer Immunity

The interaction between tumor cells and their microenvironment is crucial for cancer progression. The TME is composed of many cell types, such as immunocompetent cells, vascular endothelial cells, mesenchymal stem cells, and cancer-associated fibroblasts, together with non-cellular factors, including cell secreted proteins, cofactors and enzymes, cytokines, and hormones. The composition of TME changes during tumor evolution affecting early stages of cancer progression as well as the formation of distant metastasis.

Immune cells are critical components of the TME. They can paradoxically inhibit and promote tumor development and progression through "cancer immunoediting". This process refers to the ability of immune cells to intervene in the elimination of tumor cells (immunosurveillance) and, at the same time, shape the immunogenicity of tumors favoring their growth and progression (immunotolerance). Cancer immunoediting is a dynamic process that consists of three phases: elimination, equilibrium, and escape (Schreiber et al. 2011).

During the elimination phase, the innate and adaptive immunity cooperate to detect and eliminate tumor cells that develop due to failures in their intrinsic mechanisms of tumor suppression. The elimination can be complete, meaning no tumor cells remain, or incomplete, when only a portion of them is eradicated. In the last-mentioned case, tumor cells enter an equilibrium phase, where they evolve and accumulate changes that modulate the expression of tumor antigens. In this phase, the immune system continues to eliminate susceptible tumor clones and therefore tumor growth is limited, and, over time, it can be eradicated. However, the pressure exerted by the adaptive immunity together with increased mutational burden of tumor cells can lead to a lower immunogenicity of tumor cells and consequently the selection of tumor subclones that can evade immune recognition. These tumors enter into the escape phase, characterized by unrestrained growth in which the disease becomes clinically apparent (O'Donnell et al. 2019).

The complex interplay between the different tumor-infiltrating immune subsets that include macrophages, neutrophils, mast cells, natural killer (NK) cells, DCs, myeloid-derived suppressor cells (MDSCs), B cells, and different subtypes of T cells, as well as the secreted cytokines and chemokines, determine the balance between immunological surveillance and tolerance. The modulation of the signaling via coinhibitory or costimulatory receptors expressed on T cells may alter antitumor immune responses. Numerous evidence indicate that different factors promote immune tolerance, and one of the most prominent features is the expression of multiple coinhibitory molecules leading to the activation of immune checkpoints. The programmed death-1 (PD-1) and the cytotoxic T lymphocyte antigen 4

(CTLA-4) receptors have emerged as the most important ones. These receptors along with their respective ligands participate in the immune checkpoints, preventing tissue damage caused by excessive activation of T cells to maintain self-tolerance and avoid autoimmune reactions. However, tumor cells take advantage of these mechanisms, over-expressing the coinhibitory molecules in order to escape immunosurveillance and therefore promote tumor growth. Therefore, the specific blockade of these immune checkpoints in tumors allows the antitumor immune response to reactivate (Palleschi et al. 2020; Wang et al. 2019), and immunotherapy has been at the forefront of research in recent decades. The use of neutralizing or blocking antibodies to PD-1 and programmed death-ligand 1 (PD-L1) as immune checkpoint inhibitors is a promising and rapidly growing field with impressive success in many solid tumors, achieving durable clinical responses, but only in a fraction of patients (Alsaab et al. 2017; Chae et al. 2018; Gong et al. 2018; Kang et al. 2021; LaFleur et al. 2018; Reddy et al. 2020; Singh and Jadhav 2018).

In this section, we highlight the immunomodulatory role of histamine and histamine receptors in BC with implications in cancer immunotherapy.

BC has been originally considered poorly immunogenic because of their low mutational burden. However, in recent years the infiltration with a wide range of immune cells has been demonstrated in the TME of BC patients. However, the levels of immune cell infiltration depend on the BC subtype. The most aggressive subtypes, TNBC and HER2+ BC, are also the most immunogenic ones, as they are characterized by a higher genomic instability and tumor mutational burden than the other BC subtypes (Dieci et al. 2021). Indeed, TNBC and HER2+ BCs usually exhibit the higher percentages of tumor-infiltrating lymphocytes (TILs) (Stanton et al. 2016). The predominant TILs in BC are T cells. Among them, the presence of CD8+ cytotoxic T lymphocytes (CTLs) has been usually described as a marker of good prognosis in BC.

To date, the PD-L1 expression is the only established predictive biomarker for immunotherapy efficacy in metastatic BC. In this regard, the FDA granted accelerated approval to both atezolizumab and pembrolizumab in combination with chemotherapy for the first-line treatment of unresectable locally recurrent or metastatic TNBC patients whose tumors express PD-L1 (Emens 2018). These immunotherapeutic approaches improve progression-free and overall survival (Schmid et al. 2019), however, only a minority of patients benefit from them due to primary and acquired (secondary) resistance. Current efforts focus on developing immunotherapy combinations that can convert non-responders into responders, improve therapeutic responses, and counteract acquired therapy resistance (Reddy et al. 2020; Emens 2018).

In this scenario, histamine stands as a pivotal participant of the dynamic interaction between immune cells and tumor cells, and therefore it can tremendously impact the design of new immunomodulatory strategies to boost the immune system to fight cancer. Histamine is capable of promoting inflammatory and immunoregulatory responses that orchestrate both pathological and homeostatic conditions by balancing inflammatory reactions (Jutel et al. 2009; Stark 2013). In turn, it participates in processes related to tumorigenesis through angiogenesis (Marks et al. 1986; Moya-

García et al. 2021; Qin et al. 2013; Sörbo et al. 1994), antitumor immunity (Nicoud et al. 2020; Sterle et al. 2019), metabolic remodeling of cancer cells (Ocaña et al. 2019), epithelial-mesenchymal transition (Galarza et al. 2020; Porretti et al. 2014), and regulation of immune responses through histamine-secreting microbiota (Gao et al. 2017; Mou et al. 2021; Nejman et al. 2020; Spakowicz et al. 2020).

Previous reports have shown that histamine modulates the antigen-specific T helper (Th) cells by changing the cytokine production from a Th1 to a Th2 phenotype (Stark 2013; Shankaran et al. 2001) and stimulates the maturation of DCs from monocytes, thus reducing the growth of murine lymphoma developed with EL-4 cells (Martner et al. 2013). Histamine dihydrochloride has been used in numerous clinical trials as an adjuvant to IL-2 immunotherapy with proven clinical benefit in acute myeloid leukemia. Histamine preserves the function of T lymphocytes and NK cells by reducing the monocyte- and macrophage-induced formation and release of ROS. Histamine was generally well tolerated, demonstrating that it can be safely administered (Sarasola et al. 2021; Martner et al. 2013; Nilsson et al. 2020; Thorén et al. 2009).

In contrast, a BC model developed in HDC-deficient mice with the LM2 (adenocarcinoma mammary cells) showed a reduction of tumor size accompanied by a decreased population of regulatory T cells in the spleen compared with wild-type mice. In addition, HDC deficiency upregulated splenic T-bet+ lymphocytes and IL-12/IFNγ production (Hegyesi et al. 2007). However, in this study, the role of histamine in the local immune TME was not studied.

A recent study showed that desloratadine usage was associated with better survival for all immunogenic tumors (gastric, colorectal/anal, pancreatic, lung, breast, prostate, kidney, and bladder cancers, melanoma and Hodgkin lymphoma), but not for non-immunogenic ones (liver, uterine, ovarian, brain/CNS, thyroid cancers, and non-Hodgkin lymphoma), which drove the authors to hypothesize that these results could be related to a potential immune checkpoint inhibition of this drug (Fritz et al. 2021). In this regard, an exhaustive and original study provides new insights into combinatorial therapeutic approaches targeting H1R to improve cancer immunotherapy. Li et al., retrospectively, analyzed BC patients taking common pharmaceutical drugs during the course of anti-PD-1/PD-L1 immunotherapy. Results show that uptake of H1R antihistamines correlated with improved survival in immunotherapy-treated patients. Kaplan-Meier overall survival analysis of TNBC patients with infiltration of cytotoxic T lymphocytes showed that a high H1R expression in tumors was significantly associated with a poor survival. Increased histamine levels were further observed in blood and tumor tissue of tumor-bearing mice compared with normal counterparts. The levels of histamine in blood correlated with a lower density of tumor-infiltrating granzyme B+ cells (cytotoxic CD8+ T cells or NK cells). These effects were further demonstrated to be mediated by the histamine-induced H1R activation on tumor-associated macrophages that lead to an increased expression of the immune checkpoint VISTA and further polarization toward a M2-like phenotype. As a consequence, CD8+ T cell function is suppressed, promoting tumor growth, and leading to immunotherapy resistance. Moreover, the authors suggest that allergic reactions could decrease the effectiveness of immunotherapy because histamine, acting through the H1R in macrophages, can promote immune evasion (Li et al. 2022). However, the role of allergy in BC progression is highly controversial and remains inconclusive. Although a history of allergy may be associated with a reduced risk of BC in young women between 35 and 45 years of age, other studies suggested that allergic diseases related to skewed immune system toward Th2 might be associated with an increased risk of BC. Future studies are needed to confirm the relationship between allergic immune responses and BC risk (Hedderson et al. 2003; Sadeghi and Shirkhoda 2019).

Oral ranitidine and famotidine, but not H1R or H4R antagonists, significantly reduced the MDSC population in the spleen and bone marrow of TNBC-bearing mice. Both drugs inhibited lung metastasis in the 4T1 TNBC model, while ranitidine decreased primary tumor growth in the E0771 BC model (Vila-Leahey et al. 2016). In line with these results, oral ranitidine treatment was associated with the development of enhanced antitumor antibody response and a decrease in primary tumor growth in the E0771-GFP BC model. These effects were not observed in B cell-deficient animals (Rogers et al. 2018). A recent study evaluated the impact of ranitidine usage on immune cells in healthy patients. A total of 29 subjects received a daily high dose of ranitidine for 6 weeks (900 mg/day) and mediation analysis were performed at baseline, 3 and 6 weeks of treatment, and 12 weeks after treatment cessation. Ranitidine was well-tolerated, and no drug-related adverse events were observed. Peripheral blood immunophenotyping revealed that the number of neutrophils, basophils, and eosinophils remained unchanged in time. However, ranitidine decreased the number of B cells and modulated the CD25 expression of T lymphocytes and NK cells, which was associated with a decrease in IL-2 levels. These results indicate that H2R blockade profoundly impacts several aspects of acquired immunity that should be considered in potential immunotherapeutic strategies (Meghnem et al. 2021).

Numerous in vivo studies have shown that the H4R plays important roles in inflammation and autoimmune diseases (Thurmond et al. 2017). However, the study of its role in the immunomodulation of BC has just started. Our laboratory reported for the first time the participation of the H4R in antitumor immunity by means of a syngeneic model of BC developed orthotopically with 4T1 cells in H4R knockout (H4R-KO) and wild-type mice. Mice lacking H4R showed reduced tumor size and weight, decreased number of lung metastases, and percentage of tumor-infiltrating T helper (CD4+) cells, whereas they exhibited increased tumor-infiltrating NK cells and B (CD19+) lymphocytes. In addition, tumor-draining lymph nodes (TDLN) of H4R-KO mice showed a decreased proportion of T helper and T regulatory cells (CD4 + CD25 + FoxP3+) and increased percentages of NK cells compared with wild-type mice (Sterle et al. 2019). In the same way, the effect of systemic treatment with histamine and specific H4R pharmacological ligands JNJ7777120 (H4R antagonist) and JNJ28610244 (H4R agonist) was investigated in a TNBC model. Histamine reduced tumor weight and lung metastases more efficiently than the H4R agonist, which could be attributed to the multifaceted action of histamine on different receptors and cell types. Moreover, it may balance antitumor immunity and at the same time act directly through the H4R on 4T1 tumor cells, reducing cell proliferation. The therapeutic and immunomodulatory effects of the H4R agonist depend on the administered dose. The lowest concentration (1 mg/kg) slightly, but significantly, reduced tumor size and increased the percentage of $CD4^+$ T cells in the TDLN. On the other hand, the administration of a higher dose did not significantly alter the tumor growth probably due to the immunosuppressive effect of H4R on the TME. In line with these results, the treatment with the H4R antagonist reduced the proportion of tumor-infiltrating $CD4^+$ T cells and Tregs in the TDLN (Nicoud et al. 2020). To further highlight the immunosuppressive role of H4R, a similar reduction of Tregs infiltration was demonstrated in a model of HER2+ BC developed with LM3 cells in H4R-KO mice compared with wild-type animals (Nicoud et al. 2021).

Mast cells are the major sources of histamine in healthy tissues, which is stored in specific cytosolic granules and released in large quantities upon degranulation (Mitsuhashi et al. 1989). Histamine regulates their functions through the H1R, H2R, and H4R (Stark 2013). Their role in cancer is controversial, since mast cells tumor infiltration could be associated either with a good or a poor prognosis depending on the cancer type, tissue localization, and the ability of mast cells to interact with TME (Hanes et al. 2021; Oldford and Marshall 2015; Rigoni et al. 2018). In invasive BC, stromal mast cell infiltration is an independent good prognostic marker and correlates with long-term follow-up, particularly in the node-negative subset, supporting the key role for host mast cells and local inflammatory responses in BC progression (Dabiri et al. 2004; Rajput et al. 2008).

Presented evidence shows that histamine produces a complex regulation of the phenotype and functions of different immune cells, depending on the receptor subtype that is activated and its associated signaling pathway. Immunotherapy is revolutionizing the management of BC, and therefore, it is imperative to understand how histamine regulates immune cells within the TME to improve antitumor immunity and patient prognosis.

The differential expression of histamine receptors in the different subpopulations of immune cells, in other cells of the TME, and in tumor cells may be responsible for the dual effect of histamine in the modulation of the immune response and in the regulation of tumor cell proliferation.

Last but not least, an increasing area of research is the modulation of human microbiome function to enhance cancer immune response. It was reported that different types of tumors comprise a wide variety of microbiomes. The BC microbiome in particular has shown to be strikingly rich and diverse. Microbiome impacts both the development and the progression of cancer as well as the patient responses to cancer treatments, affecting the response to immune checkpoint inhibitors (Mou et al. 2021; Nejman et al. 2020; Spakowicz et al. 2020; Poore et al. 2020). Numerous human cells produce histamine, including immune and tumor ones, but they are not the only sources of histamine in the human body. Histamine secreted by the intestinal microbiota also influences the immune response. A multitude of bacteria can secrete and sense it to either get information about the surrounding microenvironment or self-communicate with other bacteria. Bacteria-produced histamine, as well as endogenous histamine, affects hosts' physiological activities in several ways. For instance, many histamine-secreting bacteria are found in the human gut and are thought to influence immune response through the H1R and H2R (Ferstl et al. 2014; Gao et al. 2015; Shi et al. 2019; Smolinska et al. 2014). Moreover, bacterial HDC gene levels were measured in fecal samples from asthmatic and non-asthmatic patients, and it was found that in the former group they were significantly higher compared with the healthy control group. Although the reason is still unknown, the bacterial HDC gene levels were higher in obese people with asthma than nonobese asthmatic patients (Barcik et al. 2016). Other studies on intestinal microbiota show that histamine-secreting bacteria can suppress chemokine and cytokine secretion in monocytes. In line with these findings, the administration of *Lactobacillus rhamnosus* can shape the inflammatory response by modifying the release of interleukins as well as tumor necrosis factor α, leading to the suppression of immune system activity (Frei et al. 2013). Furthermore, the immunomodulatory role of bacteria-secreted histamine not only affects local but also distant mucosal sites such as the lungs via the H2R (Barcik et al. 2019). Interestingly, Gao et al. reported that the administration of *Lactobacillus reuteri*, a histamine-producing member of the gut microbiota, reduced the recruitment of MDSCs and the production of inflammatory cytokines, preventing chemical-induced colon carcinogenesis in HDC-deficient mice (Branco et al. 2018). The role of histamine-producing bacteria in BC progression and response to therapeutics is still unknown and deserves further investigation.

## 5 Conclusions and the Future Landscape

BC is a heterogeneous disease, showing different histological types, molecular profile, and variable clinical outcomes associated with limited response and severe toxicity of the available oncological treatments. Despite outstanding advances in early detection and improved targeted therapies, BC remains the deadliest cancer for women worldwide. Therefore, there is an unmet need for new effective therapeutic approaches for BC.

Immunotherapy is now a mainstay of cancer treatment. The success in targeting immunologic checkpoints, including the PD-1/PD-L1 blockade in metastatic BC among other solid tumors, has renewed the interest in immunotherapies and combinatorial strategies to improve clinical benefits. However, a large proportion of patients do not benefit from immunotherapy, in part due to compensatory regulation of the other immune checkpoints that are activated to overcome this treatment, reflecting the complexity and redundancy of the immune system.

In the present review, the most recent research on the pathophysiological role of histamine and histamine receptors in BC immunobiology was addressed. We have already discussed the complex interactions of histamine with its receptors expressed on immune and cancer cells, which have implications in the antitumor immunity response and therapeutic approaches.

BC cells, immune cells, and other important cells within the TME, including fibroblasts, synthesize histamine and respond to this biogenic amine through the activation of different histamine receptor subtypes present in cell surface. However, contradictory roles of histamine in regulating the phenotype and functions of different immune cells and cancer cells have been reported, not only in vitro but also in vivo and in human clinical studies. Overwhelming evidence supports the key role of histaminergic system in BC development and metastasis and in the response to cancer therapies.

Many academic groups have reported several protumoral or antitumoral effects mediated by histamine in BC. In this regard, it is important to highlight that not only immune cells but also cancer cells can produce and respond to histamine in a paracrine or autocrine manner, which denotes the complexity of the histamine/histamine receptor axis modulation of the antitumor immunity. Differences in the local levels of histamine, the composition of TME, or histamine receptor subtypes present in tumor cells and in immune cells could ultimately determine the biological effects of histamine and pharmacological agents targeting histamine receptors. Therefore, these facts help to understand the controversial studies in cancer research.

The expression of histamine receptor subtypes, including H1R and H4R, was associated with clinicopathological characteristics in different BC subtypes, reinforcing the role of the histaminergic system in carcinogenesis. It is interesting to point out that at least in TNBC, the H3R-mediated effects on proliferation seem to be counterbalanced by the H4R, both histamine receptor subtypes with higher affinity for histamine compared with the H1R and H2R. In this line, analyses of TCGA show that H4R mRNA expression was lower while H3R mRNA expression was higher in the aggressive basal-like/TNBC tumors compared with the more favorable clinical outcome luminal A BC. Therefore, a direct effect of histamine through the activation of histamine receptors in cancer cells together with the histamine immune-mediated effects should be taken into consideration in order to predict tumor fate.

Histamine was the first inflammatory biogenic amine to be characterized and produces a complex regulation of the phenotype and functions of the different immune cells, depending on the activated receptor subtype and its associated signaling pathway. Experimental and clinical findings show that histamine is capable of shaping the immune TME of BC, exerting dual effects. On the one hand, some studies support the protumorigenic effects of histamine through enhancing tumor immune escape. A recent study suggests that the histamine-H1R axis could be a potential biomarker of T cell dysfunction and immunotherapy response as well as a promising therapeutic target for enhancing immune checkpoint inhibitors response. In addition, a novel immunosuppressive role of H4R was demonstrated in BC, suggesting the potential therapeutic value of H4R ligands.

On the other hand, some reports demonstrated potent antitumorigenic properties of histamine, shaping innate and adaptive immune responses to control tumor growth. In this regard, the histamine potential as an adjuvant to cancer immunotherapy has been investigated for more than a decade. For instance, clinical data support the use of histamine in combination with IL-2 in the maintenance therapy for patients with acute myeloid leukemia in remission. In addition, preclinical data show that histamine is able to enhance conventional antitumor therapies in different cancer types, supporting the rationale for the use of combination therapy with histamine in clinical settings.

Checkpoint inhibitor immunotherapy has the potential to generate tumor-specific and durable responses, revolutionizing the cancer treatment paradigms. However, improvements in terms of rational immunotherapy combinations should be made to overcome reduced clinical responses and the lack of predictive biomarkers. In this scenario, the histaminergic system provides a promising strategy for the potential therapeutic exploitation of new immunomodulatory drug targets. We hypothesize that targeting histamine receptors could produce non-redundant and complementary antitumor effects through indirect immunoregulatory activity and direct antiproliferative actions via their expression in cells within TME including immune and tumor cells.

Future double-blinded, randomized placebo-controlled prospective clinical studies are necessary for validating the use of H1R antihistamines or other histamine receptor ligands as adjuvant therapy for enhancing immunotherapy responses in BC patients, especially in hard-to-treat TNBC. Identifying new uses for histamine or approved histamine receptor antagonists will contribute to drug repurposing.

Further studies should be carried out in order to understand the role of histamine in the dynamic relationship between the microbiome and the immune system in the context of BC, which could have potential impact on the response to BC therapies.

**Nomenclature of Targets and Ligands**

Key protein targets and ligands in this article are hyperlinked to corresponding entries in https://www.guidetopharmacology.org/, the common portal for data from the IUPHAR/BPS Guide to PHARMACOLOGY (Harding et al. 2018), and are permanently archived in the Concise Guide to PHARMACOLOGY 2019/2020 (Alexander et al. 2019).

**Acknowledgements** The results shown here are in whole or part based upon data generated by the TCGA Research Network: https://www.cancer.gov/tcga

## References

Alexander SP, Christopoulos A, Davenport AP, Kelly E, Marrion NV, Peters JA et al (2017) The concise guide to pharmacology 2017/18: G protein-coupled receptors. Br J Pharmacol 174 (Suppl 1):S17–s129

Alexander SPH, Christopoulos A, Davenport AP, Kelly E, Mathie A, Peters JA et al (2019) The concise guide to pharmacology 2019/20: G protein-coupled receptors. Br J Pharmacol 176 (Suppl 1):S21–s141

Alsaab HO, Sau S, Alzhrani R, Tatiparti K, Bhise K, Kashaw SK et al (2017) PD-1 and PD-L1 checkpoint signaling inhibition for cancer immunotherapy: mechanism, combinations, and clinical outcome. Front Pharmacol 8:561

Amini A, Heidari K, Kariman H, Taghizadeh M, Hatamabadi H, Shahrami A et al (2015) Histamine antagonists for treatment of peripheral vertigo: a meta-analysis. J Int Adv Otol 11(2):138–142

Barcik W, Pugin B, Westermann P, Perez NR, Ferstl R, Wawrzyniak M et al (2016) Histamine-secreting microbes are increased in the gut of adult asthma patients. J Allergy Clin Immunol 138(5):1491–4.e7

Barcik W, Pugin B, Brescó MS, Westermann P, Rinaldi A, Groeger D et al (2019) Bacterial secretion of histamine within the gut influences immune responses within the lung. Allergy 74(5):899–909

Bäumer W, Rossbach K (2010) Histamine as an immunomodulator. J Ger Soc Dermatol 8(7): 495–504

Bens A, Dehlendorff C, Friis S, Cronin-Fenton D, Jensen MB, Ejlertsen B et al (2020) The role of H1 antihistamines in contralateral breast cancer: a Danish nationwide cohort study. Br J Cancer 122(7):1102–1108

Best CH (1929) The disappearance of histamine from autolysing lung tissue. J Physiol 67(3): 256–263

Bowrey PF, King J, Magarey C, Schwartz P, Marr P, Bolton E et al (2000) Histamine, mast cells and tumour cell proliferation in breast cancer: does preoperative cimetidine administration have an effect? Br J Cancer 82(1):167–170

Branco A, Yoshikawa FSY, Pietrobon AJ, Sato MN (2018) Role of histamine in modulating the immune response and inflammation. Mediat Inflamm 2018:9524075

Bristow MR, Cubicciotti R, Ginsburg R, Stinson EB, Johnson C (1982) Histamine-mediated adenylate cyclase stimulation in human myocardium. Mol Pharmacol 21(3):671–679

Brouckaert O, Wildiers H, Floris G, Neven P (2012) Update on triple-negative breast cancer: prognosis and management strategies. Int J Women's Health 4:511–520

Cai WK, Zhang JB, Wang NM, Wang YL, Zhao CH, Lu J et al (2015) Lack of association between rs2067474 polymorphism in histamine receptor H2 gene and Breast cancer in Chinese Han population. TheScientificWorldJournal 2015:545292

Carabajal E, Massari N, Croci M, Martinel Lamas DJ, Prestifilippo JP, Bergoc RM et al (2012) Radioprotective potential of histamine on rat small intestine and uterus. Eur J Histochem 56(4): e48

Cerami E, Gao J, Dogrusoz U, Gross BE, Sumer SO, Aksoy BA et al (2012) The cBio cancer genomics portal: an open platform for exploring multidimensional cancer genomics data. Cancer Discov 2(5):401–404

Chae YK, Arya A, Iams W, Cruz MR, Chandra S, Choi J et al (2018) Current landscape and future of dual anti-CTLA4 and PD-1/PD-L1 blockade immunotherapy in cancer; lessons learned from clinical trials with melanoma and non-small cell lung cancer (NSCLC). J Immunother Cancer 6(1):39

Chaudhry GE, Jan R, Akim A, Zafar MN, Sung YY, Muhammad TST (2021) Breast cancer: a global concern, diagnostic and therapeutic perspectives, mechanistic targets in drug development. Adv Pharm Bull 11(4):580–594

Cleator S, Heller W, Coombes RC (2007) Triple-negative breast cancer: therapeutic options. Lancet Oncol 8(3):235–244

Coogan PF, Zhang Y, Palmer JR, Strom BL, Rosenberg L (2005) Cimetidine and other histamine2-receptor antagonist use in relation to risk of breast cancer. Cancer Epidemiol Biomark Prev 14(4):1012–1015

Cricco GP, Davio CA, Martin G, Engel N, Fitzsimons CP, Bergoc RM et al (1994) Histamine as an autocrine growth factor in experimental mammary carcinomas. Agents Actions 43(1–2):17–20

Cui J, Xu G, Liu J, Pang Z, Florholmen J, Cui G (2013) The expression of non-mast histamine in tumor associated microvessels in human colorectal cancers. Pathol Oncol Res 19(2):311–316

Dabiri S, Huntsman D, Makretsov N, Cheang M, Gilks B, Bajdik C et al (2004) The presence of stromal mast cells identifies a subset of invasive breast cancers with a favorable prognosis. Mod Pathol 17(6):690–695

Dale HH, Laidlaw PP (1910) The physiological action of beta-iminazolylethylamine. J Physiol 41(5):318–344

Darvas Z, Falus A (2004) SpringMed maB. Histidine decarboxylase (HDC) enzyme and gene. 2004:31–42

Darvas Z, Sakurai E, Schwelberger HG, Hegyesi H, Rivera E, Othsu H et al (2003) Autonomous histamine metabolism in human melanoma cells. Melanoma Res 13(3):239–246

Dauvilliers Y, Bassetti C, Lammers GJ, Arnulf I, Mayer G, Rodenbeck A et al (2013) Pitolisant versus placebo or modafinil in patients with narcolepsy: a double-blind, randomised trial. Lancet Neurol 12(11):1068–1075

Davio C, Cricco G, Andrade N, Bergoc R, Rivera EJA (1993) H1 and H2 histamine receptors in human mammary carcinomas. Agents Actions 38(3):C172–C1C4

Davio CA, Cricco GP, Martin G, Fitzsimons CP, Bergoc RM, Rivera ES (1994) Effect of histamine on growth and differentiation of the rat mammary gland. Agents Actions 41 Spec No:C115–7

Davio CA, Cricco GP, Bergoc RM, Rivera ES (1995) H1 and H2 histamine receptors in N-nitroso-N-methylurea (NMU)-induced carcinomas with atypical coupling to signal transducers. Biochem Pharmacol 50(1):91–96

DeSantis CE, Ma J, Gaudet MM, Newman LA, Miller KD, Goding Sauer A et al (2019) Breast cancer statistics, 2019. CA Cancer J Clin 69(6):438–451

Dieci MV, Miglietta F, Guarneri V (2021) Immune infiltrates in breast cancer: recent updates and clinical implications. Cell 10(2)

Dimitriadou V, Rouleau A, Dam Trung Tuong M, Newlands GJ, Miller HR, Luffau G et al (1994) Functional relationship between mast cells and C-sensitive nerve fibres evidenced by histamine H3-receptor modulation in rat lung and spleen. Clin Sci (Lond) 87(2):151–163

Duggan C, Trapani D, Ilbawi AM, Fidarova E, Laversanne M, Curigliano G et al (2021) National health system characteristics, breast cancer stage at diagnosis, and breast cancer mortality: a population-based analysis. Lancet Oncol 22(11):1632–1642

Dy M, Schneider E (2004) Histamine-cytokine connection in immunity and hematopoiesis. Cytokine Growth Factor Rev 15(5):393–410

Emens LA (2018) Breast cancer immunotherapy: facts and hopes. Clin Cancer Res 24(3):511–520

Engel K, Zhou M, Wang J (2004) Identification and characterization of a novel monoamine transporter in the human brain. J Biol Chem 279(48):50042–50049

Falus A, Pós Z, Darvas Z (2010) Histamine in normal and malignant cell proliferation. Adv Exp Med Biol 709:109–123

Fernández-Nogueira P, Bragado P, Almendro V, Ametller E, Rios J, Choudhury S et al (2016) Differential expression of neurogenes among breast cancer subtypes identifies high risk patients. Oncotarget 7(5):5313–5326

Fernández-Nogueira P, Noguera-Castells A, Fuster G, Recalde-Percaz L, Moragas N, López-Plana A et al (2018) Histamine receptor 1 inhibition enhances antitumor therapeutic responses through extracellular signal-regulated kinase (ERK) activation in breast cancer. Cancer Lett 424:70–83

Ferstl R, Frei R, Schiavi E, Konieczna P, Barcik W, Ziegler M et al (2014) Histamine receptor 2 is a key influence in immune responses to intestinal histamine-secreting microbes. J Allergy Clin Immunol 134(3):744–6.e3

Frei R, Ferstl R, Konieczna P, Ziegler M, Simon T, Rugeles TM et al (2013) Histamine receptor 2 modifies dendritic cell responses to microbial ligands. J Allergy Clin Immunol 132(1): 194–204

Fritz I, Wagner P, Broberg P, Einefors R, Olsson H (2020) Desloratadine and loratadine stand out among common H(1)-antihistamines for association with improved breast cancer survival. Acta Oncol 59(9):1103–1109

Fritz I, Wagner P, Olsson H (2021) Improved survival in several cancers with use of H(1)-antihistamines desloratadine and loratadine. Transl Oncol 14(4):101029

Galarza TE, Mohamad NA, Táquez Delgado MA, Vedoya GM, Crescenti EJ, Bergoc RM et al (2016) Histamine prevents radiation-induced mesenchymal changes in breast cancer cells. Pharmacol Res 111:731–739

Galarza TE, Táquez Delgado MA, Mohamad NA, Martín GA, Cricco GP (2020) Histamine H4 receptor agonists induce epithelial-mesenchymal transition events and enhance mammosphere formation via Src and TGF-β signaling in breast cancer cells. Biochem Pharmacol 180:114177

Gao J, Aksoy BA, Dogrusoz U, Dresdner G, Gross B, Sumer SO et al (2013) Integrative analysis of complex cancer genomics and clinical profiles using the cBioPortal. Sci Signal 6(269):pl1

Gao C, Major A, Rendon D, Lugo M, Jackson V, Shi Z et al (2015) Histamine H2 receptor-mediated suppression of intestinal inflammation by probiotic lactobacillus reuteri. MBio 6(6): e01358–e01315

Gao C, Ganesh BP, Shi Z, Shah RR, Fultz R, Major A et al (2017) Gut microbe-mediated suppression of inflammation-associated colon carcinogenesis by luminal histamine production. Am J Pathol 187(10):2323–2336

García-Quiroz J, García-Becerra R, Santos-Martínez N, Barrera D, Ordaz-Rosado D, Avila E et al (2014) In vivo dual targeting of the oncogenic Ether-à-go-go-1 potassium channel by calcitriol and astemizole results in enhanced antineoplastic effects in breast tumors. BMC Cancer 14:745

García-Quiroz J, González-González ME, Díaz L, Ordaz-Rosado D, Segovia-Mendoza M, Prado-García H et al (2019) Astemizole, an inhibitor of Ether-À-Go-Go-1 potassium channel, increases the activity of the tyrosine kinase inhibitor gefitinib in breast cancer cells. Rev Invest Clin 71(3):186–194

Ghandi M, Huang FW, Jané-Valbuena J, Kryukov GV, Lo CC, McDonald ER 3rd et al (2019) Next-generation characterization of the cancer cell line encyclopedia. Nature 569(7757): 503–508

Gong J, Chehrazi-Raffle A, Reddi S, Salgia R (2018) Development of PD-1 and PD-L1 inhibitors as a form of cancer immunotherapy: a comprehensive review of registration trials and future considerations. J Immunother Cancer 6(1):8

Guevarra JT, Hiensch R, Varga AW, Rapoport DM (2020) Pitolisant to treat excessive daytime sleepiness and cataplexy in adults with narcolepsy: rationale and clinical utility. Nat Sci Sleep 12:709–719

Hanes MR, Giacomantonio CA, Marshall JS (2021) Mast cells and skin and breast cancers: a complicated and microenvironment-dependent role. Cell 10(5):986

Harbeck N, Penault-Llorca F, Cortes J, Gnant M, Houssami N, Poortmans P et al (2019) Breast cancer. Nat Rev Dis Primers 5(1):66

Harding SD, Sharman JL, Faccenda E, Southan C, Pawson AJ, Ireland S et al (2018) The IUPHAR/BPS guide to pharmacology in 2018: updates and expansion to encompass the new guide to immunopharmacology. Nucleic Acids Res 46(D1):D1091–Dd106

He GH, Lu J, Shi PP, Xia W, Yin SJ, Jin TB et al (2013) Polymorphisms of human histamine receptor H4 gene are associated with breast cancer in Chinese Han population. Gene 519(2): 260–265

He GH, Lin JJ, Cai WK, Xu WM, Yu ZP, Yin SJ et al (2014) Associations of polymorphisms in histidine decarboxylase, histamine N-methyltransferase and histamine receptor H3 genes with breast cancer. PLoS One 9(5):e97728

Hedderson MM, Malone KE, Daling JR, White E (2003) Allergy and risk of breast cancer among young women (United States). Cancer Causes Control 14(7):619–626

Hegyesi H, Colombo L, Pállinger E, Tóth S, Boer K, Molnár V et al (2007) Impact of systemic histamine deficiency on the crosstalk between mammary adenocarcinoma and T cells. J Pharmacol Sci 105(1):66–73

Hill SJ, Ganellin CR, Timmerman H, Schwartz JC, Shankley NP, Young JM et al (1997) International Union of Pharmacology. XIII. Classification of histamine receptors. Pharmacol Rev 49(3):253–278

Hoadley KA, Yau C, Hinoue T, Wolf DM, Lazar AJ, Drill E et al (2018) Cell-of-origin patterns dominate the molecular classification of 10,000 tumors from 33 types of cancer. Cell 173(2): 291–304.e6

Huang H, Li Y, Liang J, Finkelman FD (2018) Molecular regulation of histamine synthesis. Front Immunol 9:1392

Jakhar R, Paul S, Bhardwaj M, Kang SC (2016) Astemizole-histamine induces Beclin-1-independent autophagy by targeting p53-dependent crosstalk between autophagy and apoptosis. Cancer Lett 372(1):89–100

Jutel M, Akdis M, Akdis CA (2009) Histamine, histamine receptors and their role in immune pathology. Clin Exp Allergy 39(12):1786–1800

Kang J, Zhang C, Zhong WZ (2021) Neoadjuvant immunotherapy for non-small cell lung cancer: state of the art. Cancer Commun (Lond) 41(4):287–302
Kantor I, Jurkiewicz D (2006) Assessment of betahistine dihydrochloride effectiveness in the treatment of vertigo of a different aetiology based on videonystagmography test results. Pol Merkur Lekarski 20(117):318–321
Kasteleijn-Nolst Trenité D, Parain D, Genton P, Masnou P, Schwartz JC, Hirsch E (2013) Efficacy of the histamine 3 receptor (H3R) antagonist pitolisant (formerly known as tiprolisant; BF2.649) in epilepsy: dose-dependent effects in the human photosensitivity model. Epilepsy Behav 28(1): 66–70
Kollmeier AP, Barnathan ES, O'Brien C, Chen B, Xia YK, Zhou B et al (2018) A phase 2a study of toreforant, a histamine H(4) receptor antagonist, in eosinophilic asthma. Ann Allergy Asthma Immunol 121(5):568–574
LaFleur MW, Muroyama Y, Drake CG, Sharpe AH (2018) Inhibitors of the PD-1 pathway in tumor therapy. J Immunol 200(2):375–383
Lamas DJM, Nicoud MB, Sterle HA, Cricco GP, Martín GA, Cremaschi GA et al (eds) (2017) Methodological approaches to investigate the effects of histamine receptor targeting compounds in preclinical models of breast cancer
Lemos B, Davio C, Gass H, Gonzalez P, Cricco G, Martin G et al (1995) Histamine receptors in human mammary gland, different benign lesions and mammary carcinomas. Inflamm Res 44 (Suppl 1):S68–S69
Li Q, Liu KY, Liu Q, Wang G, Jiang W, Meng Q et al (2020) Antihistamine drug Ebastine inhibits cancer growth by targeting Polycomb group protein EZH2. Mol Cancer Ther 19(10):2023–2033
Li H, Xiao Y, Li Q, Yao J, Yuan X, Zhang Y et al (2022) The allergy mediator histamine confers resistance to immunotherapy in cancer patients via activation of the macrophage histamine receptor H1. Cancer Cell 40(1):36–52.e9
Lieberman P (2011) The basics of histamine biology. Ann Allergy Asthma Immunol 106(2 Suppl): S2–S5
Marks RM, Roche WR, Czerniecki M, Penny R, Nelson DS (1986) Mast cell granules cause proliferation of human microvascular endothelial cells. Lab Investig 55(3):289–294
Maršavelski A, Vianello R (2017) What a difference a methyl group makes: the selectivity of monoamine oxidase B towards histamine and N-Methylhistamine. Chemistry 23(12): 2915–2925
Martinel Lamas DJ, Croci M, Carabajal E, Crescenti EJ, Sambuco L, Massari NA et al (2013) Therapeutic potential of histamine $H_4$ receptor agonists in triple-negative human breast cancer experimental model. Br J Pharmacol 170(1):188–199
Martinel Lamas DJ, Rivera ES, Medina VA (2015a) Histamine H4 receptor: insights into a potential therapeutic target in breast cancer. Front Biosci (Schol Ed) 7:1–9
Martinel Lamas DJ, Cortina JE, Ventura C, Sterle HA, Valli E, Balestrasse KB et al (2015b) Enhancement of ionizing radiation response by histamine in vitro and in vivo in human breast cancer. Cancer Biol Ther 16(1):137–148
Martinel Lamas DJ, Nicoud MB, Sterle HA, Carabajal E, Tesan F, Perazzo JC et al (2015c) Selective cytoprotective effect of histamine on doxorubicin-induced hepatic and cardiac toxicity in animal models. Cell Death Discov 1:15059
Martinel Lamas DJ, Nicoud MB, Sterle HA, Cremaschi GA, Medina VA (2015d) Histamine: a potential cytoprotective agent to improve cancer therapy? Cell Death Dis 6(12):e2029
Martner A, Thorén FB, Aurelius J, Hellstrand K (2013) Immunotherapeutic strategies for relapse control in acute myeloid leukemia. Blood Rev 27(5):209–216
Massari NA, Nicoud MB, Medina VA (2020) Histamine receptors and cancer pharmacology: an update. Br J Pharmacol 177(3):516–538
Mathes RW, Malone KE, Daling JR, Porter PL, Li CI (2008) Relationship between histamine2-receptor antagonist medications and risk of invasive breast cancer. Cancer Epidemiol Biomark Prev 17(1):67–72

Medina VA (2013) Chapter 8 histamine in cancer. In: Holger S (ed) Histamine H4 receptor. De Gruyter Open, pp 259–308

Medina VA, Rivera ES (2010) Histamine receptors and cancer pharmacology. Br J Pharmacol 161(4):755–767

Medina V, Cricco G, Nuñez M, Martín G, Mohamad N, Correa-Fiz F et al (2006) Histamine-mediated signaling processes in human malignant mammary cells. Cancer Biol Ther 5(11): 1462–1471

Medina VA, Croci M, Mohamad NA, Massari N, Garbarino G, Cricco GP et al (2007) Mechanisms underlying the radioprotective effect of histamine on small intestine. Int J Radiat Biol 83(10): 653–663

Medina V, Croci M, Crescenti E, Mohamad N, Sanchez-Jiménez F, Massari N et al (2008) The role of histamine in human mammary carcinogenesis: H3 and H4 receptors as potential therapeutic targets for breast cancer treatment. Cancer Biol Ther 7(1):28–35

Medina VA, Croci M, Carabajal E, Bergoc RM, Rivera ES (2010) Histamine protects bone marrow against cellular damage induced by ionising radiation. Int J Radiat Biol 86(4):283–290

Medina VA, Brenzoni PG, Lamas DJ, Massari N, Mondillo C, Nunez MA et al (2011a) Role of histamine H4 receptor in breast cancer cell proliferation. Front Biosci (Elite Ed) 3:1042–1060

Medina VA, Prestifilippo JP, Croci M, Carabajal E, Bergoc RM, Elverdin JC et al (2011b) Histamine prevents functional and morphological alterations of submandibular glands induced by ionising radiation. Int J Radiat Biol 87(3):284–292

Meghnem D, Oldford SA, Haidl ID, Barrett L, Marshall JS (2021) Histamine receptor 2 blockade selectively impacts B and T cells in healthy subjects. Sci Rep 11(1):9405

Meng F, Han Y, Staloch D, Francis T, Stokes A, Francis H (2011) The H4 histamine receptor agonist, clobenpropit, suppresses human cholangiocarcinoma progression by disruption of epithelial mesenchymal transition and tumor metastasis. Hepatology 54(5):1718–1728

Mitsuhashi M, Mitsuhashi T, Payan DG (1989) Multiple signaling pathways of histamine H2 receptors. Identification of an H2 receptor-dependent Ca2+ mobilization pathway in human HL-60 promyelocytic leukemia cells. J Biol Chem 264(31):18356–18362

Mou Z, Yang Y, Hall AB, Jiang X (2021) The taxonomic distribution of histamine-secreting bacteria in the human gut microbiome. BMC Genomics 22(1):695

Moya-García AA, Pino-Ángeles A, Sánchez-Jiménez F, Urdiales JL, Medina M (2021) Histamine, metabolic remodelling and angiogenesis: a systems level approach. Biomol Ther 11(3):415

Murata Y, Song M, Kikuchi H, Hisamichi K, Xu XL, Greenspan A et al (2015) Phase 2a, randomized, double-blind, placebo-controlled, multicenter, parallel-group study of a H4 R-antagonist (JNJ-39758979) in Japanese adults with moderate atopic dermatitis. J Dermatol 42(2):129–139

Nadalin V, Cotterchio M, Kreiger N (2003) Antihistamine use and breast cancer risk. Int J Cancer 106(4):566–568

Nejman D, Livyatan I, Fuks G, Gavert N, Zwang Y, Geller LT et al (2020) The human tumor microbiome is composed of tumor type-specific intracellular bacteria. Science (New York, NY) 368(6494):973–980

Neumann J, Kirchhefer U, Dhein S, Hofmann B, Gergs U (2021) The roles of cardiovascular H(2)-histamine receptors under Normal and pathophysiological conditions. Front Pharmacol 12: 732842

Nguyen PL, Cho J (2021) Pathophysiological roles of histamine receptors in cancer progression: implications and perspectives as potential molecular targets. Biomol Ther 11(8):1232

Nicoud MB, Formoso K, Medina VA (2019) Pathophysiological role of histamine H4 receptor in cancer: therapeutic implications. Front Pharmacol 10:556

Nicoud MB, Sterle HA, Massari NA, Táquez Delgado MA, Formoso K, Herrero Ducloux MV et al (2020) Study of the antitumour effects and the modulation of immune response by histamine in breast cancer. Br J Cancer 122(3):348–360

Nicoud MB, Táquez Delgado MA, Sarasola MP, Vidal A, Speisky D, Cremaschi GA et al (2021) Impact of histamine H4 receptor deficiency on the modulation of T cells in a murine breast cancer model. Cancer Immunol Immunother 70(1):233–244

Nilsson MS, Hallner A, Brune M, Nilsson S, Thorén FB, Martner A et al (2020) Immunotherapy with HDC/IL-2 may be clinically efficacious in acute myeloid leukemia of normal karyotype. Hum Vaccin Immunother 16(1):109–111

Ocaña MC, Martínez-Poveda B, Quesada AR, Medina M (2019) Metabolism within the tumor microenvironment and its implication on cancer progression: an ongoing therapeutic target. Med Res Rev 39(1):70–113

O'Donnell JS, Teng MWL, Smyth MJ (2019) Cancer immunoediting and resistance to T cell-based immunotherapy. Nat Rev Clin Oncol 16(3):151–167

Oldford SA, Marshall JS (2015) Mast cells as targets for immunotherapy of solid tumors. Mol Immunol 63(1):113–124

Olsson HL (2018) Use of certain antihistamines among breast cancer patients confers survival benefit. AACR

Olsson HL, Einefors R, Broberg P (2015) Second generation antihistamines after breast cancer diagnosis to improve prognosis both in patients with ER+ and ER-breast cancer. American Society of Clinical Oncology

O'Mahony L, Akdis M, Akdis CA (2011) Regulation of the immune response and inflammation by histamine and histamine receptors. J Allergy Clin Immunol 128(6):1153–1162

Palleschi M, Maltoni R, Sarti S, Melegari E, Bravaccini S, Rocca A (2020) Immunotherapy: the end of the "dark age" for metastatic triple-negative breast cancer? Breast J 26(4):739–742

Panula P, Chazot PL, Cowart M, Gutzmer R, Leurs R, Liu WL et al (2015) International Union of Basic and Clinical Pharmacology. XCVIII. Histamine receptors. Pharmacol Rev 67(3):601–655

Parshad R, Hazrah P, Kumar S, Gupta SD, Ray R, Bal S (2005) Effect of preoperative short course famotidine on TILs and survival in breast cancer. Indian J Cancer 42(4):185–190

Poore GD, Kopylova E, Zhu Q, Carpenter C, Fraraccio S, Wandro S et al (2020) Microbiome analyses of blood and tissues suggest cancer diagnostic approach. Nature 579(7800):567–574

Porretti JC, Mohamad NA, Martín GA, Cricco GP (2014) Fibroblasts induce epithelial to mesenchymal transition in breast tumor cells which is prevented by fibroblasts treatment with histamine in high concentration. Int J Biochem Cell Biol 51:29–38

Prat A, Pineda E, Adamo B, Galván P, Fernández A, Gaba L et al (2015) Clinical implications of the intrinsic molecular subtypes of breast cancer. Breast 24(Suppl 2):S26–S35

Qin L, Zhao D, Xu J, Ren X, Terwilliger EF, Parangi S et al (2013) The vascular permeabilizing factors histamine and serotonin induce angiogenesis through TR3/Nur77 and subsequently truncate it through thrombospondin-1. Blood 121(11):2154–2164

Rajput AB, Turbin DA, Cheang MC, Voduc DK, Leung S, Gelmon KA et al (2008) Stromal mast cells in invasive breast cancer are a marker of favourable prognosis: a study of 4,444 cases. Breast Cancer Res Treat 107(2):249–257

Reddy SM, Carroll E, Nanda R (2020) Atezolizumab for the treatment of breast cancer. Expert Rev Anticancer Ther 20(3):151–158

Rigoni A, Colombo MP, Pucillo C (2018) Mast cells, basophils and eosinophils: from allergy to cancer. Semin Immunol 35:29–34

Rivera E, Davio C, Cricco G, Bergoc RJAITBO (1993) Histamine regulation of tumour growth. Role of H~ 1 and H~ 2 receptors. 89:299

Rivera ES, Cricco GP, Engel NI, Fitzsimons CP, Martín GA, Bergoc RM (2000) Histamine as an autocrine growth factor: an unusual role for a widespread mediator. Semin Cancer Biol 10(1): 15–23

Rogers D, Vila-Leahey A, Pessôa AC, Oldford S, Marignani PA, Marshall JS (2018) Ranitidine inhibition of breast tumor growth is B cell dependent and associated with an enhanced antitumor antibody response. Front Immunol 9:1894

Rouleau A, Stark H, Schunack W, Schwartz JC (2000) Anti-inflammatory and antinociceptive properties of BP 2-94, a histamine H(3)-receptor agonist prodrug. J Pharmacol Exp Ther 295(1): 219–225

Sadeghi F, Shirkhoda M (2019) Allergy-related diseases and risk of breast cancer: the role of skewed immune system on this association. Allergy & Rhinol (Providence) 10: 2152656719860820

Sarasola MP, Táquez Delgado MA, Nicoud MB, Medina VA (2021) Histamine in cancer immunology and immunotherapy. Current status and new perspectives. Pharmacol Res Perspect 9(5): e00778

Schaper-Gerhardt K, Rossbach K, Nikolouli E, Werfel T, Gutzmer R, Mommert S (2020) The role of the histamine H(4) receptor in atopic dermatitis and psoriasis. Br J Pharmacol 177(3): 490–502

Schmid P, Adams S, Rugo HS, Schneeweiss A, Barrios CH, Iwata H et al (2019) Impassion130: updated overall survival (OS) from a global, randomized, double-blind, placebo-controlled, phase III study of atezolizumab (atezo)+ nab-paclitaxel (nP) in previously untreated locally advanced or metastatic triple-negative breast cancer (mTNBC). American Society of Clinical Oncology

Schreiber RD, Old LJ, Smyth MJ (2011) Cancer immunoediting: integrating immunity's roles in cancer suppression and promotion. Science (New York, NY) 331(6024):1565–1570

Schwelberger HG, Sánchez-Jiménez F (eds) (2013) Chapter 3 Histamine metabolism

Seifert R, Strasser A, Schneider EH, Neumann D, Dove S, Buschauer A (2013) Molecular and cellular analysis of human histamine receptor subtypes. Trends Pharmacol Sci 34(1):33–58

Shahid M, Tripathi T, Sobia F, Moin S, Siddiqui M, Khan RA (2009) Histamine, histamine receptors, and their role in immunomodulation: an updated systematic review. Open Immunol J 2:32

Shankaran V, Ikeda H, Bruce AT, White JM, Swanson PE, Old LJ et al (2001) IFNgamma and lymphocytes prevent primary tumour development and shape tumour immunogenicity. Nature 410(6832):1107–1111

Shi Z, Fultz RS, Engevik MA, Gao C, Hall A, Major A et al (2019) Distinct roles of histamine H1- and H2-receptor signaling pathways in inflammation-associated colonic tumorigenesis. Am J Physiol Gastrointest Liver Physiol 316(1):G205–Gg16

Sieja K, Stanosz S, von Mach-Szczypiński J, Olewniczak S, Stanosz M (2005) Concentration of histamine in serum and tissues of the primary ductal breast cancers in women. Breast 14(3): 236–241

Singh M, Jadhav HR (2018) Targeting non-small cell lung cancer with small-molecule EGFR tyrosine kinase inhibitors. Drug Discov Today 23(3):745–753

Smolinska S, Jutel M, Crameri R, O'Mahony L (2014) Histamine and gut mucosal immune regulation. Allergy 69(3):273–281

Sörbo J, Jakobsson A, Norrby K (1994) Mast-cell histamine is angiogenic through receptors for histamine1 and histamine2. Int J Exp Pathol 75(1):43–50

Sotiriou C, Neo SY, McShane LM, Korn EL, Long PM, Jazaeri A et al (2003) Breast cancer classification and prognosis based on gene expression profiles from a population-based study. Proc Natl Acad Sci USA 100(18):10393–10398

Spakowicz D, Hoyd R, Muniak M, Husain M, Bassett JS, Wang L et al (2020) Inferring the role of the microbiome on survival in patients treated with immune checkpoint inhibitors: causal modeling, timing, and classes of concomitant medications. BMC Cancer 20(1):383

Speisky D, Taquez Delgado MA, Lotti A, Nicoud M, Ospital I, Vigovich F et al (2022) Histamine h4 receptor expression in triple negative breast cancer: an exploratory study. J Histochem Cytochem 70(4):311–322

Stanton SE, Adams S, Disis ML (2016) Variation in the incidence and magnitude of tumor-infiltrating lymphocytes in breast cancer subtypes: a systematic review. JAMA Oncol 2(10): 1354–1360

Stark H (2013) De Gruyter Open Poland
Sterle HA, Nicoud MB, Massari NA, Táquez Delgado MA, Herrero Ducloux MV, Cremaschi GA et al (2019) Immunomodulatory role of histamine H4 receptor in breast cancer. Br J Cancer 120(1):128–138
Strasser A, Wittmann HJ, Buschauer A, Schneider EH, Seifert R (2013) Species-dependent activities of G-protein-coupled receptor ligands: lessons from histamine receptor orthologs. Trends Pharmacol Sci 34(1):13–32
Sung H, Ferlay J, Siegel RL, Laversanne M, Soerjomataram I, Jemal A et al (2021) Global cancer statistics 2020: GLOBOCAN estimates of incidence and mortality worldwide for 36 cancers in 185 countries. CA Cancer J Clin 71(3):209–249
Szakacs Z, Dauvilliers Y, Mikhaylov V, Poverennova I, Krylov S, Jankovic S et al (2017) Safety and efficacy of pitolisant on cataplexy in patients with narcolepsy: a randomised, double-blind, placebo-controlled trial. Lancet Neurol 16(3):200–207
Tanaka S, Sakaguchi M, Yoneyama H, Usami Y, Harusawa S (2016) Histamine H(3) receptor antagonist OUP-186 attenuates the proliferation of cultured human breast cancer cell lines. Biochem Biophys Res Commun 480(3):479–485
Thorén FB, Romero AI, Brune M, Hellstrand K (2009) Histamine dihydrochloride and low-dose interleukin-2 as post-consolidation immunotherapy in acute myeloid leukemia. Expert Opin Biol Ther 9(9):1217–1223
Thurmond RL, Chen B, Dunford PJ, Greenspan AJ, Karlsson L, La D et al (2014) Clinical and preclinical characterization of the histamine H(4) receptor antagonist JNJ-39758979. J Pharmacol Exp Ther 349(2):176–184
Thurmond RL, Greenspan A, Radziszewski W, Xu XL, Miao Y, Chen B et al (2016) Toreforant, a histamine H4 receptor antagonist, in patients with active rheumatoid arthritis despite methotrexate therapy: results of 2 phase II studies. J Rheumatol 43(9):1637–1642
Thurmond RL, Venable J, Savall B, La D, Snook S, Dunford PJ et al (2017) Clinical development of histamine H(4) receptor antagonists. In: Handbook of experimental pharmacology, vol 241, pp 301–320
Tiligada E, Ennis M (2020) Histamine pharmacology: from Sir Henry Dale to the 21st century. Br J Pharmacol 177(3):469–489
Tsang JYS, Tse GM (2020) Molecular classification of breast cancer. Adv Anat Pathol 27(1):27–35
Vila-Leahey A, Oldford SA, Marignani PA, Wang J, Haidl ID, Marshall JS (2016) Ranitidine modifies myeloid cell populations and inhibits breast tumor development and spread in mice. Onco Targets Ther 5(7):e1151591
von Mach-Szczypiński J, Stanosz S, Sieja K, Stanosz M (2009) Metabolism of histamine in tissues of primary ductal breast cancer. Metab Clin Exp 58(6):867–870
Wang X, Qi Y, Kong X, Zhai J, Li Y, Song Y et al (2019) Immunological therapy: a novel thriving area for triple-negative breast cancer treatment. Cancer Lett 442:409–428
Wendell SG, Fan H, Zhang C (2020) G protein-coupled receptors in asthma therapy: pharmacology and drug action. Pharmacol Rev 72(1):1–49
Werfel T, Layton G, Yeadon M, Whitlock L, Osterloh I, Jimenez P et al (2019) Efficacy and safety of the histamine H(4) receptor antagonist ZPL-3893787 in patients with atopic dermatitis. J Allergy Clin Immunol 143(5):1830–1837.e4

# Nutrient Transporters: New Molecular Targets for Triple Negative Breast Cancer in Type 2 Diabetics

Cláudia Silva and Fátima Martel

**Abstract**

Breast cancer is the most common malignancy among women worldwide and is also the leading cause of cancer-related death in this gender. Type 2 diabetes mellitus (T2D) is one of the most common chronic diseases worldwide. There is a positive association between breast cancer and T2D, especially evident for triple-negative breast cancer (TNBC). In this chapter, we discuss current knowledge regarding the effect of T2D-associated characteristics upon glucose and glutamine uptake by TNBC cells, and whether alterations in the main transporters of these nutrients (GLUT1 and ASCT2, respectively) can contribute to TNBC initiation, promotion and progression in T2D patients. Insulin, leptin, interferon-γ and an oxidative stress inducer were found to create a microenvironment favouring TNBC development by upregulating GLUT1-mediated glucose and/or ASCT2-mediated glutamine transport. Based on the existing evidences, we conclude that GLUT1 and ASCT2 may constitute a mechanism by which T2D participates in malignancy of TNBC tumours. So, it is possible that these two nutrient transporters can be used in the future as molecular targets for TNBC therapy and as biomarkers for the risk of TNBC development in T2D patients.

**Keywords**

Breast cancer · Glucose uptake · Glutamine uptake · Type 2 diabetes

C. Silva · F. Martel (✉)
Unit of Biochemistry, Department of Biomedicine, Faculty of Medicine, University of Porto, Porto, Portugal

Instituto de Investigação e Inovação em Saúde (i3S), University of Porto, Porto, Portugal
e-mail: fmartel@med.up.pt

 
Interdisciplinary Cancer Research, https://doi.org/10.1007/16833_2022_95
Published online: 16 December 2022

# 1 Introduction

Breast cancer is the most common malignancy among women worldwide and is also the leading cause of cancer-related death in this gender (Sung et al. 2021). About 25% of women with breast cancer are also diagnosed with type 2 diabetes mellitus (T2D) (Larsson et al. 2007; Boyle et al. 2012; Hardefeldt et al. 2012; Samuel et al. 2018) and this association is especially evident for triple-negative breast cancer (TNBC) (Chen et al. 2019). Indeed, studies have specifically linked T2D to the highly aggressive TNBCs, which do not respond to conventional hormonal/human epidermal growth factor receptor 2 (HER2) targeted interventions, have chances of early recurrence, metastasize, tend to be more invasive in nature and develop drug resistance.

The metabolic environment occurring in T2D appears to promote, directly or indirectly, breast tumour initiation, promotion, progression, and/or response to therapy. Actually, metabolic alterations associated with T2D, namely insulin resistance (hyperinsulinemia and hyperglycaemia), altered adipokine levels (higher leptin and lower adiponectin), chronic low-grade inflammation (higher tumour necrosis factor alpha (TNF-$\alpha$), interferon gamma (IFN-$\gamma$), interleukin-6 (IL-6) and interleukin-1 beta (IL-1$\beta$) levels) and increased oxidative stress levels have been found to contribute to the link between T2D and breast cancer (Martin and McGee 2018; Samuel et al. 2018).

It is firmly established that metabolic reprogramming, a recognized hallmark of cancer cells, is fundamental for the development, rapid proliferation, invasion and survival of cancer cells (Hanahan and Weinberg 2011; DeBerardinis and Chandel 2016). A classic example of a reprogrammed metabolic pathway in cancer cells is the "Warburg effect" (Warburg et al. 1927). In addition, many cancer cells also display a reliance on glutamine, a phenomenon known as "glutamine addiction" (Vander Heiden 2011; Kato et al. 2018; Schiliro and Firestein 2021). The rewiring of metabolism in breast cancer cells, which results in higher glucose and glutamine cellular demands, depends on an upregulation of nutrient transporters (Ogrodzinski et al. 2017). In line with this, higher expression of the main glucose and glutamine transporters in cancer cells, namely the glucose facilitative transporter 1 (GLUT1; encoded by SLC2A1) and the alanine, serine, cysteine transporter 2 (ASCT2; encoded by SLC1A5), respectively, has been associated with cancers of higher grade, increased proliferation, malignant potential and poor prognosis (Younes et al. 1995; Pinheiro et al. 2011; Liu et al. 2018; Teixeira et al. 2021). Therefore, these genes have been proposed as oncogenes and glucose and glutamine transporters are currently being investigated as potential therapeutic targets in cancer treatment (Kato et al. 2018; Barbosa and Martel 2020).

In this chapter, we will discuss the effect of the T2D-associated characteristics (hyperglycaemia, hyperinsulinemia, hyperleptinemia, chronic low-grade inflammation and increased levels of oxidative stress) upon glucose and glutamine uptake by TNBC cells, and whether alterations in the main transporters of these nutrients (GLUT1 and ASCT2, respectively) can contribute to the development of TNBC in T2D patients.

## 2 Metabolic Rewiring in Triple-Negative Breast Cancer Cells

TNBC is a heterogeneous group of breast cancers that lack expression of oestrogen receptor (ER), progesterone receptor (PR) and HER2 amplification (Feng et al. 2018; Harbeck et al. 2019). TNBC accounts for 10–20% of all breast cancers and can be classified into six molecularly distinct subtypes: two basal-like (BL1 and BL2), an immunomodulatory (IM), a mesenchymal (M), a mesenchymal stem-like (MSL), and a luminal AR (LAR) subtype (Lehmann et al. 2015). TNBC is associated with high grade, aggressive phenotype and a poor prognosis; it is common in women with BRCA1 gene mutations and, to a less extent, BRCA2 mutations (Harbeck et al. 2019; Sun et al. 2020; Tsang and Tse 2020; Loibl et al. 2021).

In order to sustain the increase in proliferation and rapid growth associated with tumorigenesis, cancer cells consume alternative nutrients and increase the production of macromolecules that contribute to cellular biomass and support generation of new cells (Hanahan and Weinberg 2011). As a result, cancer cells must undergo reprogramming of their metabolic pathways in order to obtain huge amounts of energy and materials (DeBerardinis and Chandel 2016). These altered metabolic processes are recognized as an important hallmark of cancer, which is critical for successful neoplastic transformation and continued tumour cell growth (Hanahan and Weinberg 2011; Pavlova and Thompson 2016). A classic example of a reprogrammed metabolic pathway in cancer cells is the "Warburg effect" or aerobic glycolysis (Warburg et al. 1927). In addition, many cancer cells also display reliance on glutamine, termed "glutamine addiction" (Vander Heiden 2011; Kato et al. 2018; Schiliro and Firestein 2021).

Metabolic reprogramming varies significantly between different breast cancers subtypes, that display distinct features and metabolic alterations (Ogrodzinski et al. 2017). Importantly, in TNBC cells, an increase in aerobic glycolysis and glutamine addiction is observed, in comparison with other breast cancer subtypes. For instance, in luminal breast cancers, the changes in glucose metabolism are inconsistent and a decrease in glutamine catabolism is observed (Robey et al. 2005; van Geldermalsen et al. 2016; Ogrodzinski et al. 2017). So, especially for TNBC tumours, understanding the mechanisms involved in their metabolic rewiring could help to discover new vulnerabilities and therapeutic opportunities.

### 2.1 Glucose Metabolism in Triple-Negative Breast Cancer

Glucose metabolism is considered one of the most important aspects of cancer cell metabolism, because it supplies intermediates and precursors for several other key metabolic pathways, including the generation of amino acids, nucleotides and lipids (Cassim et al. 2020; Schiliro and Firestein 2021), and the generation of lactate to maintain the nicotinamide adenine dinucleotide ($NAD^+$)/reduced nicotinamide adenine dinucleotide (NADH) redox balance (DeBerardinis and Chandel 2016; Cassim et al. 2020).

This process, termed glycolysis, comprises several enzymatic reactions where glucose is converted into pyruvate, being the energy released conserved in the form of adenosine triphosphate (ATP) and NADH (Annibaldi and Widmann 2010; Kalyanaraman 2017). In well-oxygenated tissues, pyruvate enters the mitochondria and is completely oxidized to $CO_2$. Under oxygen deprivation, pyruvate is converted to lactate into cytoplasm (Shiratori et al. 2019). Contrary to most normal cells, that obtain ATP both from glycolysis and oxidative phosphorylation (OXPHOS), most cancer cells mainly obtain ATP from glycolysis, even when there is sufficient oxygen (Warburg et al. 1927; Shiratori et al. 2019) This metabolic change is one of the first recognized biochemical hallmarks of cancer cells (Hanahan and Weinberg 2011; Pavlova and Thompson 2016) and is called the Warburg effect or aerobic glycolysis (Warburg et al. 1927).

Glucose metabolic reprogramming varies significantly between different breast cancers subtypes (Ogrodzinski et al. 2017). TNBC cells generally display the Warburg effect, having elevated glucose uptake and metabolism and increased lactate secretion relative to other breast cancer subtypes (Ogrodzinski et al. 2017). So, TNBC cells present a high metabolic status and intensive glucose dependence, being this a fundamental trait of this breast cancer subtype (Robey et al. 2005; Shin and Koo 2021; Silva et al. 2021a). Consistent with the Warburg effect, TNBC cells present elevated glucose uptake and lactate production (Kim et al. 2013a, b; Ogrodzinski et al. 2017) and high expression of glycolysis-related enzymes, namely hexokinase 2 (HK2), pyruvate kinase 2 (PKM2) and lactate dehydrogenase (LDH). HK2 catalyse the first and the foremost important step of glycolysis (Hay 2016). Upregulation of HK2 expression contributes to aerobic glycolysis, and thus is a critical enzyme in the Warburg effect, favouring rapid proliferation and apoptosis resistance and appears also to induce chemotherapy resistance (Peng et al. 2021). The miR-155 facilitates HK2 expression, leading to an enhanced glycolytic phenotype in TNBC cells (Jiang et al. 2012). PKM2, considered a gatekeeper of pyruvate flux, is responsible for the last committed step of glycolysis, catalysing the conversion of phosphoenolpyruvate to pyruvate (PEP) (Hay 2016). This enzyme plays an important role in inducing chemotherapy resistance in TNBC. In TNBC cell lines and orthotopic breast models, PKM2 confers some resistance to doxorubicin treatment (Wang and Yang 2014). Additionally, it was reported that PKM2 knockdown has anticancer effects against TNBC cells (Ma et al. 2019). In anaerobic conditions, LDH generates lactate from the glycolytic end product pyruvate (Hay 2016). Higher LDH levels and lower oxygen consumption rates have been reported in TNBC compared to luminal breast cancer (Huang et al. 2016). Moreover, LDH overexpression is closely associated with poor prognosis and tumour aggressiveness in patients with TNBC (Dong et al. 2017). The lactate transporters monocarboxylate transporter 1 (MCT1) and MCT4 are also upregulated in TNBC (Shin and Koo 2021) and silencing of lactate transports in TNBC impairs cell proliferation and tumour growth in vitro and in vivo (Morais-Santos et al. 2015).

The high rate of glucose uptake observed in cancer cells is often associated with deregulated and increased expression of glucose carriers, particularly GLUT1

(Barbosa and Martel 2020). High GLUT1 expression is observed in TNBC patient samples compared to other subtypes of breast cancer, facilitating a greater subsequent glucose utilization by this subtype of breast cancer (Choi et al. 2013). Furthermore, GLUT1 upregulation in TNBC correlates with increased cell proliferation, migration and invasion, as well as with a worse prognosis and chemoresistance (Hussein et al. 2011; Oh et al. 2017). Consistent with this, GLUT1 silencing or pharmacological inhibition severely reduces glucose cellular uptake, impairing the growth and proliferation and increasing apoptosis in TNBC cells (Wu et al. 2020; Silva et al. 2021a). The stimulatory effect of GLUT1 upon cell proliferation, migration, invasion and colony formation appears to involve stimulation of the mitogen-activated protein kinase (MAPK) signalling pathway (Oh et al. 2017).

To conclude, TNBC cells are more dependent on glycolysis than the other breast cancer subtypes, displaying a high expression of GLUT1 and a concomitant increased glucose uptake. So, glucose metabolism may represent an alternative therapeutic strategy for reducing or preventing TNBC.

## 2.2 Glutamine Metabolism in Triple-Negative Breast Cancer

Although it is widely accepted that glucose is the dominant energy fuel for most cancers (Zheng 2012), another energy fuel of many cancer cells is glutamine (El Ansari et al. 2018), which is the amino acid most utilized by transformed cells (Schiliro and Firestein 2021). Glutamine has a versatile role in cell metabolism. First, glutamine participates as a source of nitrogen, which is required for the synthesis of nucleotides and non-essential amino acids (Cory and Cory 2006; Cha et al. 2018; El Ansari et al. 2018), which are required for rapidly proliferating cancer cells (Choi and Park 2018). A further role for glutamine is that it provides a carbon source for tricarboxylic acid (TCA) cycle and fatty acid synthesis via α-ketoglutarate production (DeBerardinis et al. 2007; Metallo et al. 2011; Cha et al. 2018; El Ansari et al. 2018), and cancer cells require large amounts of lipids during cell division (Choi and Park 2018). Alternatively, glutamine is involved in glutathione synthesis, which reinforces the antioxidant response and helps to maintain the redox balance, apoptosis and autophagy (Yudkoff et al. 1990; Vaughn and Deshmukh 2008; Cha et al. 2018; El Ansari et al. 2018). Last, glutamine activates the mammalian target of rapamycin (mTOR) pathway by enhancing the influx of amino acids, including leucine, which maintains activation of the mammalian target of rapamycin complex 1 (mTORC1). This, in turn, regulates protein translation and prevents apoptosis in cancer cells (Yanagida et al. 2001; Duran et al. 2012; Cha et al. 2018; El Ansari et al. 2018).

Many cancer cells, including TNBC cells, are highly dependent on glutamine uptake, being thus considered glutamine-sensitive (van Geldermalsen et al. 2016; Silva et al. 2021b). Increased glutaminolysis, the process by which cells convert glutamine into TCA cycle metabolites through the activity of numerous enzymes (Yang et al. 2017), is observed in some breast cancer subtypes (van Geldermalsen et al. 2016, Silva et al. 2021b). Compared to luminal breast cancer cells, TNBC cells present a greater glutamine addiction, increased glutaminolysis and are sensitive to glutamine depletion (van Geldermalsen et al. 2016; Ogrodzinski et al. 2017). Glutaminase, which catalyses the first step of glutaminolysis by converting glutamine into glutamate (Yang et al. 2017; Yoo et al. 2020), is overexpressed in TNBC cells (Gross et al. 2014) and is associated with a higher disease stage and a poor prognosis in this breast cancer subtype (Kim et al. 2017). Interestingly, glutaminase inhibitors reduce cell proliferation in TNBC cell lines and xenografts (Gross et al. 2014) and loss or depletion of glutaminase leads to dysregulated glutaminolysis and a reduction in tumour growth in vitro and in vivo (Kung et al. 2011). The high dependence of TNBC cells on exogenous glutamine may be related to a lower expression level of glutamine synthetase, the enzyme that allows synthesis of glutamine by the cells (Kung et al. 2011).

ASCT2 is recognized as a primary transporter of glutamine (Teixeira et al. 2021), and is known to be upregulated in breast cancer (Kim et al. 2013a, b; van Geldermalsen et al. 2016), but ASCT2 expression levels seem to vary depending on the molecular subtype (van Geldermalsen et al. 2016; Cha et al. 2018). Indeed, high ASCT2 expression is observed in highly proliferative subtypes (TNBC) (Kim et al. 2013a, b, van Geldermalsen et al. 2016) and a lower expression in less aggressive tumours (luminal) (van Geldermalsen et al. 2016). This is in concordance with lower levels of glutamine observed in TNBC tumours compared with luminal A tumours, suggesting an increase in glutamine consumption in TNBC cells (Cao et al. 2014; Kanaan et al. 2014). It is also in agreement with the fact that TNBC cells display significantly increased ASCT2-mediated glutamine uptake, compared to $ER^+$ tumours (van Geldermalsen et al. 2016, Silva et al. 2021b). Importantly, in these cells, pharmacological inhibition of ASCT2 or ASCT2 knockdown significantly reduced glutamine uptake and cell proliferation (van Geldermalsen et al. 2016, Silva et al. 2021b), interfering with mTORC1 signalling (van Geldermalsen et al. 2016).

So, TNBC display high ASCT2 expression and increased glutamine uptake to sustain rapid proliferation, suggesting that a therapeutic strategy against glutamine metabolism can be effective in TNBC.

## 3 Linking Metabolic Alterations That Occur in Type 2 Diabetes to Triple-Negative Breast Cancer

T2D is one of the most common chronic diseases worldwide (Zheng et al. 2018) and is characterized by insulin resistance and glucose intolerance (Samuel et al. 2018). It results from the body's inability to use insulin and is responsible for 90% of patients with diabetes around the world (Samuel et al. 2018; Zheng et al. 2018).

T2D and breast cancer are diagnosed in the same individual more frequently than would be expected (Ferroni et al. 2015; Martin and McGee 2018; Samuel et al. 2018; Talib et al. 2021). This suggests that the metabolic alterations that occur in T2D provide a favourable metabolic environment for the development of breast cancer, mainly TNBC (Martin and McGee 2018). It is known that hyperglycaemia and hyperinsulinemia act as triggers to cancerous phenotypes in T2D patients. Additionally, altered adipokine levels (higher leptin and lower adiponectin levels) also seem to be implicated in cancer pathophysiology in T2D patients. Moreover, the metabolic dysregulation observed in T2D is responsible for a chronic low-grade proinflammatory condition (higher IL-6, IL-1$\beta$, TNF-$\alpha$ and IFN-$\gamma$ levels), leading to increased oxidative stress levels (Martin and McGee 2018; Samuel et al. 2018; Silva et al. 2021a, b, 2022). These metabolic changes appear to interfere with the activity of GLUT1 and ASCT2, which are the main glucose and glutamine transporters in breast cancer cells, respectively, as shown next.

## 3.1 Type 2 Diabetes-Associated Characteristics as GLUT1 and ASCT2 Regulators in Triple-Negative Breast Cancer

Breast cancer in women with T2D is often diagnosed at an advanced stage compared to women without T2D (Bronsveld et al. 2017). Curiously, in T2D women, the most frequently found breast cancer subtype is TNBC (Chen et al. 2019). This is of special interest because glucose (Robey et al. 2005) and glutamine (van Geldermalsen et al. 2016) dependence is particularly evident in TNBC. Interestingly, GLUT1 and ASCT2 transporters are highly expressed in TNBC cells, and their expression levels associate with increased proliferation, higher grade, malignant potential and poor prognosis (Choi et al. 2013; van Geldermalsen et al. 2016; Silva et al. 2021a, b). Therefore, these genes have been proposed as oncogenes (Kato et al. 2018; Barbosa and Martel 2020). So, the dependence of TNBC cells in fuels such as glucose and glutamine may constitute a tumour-specific metabolic vulnerability, and thus, may create a new therapeutic window to eradicate TNBC cells in patients with T2D.

GLUT1 and ASCT2 are regulated through several molecular mechanisms. In TNBC, hypoxia-inducible factor 1-alpha (HIF-1$\alpha$), the c-MYC oncogene, as well as the mutated gene tumour protein 53 (p53) or the retinoblastoma (Rb) have been reported as mechanisms regulating GLUT1 and ASCT2 activity (Cha et al. 2018; Liu et al. 2018; Barbosa and Martel 2020; Shin and Koo 2021; Teixeira et al. 2021). However, other mechanisms are believed to regulate these two transporters. Given the correlation between T2D and TNBC, it is possible that the metabolic changes associated with T2D interfere with the activity of these two transporters. In this context, we will next describe the effects of some T2D-associated characteristics on GLUT1 and ASCT2, and how these alterations may contribute to TNBC initiation, promotion and progression.

### 3.1.1 Glucose

The effects of glucose upon TNBC cells seems to vary depending on the glucose concentration. It seems consensual that a high-glucose condition (25–30mM) significantly increases cell proliferation and migration in basal-like TNBC cell lines (MDA-MB-231) compared to a normal-glucose (5–5.6mM) condition (Gupta and

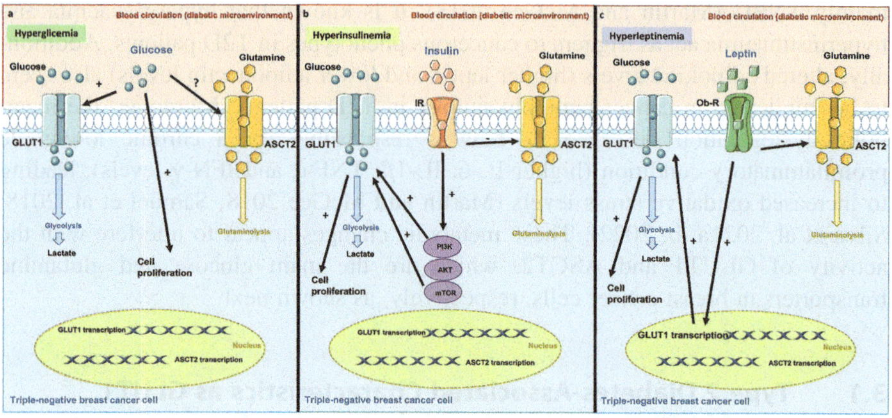


**Fig. 1** Schematic depiction of the effect of hyperglycaemia, hyperinsulinemia and hyperleptinemia on glutamine and glucose uptake and metabolism in triple-negative breast cancer cells. (**a**) Hyperglycaemia stimulates cell proliferation and GLUT1-mediated glucose uptake; the effects of glucose upon glutamine uptake, metabolism and ASCT2 expression are not yet known in TNBC. (**b**) Hyperinsulinemia stimulates cell proliferation, dependent on an increase in GLUT1-mediated glucose uptake, which is mediated by the PI3K and mTOR signalling pathways. Contrariwise, it reduces ASCT2-mediated glutamine uptake. (**c**) Hyperleptinemia causes a pro-proliferative effect in TNBC cells, which is associated with enhanced GLUT1 transcription rates and an increase in GLUT1-mediated glucose transport; on the other hand, it reduces ASCT2-mediated glutamine transport. Abbreviations: *GLUT1* facilitative glucose transporters 1, *ASCT2* alanine, serine, cysteine transporter, *IR* insulin receptor, *Ob-R* leptin receptor, *MAPK* mitogen-activated protein kinase, *PI3K* phosphatidylinositol 3-kinase, *AKT* protein kinase B, *mTOR* mammalian target of rapamycin

Tikoo 2013; Flores-Lopez et al. 2016; Hou et al. 2017; Silva et al. 2021a). Additionally, Venturelli et al. demonstrated that the relative amount of GLUT1 in MDA-MB-231 cells was about six times higher with a high-concentration glucose (25 mM) than with a lower-concentration glucose (5 mM) (Venturelli et al. 2016) (Fig. 1).

It is well established that glutamine provides a survival advantage for TNBC tumour cells, and ASCT2 is highly expressed in this breast cancer subtype (van Geldermalsen et al. 2016, Silva et al. 2021b). Although no studies have yet reported the effect of high glucose levels upon ASCT2 in TNBC cells, a cross-talk between glutamine and glucose metabolism has become evident in either physiological or pathological conditions. For example, in breast cancer cells, receptor peroxisome proliferator activated receptor delta (PPARδ) has been shown to promote tumour progression by activating both ASCT2 and GLUT1 expression (Zhang et al. 2017) and high glucose levels seem to activate PPARδ (Lee et al. 2006).

To conclude, elevated blood glucose levels appear to cause an increase in GLUT1-mediated glucose uptake by TNBC cells, but the effects of glucose upon glutamine metabolism and ASCT2 expression are not yet known in these tumours (Fig. 1).

### 3.1.2 Insulin

The ability of insulin to stimulate glucose uptake in different types of cancers is well accepted. Interestingly, TNBC cells have an elevated insulin receptor (IR) content as they carry an uncommon IR gene amplification (Papa et al. 1997), indicating that insulin may play an important role in the metabolism in this subtype of breast cancer. Although there is some inconsistency in the literature, high insulin levels have been reported to promote proliferation (Gupta and Tikoo 2013; Silva et al. 2021a) migration and angiogenesis (Silva et al. 2021a) of MDA-MB-231 cells. In addition, in MDA-MB-231 cells, 50 nM insulin increased the ability of these cells to incorporate glucose as well as to produce lactate (Silva et al. 2021a). So, insulin seems to promote the Warburg effect in TNBC cells. In the same study, by using a selective GLUT1 inhibitor (BAY-876), it was also shown that insulin stimulates GLUT1-mediated uptake, and that insulin acts post-transcriptionally, as it did not interfere with GLUT1 transcription rates. Finally, it was also concluded that the stimulatory effect of insulin upon cell proliferation and viability is dependent on GLUT1 stimulation (Silva et al. 2021a). These effects may be partly dependent on insulin-stimulation of phosphatidylinositol 3-kinase (PI3K) and mTOR signalling pathway (Silva et al. 2022) (Fig. 1).

Regarding glutamine metabolism, high insulin concentrations appear to decrease ASCT2-mediated glutamine uptake in a basal-like TNBC cell line (MDA-MB-231) (Silva et al. 2021b) (Fig. 1).

Overall, these observations suggest that insulin promotes TNBC cell proliferation by stimulating, via PI3K and mTOR intracellular pathways, GLUT1-dependent glucose uptake.

### 3.1.3 Leptin

Most reports indicate that high leptin serum levels are associated with advanced stage breast cancer (Yi et al. 2018; Ando et al. 2019; Sanchez-Jimenez et al. 2019). Some studies comparing the influence of leptin in different breast cancer subtypes found significant differences in leptin levels among breast cancer patients, with levels of leptin in TN > $HER2^+$ > $ER^+$ patients (Yuan et al. 2017; Kong et al. 2019). In a basal-like TNBC cell line (MDA-MB-231 cells), leptin was recently found to present pro-proliferative, pro-migratory and pro-angiogenic effects and to promote glucose uptake, associated with an increase of GLUT1 expression (Silva et al. 2021a).

In relation to glutamine metabolism, leptin seems to decrease ASCT2-mediate glutamine uptake in MDA-MB-231 cells (Silva et al. 2021b).

These observations thus show that the pro-proliferative, pro-migratory and pro-angiogenic effects of leptin are associated with an increase in GLUT1-mediated glucose uptake, but not with an increase in ASCT2-mediated glutamine uptake (Fig. 1).

### 3.1.4 Interferon-γ

IFN-γ, an important pro-inflammatory cytokine involved in the inflammatory process in T2D, can create a local microenvironment conducive to enhanced TNBC

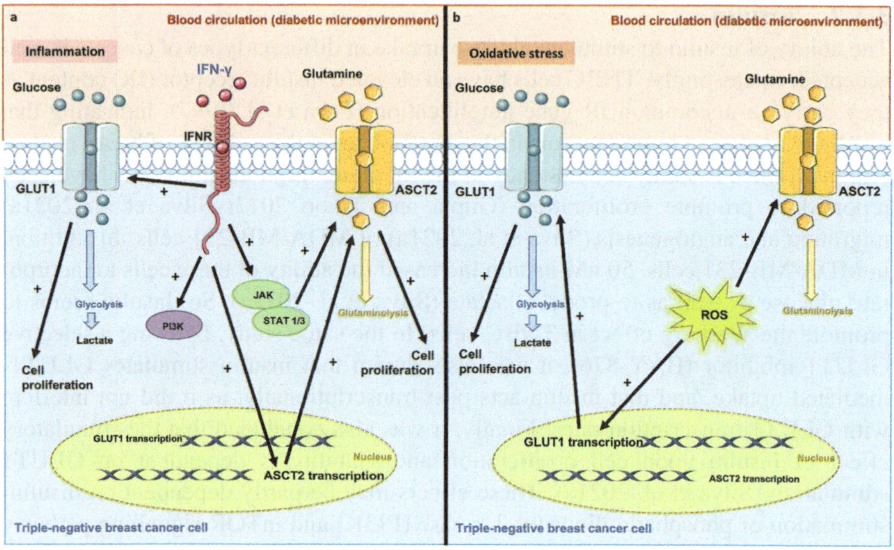


**Fig. 2** Schematic depiction of the effect of IFN-γ and oxidative stress on glutamine and glucose uptake and metabolism in triple-negative breast cancer cells. (**a**) A high inflammatory status (IFN-γ) has a pro-proliferative effect in TNBC cells, which is dependent on an increase in GLUT1-mediated and ASCT2-mediated glucose and glutamine uptake, respectively. IFN-γ upregulates ASCT2 transcription and translation rates and its effect upon glutamine uptake involves activation of the PI3K, STAT3 and STAT1 intracellular signalling pathways. (**b**) Oxidative stress stimulates GLUT1-mediated and ASCT2-mediated glucose and glutamine uptake, respectively, increases GLUT1 transcription rates and stimulates cell proliferation, which is dependent on an increase in GLUT1-mediated glucose uptake. Abbreviations: *GLUT1* facilitative glucose transporters 1, *ASCT2* alanine, serine, cysteine transporter, *IFN- γ* interferon gamma, *INFR* interferon receptor, *ROS* reactive oxygen species, *PI3K* phosphatidylinositol 3-kinase, *JAK* janus kinase, *STAT* signal transducer and activator of transcription

survival and proliferation, and to the promotion of angiogenesis and metastatic dissemination (Mantovani et al. 2008; Crespi et al. 2016; Silva et al. 2021a). However, the direct effect of an inflammatory microenvironment on GLUT1 and ASCT2 expression/activity in TNBC cells is practically unknown. In this context, Silva et al. recently examined the effect of IFN-γ on glucose uptake by MDA-MB-231 cells. It was found that treatment of MDA-MB-231 cells with IFN-γ results in a significant increase in GLUT1-mediated glucose uptake and lactate production. Furthermore, the effect of IFN-γ on cell proliferation and viability was shown to be dependent on interference with GLUT1-mediated glucose transport (Silva et al. 2021a) (Fig. 2).

The effects of IFN-γ on glutamine metabolism in TNBC cells were also recently investigated. In the MDA-MB-231 cell line, IFN-γ was found to increase $Na^+$-dependent glutamine uptake. Pharmacological inhibition of ASCT2 with GPNA or ASCT2 knockdown was able to abolish the stimulatory effect of IFN-γ upon glutamine uptake, suggesting that IFN-γ promotes ASCT2-mediated glutamine uptake. It was also concluded that IFN-γ upregulates ASCT2 transcription and translation rates, and that the stimulatory effect of IFN-γ upon ASCT2-mediated glutamine transport involves activation of signal transducer and activator of transcription 1 (STAT1), STAT3 and PI3K intracellular signalling pathways (Silva et al. 2021b). Further, the pro-proliferative effect of this pro-inflammatory cytokine in MDA-MB-231 cells appears to be ASCT2-mediated (Silva et al. 2021b) (Fig. 2).

All these results thus strongly support the conclusion that GLUT1 and ASCT2 are mediators of the effects of IFN-γ in promoting TNBC cell proliferation.

### 3.1.5 Oxidative Stress

Reactive oxygen species (ROS) have a controversial role in breast cancer (Sarmiento-Salinas et al. 2019). In high oxidative stress environments, they may cause deleterious cellular effects (anti-proliferative, pro-necrotic and pro-apoptotic), while in low or medium oxidative stress environments, ROS may induce DNA damage, inflammation and cell proliferation, eventually promoting carcinogenesis (Isnaini et al. 2018). Interestingly, an oxidative stress inducer (*tert*-butylhydroperoxide (TBH)) presented pro-proliferative, pro-migratory, pro-angiogenic, pro-apoptotic and cytotoxic effects in MDA-MB-231 cells (Silva et al. 2021a). TBH appears to potentiate the Warburg effect in TNBC cells. In fact, exposure of MDA-MB-231 cells to TBH stimulated GLUT1-mediated glucose uptake by increasing GLUT1 transcription rates, with a concomitant increase in lactate production. It was also shown that the pro-proliferative and cytotoxic effects of TBH upon TNBC cells are dependent on GLUT1 stimulation (Silva et al. 2021a) (Fig. 2).

TBH was also found to increase total and $Na^+$-dependent glutamine uptake in MDA-MB-231 cells, although it did not interfere with ASCT2 transcription and translation rates. Moreover, the ASCT2 inhibitor GPNA was able to abolish the stimulatory effect of TBH upon total and $Na^+$ dependent glutamine uptake. Together, these results indicate that the effect of TBH upon glutamine uptake in TNBC cells is GPNA-sensitive and probably ASCT2-mediated and involves a mechanism other than regulation of gene or protein expression (Silva et al. 2021b). So, oxidative stress can exacerbate aerobic glycolysis and glutaminolysis in TNBC tumours in a GLUT1- and ASCT2-dependent manner, respectively, and the pro-proliferative and cytotoxic effects of TBH are dependent on GLUT1 stimulation (Fig. 2).

## 4 Conclusions

Despite the fact that T2D and breast cancer are distinct diseases, glucose and glutamine metabolism appear to play a central role in both diseases. TNBC is one of the most challenging breast cancers and it is clear that the Warburg effect and glutamine addiction support the malignancy of TNBC tumours. Based on the presented evidences, it seems recognizable that most T2D-associated conditions can create a microenvironment auspicious to TNBC development, which is dependent on an increase in GLUT1-mediated glucose and/or ASCT2-mediated glutamine uptake. Therefore, the glucose transporter GLUT1 and the glutamine transporter ASCT2 may constitute mechanisms by which T2D participates in malignancy of TNBC tumours. So, it is possible that these two nutrient transporters can be used in the future as molecular targets for TNBC therapy and as a biomarker for the risk of TNBC development in T2D patients.

**Conflict of Interest Statement** The authors declare that there are no conflicts of interest.

**Funding** This work was supported by I3S (UID/BIM/04293/2013) to FM and CS was partially funded by the project NORTE-08-5369-FSE-000018, supported by Norte Portugal Regional Programme (Norte 2020), under the PORTUGAL 2020 Partnership Agreement, through the European Regional Development Fund (ERDF) and Faculty of Medicine, University of Porto (FMUP).

## References

Ando S, Gelsomino L, Panza S, Giordano C, Bonofiglio D, Barone I, Catalano S (2019) Obesity, leptin and breast cancer: epidemiological evidence and proposed mechanisms. Cancers (Basel) 11:62. https://doi.org/10.3390/cancers11010062

Annibaldi A, Widmann C (2010) Glucose metabolism in cancer cells. Curr Opin Clin Nutr Metab Care 13:466–470. https://doi.org/10.1097/MCO.0b013e32833a5577

Barbosa AM, Martel F (2020) Targeting glucose transporters for breast cancer therapy: the effect of natural and synthetic compounds. Cancers (Basel) 12:154. https://doi.org/10.3390/cancers12010154

Boyle P, Boniol M, Koechlin A, Robertson C, Valentini F, Coppens K, Fairley LL, Boniol M, Zheng T, Zhang Y, Pasterk M, Smans M, Curado MP, Mullie P, Gandini S, Bota M, Bolli GB, Rosenstock J, Autier P (2012) Diabetes and breast cancer risk: a meta-analysis. Br J Cancer 107: 1608–1617. https://doi.org/10.1038/bjc.2012.414

Bronsveld HK, Jensen V, Vahl P, De Bruin ML, Cornelissen S, Sanders J, Auvinen A, Haukka J, Andersen M, Vestergaard P, Schmidt MK (2017) Diabetes and breast cancer subtypes. PLoS One 12:e0170084. https://doi.org/10.1371/journal.pone.0170084

Cao MD, Lamichhane S, Lundgren S, Bofin A, Fjosne H, Giskeodegard GF, Bathen TF (2014) Metabolic characterization of triple negative breast cancer. BMC Cancer 14:941. https://doi.org/10.1186/1471-2407-14-941

Cassim S, Vucetic M, Zdralevic M, Pouyssegur J (2020) Warburg and beyond: the power of mitochondrial metabolism to collaborate or replace fermentative glycolysis in cancer. Cancers (Basel) 12:1119. https://doi.org/10.3390/cancers12051119

Cha YJ, Kim ES, Koo JS (2018) Amino acid transporters and glutamine metabolism in breast cancer. Int J Mol Sci 19:907. https://doi.org/10.3390/ijms19030907

Chen H, Cook LS, Tang MC, Hill DA, Wiggins CL, Li CI (2019) Relationship between diabetes and diabetes medications and risk of different molecular subtypes of breast cancer. Cancer Epidemiol Biomark Prev 28:1802–1808. https://doi.org/10.1158/1055-9965.EPI-19-0291

Choi YK, Park KG (2018) Targeting glutamine metabolism for cancer treatment. Biomol Ther (Seoul) 26:19–28. https://doi.org/10.4062/biomolther.2017.178

Choi J, Jung WH, Koo JS (2013) Metabolism-related proteins are differentially expressed according to the molecular subtype of invasive breast cancer defined by surrogate immunohistochemistry. Pathobiology 80:41–52. https://doi.org/10.1159/000339513

Cory JG, Cory AH (2006) Critical roles of glutamine as nitrogen donors in purine and pyrimidine nucleotide synthesis: asparaginase treatment in childhood acute lymphoblastic leukemia. In Vivo 20:587–589

Crespi E, Bottai G, Santarpia L (2016) Role of inflammation in obesity-related breast cancer. Curr Opin Pharmacol 31:114–122. https://doi.org/10.1016/j.coph.2016.11.004

DeBerardinis RJ, Chandel NS (2016) Fundamentals of cancer metabolism. Sci Adv 2:e1600200. https://doi.org/10.1126/sciadv.1600200

DeBerardinis RJ, Mancuso A, Daikhin E, Nissim I, Yudkoff M, Wehrli S, Thompson CB (2007) Beyond aerobic glycolysis: transformed cells can engage in glutamine metabolism that exceeds the requirement for protein and nucleotide synthesis. Proc Natl Acad Sci U S A 104:19345–19350. https://doi.org/10.1073/pnas.0709747104

Dong T, Liu Z, Xuan Q, Wang Z, Ma W, Zhang Q (2017) Tumor LDH-A expression and serum LDH status are two metabolic predictors for triple negative breast cancer brain metastasis. Sci Rep 7:6069. https://doi.org/10.1038/s41598-017-06378-7

Duran RV, Oppliger W, Robitaille AM, Heiserich L, Skendaj R, Gottlieb E, Hall MN (2012) Glutaminolysis activates Rag-mTORC1 signaling. Mol Cell 47:349–358. https://doi.org/10.1016/j.molcel.2012.05.043

El Ansari R, McIntyre A, Craze ML, Ellis IO, Rakha EA, Green AR (2018) Altered glutamine metabolism in breast cancer; subtype dependencies and alternative adaptations. Histopathology 72:183–190. https://doi.org/10.1111/his.13334

Feng Y, Spezia M, Huang S, Yuan C, Zeng Z, Zhang L, Ji X, Liu W, Huang B, Luo W, Liu B, Lei Y, Du S, Vuppalapati A, Luu HH, Haydon RC, He TC, Ren G (2018) Breast cancer development and progression: risk factors, cancer stem cells, signaling pathways, genomics, and molecular pathogenesis. Genes Dis 5:77–106. https://doi.org/10.1016/j.gendis.2018.05.001

Ferroni P, Riondino S, Buonomo O, Palmirotta R, Guadagni F, Roselli M (2015) Type 2 diabetes and breast cancer: the interplay between impaired glucose metabolism and oxidant stress. Oxidative Med Cell Longev 2015:183928. https://doi.org/10.1155/2015/183928

Flores-Lopez LA, Martinez-Hernandez MG, Viedma-Rodriguez R, Diaz-Flores M, Baiza-Gutman LA (2016) High glucose and insulin enhance uPA expression, ROS formation and invasiveness in breast cancer-derived cells. Cell Oncol (Dordr) 39:365–378. https://doi.org/10.1007/s13402-016-0282-8

Gross MI, Demo SD, Dennison JB, Chen L, Chernov-Rogan T, Goyal B, Janes JR, Laidig GJ, Lewis ER, Li J, Mackinnon AL, Parlati F, Rodriguez ML, Shwonek PJ, Sjogren EB, Stanton TF, Wang T, Yang J, Zhao F, Bennett MK (2014) Antitumor activity of the glutaminase inhibitor CB-839 in triple-negative breast cancer. Mol Cancer Ther 13:890–901. https://doi.org/10.1158/1535-7163.MCT-13-0870

Gupta C, Tikoo K (2013) High glucose and insulin differentially modulates proliferation in MCF-7 and MDA-MB-231 cells. J Mol Endocrinol 51:119–129. https://doi.org/10.1530/JME-13-0062

Hanahan D, Weinberg RA (2011) Hallmarks of cancer: the next generation. Cell 144:646–674. https://doi.org/10.1016/j.cell.2011.02.013

Harbeck N, Penault-Llorca F, Cortes J, Gnant M, Houssami N, Poortmans P, Ruddy K, Tsang J, Cardoso F (2019) Breast cancer. Nat Rev Dis Primers 5:66. https://doi.org/10.1038/s41572-019-0111-2

Hardefeldt PJ, Edirimanne S, Eslick GD (2012) Diabetes increases the risk of breast cancer: a meta-analysis. Endocr Relat Cancer 19:793–803. https://doi.org/10.1530/ERC-12-0242

Hay N (2016) Reprogramming glucose metabolism in cancer: can it be exploited for cancer therapy? Nat Rev Cancer 16:635–649. https://doi.org/10.1038/nrc.2016.77

Hou Y, Zhou M, Xie J, Chao P, Feng Q, Wu J (2017) High glucose levels promote the proliferation of breast cancer cells through GTPases. Breast Cancer (Dove Med Press) 9:429–436. https://doi.org/10.2147/BCTT.S135665

Huang X, Li X, Xie X, Ye F, Chen B, Song C, Tang H, Xie X (2016) High expressions of LDHA and AMPK as prognostic biomarkers for breast cancer. Breast 30:39–46. https://doi.org/10.1016/j.breast.2016.08.014

Hussein YR, Bandyopadhyay S, Semaan A, Ahmed Q, Albashiti B, Jazaerly T, Nahleh Z, Ali-Fehmi R (2011) Glut-1 expression correlates with basal-like breast cancer. Transl Oncol 4:321–327. https://doi.org/10.1593/tlo.11256

Isnaini I, Permatasari N, Mintaroem K, Prihardina B, Widodo MA (2018) Oxidants-antioxidants profile in the breast cancer cell line MCF-7. Asian Pac J Cancer Prev 19:3175–3178. https://doi.org/10.31557/APJCP.2018.19.11.3175

Jiang S, Zhang LF, Zhang HW, Hu S, Lu MH, Liang S, Li B, Li Y, Li D, Wang ED, Liu MF (2012) A novel miR-155/miR-143 cascade controls glycolysis by regulating hexokinase 2 in breast cancer cells. EMBO J 31:1985–1998. https://doi.org/10.1038/emboj.2012.45

Kalyanaraman B (2017) Teaching the basics of cancer metabolism: developing antitumor strategies by exploiting the differences between normal and cancer cell metabolism. Redox Biol 12:833–842. https://doi.org/10.1016/j.redox.2017.04.018

Kanaan YM, Sampey BP, Beyene D, Esnakula AK, Naab TJ, Ricks-Santi LJ, Dasi S, Day A, Blackman KW, Frederick W, Copeland RL Sr, Gabrielson E, Dewitty RL Jr (2014) Metabolic profile of triple-negative breast cancer in African-American women reveals potential biomarkers of aggressive disease. Cancer Genomics Proteomics 11:279–294

Kato Y, Maeda T, Suzuki A, Baba Y (2018) Cancer metabolism: new insights into classic characteristics. Jpn Dent Sci Rev 54:8–21. https://doi.org/10.1016/j.jdsr.2017.08.003

Kim S, Kim DH, Jung WH, Koo JS (2013a) Expression of glutamine metabolism-related proteins according to molecular subtype of breast cancer. Endocr Relat Cancer 20:339–348. https://doi.org/10.1530/ERC-12-0398

Kim S, Kim DH, Jung WH, Koo JS (2013b) Metabolic phenotypes in triple-negative breast cancer. Tumour Biol 34:1699–1712. https://doi.org/10.1007/s13277-013-0707-1

Kim JY, Heo SH, Choi SK, Song IH, Park IA, Kim YA, Park HS, Park SY, Bang WS, Gong G, Lee HJ (2017) Glutaminase expression is a poor prognostic factor in node-positive triple-negative breast cancer patients with a high level of tumor-infiltrating lymphocytes. Virchows Arch 470: 381–389. https://doi.org/10.1007/s00428-017-2083-5

Kong Y, Dong Q, Ji H, Sang M, Ding Y, Zhao M, Yang H, Geng C (2019) The effect of the leptin and leptin receptor expression on the efficacy of neoadjuvant chemotherapy in breast cancer. Med Sci Monit 25:3005–3013. https://doi.org/10.12659/MSM.915368

Kung HN, Marks JR, Chi JT (2011) Glutamine synthetase is a genetic determinant of cell type-specific glutamine independence in breast epithelia. PLoS Genet 7:e1002229. https://doi.org/10.1371/journal.pgen.1002229

Larsson SC, Mantzoros CS, Wolk A (2007) Diabetes mellitus and risk of breast cancer: a meta-analysis. Int J Cancer 121:856–862. https://doi.org/10.1002/ijc.22717

Lee CH, Olson P, Hevener A, Mehl I, Chong LW, Olefsky JM, Gonzalez FJ, Ham J, Kang H, Peters JM, Evans RM (2006) PPARdelta regulates glucose metabolism and insulin sensitivity. Proc Natl Acad Sci U S A 103:3444–3449. https://doi.org/10.1073/pnas.0511253103

Lehmann BD, Pietenpol JA, Tan AR (2015) Triple-negative breast cancer: molecular subtypes and new targets for therapy. Am Soc Clin Oncol Educ Book 35:e31–e39. https://doi.org/10.14694/EdBook_AM.2015.35.e31

Liu Y, Zhao T, Li Z, Wang L, Yuan S, Sun L (2018) The role of ASCT2 in cancer: a review. Eur J Pharmacol 837:81–87. https://doi.org/10.1016/j.ejphar.2018.07.007

Loibl S, Poortmans P, Morrow M, Denkert C, Curigliano G (2021) Breast cancer. Lancet 397: 1750–1769. https://doi.org/10.1016/S0140-6736(20)32381-3

Ma C, Zu X, Liu K, Bode AM, Dong Z, Liu Z, Kim DJ (2019) Knockdown of pyruvate kinase M inhibits cell growth and migration by reducing NF-kB activity in triple-negative breast cancer cells. Mol Cells 42:628–636. https://doi.org/10.14348/molcells.2019.0038

Mantovani A, Allavena P, Sica A, Balkwill F (2008) Cancer-related inflammation. Nature 454:436–444. https://doi.org/10.1038/nature07205

Martin SD, McGee SL (2018) Metabolic reprogramming in type 2 diabetes and the development of breast cancer. J Endocrinol 237:R35–R46. https://doi.org/10.1530/JOE-18-0037

Metallo CM, Gameiro PA, Bell EL, Mattaini KR, Yang J, Hiller K, Jewell CM, Johnson ZR, Irvine DJ, Guarente L, Kelleher JK, Vander Heiden MG, Iliopoulos O, Stephanopoulos G (2011) Reductive glutamine metabolism by IDH1 mediates lipogenesis under hypoxia. Nature 481: 380–384. https://doi.org/10.1038/nature10602

Morais-Santos F, Granja S, Miranda-Goncalves V, Moreira AH, Queiros S, Vilaca JL, Schmitt FC, Longatto-Filho A, Paredes J, Baltazar F, Pinheiro C (2015) Targeting lactate transport suppresses in vivo breast tumour growth. Oncotarget 6:19177–19189. https://doi.org/10.18632/oncotarget.3910

Ogrodzinski MP, Bernard JJ, Lunt SY (2017) Deciphering metabolic rewiring in breast cancer subtypes. Transl Res 189:105–122. https://doi.org/10.1016/j.trsl.2017.07.004

Oh S, Kim H, Nam K, Shin I (2017) Glut1 promotes cell proliferation, migration and invasion by regulating epidermal growth factor receptor and integrin signaling in triple-negative breast cancer cells. BMB Rep 50:132–137. https://doi.org/10.5483/bmbrep.2017.50.3.189

Papa V, Milazzo G, Goldfine ID, Waldman FM, Vigneri R (1997) Sporadic amplification of the insulin receptor gene in human breast cancer. J Endocrinol Investig 20:531–536. https://doi.org/10.1007/BF03348015

Pavlova NN, Thompson CB (2016) The emerging hallmarks of cancer metabolism. Cell Metab 23: 27–47. https://doi.org/10.1016/j.cmet.2015.12.006

Peng J, Cui Y, Xu S, Wu X, Huang Y, Zhou W, Wang S, Fu Z, Xie H (2021) Altered glycolysis results in drug-resistant in clinical tumor therapy. Oncol Lett 21:369. https://doi.org/10.3892/ol.2021.12630

Pinheiro C, Sousa B, Albergaria A, Paredes J, Dufloth R, Vieira D, Schmitt F, Baltazar F (2011) GLUT1 and CAIX expression profiles in breast cancer correlate with adverse prognostic factors and MCT1 overexpression. Histol Histopathol 26:1279–1286. https://doi.org/10.14670/HH-26.1279

Robey IF, Lien AD, Welsh SJ, Baggett BK, Gillies RJ (2005) Hypoxia-inducible factor-1alpha and the glycolytic phenotype in tumors. Neoplasia 7:324–330. https://doi.org/10.1593/neo.04430

Samuel SM, Varghese E, Varghese S, Busselberg D (2018) Challenges and perspectives in the treatment of diabetes associated breast cancer. Cancer Treat Rev 70:98–111. https://doi.org/10.1016/j.ctrv.2018.08.004

Sanchez-Jimenez F, Perez-Perez A, de la Cruz-Merino L, Sanchez-Margalet V (2019) Obesity and breast cancer: role of leptin. Front Oncol 9:596. https://doi.org/10.3389/fonc.2019.00596

Sarmiento-Salinas FL, Delgado-Magallon A, Montes-Alvarado JB, Ramirez-Ramirez D, Flores-Alonso JC, Cortes-Hernandez P, Reyes-Leyva J, Herrera-Camacho I, Anaya-Ruiz M, Pelayo R, Millan-Perez-Pena L, Maycotte P (2019) Breast cancer subtypes present a differential production of reactive oxygen species (ROS) and susceptibility to antioxidant treatment. Front Oncol 9: 480. https://doi.org/10.3389/fonc.2019.00480

Schiliro C, Firestein BL (2021) Mechanisms of metabolic reprogramming in cancer cells supporting enhanced growth and proliferation. Cell 10:1056. https://doi.org/10.3390/cells10051056

Shin E, Koo JS (2021) Glucose metabolism and glucose transporters in breast cancer. Front Cell Dev Biol 9:728759. https://doi.org/10.3389/fcell.2021.728759

Shiratori R, Furuichi K, Yamaguchi M, Miyazaki N, Aoki H, Chibana H, Ito K, Aoki S (2019) Glycolytic suppression dramatically changes the intracellular metabolic profile of multiple cancer cell lines in a mitochondrial metabolism-dependent manner. Sci Rep 9:18699. https://doi.org/10.1038/s41598-019-55296-3

Silva C, Andrade N, Guimaraes JT, Patricio E, Martel F (2021a) The in vitro effect of the diabetes-associated markers insulin, leptin and oxidative stress on cellular characteristics promoting breast cancer progression is GLUT1-dependent. Eur J Pharmacol 898:173980. https://doi.org/10.1016/j.ejphar.2021.173980

Silva C, Andrade N, Rodrigues I, Ferreira AC, Soares ML, Martel F (2021b) The pro-proliferative effect of interferon-gamma in breast cancer cell lines is dependent on stimulation of ASCT2-mediated glutamine cellular uptake. Life Sci 286:120054. https://doi.org/10.1016/j.lfs.2021.120054

Silva C, Andrade N, Guimaraes JT, Cardoso E, Meireles C, Pinto V, Paiva J, Martel F (2022) The pro-proliferative effect of insulin in human breast epithelial DMBA-transformed and non-transformed cell lines is PI3K-, mTOR- and GLUT1-dependent. Cell Biochem 40:127–137 Funct. https://doi.org/10.1002/cbf.3681

Sun X, Wang M, Wang M, Yu X, Guo J, Sun T, Li X, Yao L, Dong H, Xu Y (2020) Metabolic reprogramming in triple-negative breast cancer. Front Oncol 10:428. https://doi.org/10.3389/fonc.2020.00428

Sung H, Ferlay J, Siegel RL, Laversanne M, Soerjomataram I, Jemal A, Bray F (2021) Global cancer statistics 2020: GLOBOCAN estimates of incidence and mortality worldwide for 36 cancers in 185 countries. CA Cancer J Clin 71:209–249. https://doi.org/10.3322/caac.21660

Talib WH, Mahmod AI, Abuarab SF, Hasen E, Munaim AA, Haif SK, Ayyash AM, Khater S, Al-Yasari IH, Kury LTA (2021) Diabetes and cancer: metabolic association, therapeutic challenges, and the role of natural products. Molecules 26:2179. https://doi.org/10.3390/molecules26082179

Teixeira E, Silva C, Martel F (2021) The role of the glutamine transporter ASCT2 in antineoplastic therapy. Cancer Chemother Pharmacol 87:447–464. https://doi.org/10.1007/s00280-020-04218-6

Tsang JYS, Tse GM (2020) Molecular classification of breast cancer. Adv Anat Pathol 27:27–35. https://doi.org/10.1097/PAP.0000000000000232

van Geldermalsen M, Wang Q, Nagarajah R, Marshall AD, Thoeng A, Gao D, Ritchie W, Feng Y, Bailey CG, Deng N, Harvey K, Beith JM, Selinger CI, O'Toole SA, Rasko JE, Holst J (2016) ASCT2/SLC1A5 controls glutamine uptake and tumour growth in triple-negative basal-like breast cancer. Oncogene 35:3201–3208. https://doi.org/10.1038/onc.2015.381

Vander Heiden MG (2011) Targeting cancer metabolism: a therapeutic window opens. Nat Rev Drug Discov 10:671–684. https://doi.org/10.1038/nrd3504

Vaughn AE, Deshmukh M (2008) Glucose metabolism inhibits apoptosis in neurons and cancer cells by redox inactivation of cytochrome c. Nat Cell Biol 10:1477–1483. https://doi.org/10.1038/ncb1807

Venturelli L, Nappini S, Bulfoni M, Gianfranceschi G, Dal Zilio S, Coceano G, Del Ben F, Turetta M, Scoles G, Vaccari L, Cesselli D, Cojoc D (2016) Glucose is a key driver for GLUT1-mediated nanoparticles internalization in breast cancer cells. Sci Rep 6:21629. https://doi.org/10.1038/srep21629

Wang F, Yang Y (2014) Inhibition of PKM2 sensitizes triple-negative breast cancer cells to doxorubicin. Biochem Biophys Res Commun 454:465–470. https://doi.org/10.1016/j.bbrc.2014.10.109

Warburg O, Wind F, Negelein E (1927) The metabolism of tumors in the body. J Gen Physiol 8:519–530. https://doi.org/10.1085/jgp.8.6.519

Wu Q, Ba-Alawi W, Deblois G, Cruickshank J, Duan S, Lima-Fernandes E, Haight J, Tonekaboni SAM, Fortier AM, Kuasne H, McKee TD, Mahmoud H, Kushida M, Cameron S, Dogan-Artun N, Chen W, Nie Y, Zhang LX, Vellanki RN, Zhou S, Prinos P, Wouters BG, Dirks PB, Done SJ, Park M, Cescon DW, Haibe-Kains B, Lupien M, Arrowsmith CH (2020) GLUT1 inhibition blocks growth of RB1-positive triple negative breast cancer. Nat Commun 11:4205. https://doi.org/10.1038/s41467-020-18020-8

Yanagida O, Kanai Y, Chairoungdua A, Kim DK, Segawa H, Nii T, Cha SH, Matsuo H, Fukushima J, Fukasawa Y, Tani Y, Taketani Y, Uchino H, Kim JY, Inatomi J, Okayasu I, Miyamoto K, Takeda E, Goya T, Endou H (2001) Human L-type amino acid transporter 1 (LAT1): characterization of function and expression in tumor cell lines. Biochim Biophys Acta 1514:291–302. https://doi.org/10.1016/s0005-2736(01)00384-4

Yang L, Venneti S, Nagrath D (2017) Glutaminolysis: a Hallmark of cancer metabolism. Annu Rev Biomed Eng 19:163–194. https://doi.org/10.1146/annurev-bioeng-071516-044546

Yi F, Diao S, Yuan XL, Li JY (2018) Association of plasma leptin levels and soluble leptin receptor with breast cancer. Zhonghua Yu Fang Yi Xue Za Zhi 52:253–259. https://doi.org/10.3760/cma.j.issn.0253-9624.2018.03.007

Yoo HC, Yu YC, Sung Y, Han JM (2020) Glutamine reliance in cell metabolism. Exp Mol Med 52: 1496–1516. https://doi.org/10.1038/s12276-020-00504-8

Younes M, Brown RW, Mody DR, Fernandez L, Laucirica R (1995) GLUT1 expression in human breast carcinoma: correlation with known prognostic markers. Anticancer Res 15:2895–2898

Yuan XL, Xu ZP, Liu CR, Yan LP, Tao P, Xiong P, Li Q, Zhou M, Li H, Zhao M, Li JY (2017) Study of the association between polymorphism of persistent obesity, human leptin gene/leptin receptor gene and molecular subtypes of breast cancer. Zhonghua Yu Fang Yi Xue Za Zhi 51: 533–538. https://doi.org/10.3760/cma.j.issn.0253-9624.2017.06.015

Yudkoff M, Pleasure D, Cregar L, Lin ZP, Nissim I, Stern J, Nissim I (1990) Glutathione turnover in cultured astrocytes: studies with [15N]glutamate. J Neurochem 55:137–145. https://doi.org/10.1111/j.1471-4159.1990.tb08831.x

Zhang W, Xu Y, Xu Q, Shi H, Shi J, Hou Y (2017) PPARdelta promotes tumor progression via activation of Glut1 and SLC1-A5 transcription. Carcinogenesis 38:748–755. https://doi.org/10.1093/carcin/bgx035

Zheng J (2012) Energy metabolism of cancer: glycolysis versus oxidative phosphorylation (Review). Oncol Lett 4:1151–1157. https://doi.org/10.3892/ol.2012.928

Zheng Y, Ley SH, Hu FB (2018) Global aetiology and epidemiology of type 2 diabetes mellitus and its complications. Nat Rev Endocrinol 14:88–98. https://doi.org/10.1038/nrendo.2017.151

# Holistic Nursing Approach: A Case Study on Integrating Traditional Chinese Medicine into Nursing Care for a Patient with Upper Limb Lymphedema and Coexisting Pressure Injury Post-breast Cancer Surgery

Haiyan Y. Zhao, Ye Zhou, Chong Chin Che, Mei Chan Chong, Ying Ji, Yantao T. Zhu, Canjuan J. Chen, and Ming Chen

**Abstract**

This case study describes the nursing care of a patient following breast cancer surgery, who experienced a complication of a pressure injury in the elbow joint. The patient underwent comprehensive decongestive therapy (CDT), which included skin care to prevent and mitigate pressure injuries, manual lymphatic drainage to facilitate lymphatic flow, intermittent bandaging with pressure to sustain the edema-reducing impact, and integration of traditional Chinese medicine into the nursing care to enhance clinical efficacy. Additionally, the treatment emphasized strengthening nutritional support. Patient education on self-lymphatic drainage and strategies to

The original version of the chapter has been revised. Errors in the author names and affiliations were corrected. A correction to this chapter can be found at https://doi.org/10.1007/978-3-031-65827-3_282

Haiyan Y. Zhao and Ye Zhou have contributed equally to this work and should be considered co-first author.

H. Y. Zhao · Y. Ji · Y. T. Zhu · C. J. Chen · M. Chen
Nursing Department, Jingjiang People's Hospital Affiliated to Yangzhou University, Taizhou City, Jiangsu Province, China

Y. Zhou
Editorial Department of IJNSS, Chinese Nursing Journals Publishing House Co., Ltd., Beijing, China

Department of Nursing Science, Faculty of Medicine, Universiti Malaya, Kuala Lumpur, Malaysia

C. C. Che (✉) · M. C. Chong
Department of Nursing Science, Faculty of Medicine, Universiti Malaya, Kuala Lumpur, Malaysia
e-mail: chechongchin@um.edu.my



Interdisciplinary Cancer Research, https://doi.org/10.1007/16833_2024_249
Published online: 16 May 2024

improve patient compliance were also key components. After 4 weeks of meticulous treatment and care, the pressure injury completely healed, and the circumference difference between the bilateral arms was reduced to less than 1 cm, accompanied by the elimination of the sensation of heaviness.

**Keywords**

Breast cancer · Holistic nursing · Lymphedema · Pressure injury · Traditional Chinese medicine

## 1 Introduction

Breast cancer–related lymphedema (BCRL) is characterized by impaired lymphatic drainage and venous return in the upper limb on the affected side. This impairment results from factors such as the removal of axillary lymph nodes, ligation of the thoracic duct, occlusion of the axillary vein, local fluid accumulation, or infection following surgery for breast cancer (Li & Lu 2020). BCRL is a serious potential complication of breast cancer surgery, with an incidence ranging from 5% to 50% (Leppäpuska 2022; Tracey 2013). It impacts limb function, leading to reduced mobility, joint contractures, weakened muscle strength, and restricted movement. Moreover, decreased upper limb mobility further increases the risk of lymphedema, creating a vicious cycle. Recurrent episodes of lymphedema can lead to secondary infections like cellulitis, and in some cases, persistent conditions may develop into elephantiasis, significantly affecting patients' quality of life (Li & Que 2016). BCRL thus becomes a significant, ongoing issue for postoperative breast cancer patients, particularly those with long-term survival (Wei et al. 2022). Comprehensive decongestive therapy (CDT) is currently considered the "gold standard" for lymphedema treatment. It involves an intensive treatment phase led by therapists and a home self-maintenance phase led by patients and/or caregivers (Zhang et al. 2020). While BCRL cannot be completely cured, proactive decongestive therapy, professional care, and long-term rehabilitation can provide lasting relief from edema symptoms.

Traditional Chinese medicine (TCM) views BCRL as one of the manifestations of "edema," caused by qi deficiency and blood stasis (Bao 2018). Some scholars also believe that upper limb lymphedema results from blood stasis, insufficient qi and blood, and retention of fluids. Its pathogenesis is related to deficiencies in spleen-lung qi, spleen-kidney yang, and stagnation of cold-congealed blood (Jin et al. 2017). Routine care in comprehensive hospitals typically adopts a Western medicine perspective, often unilateral. The implementation of various measures, usually carried out by nurses based on doctors' orders and experience, may not consider the overall picture, thereby affecting nursing effectiveness (Jin et al. 2017). TCM treatment for BCRL has unique advantages. Research (Zhang et al. 2022) suggests that incorporating TCM nursing into CDT can more effectively alleviate edema, improve upper limb vein blockage, ensure timely lymphatic fluid reflux, and reduce edema recurrence.

On April 20, 2023, our department admitted a patient with moderate upper limb edema following breast cancer surgery. Seeking further treatment, and after a multidisciplinary consultation, protective measures were taken, and intermittent complex decongestive therapy (CDT) was implemented. After 4 weeks, the patient's arm circumference significantly reduced, and discomforts such as heaviness, numbness, and tightness were effectively relieved. The following report details this nursing experience.

## 2 Clinical Presentation of the Patient

The patient is a 48-year-old female with a BMI of 31.3. On September 30, 2022, she underwent a "left breast cancer–modified radical mastectomy" under general anesthesia. During the surgery, an irregular, hard mass measuring approximately 3.0 × 2.0 cm was found in the outer upper quadrant of the left breast. The surgery was successful, and the postoperative recovery was good. Pathological findings revealed grade 3 infiltrating ductal carcinoma with necrosis, with tumor sizes of 1.5 × 1 × 1 cm and 3 × 2 × 2 cm. Multiple cancer emboli were observed in blood vessels, and there was no nerve invasion (−). The modified radical mastectomy of the left breast was performed without involving the nipple. In the adjacent area of the original surgical site, high-grade ductal carcinoma in situ was found. Multiple cancer emboli were observed in blood vessels, and the margins were negative (−). Cancer metastasis was found in 13 out of 18 axillary lymph nodes and 1 out of 1 lymph node under the left clavicle. Postoperatively, the patient received the EC-T regimen (liposomal doxorubicin 60 mg ivgtt + cyclophosphamide 1.1 g iv + docetaxel 140 mg ivgtt) for chemotherapy. Gastroprotective and supportive treatments with azasetron and famotidine were also administered, and the process was smooth.

On April 20, 2023, the patient complained of heaviness, numbness, and significant swelling in the left upper limb, which did not subside on its own after rest. No redness or fever was reported. Physical examination revealed swelling in the left upper limb with no redness, normal skin temperature, taut skin, surface hardness, and increased fibrosis of subcutaneous fibrous tissue. Some movement of the left upper limb was restricted. Laboratory indicators showed a white blood cell count of $7.3 \times 10^9$/L, red blood cell count of $4.72 \times 10^9$/L, platelet count of $297 \times 10^9$/L, hemoglobin of 144 g/L, and absolute neutrophil count of $0.55 \times 10^9$/L. Auxiliary examinations revealed no venous thrombosis in the left upper limb on vascular ultrasound. Magnetic resonance imaging showed soft tissue swelling from the distal end of the left brachial plexus to the axillary region, as well as subcutaneous and deep tissue swelling in the left shoulder and axillary regions. Based on the progression stage and severity grading of lymphedema (as measured by the weekly measurement method), the patient's left upper limb showed a significant volume increase with obvious tissue fibrosis, classified as stage II. The affected limb was 5 cm larger than the healthy limb, measured by the circumference method, indicating moderate lymphedema, as shown in Fig. 1. During the home care of the patient, when the patient's family members relieved the swelling pressure and reapplied the bandage, the limb developed pressure sores at the elbow due to the tightness of the pressure bandage. It was a stage 1 pressure sore with a wound size of 1 × 3 cm; the skin is not broken but appears red. After releasing the pressure for half an hour, the skin color did not recover.

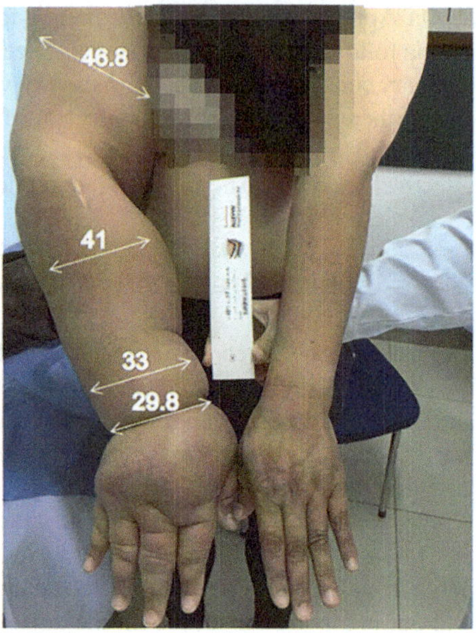


**Fig. 1** Lymphedema in the limb

## 3 Therapeutic Nursing Intervention

### 3.1 Comprehensive Decongestive Therapy

The assessment and implementation of comprehensive decongestive therapy (CDT) for the patient are under the complete supervision of an international lymphedema therapist. CDT is recognized as the standard comprehensive treatment for lymphedema and is the most widely applied and effective conservative treatment globally. It includes four components: skin care, manual lymphatic drainage, low-stretch bandaging, and functional exercises. These components are administered by trained and qualified lymphedema therapists (Tian et al. 2021; Brandão et al. 2020; Wu et al. 2021). After a 4-week CDT course, the patient's left upper arm circumference noticeably reduced, with less than a 1 cm difference in bilateral arm measurements. The sensation of heaviness disappeared, the pressure injuries completely healed, and the patient reported a reduction in numbness and swelling in the affected limb.

#### 3.1.1 Skin Care

Skin care is an essential aspect of CDT for lymphedema, playing a significant role in optimizing skin structure, preventing and reducing infections, and decelerating edema progression (Lu et al. 2022). The patient's left upper limb presented with dry skin. Before daily treatment, it is recommended to cleanse the skin with warm water and wipe it with a soft cotton towel. SkinRun lotion, applied in the direction of the skin texture, promotes absorption. SkinRun is a skin-protective liquid dressing, comprising about 99% linoleic acid, alpha-linolenic acid, and 1% fennel as an additive (Xu 2018). It forms a lipid protective film on the skin, improving local microcirculation and blood flow. This nourishes the skin, maintains moisture, enhances local oxygen supply, and accelerates the repair of damaged areas. It also alleviates symptoms like skin dryness, with a high level of clinical safety. After 1 week of care, the patient's skin dryness was effectively relieved.

### 3.1.2 Manual Lymphatic Drainage (MLD)

Manual lymphatic drainage (MLD) is a conservative therapeutic technique aligned with the anatomy and physiology of the lymphatic system. It involves massaging the lymphatic system in a specific sequence. Research (Lu et al. 2022; Thompson et al. 2021; Chen et al. 2022; Xiao et al. 2023) indicates that MLD can stimulate lymphatic circulation, thereby increasing lymphatic drainage, alleviating patient pain, and improving symptoms of skin and pressure injuries. Due to the patient's severe limb swelling and lower hemoglobin levels, a gentle approach is recommended.

Trained nurses performed lymphatic drainage and instructed the patient to take deep breaths before the massage. As no back issue was present, the patient combined the massage with strong abdominal muscle contractions. The massage sequence followed three independent lymphatic pathways (chest, trunk, and affected limb) with specific techniques:

(a) Aiding in the drainage of lymphatic fluid from the affected limb to the healthy armpit lymph nodes. This involves sequential massaging of the neck and armpit lymph nodes, followed by the sternum and the upper chest from the affected to the healthy side.
(b) Assisting in draining lymphatic fluid from the affected armpit to the groin area. It is started by massaging the affected groin area (lying down or standing), then massaging the abdominal, lower back, and upper chest areas, moving from the affected side to the groin area.
(c) Facilitating drainage from the affected shoulder and back to the neck lymph nodes (not the armpit nodes) massaging the shoulder, outer upper arm to the neck lymph nodes, inner upper arm to the back of the shoulder, and front of the upper arm to the neck lymph nodes on the affected side.
(d) Finally, massaging the neck, armpit, groin area, and lymph nodes above and below the elbow, and massaging the inner and outer sides of the affected limb from the forearm to the upper arm. Repeat each motion 5–10 times, twice a day, for 30 min per session (Zhao et al. 2018).

The massage was gentle enough to move the skin without exerting excessive pressure on the superficial lymphatic vessels under the fascia. After 4 weeks of treatment, reductions in limb lymphedema, relief of shoulder joint dysfunction, and alleviation of self-reported symptoms (pain, swelling, and skin tightness) were observed.

### 3.1.3 Pressure Therapy

Pressure therapy primarily involves the use of short-stretch bandages, aiming to maintain the effects achieved during manual lymphatic drainage (MLD) treatment and prevent the re-accumulation of lymphatic fluid. A significant challenge in patient care is managing pressure injuries around the patient's elbow. Applying pressure during bandaging can easily exacerbate existing injuries or lead to new pressure injuries in other areas. Biatain dressing is a sterile polyurethane foam dressing, primarily consisting of polyurethane foam and a hydrocolloid adhesive. It includes an upper polyurethane film, a middle polyurethane foam layer, and a lower hydrocolloid adhesive layer. This foam dressing supports and reduces pressure, exhibits excellent fluid absorption properties, conforms well to the skin, and provides a comfortable pressure-relieving sensation for the patient. It does not stick to wounds, making it easier for wound cleansing and maintaining dryness. Available in various shapes and sizes, it can be applied to different anatomical sites. The foam dressing, cut into a 3 × 6 cm rectangular shape, was applied to cover the elbow, followed by bandaging to effectively achieve a pressure-relieving effect.

Prior to each bandaging session, a thorough examination of the affected limb was conducted. If redness or mild friction on the skin is observed, the dressing is promptly cut to the appropriate shape and size for coverage and facilitate early detection and prevention. Fibrosis and significant edema were evident in the lower third of the inner forearm of the patient. When a large bandage was used for pressure wrapping at the elbow joint, the bandage was loosening. To address this, an 8 cm bandage was spirally wound using the figure-eight method at the elbow joint to maintain its functionality. A 10 cm bandage was then spirally wrapped around the outer layer of the elbow joint to ensure continued pressure, as shown in Fig. 2. Intermittent bandaging with pressure is applied to alleviate pressure injuries in the patient's elbow region.

### 3.1.4 Functional Exercise

Research by Zhao et al. (2020) indicates that resistance and aerobic exercises can stimulate muscle contraction, enhance aerobic metabolism, and are beneficial for reducing edema. Wei et al. (2022) also confirm that appropriate exercise not only strengthens the activity capacity of the affected limb but also promotes lymphatic circulation through muscle contractions, serving as an effective measure for preventing and controlling lymphedema in the patient. Lymphedema therapists instructed the patient in deep diaphragmatic breathing to facilitate the entry of lymphatic fluid into the bloodstream. They guided the patient to gradually engage in functional exercises, including tai chi, yoga, qigong, and basic limb movement training, and integrate basic limb movement training into daily activities such as walking, climbing stairs, hanging clothes, and performing household chores. The exercises were less intense, progressive, and adjusted as needed based on the patient's tolerance.

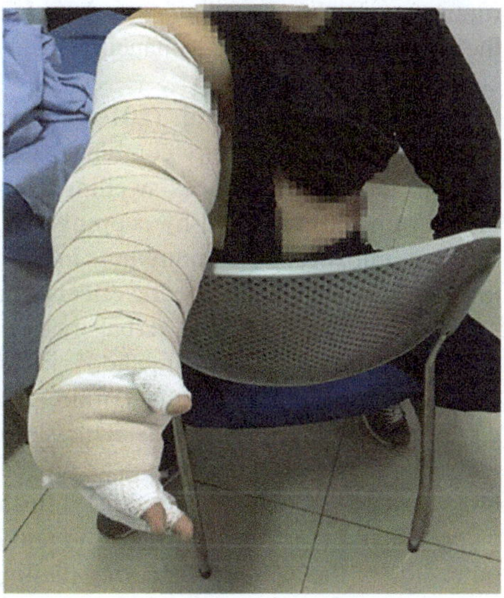

**Fig. 2** Pressure therapy

## 3.2 Traditional Chinese Medicine Nursing

Traditional Chinese nursing is a nursing theory and technique that follows the theoretical system of traditional Chinese medicine, embodying characteristics of nursing theory and techniques based on traditional Chinese medical diagnostic and therapeutic principles. In traditional Chinese medical practice, the emphasis on differential diagnosis and treatment extends to nursing care, wherein interventions are determined based on the foundational theories of traditional Chinese medicine. The basic principles of nursing care are derived from differential diagnosis, and specific measures are formulated in accordance with the principle of "tailoring care to the individual based on the time, place, and person" (Sun & Zhang 2022).

### 3.2.1 Fire Needle Combined with Cupping to Unblock Stagnation

The method of bloodletting (vessel puncture) is first mentioned in the "Huangdi Neijing": "Piercing the vessels means puncturing the blood vessels of small collaterals." The use of cupping (angular method) is first seen in the "Fifty-Two Medical Formulas: Mu Zhi": "Mu Zhi resides next to the orifice...using small angular cups." The combination of vessel puncture and cupping is first mentioned in "Emergency Formulas Behind the Elbow," specifically the term "needle angular," which refers to vessel puncture combined with cupping. By using tools such as a skin needle, triangular needle, and plum blossom needle to puncture and bleed, cupping after bloodletting can enhance the effect of vessel puncture (Sun & Zhang 2022). This practice initiates neurohumoral regulation, improves blood circulation around the Ashi point, and accelerates the removal of "cold," "toxins," and "dampness" within tissues along with the blood flow. Cupping, with its warm and negative pressure effects, can change local capillary permeability, cause local capillary rupture, and release mediators such as histamine and serotonin, thereby activating the immune system and enhancing local tolerance and overall resistance (Zhang et al. 2015).

A study by Wang et al. (2019) suggests that treating upper limb edema with localized vessel puncture combined with cupping promotes blood circulation, resolves stasis, and reduces swelling. Treating a selected Ashi point with regular needle puncture followed by cupping for 7 days results in reduced swelling in the affected limb, occasional numbness in the left upper limb, and significant pain relief, indicating favorable therapeutic effects. Liang et al. (2021) proposes that the combination of acupuncture and herbal medicine, emphasizing the treatment of edema using warming and diuretic methods, favors the use of fire needles. Fire, being yang in nature, employs the "warmth" of fire needles to invigorate yang qi, dispel cold and dampness, and open meridians. This stimulates the body's yang qi, harmonizes qi and blood, and produces more pronounced effects than regular fine needles or plum blossom needles (Chen et al. 2021). In this case, the patient underwent treatment using the combination of fire needle and cupping, selecting the Ashi point on the outer side of the left calf. After bloodletting with a fire needle, cupping was applied and repeated four times during the course of treatment.

### 3.2.2 Traditional Chinese Dietary Care for Postoperative Lymphedema in Breast Cancer

The main factors contributing to the occurrence of lymphedema in the upper limbs after breast cancer surgery are categorized as "deficiency, toxicity, stasis, water, phlegm, and heat." These are further classified into deficiency, including qi deficiency and yang deficiency, and excess, including heat toxin, blood stasis, and water retention (Liu et al. 2023). Based on the patient's actual condition, an oral blood-activating and spleen-invigorating diuretic formula was recommended. The main ingredients are 15 g of danggui (Chinese angelica), 10 g of shaoyao (peony), 15 g of huangqi (astragalus), 15 g of fuling (poria), 15 g of baizhu (white atractylodes), 15 g of guizhi (cinnamon twig), and 5 g of gancao (licorice). This formula was taken twice a day for a duration of 14 days per treatment cycle and decocted before consumption. Following a consultation with the nutrition department, an individualized plan was formulated based on the patient's specific circumstances. For meat consumption, it was recommended to include protein-rich foods such as carp, crucian carp, beef, large shrimp, and silkworm pupa to enhance the body's resistance and prevent infection. In the vegetable category, the patient was advised to consume carrots, winter melon, spinach, onions, and bitter gourd, along with incorporating anti-edema recipes like Wuhong Tang, adzuki bean and coix seed porridge, and winter melon soup.

### 3.3 Home Monitoring for Breast Cancer–Related Lymphedema (BCRL)

Due to the chronic and progressive nature of lymphedema, patients require long-term self-management even after discharge (Wei et al. 2022). Caregivers also play a crucial role in the self-care of BCRL patients, offering daily life assistance, symptom management, and psychological support (Wang & Yu 2023). Therefore, active communication with patients and their families is essential, encouraging family members to actively participate in lymphedema treatment. During treatment intervals, lymphedema therapists use WeChat to continuously monitor patients' home situations and teach both patients and their family caregivers the way of self-maintenance. Specific methods include the following: (1) Providing patients with a lymphedema management manual and guiding them to monitor and record arm circumference daily. (2) Instructing patients in self-lymphatic drainage, emphasizing twice daily for 30 min each time. (3) Paying close attention to the skin at the elbow pressure injury site to prevent compression again. (4) Guiding patients in progressive functional exercises, with the option to wear pressure sleeves when necessary. (5) In daily life, emphasizing the avoidance of mosquito bites and prohibiting intravenous procedures and blood collection on the affected limb. (6) Actively participating in patients' daily lives and self-monitoring management, helping patients build confidence in recovery, and encouraging them to adhere to self-care during home rehabilitation.

## 4 Discussion

CDT is recognized as an effective conservative treatment for breast cancer–related lymphedema (BCRL). It primarily includes skin care, manual lymphatic drainage (MLD), pressure therapy, and functional exercise. This study, through intervention techniques, aims to promote the reflux of lymphatic fluid, reduce the degree of edema and pain in patients, and improve their quality of life. In clinical nursing, caregivers need to master the essentials of each drainage technique, conduct a comprehensive assessment of the patient's surgical wounds and shoulder joint mobility, create personalized exercise plans, and educate them on home monitoring for BCRL.

BCRL falls within the realm of traditional Chinese medicine (TCM) edema. If it presents with cord-like structures and pain, it is classified as "mài bì" (Feng & Hou 2022). The earliest TCM theoretical records on edema are found in the Neijing, stating, "All damp swellings belong to the spleen. When the spleen's qi is deficient, the earth cannot control water, and water qi overflows the muscles and skin, causing swelling" (Huangdi 2001). The Jin Kui Yao Lue records, "Poor blood circulation results in water retention" (Zhang & Jin 2016). Sun Simiao in Qian Jin Yao Fang and Qian Jin Yi Fang elaborated on the types of edema, classifying them into 10 categories based on symptoms and causes: blue, red, yellow, white, black, mysterious, wind, stone, fruit, and qi edema (Huang 2021). TCM has a long history of understanding and treating edema, with rich clinical experience. Its literature documents methods for treating edema, including internal and external medication, acupuncture, moxibustion, diet therapy, and nondrug therapies, focusing on nourishing vitality; promoting blood circulation; facilitating fluid elimination and detoxification; and promoting the resolution of edema. Meanwhile, TCM nursing has developed professional theories and teams in a relatively short time, yet its scope of work needs further definition (Zheng et al. 2020). The management of postoperative breast cancer lymphedema involving only nursing personnel cannot fully leverage the advantages of TCM treatment. In the future, a multidisciplinary cooperation platform should be established to promote collaboration among multiple disciplines, develop TCM nursing plans for BCRL, and enhance treatment outcomes.

Literature (Wang 2017; Li et al. 2018) suggests that when implementing moxibustion for CBRL patients, caregivers must prevent skin burns and avoid damaging the integrity of the affected limb's skin, which may increase the risk of infection. Therefore, caregivers need to have knowledge of acupoints, assess the timing of moxibustion treatment, and control the moxibustion temperature. TCM nursing measures for CBRL patients require a high level of professionalism, placing specific demands on caregivers. However, many TCM nursing professionals in recent years have graduated from Western medical schools, resulting in weak knowledge of acupoints and meridians. There is a need to strengthen caregiver training, reinforce the learning of TCM basic knowledge, and improve treatment outcomes while preventing adverse reactions.

## 5 Implications for Nursing

The development trend of traditional Chinese nursing primarily revolves around the integration of traditional Chinese and Western medicine. Utilizing traditional Chinese medicine (TCM) theories, nursing practices are guided to combine both traditions. Incorporating TCM nursing with Western medical treatments aims to shorten the duration of illnesses. Traditional Chinese medicine emphasizes a holistic approach to health, and nurses can integrate these principles of holistic care into their practices. Adapting to the changes of the four seasons and providing targeted care in terms of activities and diet, as per TCM health preservation theory, is crucial for disease prevention and treatment. Nurses can use TCM theories such as the compatibility and incompatibility of drugs and foods, as well as the theory of the four seasons, to guide patients in dietary choices. This promotes early recovery from diseases, adjusts physiological mechanisms, and achieves preventive effects by treating diseases before they manifest.

## 6 Conclusion

In addition to skin care, manual lymphatic drainage (MLD), pressure therapy, functional exercise, and home monitoring for breast cancer–related lymphedema (BCRL) have beneficial effects on patients with BCRL and concomitant elbow joint pressure injuries; this chapter also explores distinctive traditional Chinese medicine (TCM) techniques like fire needling combined with cupping. Based on the holistic concept of TCM, this approach focuses on reducing swelling, alleviating pain, promoting meridian circulation, and instructing patients to consume foods that regulate qi, transform dampness, and reduce swelling. The application of TCM-specific nursing interventions in the country is currently fragmented and lacks a standardized scientific approach for lymphedema in breast cancer patients. Therefore, future efforts are necessary to consolidate and summarize these interventions, establishing evidence-based guidelines to guide clinical practice.

**Acknowledgments** We would like to acknowledge the patient who accepted the nursing in this study.

**Consent for Publication** All the authors agree to publication.

**Conflicts of Interest** The authors declare no potential conflicts of interest with respect to the research, authorship, and/or publication of this chapter.

**Data Availability Statement** Data from this study are available from the first author upon reasonable request.

**Author Contributions** Haiyan Y. Zhao, Ye Zhou, Chong Chin Che, and Mei Chan Chong participated mainly in the research design and in the drafting and critical revision of the manuscript. Haiyan Y. Zhao, Ye Zhou, Ying Ji, Yantao T. Zhu, Canjuan J. Chen, and Ming Chen conducted the holistic therapy and nursing.

**Funding Information** This research was funded by Taizhou Traditional Chinese Medicine Technology Development Project grant number TZ202106.

## References

Bao T (2018) Acupuncture for breast cancer-related lymphedema: a randomized controlled trial. Breast Cancer Res Treat 170(1):77–87

Brandão ML, Soares HPDS, Andrade MDA et al (2020) Efficacy of complex decongestive therapy for lymphedema of the lower limbs: a systematic review. J Vasc Bras 29:e20190074

Chen P, Liu H, Wang S et al (2021) Treatment of 27 cases of primary Parkinson's disease with fire needle at the base of the skull combined with body needle. Chin Acup 41(1):71–72

Chen Y, Shi S, Li L et al (2022) The preventive effect of manual lymphatic drainage combined with pneumatic treatment on upper limb lymphedema after breast cancer surgery. Clin Nurs J 21(5): 45–44

Feng K, Hou J (2022) Treatment of upper limb lymphedema after breast cancer. Hebei Trad Chin Med 44(6):1052–1106
Huang B (2021) Analysis on the treatment of water and gas disease in Medical Jian. Guangzhou University of Trad Chin Med. https://doi.org/10.27044/d.cnki.ggzzu.2021.00761
Huangdi (2001) The Huangdi Neijing. Zhonghua Book Company Press, Beijing
Jin M, Wang L, Shen X (2017) Progress of TCM external treatment for breast cancer-related upper limb lymphedema. Chin Med 26(3):464–466
Leppäpuska I (2022) Phase 1 lymfactin study: 24-month efficacy and safety results of combined adenoviral VEGF-C and lymph node transfer treatment for upper extremity lymphedema. J Plas, Reconst Aesthet Surg 75(11):3938–3945
Li LZ, Lu Q (2020) Surgical nursing, 6th edn. People's Health Publishing House, Beijing
Li S, Que H (2016) Huafa Que's experience in the treatment of upper limb lymphedema after breast cancer surgery. Shanghai J Trad Chin Med 50(8):3
Li H, Yang Y, Sultan et al (2018) A meta-analysis of the preventive effects of different preventive behaviors on postoperative lymphedema in breast cancer. Mod Prev Med 45(5):949–960
Liang R, Wang L, Cheng H (2021) Treatment of edema with the combined use of acupuncture and medicine by Cheng Haiying's method of warming yang and promoting diuresis. Beijing Trad Chin Med 40(12):107–1309
Liu Z, Li Q, Li W et al (2023) Chinese expert consensus on the diagnosis and treatment of postoperative lymphedema after breast cancer surgery. Trauma Crit Care Med 1(1):1–8
Lu Y, Chai C, Yang X et al (2022) Nursing care of a patient with stage III lymphedema of both lower limbs after cervical cancer surgery combined with stage II lymphedema of the lower abdomen and vulva. Chin J Nurs 57(20):2522–2526
Sun Q, Zhang D (2022) Application of bloodletting cupping therapy in the treatment of yin syndrome with breast disease as an example. Chin Med Drug Guide 28(6):78–81
Thompson B, Gaitatzis K, Janse DJX et al (2021) Manual lymphatic drainage treatment for lymphedema: a systematic review of the literature. J Cancer Surv 1(2):244–258
Tian C, Zhang J, Yang Z et al (2021) Application of the "Internet+" model in comprehensive decongestive therapy for lower limb lymphedema in gynecologic tumor patients. Chin Nurs Manag 21(2):171–175
Tracey D (2013) Incidence of unilateral arm lymphedema after breast cancer: a systematic review and meta-analysis. Lanct Oncol 14(6):500–515
Wang H (2017) Preliminary construction of clinical practice guidelines for postoperative lymphedema prevention behavior in breast cancer. Anhui Meical University, Hefei
Wang W, Yu J (2023) Research progress on the involvement of caregivers in self-care of breast cancer-related lymphedema patients. J Nurs 38(6):19–22
Wang C, Sun Y, Li Q et al (2019) Experience of Pei Xiaohua in the treatment of upper limb lymphedema after breast cancer surgery. Chin J Trad Chin 34(3):513–516
Wei X, Fu X, Shen A et al (2022) Summary of evidence for self-management of lymphedema in breast cancer patients. Chin J Nurs 57(2):237–244
Wu X, Liu Y, Zhu D et al (2021) Early prevention of complex decongestive therapy and rehabilitation exercise for prevention of lower extremity lymphedema after operation of gynecologic cancer. Asian J Surg 44(1):111–115
Xiao Q, Mi P, Chen W (2023) Effect analysis of manual lymphatic drainage nursing on relieving postoperative lymphedema and axillary web syndrome in breast cancer patients. Clin Electc J Cinic Surg 11(1):64–68
Xu T (2018) Analysis of the advantages of Saifurun in the treatment of pressure sores. Health Wkly. https://doi.org/10.3969/j.issn2096-1499.2018.08.025
Zhang Z, Jin K (2016) Jin Kui Yao Lue. China Medical Science and Technology Press, Beijing
Zhang Y, Pei X, Zhang S et al (2015) Clinical application of bloodletting cupping in traditional Chin surgery. World J Trad Chin Med 10(10):1465–1468

Zhang X, Oliveri JM, Paskett ED (2020) Features, predictors, and treatment of breast cancer-related lymphedema. Curr Breast Cancer Rep 12(4):244–254

Zhang H, Yu Q, Bai Z et al (2022) Progress in the pathogenesis of breast cancer-related lymphedema and TCM intervention. J Nanjing University of Trad Chin Med 38(3):258–264. https://doi.org/10.1414/j.issn.1672-0482.2022.0258

Zhao H, Bao Z, Du J et al (2018) Application of lymphatic drainage technique in patients with breast cancer-related upper limb lymphedema. Intern J Nurs 37(5):5. https://doi.org/10.3760/cma.j.isn.1673-4351.2018.05.042

Zhao H, Zhou C, Wu Y et al (2020) Summary of evidence-based exercise guidance for patients with breast cancer-related lymphedema. Chin J Nurs 55(5):779–785

Zheng H, Zhao S, Zheng Q (2020) Analysis and countermeasures of the current situation of TCM nursing talents training. J Trad Chin Med Manag 28(20):67–68

# Correction to: Holistic Nursing Approach: A Case Study on Integrating Traditional Chinese Medicine into Nursing Care for a Patient with Upper Limb Lymphedema and Coexisting Pressure Injury Post-breast Cancer Surgery

Haiyan Y. Zhao, Ye Zhou, Chong Chin Che, Mei Chan Chong, Ying Ji, Yantao T. Zhu, Canjuan J. Chen, and Ming Chen

**Correction to:**
**Chapter 249 in: H. Y. Zhao et al., Interdisciplinary Cancer Research, https://doi.org/10.1007/16833_2024_249**

The original version of the chapter was inadvertently published with the below listed errors, which have now been corrected.

1. First and last name of all the authors were interchanged.
2. Middle initials were added to the authors Haiyan Y. Zhao, Yantao T. Zhu, and Canjuan J. Chen.
3. Error in the author name Mei Chan Chong was corrected.
4. A new affiliation was added for the author Ye Zhou.
5. Affiliations of all the authors were corrected.

The updated version of this chapter can be found at
https://doi.org/10.1007/16833_2024_249


Interdisciplinary Cancer Research, https://doi.org/10.1007/978-3-031-65827-3_282

# Index




Interdisciplinary Cancer Research, https://doi.org/10.1007/978-3-031-65827-3

SPRINGER NATURE

# GPSR Compliance

*The European Union's (EU) General Product Safety Regulation (GPSR) is a set of rules that requires consumer products to be safe and our obligations to ensure this.*

*If you have any concerns about our products, you can contact us on ProductSafety@springernature.com*

In case Publisher is established outside the EU, the EU authorized representative is:

Springer Nature Customer Service Center GmbH
Europaplatz 3
69115 Heidelberg, Germany